Probability and Statistical Models with Applications

Probability and Statistical Models with Applications

Edited by

Ch.A. Charalambides
Markos V. Koutras
N. Balakrishnan

CRC Press
Taylor & Francis Group
Boca Raton London New York

CRC Press is an imprint of the
Taylor & Francis Group, an **informa** business

A CHAPMAN & HALL BOOK

Chapman & Hall/CRC
Taylor & Francis Group
6000 Broken Sound Parkway NW, Suite 300
Boca Raton, FL 33487-2742

First issued in paperback 2020

ISBN 13: 978-0-367-57892-3 (pbk)
ISBN 13: 978-1-58488-124-7 (hbk)

Library of Congress Card Number 00-059049

Library of Congress Cataloging-in-Publication Data

Probability and Statistical Models with Applications / editors, Ch. A. Charalambides, Markos V. Koutras, N. Balakrishnan.
      p. cm.
   Includes bibliographical references and index.
   ISBN 1-58488-124-0
    1. Distribution (Probability theory) 2. Mathematical statistics. I. Charalambides, Ch.
  A. II. Koutras, Markos V. III. Balakrishnan, N., 1956-
QA273.6 .S694 2000
519.5--dc21   00-059049

Visit the Taylor & Francis Web site at
http://www.taylorandfrancis.com

and the CRC Press Web site at
http://www.crcpress.com

Preface

Theophilos Cacoullos has made significant contributions to many areas of probability and statistics including Discriminant Analysis, Estimation for Compound and Truncated Power Series Distributions, Characterizations, Differential Variance Bounds, Variational Type Inequalities, and Limit Theorems. This is reflected in his lifetime publications list and the numerous citations his publications have received over the past three decades.

We have been associated with him on different levels professionally and personally. We have benefited greatly from him on both these grounds. We have enjoyed his poetry and his knowledge of history, and admired his honesty, sincerity, and dedication.

This volume has been put together in order to (i) review some of the recent developments in Statistical Science, (ii) highlight some of the new noteworthy results and illustrate their applications, and (iii) point out possible new directions to pursue. With these goals in mind, a number of authors actively involved in theoretical and applied aspects of Statistical Science were invited to write an article for this volume. The articles so collected have been carefully organized into this volume in the form of 38 chapters. For the convenience of the readers, the volume has been divided into following six parts:

- APPROXIMATIONS, BOUNDS, AND INEQUALITIES
- PROBABILITY AND STOCHASTIC PROCESSES
- DISTRIBUTIONS, CHARACTERIZATIONS, AND APPLICATIONS
- TIME SERIES, LINEAR, AND NON-LINEAR MODELS
- INFERENCE AND APPLICATIONS
- APPLICATIONS TO BIOLOGY AND MEDICINE

From the above list, it should be clear that recent advances in both theoretical and applied aspects of Statistical Science have received due attention in this volume. Most of the authors who have contributed to this volume were present at the conference in honor of Professor Theophilos Cacoullos that was organized by us at the University of Athens, Greece during June 3–6, 1999. However, it should be stressed here that this volume is **not** a proceedings of this conference, but rather a volume comprised of carefully collected articles with specific editorial goals (mentioned earlier) in mind. The articles in this volume were refereed, and we express our sincere thanks to all the referees for their diligent work and commitment.

It has been a very pleasant experience corresponding with all the authors involved, and it is with great pleasure that we dedicated this volume to **Theophilos Cacoullos**. We sincerely hope that this work will be of interest to mathematicians, theoretical and applied statisticians, and graduate students.

Our sincere thanks go to all the authors who have contributed to this volume, and provided great support and cooperation throughout the course of this project. Special thanks go to Mrs. Debbie Iscoe for the excellent typesetting of the entire volume. Our final gratitude goes to Mr. Robert Stern (Editor, CRC Press) for the invitation to take on this project and providing constant encouragement.

Ch. A. Charalambides   **Athens, Greece**
Markos V. Koutras   **Athens, Greece**
N. Balakrishnan   **Hamilton, Canada**

April 2000

Contents

Part I. Approximation, Bounds, and Inequalities

II. Probability and Stochastic Processes

III. Distributions, Characterizations, and Applications

IV. Time Series, Linear, and Non-Linear Models

V. Inference and Applications

VI. Applications to Biology and Medicine

List of Contributors

M. Albassam, School of Mathematics, Department of Probability and Statistics, University of Sheffield, Sheffield, UK
M.Al-Bassam@sheffield.ac.uk

T. W. Anderson, Department of Statistics, Stanford University, Stanford, CA 94305, USA
twa@stat.stanford.edu

Barry C. Arnold, Department of Statistics, University of California, Riverside, CA 92521-0138, USA
barnold@ucrstat2.ucr.edu

Vilijandas Bagdonavičius, MI2S, BP-69, Université de Bordeaux-2, Bordeaux 33076, France; Department of Statistics, University of Vilnius, Vilnius, Lithuania

N. Balakrishnan, Department of Mathematics and Statistics, McMaster University, 1280 Main Street West, Hamilton, Ontario L8S 4K1, Canada
bala@mcmail.cis.mcmaster.ca

A. D. Barbour, Abt. Angewandte Mathematik, University Zürich-Irchel, Winterthurerstr. 190, 8057 Zürich, Switerland
adb@amath.unizh.ch

Kaye Basford, School of Land and Food, The University of Queensland, Brisbane, Queensland 4304, Australia
k.e.basford@mailbox.uq.edu.au

Niels G. Becker, School of Mathematical Sciences, Australian National University, Canberra ACT 0200, Australia

Michael V. Boutsikas, Department of Mathematics, University of Athens, Panepistemiopolis 15784, Greece
mbouts@math.uoa.gr

Enrique Castillo, Departamento de Economia, Universidad de Cantabria, Avda. de los Castros, 39005 Santander, Spain

Ch. A. Charalambides, Department of Mathematics, University of Athens, Panepistemiopolis 15784, Greece
ccharal@math.uoa.gr

Louis H. Y. Chen, Department of Mathematics, National University of Singapore, Lower Kent Ridge Road, Singapore 0511, Singapore
matchyl@leonis.nus.sg

Herman Chernoff, Department of Statistics, Science Center, Harvard University, 1 Oxford Street, Cambridge, MA 02138, USA
chernoff@stat.harvard.edu

Tasos C. Christofides, Department of Mathematics and Statistics, University of Cyprus, P.O. Box 537 1678, Nicosia, Cyprus
tasos@pythagoras.mas.ucy.ac.cy

O. Chryssaphinou, Department of Mathematics, University of Athens, Panepistemiopolis 15784, Greece
ocrysaf@math.uoa.gr

Arthur Cohen, Department of Statistics, Rutgers University, New Brunswick, NJ 08903, USA
artcohen@rci.rutgers.edu

Somesh Das Gupta, Division of Statistics/Mathematics, Indian Statistical Institute, 203 Barrackpore Trunk Road, Calcutta 700 035, India
sdg@isical.ernet.in

Victor H. de la Peña, Department of Statistics, Columbia University, New York, NY 10027, USA
vp@stat.columbia.edu

N. Flournoy, Department of Mathematics and Statistics, The American University, 4400 Massachusetts Avenue, Washington, D.C. 20016, USA
flournoy@american.edu

J. C. Fu, Department of Statistics, University of Manitoba, Winnipeg, Manitoba R3T 2N2, Canada
fu@ccm.umanitoba.ca

Joseph Glaz, Department of Statistics, The College of Liberal Arts and Sciences, University of Connecticut, 196 Auditorium Road U-120 MSB

428, Storrs, CT 06269-3120, USA
glaz@uconnvm.uconn.edu

Anastasia Ivanova, Department of Biostatistics, University of North Carolina, Chapel Hill, NC 27599, USA

M. C. Jones, Department of Statistics, The Open University, Walton Hall, Milton Keynes MK7 6AA, UK
m.c.jones@open.ac.uk

Adrienne W. Kemp, Mathematical Institute, University of St. Andrews, North Haugh, St. Andrews KY16 9SS, Scotland
cdk@st-and.ac.uk

V. S. Korolyuk, Institute of Mathematices, National Academy of Sciences of Ukraine, Tereschenkovska 3, 252601 Kiev, Ukraine

S. Kourouklis, Department of Mathematics, University of Patras, Patras 26110, Greece
stavros@math.upatras.gr

M. V. Koutras, Department of Mathematics, University of Athens, Panepistemiopolis 15784, Greece
mkoutras@cc.uoa.gr

Tze Leung Lai, Department of Statistics, Stanford University, Stanford, CA 94305, USA

Claude Lefèvre, Institute de Statistique, Université Libre de Bruxelles, Campus Plan, CP 210 B-1050, Bruxelles, Belgique
clefevre@ulb.ac.be

N. Limnios, Division Mathématiques Appliquées, Université de Technologie de Compiègne, Compiègne Cedex, France
nikolaos.limnois@uct.fr

Yimin Ma, Department of Mathematics and Statistics, McMaster University, 1280 Main Street West, Hamilton, Ontario L8S 4K1, Canada

Marianthi Markatou, Columbia University, 615 Mathematics Bldg., New York, NY 10027, USA
markat@stat.columbia.edu

A. M. Mathai, Department of Mathematics, McGill University, 805 Sherbrooke Street West, Montreal, Quebec H3A 2K6, Canada
mathai@math.utep.edu

S. Morgenthaler, Département de Mathématiques, Swiss Federal Institute of Technology, Lausanne CH-1015, Switzerland
Stephan.Morgenthaler@epfl.ch

P. G. Moschopoulos, Department of Mathematical Sciences, University of Texas at El Paso, El Paso, TX 79968-0001, USA
peter@math.utep.edu

Marcel F. Neuts, Department of Mathematics, University of Arizona, Tucson, AZ 85721-0001, USA
marcel@sie.arizona.edu

Mikhail Nikulin, Mathématiques, Informatiques et Sciences Sociales, University Victor Segalen, Bordeaux 2 U.F.R. M.I. 2S, 146 rue Léo Saignat, 33076 Bordeaux, France
M.S.Nikouline@u-bordeaux2.fr

I. Olkin, Department of Statistics and Education, Stanford University, Sequoia Hall, Stanford, CA 94305, USA
iolkin@stat.stanford.edu

N. Papadatos, Department of Mathematics and Statistics, University of Cyprus, P.O. Box 537 1678, Nicosia, Cyprus
npapadat@pythagoras.mas.ucy.ac.cy

Takis Papaioannou, Department of Mathematics, University of Ioannina, Ioannina 45110, Greece
takpap@unipi..gr

Fredos Papangelou, Department of Mathematics, University of Manchester, Manchester M13 9PL, UK
fredos.papangelou@man.ac.uk

Efstathios Paparoditis, Department of Mathematics and Statistics, University of Cyprus, P.O. Box 537 1678, Nicosia, Cyprus
stathisp@pythagoras.mas.ucy.ac.cy

V. Papathanasiou, Department of Mathematics, University of Athens, Panepistemiopolis 15784, Greece
bpapath@math.uoa.gr

Philippe Picard, 14 Quai de Serbie, 69006 Lyon, France
picard@univ-lyon1.fr

Dimitris N. Politis, Department of Mathematics, University of California at San Diego, La Jolla, CA 92093-0112, USA
politis@mathgrad.ucsd.edu

C. R. Rao, Department of Statistics, Pond Laboratory, Penn State University, University Park, PA 16802, USA
crr1@psuvm.psu.edu

Wolf-Dieter D. Richter, Fachbereich Mathematik, Universität Rostock, Universitätsplatz 1 D-18051, Rostock, Germany
richter@likeli-uni-rostock.de

H. B. Sackrowitz, Department of Applied Mathematics and Statistics, Rutgers University, Hill Center, Busch Campus, 110 Frelinghuysen Road, Piscataway, NJ 08855, USA
sackrowi@rci.rutgers.edu

Antonio I. Sanhueza, Universidad de La Frontera, Temuco, Chile

José Maria Sarabia, Departamento de Economia, Universidad de Cantabria, Avda. de los Castros, 39005 Santander, Spain

Pranab K. Sen, Department of Biostatistics, University of North Carolina, CB #7400 3101, McGavran-Greenberg Hall, Chapel Hill, NC 27599-7400, USA
pksen@bios.unc.edu

D. N. Shanbhag, School of Mathematics, Department of Probability and Statistics, University of Sheffield, Sheffield, UK
D.Shanbhag@sheffield.ac.uk

Milton Sobel, Department of Mathematics, University of California at Santa Barbara, Santa Barbara, CA 93017, USA
sobel@pstat.ucsb.edu

H. Tsakiridou, Department of Mathematics, Section of Statistics and O.R., Aristotle University of Thessaloniki, 54006 Thessaloniki, Greece
heltsake@auth.gr

John W. Tukey Princeton University, Washington Road, Princeton, NJ 08544-1000, USA
eo@math.princeton.edu

Sergei Utev, Department of Mathematics and Statistics, School of Statistical Sciences, La Trobe Unviersity, Bundoora, VIC 3083, Australia
stasu@lure.latrobe.edu.au

E. Vaggelatou, Department of Mathematics, University of Athens, Panepistemiopolis 15784, Greece
evagel@math.uoa.gr

P.-C. G. Vassiliou, Department of Mathematics, Section of Statistics and O.R., Aristotle University of Thessaloniki, 54006 Thessaloniki, Greece
Vasiliou@ccf.auth.gr

List of Tables

List of Figures

Theophilos N. Cacoullos—A View of his Career

Theophilos N. Cacoullos was born on April 5, 1932 in Pachna, a village in the Limassol district of Cyprus. He is the eldest son of Nicolas and Galatea Cacoullos. After his secondary education at Limassol Lanitio Gymnasium, he studied mathematics at the University of Athens, receiving his diploma (B.Sc.) in 1954. He then returned to Cyprus and worked for two years as a teacher of mathematics in his alma mater high school. In 1957 he went to the USA, on a scholarship from the Greek Scholarship Foundation, for graduate studies. He received his M.A. in 1960 and his Ph.D. in 1962 in Mathematical Statistics from Columbia University, New York. His M.A. thesis, with Alan Birnbaum, on an almost forgotten topic of *median-unbiased estimation*, prompted his combinatorial derivation of the distribution of the n-fold convolution of the zero truncated Poisson distribution, published in the *Annals of Mathematical Statistics* of 1965. His doctoral thesis on *comparing Mahalanobis Distances between normal populations with unknown means*, published in two extensive papers in *Sankhyā* of 1965, was written under the supervision of the late Professor Rosedith Sitgreaves and Professor T. W. Anderson. Cacoullos says that while many colleagues think he is a multivariate analysis expert, the truth is "*I am a multitopic visitor; as in everyday life I am a random variable, a random walker; I am an actual stochastician.*" He revisited the topic of discriminant analysis with his paper in the *Journal of Multivariate Analysis* of 1992; he characterized normality under spherical symmetry by the admissibility of Fisher's linear discriminant functions and showed that, within the family of spherical normal scale mixtures, the normal maximizes the minimax probability of correct classification provided the Mahalanobis distance between the two populations is the same.

While at Columbia University (1957–1961), Cacoullos met and married Ann Rossettos, a graduate student in the Department of Philosophy. They have three daughters, Rena, Nike, and Galatea, and a granddaughter, Annan Torres, all born in the USA. Ann Rossettos-Cacoullos is Professor in the School of Philosophy of the University of Athens. Rena Cacoullos-Torres is an Assistant Professor of Linguistics and Spanish at the University of Florida; Nike, a biologist, is a researcher at the Greek Cancer Institute in Athens; Galatea, also a Columbia alumna, works for the Alpha Credit Bank of Greece. Cacoullos himself has always been interested in Greek linguistics, philosophy and stochasticity, as well as poetry.

In 1961–1962 he worked as Research Associate at Stanford University, meeting Herman Chernoff, Ingram Olkin, Lincoln Moses, Charles Stein, Gerald Lieberman and Samuel Karlin, among others. He worked with Olkin, producing their joint 1965 *Biometrika* paper on the bias of the characteristic roots of a random matrix. In May 1962 he returned to Cyprus accepting the position of the Director of the Department of Statistics and Research of the new independent Republic of Cyprus. Unable to adjust to the routine work of a small Government Statistics Office, he returned to the States, accepting an Assistant Professorship at the University of Minessota, Minneapolis. Yet, during his 15-month bureaucratic work in Cyprus, he prepared his paper on nonparametric multivariate density estimation, published in 1966 in the *Annals of the Institute of Statistical Mathematics*. In Minneapolis he worked on characterizations of normality by constant regression of quadratic and linear forms; a paper by Geisser motivated the interesting relation between the t and F distributions, published in 1965 in the *Journal of the American Statistical Association*. He also worked with Milton Sobel on inverse-sampling procedures for ranking multinomial distributions. In 1965 he accepted an appointment as Associate Professor in the Department of Industrial Engineering and Operations Research of New York University.

In April 1967, Cacoullos was elected Professor in the newly created Chair of Probability and Statistics of the University of Athens, and in 1968 he returned to Greece, beginning his 31 year long career in the Faculty of Mathematical and Physical Sciences. An excellent teacher, never using notes, with spontaneous humor, his inspired lectures influenced many of his students. As the first professor of Mathematical Statistics in Greece, he definitely shaped the teaching of Statistics in Greek Universities. In addition to standard courses of statistics, he introduced combinatorics, linear programming, stochastic processes and recently actuarial mathematics. Equally important and decisive for the future of research in mathematical statistics in Greece was his supervising of young mathematicians. Two of the editors of this volume (Ch. A. Charalambides and Markos V. Koutras) were honored to be among his first doctoral students. Several other people received their Ph.D. in Mathematical Statistics under his supervision or joint supervision. Among them are Vassilis Papathanasiou, and Ourania Chryssaphinou, now associate professors of the Department of Mathematics of the University of Athens. His guidance in widening his doctoral students' research interests and his encouragement to take advantage of a leave of absence to do research in other Universities abroad were invaluable. The international recognition of his research work and his association with top rank Statistics Departments were instrumental in obtaining positions for his doctoral advisees.

Cacoullos himself visited several Institutions to teach, do research or give seminar talks. He visited McGill University, Montreal, Canada, in 1972–1973 and 1975–1976, always enjoying the collaboration and friendship of Harold Ruben, George Styan, and Arak Mathai. At Stanford he taught in the summer of 1974. In the fall of 1983, at the invitation of Chernoff, he gave a series of lectures under Assorted Topics at the Statistics Center of M.I.T.; in the spring of 1984 he visited the University of Arizona, Tucson. In 1986–1987 he spent the whole academic year at Columbia University. In the spring of 1989 he visited the Center for Multivariate Analysis under C. R. Rao, then at the University of Pittsburgh.

Cacoullos has published about 60 papers in international journals. His main research contribution since the early eighties has been the derivation of differential upper (Chernoff-type) and lower variance bounds for functions of random variables. Related characterizations of distributions as well as elementary proofs of the classical central limit theorem have also been obtained in a series of papers with his collaborators. Recently he has been working on reducing tests of homoscedasticity to testing independence under multinormality. He wrote seven textbooks on probability and statistics in Greek and edited the *Discriminant Analysis and Applications*, published in 1973 by Academic Press. In 1989 Springer published his book *Exercises in Probability*. He also translated into Greek, Stephenson's *Mathematical Methods for Science Students*, Gass' *Linear Programming*, and *Formulae and Tables for Statistical Work* by C. R. Rao *et al.*

Ch. A. Charalambides
M. V. Koutras
N. Balakrishnan

Publications of Theophilos N. Cacoullos

Books

1. *Discriminant Analysis and Applications* (Editor), Academic Press, New York. 1973.

2. *Exercises in Probability*. Springer-Verlag, New York, 1989. Also published by Narosa Publishing House, New Delhi as a Springer International Student Edition, 1993.

Books translated into Greek

1. G. Stephenson: *Mathematical Methods for Science Students*, Longman, London, 1973.

2. S. Gass: *Linear Programming*, McGraw Hill, New York, 1974.

3. C. R. Rao, S. K. Mitra, A. Mathai, K. G. Ramamurthy, *Formulae and Tables for Statistical Work*, Statistical Publishing Society Calcutta, 1974.

Books in Greek

1. *A Course in Probability Theory*, Athens, 1969.

2. *Elements of Stochastic Processes*, Athens, 1970.

3. *Exercises in Probability with Solutions and Elements of Theory*, Vol. 1, 2nd edition, Athens, 1986.

4. *Probability Theory: Exercises and Problems with Solution and Elements of Theory*, Vol. 2, 3rd edition, Athens, 1986.

5. *Statistics, Theory and Applications*, Athens, 1972.

6. *Stochastic Processes*, Athens, 1978.

7. *Actuarial Science*, Vol. I: *Risk Theory and Probability*, Athens, 1994.

Articles

1. A combinatorial derivation of the distribution of the truncated Poisson sufficient statistic. *Annals of Mathematical Statistics* **32** (1961), 904–905.

2. Comparing Mahalanobis Distances I: Comparing distances between k known normal populations and another unknown. *Sankhyā, Series A* **27**(1965), 1–22.

3. Comparing Mahalanobis Distances II: Bayes procedures when the mean vectors are unknown. *Sankhyā, Series A* **27**(1965), 23–32.

4. On the bias of functions of characteristic roots of a random matrix (with Ingram Olkin). *Biometrika* **52**(1965), 87–94.

5. A relation between t- and F-distributions. *Journal of the American Statistical Association* **60**(1965), 179–182, 1249.

6. Estimation of a multivariate density. *Annals of the Institute of Statistical Mathematics, Tokyo* **18**(1966), 179–189.

7. An inverse-sampling procedure for selecting the most probable event in a multinomial distribution (with Milton Sobel). In *Multivariate Analysis* (Ed., P. R. Krishnaiah) (1966), 423–455, Academic Press, New York.

8. On a class of admissible partitions. *Annals of Mathematical Statistics* **37**(1966), 189–195.

9. Asymptotic distribution for a generalized Banach match-box problem. *Journal of the American Statistical Association* **62**(1967), 1252–57.

10. Characterizations of normality by constant regression of linear statistics on another linear statistic. *Annals of Mathematical Statistics* **38**(1967), 1894–1898.

11. On the distribution of the bivariate range (with Henry DeCicco). *Technometrics* **9**(1967), 476–480.

12. Some characterizations of normality. *Sankhyā, Series A* **29**(1967), 399–404.

13. A sequential scheme for detecting outliers. *Bull. de la Soc. Math. de Grece, Nouvelle Serie* **9**(1968), 113–123.

14. Some properties of the incomplete beta-function ratio. *Bull. de la Soc. Math. de Grece, Nouvelle Serie* **11**(1970), 132–138.

15. Some remarks on topothetical procedures. *Proceedings, ISI 38th Session*, Washington, DC, 70–73. *Bull. Int. Statist. Inst.* **1**(1971), 128–131.

16. MVUE for a truncated Poisson process (with Ch. Charalambides). *Proceedings of the 4th Session of the Greek Mathematics Society.* Patras (1971), 73–77.

17. On fixed-point solutions of operators and some reliability problems (with N. Apostolatos). *Research Chronicles*, University of Athens, Volume 3(1972), 306– 312.

18. Some remarks on the convolution of Poisson distributions with the zero class missing. *The American Statistician* **83**(1972).

19. A bibliography of discriminant analysis (with G.P.H. Styan). In *Discriminant Analysis and Applications* (Ed., T. Cacoullos), (1973), 375–435, Academic Press, New York.

20. Distance, discrimination and error. In *Discriminant Analysis and Applications* (Ed., T. Cacoullos), (1973), 61–75, Academic Press, New York.

21. MVUE for truncated discrete distributions (with Ch. Charalambides). Invited paper, *European IMS Meeting*, Budapest, August 31-September 5, 1972, in *Progress in Statistics* (Eds., J. Gani *et al.*), Vol. **1**(1974), 133–144, *Colloquia Mathematica Societatis Janos Bolyai*, North-Holland.

22. On minimum variance unbiased estimation for truncated binomial and negative binomial distributions (with Ch. Charalambides). *Annals of the Institute of Statistical Mathematics*, **27**(1975), 133–144.

23. Multiparameter Stirling and C-type distributions. In *Statistical Distributions in Scientific Work* **1**(1975): Models and Structures (Eds., G.P. Patil *et al.*), pp. 19–30. Reidel, Dordrecht.

24. Best estimation for multiply truncated PSD's. *Scandinavian Journal of Statistics* **4**(1977), 159–164.

25. On some bivariate probability models applicable to traffic accidents and fatalities (with H. Papageorgiou). *International Statistical Review* **48**(1980), 345–356.

26. On bivariate discrete distributions generated by compounding (with H. Papageorgiou). In *Statistical Distributions in Scientific Work* **4**(1981), *Models, Structures and Characterizations* (Eds:, C. Taillie *et al.*), pp. 197–212. Reidel, Dordrecht.

27. On upper and lower bounds for the variance of a function of a random variable. *Annals of Probability* **10**(1982), 799–809.

28. Bivariate negative binomial-Poisson and negative binomial-Bernoulli models with an application to accident data (with H. Papageorgiou). In *Statistics and Probability: Essays in honor of C.R. Rao* (Eds., G. Kallianpur *et al.*), (1982), pp. 155– 168, North-Holland.

29. Characterizing the negative binomial distribution (with H. Papageorgiou), letter to the Editor. *Journal of Applied Probability* **19**(1982), 742–743.

30. Characterizations of discrete distributions by a conditional distribution and a regression function (with H. Papageorgiou). *Annals of the Institute of Statistical Mathematics*, **35**(1983), 95–103.

31. Characterizations of mixtures of discrete distributions by a regression point. *Statistics & Probability Letters* **1**(1983), 269–272.

32. Multiparameter Stirling and C-numbers; recurrences and applications (with H. Papageorgiou). *The Fibonacci Quarterly* **22**(1984), 119–133.

33. Characterizations of mixtures of continuous distributions by their posterior means (with H. Papageorgiou). *Scandinavian Actuarial Journal*, (1984), 23–30.

34. Quadratic forms in spherical random variables: generalized noncentral 2- distributions (with M. Koutras). *Naval Research Logistics Quarterly* **31**(1984), 447–461.

35. On upper bounds for the variance of functions of random variables (with V. Papathanasiou). *Statistics & Probability Letters* **3**(1985), 175–184.

36. Minimum distance discrimination for spherical distributions (with M. Koutras). In *Statistical Theory and Data Analysis*, (Ed., K. Matusita), (1985), pp. 91–102, North-Holland.

37. Bounds for the variance of functions of random variables by orthogonal polynomials and Bhattacharya bounds (with V. Papathanasiou). *Statistics & Probability Letters* **4**(1986), 21–23.

38. Characterizing priors by posterior expectations in multiparameter exponential families. *Annals of the Institute of Statistical Mathematics*, **39**(1987), 399–405.

39. Characterization of generalized multivariate discrete distributions by a regression point. *Statistics & Probability Letters* **5**(1987), 39–42.

40. On minimum-distance location discrimination for isotropic distributions. *Proceedings DIANA II*, Praha, (1987) 1–16.

41. Characterization of distributions by variance bounds (with V. Papathanasiou). *Statistics & Probability Letters* **7**(1989), 351–356.

42. Dual Poincaré-type inequalities via the Cramer-Rao and the Cauchy-Schwarz inequalities and related characterizations. In C. R. Rao's honorary 70th birthday festschrift *Data Analysis and Inference* (Eds., Y. Dodge), (1989), pp. 239–250, North-Holland.

43. Characterizations of distributions within the elliptical class by a gamma distributed quadratic form (with C.G. Khatri). In Khatri's Memorial Volume. *Gujarat Statistics Review*, (1990), 89–98.

44. Correcting Remarks on "Characterization of normality within the elliptical contoured class" (with C.G. Khatri). *Statistics & Probability Letters* **11**(1990), 551–552.

45. The optimality of linear discriminant functions under spherical symmetry. *Proceedings DIANA III*, (1990), pp. 20–33, Czechoslovak Academy of Sciences, Praha.

46. Another characterization of the normal law and a proof of the central limit thoerem connected with it (with V. Papathanasiou and S. Utev). *Theory of Probability and its Applications* **37**(1992), 648–657 (in Russian). Also in the English translation of the Russian journal, pp. 581–588, 1993.

47. Two LDF Characterizations of the Normal as a Spherical Distribution. *Journal of Multivariate Analysis* **40**(1992), 205–212.

48. Lower variance bounds and a new proof of the multivariate central limit theorem (with V. Papathanasiou). *Journal of Multivariate Analysis* **43**(1992), 173–184.

49. Variance Inequalities, Characterizations and a Proof of the Central Limit Theorem, Festschrift in honor of F. Eicker. *Data Analysis and Statistical Inference* (Eds., S. Schach and G. Trenkler), (1992), pp. 27–32, Joseph Eul, Verlag.

50. Variational inequalities with examples and an application to the central limit theorem (with V. Papathanasiou and S. Utev). *Annals of Probability* **23**(1994), 1607–1618.

51. A generalizattion of a covariance identity and related characterizations (with V. Papathanasiou). *Mathematical Methods of Statistics* **4**(1995), 105–113.

52. On the Performance of Minimum-Distance Classification Rules for Kotz-type Elliptical Distributions (with M. Koutras). In *Advances in the Theory and Practice of Statistics: A Volume in Honor of Samuel Kotz* (Eds., Norman L. Johnson and N. Balakrishnan), (1996), pp. 209–224, John Wiley & Sons, New York.

53. Characterizations of Distributions by Generalized Variance Bounds and a Simple Proof of the CLT (with V. Papathanasiou). *Statistical Planning and Inference*, Special Volume in honor of C. R. Rao's 75th birthday, **61**(1997), 157–171.

54. Variance inequalities for covariance kernels and applications to central limit theorems (with N. Papadatos and V. Papathanasiou). *Theory of Probability and its Applications* **42**(1997), 195–201.

55. The Statisticity and Stochasticity of the Greeks, ISI Session 51, L VII, Contributed Papers, Book 1(1997), 339–340.

56. Three elementary proofs of the central limit theorem with applications to random sums (with N. Papadatos and V. Papathanasiou). In *Stochastic Processes and Related Topics In Memory of S. Cam-*

banis, (Eds., I. Karatzas, B. Rajput, M. Taqqu), (1998), pp. 15–23, Birkhauser, Boston.

57. Il statisticità et stochasticità dei Greci, *Induzioni* **16**(1998), 23–31.

58. An application of an density transform, and the local limit theorem (with N. Papadatos and V. Papathanasiou). (2000). Submitted for publication.

59. Testing homoscedasticity via independence, under joint normality. (2000). Submitted for publication.

60. The F-test of homoscedasticity for correlated normal variables. (2000). *Statistics & Probability Letters*, to appear.

The Ten Commandments for a Statistician

Theophilos Cacoullos

1. You must not have any other masters beside Chance.

2. You must not construct for yourself any model that is biased either for your employer above or your client below or that is in the service of the underworld.

   You must not pay homage to them, nor serve them; for I, Chance, your master am a source of error, propagating sample bias to third and fourth phase subsamples of those who ignore me, but showing consistency to the thousandth order of those that pay homage to me and follow my distributions.

3. You must not invoke the name of Chance, your Master, to misinformation; for Chance will not hold him right that uses its name in vain, by commission or omission.

4. Remember the seventh stage to stochasticize.

   Six stages you are to labor and do all your work, but in the seventh stage, a meditation on Chance, you must not do any work, neither quantify, nor enumerate, nor sample, nor tabulate or calculate, by hand or machine, nor analyse.

5. Adhere to the principles and code of your Society that you make right inferences on the population that Chance is giving you.

6. You must not conceal nor distort information.

7. You must not abuse information acquired in confidence.

8. You must not plagiarize.

9. You must not bring a false charge against your fellow.

10. You must not covet your colleague's art; you must not covet your colleague's company, nor his software or hardware, nor his algorithm, nor his program, nor anything that is your colleague's.

Sources

1. Exodus, Chapter 20, *The Complete Bible, The Old Testament* (J. M. Powis Smith Transl.) The University of Chicago Press.

2. JSI (1985) Declaration on Professional Ethics, *International Statistical Review* 1986, 54, pp. 227–242.

3. *Professional Code of Conduct*, The Institute of Statisticians, Preston, Lancashire, England.

4. JCC/E.S.O.M.A.R., *International Code of Marketing and Social Research Practice*, 1976.

Theophilos Cacoulos

PART I

Approximation, Bounds, and Inequalities

1

Nonuniform Bounds in Probability Approximations Using Stein's Method

Louis H. Y. Chen
National University of Singapore, Republic of Singapore

ABSTRACT Most of the work on Stein's method deals with uniform error bounds. In this paper, we discuss non-uniform error bounds using Stein's method in Poisson, binomial, and normal approximations.

Keywords and phrases Stein's method, non-uniform bounds, probability approximations, Poisson approximation, binomial approximation, normal approximation, concentration inequality approach, binary expansion of a random integer

1.1 Introduction

In 1972 Stein introduced a method of normal approximation which does not depend on Fourier analysis but involves solving a differential equation. Although his method was for normal approximation, his ideas are applicable to other probability approximations. The method also works better than the Fourier analytic method for dependent random variables, particularly if the dependence is local or of a combinatorial nature. Since the publication of this seminal work of Stein, numerous papers have been written and Stein's ideas applied in many different contexts of probability approximation. Most notable of these works are in normal approximation, Poisson approximation, Poisson process approximation, compound Poisson approximation and binomial approximation. An account of Stein's method and a brief history of its developments can be found in Chen (1998).

In this paper we discuss another aspect of the application of Stein's method, not in terms of the approximating distribution but in terms of the nature of the error bound. Most of the papers on Stein's method deal with uniform error bounds. We show that Stein's method can also

3

be applied to obtain non-uniform error bounds and of the best possible order. Roughly speaking, a uniform bound is one on a metric between two distributions. Whereas a non-uniform bound on the discrepancy between two distributions, $\mathcal{L}(W)$ and $\mathcal{L}(\mathcal{Z})$, is one on $|Eh(W) - Eh(Z)|$, which depends on h for every h in a separating class. We will consider non-uniform bounds in three different contexts, Poisson, binomial, and normal. In the exposition below, we will focus more on ideas than on technical details.

1.2   Poisson Approximation

Poisson approximation using Stein's method was first investigated by Chen (1975a). Since then many developments have taken place and Poisson approximation has been applied to such diverse fields as random graphs, molecular biology, computer science, probabilistic number theory, extreme value theory, spatial statistics, and reliability theory, where many problems can be phrased in terms of dependent events. See for example Arratia, Goldstein, and Gordon (1990), Barbour, Holst, and Janson (1992) and Chen (1993). All these results of Poisson approximation concern error bounds on the total variation distance between the distribution of a sum of dependent indicator random variables and a Poisson distribution. These bounds are therefore uniform bounds.

The possibility of nonuniform bounds in Poisson approximation using Stein's method was first mentioned in Chen (1975b). For independent indicator random variables, nonuniform bounds were first obtained for small and moderate λ by Chen and Choi (1992) and for unrestricted λ with improved results by Barbour, Chen, and Choi (1995). To explain the ideas behind obtaining nonuniform bounds, we first illustrate how a uniform bound is obtained in the context of independent indicator random variables.

Let $X_1, \ldots, X_n$ be independent indicator random variables with $P(X_i = 1) = 1 - P(X_i = 0) = p_i$, $i = 1, \ldots, n$. Define $W = \sum_{i=1}^{n} X_i$, $W^{(i)} = W - X_i$, $\lambda = \sum_{i=1}^{n} p_i$ and Z to be a Poisson random variable with mean λ. Let f_h be the solution (which is unique except at 0) of the Stein equation

$$\lambda f(w+1) - wf(w) = h(w) - Eh(Z)$$

where h is a bounded real-valued function defined on $\mathbf{Z}^{+} = \{0, 1, 2, \ldots\}$. Then we have

$$Eh(W) - Eh(Z) = E\{h(W) - Eh(Z)\}$$
$$= E\{\lambda f_h(W+1) - Wf_h(W)\}$$
$$= \sum_{i=1}^{n} p_i^2 E\triangle f_h(W^{(i)}+1) \qquad (1.2.1)$$

where $\triangle f(w) = f(w+1) - f(w)$. A result of Barbour and Eagleson (1983) states that $\|\triangle f_h\|_\infty \leq 2(1 \wedge \lambda^{-1})\|h\|_\infty$. Applying this result, we obtain

$$d_{TV}(\mathcal{L}(W), \mathcal{L}(Z)) = \sup_A |P(W \in A) - P(Z \in A)|$$
$$= (1/2) \sup_{|h|=1} |Eh(W) - Eh(Z)|$$
$$\leq (1 \wedge \lambda^{-1}) \sum_{i=1}^{n} p_i^2 \qquad (1.2.2)$$

where d_{TV} denotes the total variation distance. It is known that the absolute constant 1 is best possible and the factor $(1 \wedge \lambda^{-1})$ has the correct order for both small and large values of λ. The significance of the factor $(1 \wedge \lambda^{-1})$ is explained in Chapter 1 of Barbour, Holst, and Janson (1992).

To obtain a nonuniform bound, we let

$$A_i(r) = \frac{P(W^{(i)} = r)}{P(Z = r)}.$$

Then (1.2.1) can be rewritten as

$$Eh(W) - Eh(Z) = \sum_{i=1}^{n} p_i^2 E A_i(Z) \triangle f_h(Z+1) \qquad (1.2.3)$$

where h is no longer assumed to be bounded.

Let $C^* = \sup_{1<i<n} \sup_{r \geq 0} A_i(r)$. Then

$$|Eh(W) - Eh(Z)| \leq C^* \left(\sum_{i=1}^{n} p_i^2\right) E|\triangle f_h(Z+1)|.$$

What remains to be done is to calculate or bound C^* and $E|\triangle f_h(Z+1)|$. In Barbour, Chen and Choi (1995), it is shown that for $\max_{1 \leq i \leq n} p_i \leq 1/2$, $C^* \leq 4e^{13/12}\sqrt{\pi}$ and the following theorem was proved.

THEOREM 1.2.1   *[Theorem 3.1 in Barbour, Chen, and Choi (1995)]*
*Let h be a real-valued function defined on $\mathbf{Z}^+$ such that $EZ^2|h(Z)| < \infty$.
We have*

$$|Eh(W) - Eh(Z)|$$

$$\leq C^* \sum_{i=1}^{n} p_i^2 [4(1 \wedge \lambda^{-1}) E|h(Z+1)| + E|h(Z+2)|$$

$$-2E|h(Z+1)| + E|h(Z)|]/2. \tag{1.2.4}$$

If $|h| = 1$, then we have

$$d_{TV}(\mathcal{L}(W), \mathcal{L}(Z)) = (1/2) \sup_{|h|=1} |Eh(W) - Eh(Z)| \leq C^*(1 \wedge \lambda^{-1}) \sum_{i=1}^{n} p_i^2$$

where the upper bound has the same order as that of (1.2.2), but it
has a larger absolute constant. However, the bound in (1.2.4) allows
a very wide choice of possible functions h, and therefore contains more
information than the total variation distance bound in (1.2.2).

By iterating (1.2.1), we obtain

$$Eh(W) - Eh(Z) = \sum_{i=1}^{n} p_i^2 E\triangle f_h(Z+1) + \text{second order terms}$$

$$= -\frac{1}{2} \sum_{i=1}^{n} p_i^2 E\triangle^2 h(Z) + \text{second order terms}$$

where $E\triangle f_h(Z+1) = -(1/2)E\triangle^2 h(Z)$ (see, for example, Chen and
Choi (1992), p.1871).

In Barbour, Chen, and Choi (1995), a more refined result (Theorem
3.2) was obtained by bounding the second order error terms in the same
way the first order error terms were bounded. From this theorem, a large
deviation result (Theorem 4.2) was proved which produces the following
corollary.

COROLLARY 1.2.2
*Let $z = \lambda + \xi\sqrt{\lambda}$. Suppose $\max_{1 \leq i \leq n} p_i \to 0$ and $\xi = o\left([\lambda/\sum_{i=1}^{n} p_i^2]^{1/2}\right)$
as $n \to \infty$. Then, as n, z and $\xi \to \infty$,*

$$\frac{P(W \geq z)}{P(Z \geq z)} - 1 \sim -\frac{\xi^2}{2\lambda} \sum_{i=1}^{n} p_i^2.$$

The following asymptotic result was also deduced.

THEOREM 1.2.3

Let N be a standard normal random variable. Let h be a nonnegative function defined on $\mathbf{R}$ which is continuous almost everywhere and not identically zero. Suppose that $\left\{(\frac{Z-\lambda}{\sqrt{\lambda}})^4 h(\frac{Z-\lambda}{\sqrt{\lambda}}) : \lambda \geq 1\right\}$ is uniformly integrable. Then as $\lambda \to \infty$ such that $\max_{1 \leq i \leq n} p_i \to 0$,

$$\sum_{r=0}^{\infty} h(\frac{r-\lambda}{\sqrt{\lambda}})|P(W=r) - P(Z=r)| \sim \frac{1}{2\lambda}(\sum_{i=1}^{n} p_i^2)E|N^2 - 1|h(N).$$

By letting $h \equiv 1$, $E|N^2 - 1|h(N) = E|N^2 - 1| = 2\sqrt{2/(\pi e)}$, and Theorem 1.2.3 yields a result of Barbour and Hall (1984a, p. 477) and Theorem 1.2 of Deheuvels and Pfeifer (1986).

Nonuniform bounds in compound Poisson approximation on a group for small and moderate λ were first obtained by Chen (1975b) and later generalized and refined by Chen and Roos (1995). In these papers, the techniques were inspired by Stein's method. The first paper on compound Poisson approximation using Stein's method directly was by Barbour, Chen, and Loh (1992).

1.3 Binomial Approximation: Binary Expansion of a Random Integer

In his monograph, Stein (1986) considered the following problem. Let n be a natural number and let X denote a random variable uniformly distributed over the set $\{0, 1, , n - 1\}$. Let W denote the number of ones in the binary expansion of X and let Z be a binomial random variable with parameters $(k, 1/2)$, where k is the unique integer such that $2^{k-1} < n \leq 2^k$. If $n = 2^k$, then W has the same distribution as Z, otherwise it is a sum of dependent indicator random variables.

By using the solution of the Stein equation

$$(k - x)f(x) - xf(x - 1) = h(x) - Eh(Z) \tag{1.3.1}$$

where $h = I_{\{r\}}$ and $r = 0, 1, \ldots, k$, Stein (1986) proved that

$$\sup_{0 \leq r \leq k} |P(W = r) - P(Z = r)| \leq 4/k.$$

Diaconis (1977), jointly with Stein, proved a normal approximation result for W with an error bound of order $1/\sqrt{k}$. A combination of this result with the normal approximation to the binomial distribution shows that $\sup_{0 \le r < k} |P(W \le r) - P(Z \le r)|$ is of the order of $1/\sqrt{k}$.

Loh (1992) obtained a bound on the solution of a multivariate version of (1.3.1) using the probabilistic approach of Barbour (1988). Using this result of Loh and arguments in Stein (1986), we can obtain a bound of order $1/\sqrt{k}$ on the total variation distance between $\mathcal{L}(W)$ and $\mathcal{L}(Z)$.

In an unpublished work of Chen and Soon (1994) which was based on the Ph.D. dissertation of the latter, the method of obtaining nonuniform bounds in Poisson approximation was applied to the approximation of $\mathcal{L}(W)$ by $\mathcal{L}(Z)$. Apart from proving other results, this work shows that the total variation distance between $\mathcal{L}(W)$ and $\mathcal{L}(Z)$ is, in many instances, of much small order than $1/\sqrt{k}$.

Let $X = \sum_{i=1}^{k} X_i 2^{k-i}$ for the binary expansion of X and $W = \sum_{i=1}^{k} X_i$. In Stein (1986, pp. 44–45), it is shown that

$$Eh(W) - Eh(Z) = EQf_h(W) \tag{1.3.2}$$

where $Q = |\{j : X_j = 0 \text{ or } X + 2^{k-j} \ge n\}|$ and f_h is the solution of (1.3.1) with h being a real-valued function defined on $\{0, 1, \ldots, k\}$. Define

$$\psi(r) = E[Q|W = r] \quad \text{and} \quad A(r) = \frac{P(W = r)}{P(Z = r)}.$$

Then (1.3.2) can be written as

$$Eh(W) - Eh(Z) = E\psi(Z)A(Z)f_h(Z). \tag{1.3.3}$$

Let l_k be the number of consecutive 1s, starting from the beginning in the binary expansion of $n - 1$. The relationship between $n - 1$ and l_k is given by

$$n - 1 = \sum_{i=1}^{l_k} 2^{k-i} + m$$

where $0 \le m < 2^{k-l_k-1}$. It is shown in Chen and Soon (1994) that for $0 \le r \le k - 1$, $l_k/k \le A(r) \le 2$. By obtaining upper and lower bounds on the right hand side of (1.3.3), the following theorem was proved.

THEOREM 1.3.1
Assume that $2^{k-1} < n < 2^k$.
  (i) If $\lim_{k \to \infty} \frac{l_k}{\sqrt{k}} = \infty$, then

$$d_{TV}(\mathcal{L}(W), \mathcal{L}(Z)) \asymp 2^{-l_k}.$$

  (ii) If $\limsup_{k \to \infty} \frac{l_k}{\sqrt{k}} < \infty$, then

$$d_{TV}(\mathcal{L}(W), \mathcal{L}(Z)) \asymp 2^{-l_k} \frac{l_k}{\sqrt{k}}$$

where $x_k \asymp y_k$ means that there exist positive constants $a < b$ such that $a \leq x_k/y_k \leq b$ for sufficiently large k.

From this theorem it follows that

$$d_{TV}(\mathcal{L}(W), \mathcal{L}(Z)) \asymp \frac{1}{\sqrt{k}}$$

if and only if

$$0 < \liminf_{k \to \infty} l_k \leq \limsup_{k \to \infty} l_k < \infty.$$

The following theorems were also proved.

THEOREM 1.3.2

$$|Eh(W) - Eh(Z)|$$

$$\leq \frac{13}{\sqrt{k}} E \left\{ \left| \frac{Z - [k/2] - 1}{\sqrt{k/4}} \right| (|h(Z)| + |h(Z+1)| + 2|Eh(Z)|) \right\}.$$

THEOREM 1.3.3
Let $a = [k/2] + b_k$ where $b_k/\sqrt{k} \to \infty$ and $b_k/k \to 0$ as $k \to \infty$. If $l_k = l$ for all sufficiently large k, then

$$\frac{P(W \geq a)}{P(Z \geq a)} - 1 \sim -2\psi\left(\left[\frac{k}{2}\right]\right) \frac{b_k}{k}$$

as $k \to \infty$, where $l(1/2 - (l-1)/[2(k-l+1)])^{l+1} < \psi([k/2]) \leq 3$.

Theorem 1.3.3 is in fact a corollary of a more general large deviation theorem.

1.4 Normal Approximation

Let $X_1, \ldots, X_n$ be independent random variables with $EX_i = 0$, $\text{var}(X_i) = \sigma_i^2$, $E|X_i|^3 = \gamma_i < \infty$ and $\sum_{i=1}^{n} \sigma_i^2 = 1$. Let F be the distribution function of $\sum_{i=1}^{n} X_i$ and let Φ be the standard normal distribution function. The Berry-Esseen Theorem states that

$$\sup_{-\infty < x < \infty} |F(x) - \Phi(x)| \leq C \sum_{i=1}^{n} \gamma_i$$

where C is an absolute constant. The smallest value of C, obtained so far by Van Beek (1972) (without using computers), is 0.7975.

If $X_1, \ldots, X_n$ are independent and identically distributed, then

$$\sup_{-\infty < x < \infty} |F(x) - \Phi(x)| \leq Cn\gamma$$

where $\gamma = \gamma_i$ for $i = 1, \ldots, n$. Nonuniform bounds were first obtained by Esseen (1945) who proved that for the i.i.d. case

$$|F(x) - \Phi(x)| \leq \frac{\lambda \log n}{\sqrt{n}(1 + x^2)}$$

and

$$|F(x) - \Phi(x)| \leq \frac{\lambda \log(2 + |x|)}{\sqrt{n}(1 + x^2)}$$

where λ depends on $n^{3/2}\gamma$. Nagaev (1965) improved the upper bounds to $Cn\gamma/(1 + |x|^3)$, also for the i.i.d. case. This was generalized by Bikelis (1966) who proved that, for independent and not necessarily identically distributed random variables,

$$|F(x) - \Phi(x)| \leq \frac{C \sum_{i=1}^n \gamma_i}{1 + |x|^3}$$

where C is an absolute constant. Paditz (1977) calculated C to be 114.7 and Michel (1981) reduced it to 30.54 for the i.i.d. case. All the above proofs used the Fourier analytic method. Chen and Shao (2000) used Stein's method to prove the following more general result:

$$|F(x) - \Phi(x)| \leq C \sum_{i=1}^n \left\{ \frac{EX_i^2 I(|X_i| > 1 + |x|)}{(1 + |x|)^2} + \frac{E|X_i|^3 I(|X_i| \leq 1 + |x|)}{(1 + |x|)^3} \right\}$$

where the existence of third moments is no longer assumed. Their proof is based on truncation and the concentration inequality approach. The concentration inequality approach was originally used by Stein for the i.i.d. case (see Ho and Chen (1978)). It was extended by Chen (1986) to dependent and non-identically distributed random variables with arbitrary index set. A proof of the Berry-Esseen Theorem for independent and non-identically distributed random variables using the concentration inequality approach is given in Section 2 of Chen (1998).

The concentration inequality approach is not the only approach for obtaining Berry-Esseen bounds using Stein's method. Another approach based on inductive arguments has been used by Barbour and Hall (1984b), Bolthausen (1984) and Stroock (1993).

We would like to mention in passing that Stein's method has also been applied to obtain bounds on the total variation distances between the standard normal distribution and distributions satisfying certain variational inequalities. See Utev (1989) and Cacoullos, Papathanasiou, and Utev (1994).

1.5 Conclusion

We would like to conclude by saying that there is much more to be done in the direction of nonuniform bounds, particularly for dependent random variables both in Poisson approximation and normal approximation. The large deviation results referred to in the above sections are actually those of moderate deviation. A related question therefore is how Stein's method can be applied to obtain results which cover both moderate and really large deviations.

Acknowledgement This work is partially supported by grant RP3982719 at the National University of Singapore. I would like to thank K. P. Choi and Qi-Man Shao for their help in preparing the manuscript and for their helpful comments.

References

1. Arratia, R., Goldstein, L., and Gordon, L. (1990). Poisson approximation and the Chen-Stein method. *Statistical Science* **5**, 403–434.

2. Barbour, A. D. (1988). Stein's method and Poisson process convergence. *Journal of Applied Probability* **25 (A)**, 175–184.

3. Barbour, A. D., Chen, L. H. Y., and Choi, K. P. (1995). Poisson approximation for unbounded functions, I: independent summands. *Statistica Sinica* **5**, 749–766.

4. Barbour, A. D., Chen, L. H. Y., and Loh, W. L. (1992). Compound Poisson approximation for nonnegative random variables via Stein's method. *Annals of Probability* **20**, 1843–1866.

5. Barbour, A. D. and Eagleson, G. (1983). Poisson approximation for some statistics based on exchangeable trials. *Advances in Applied Probability* **15**, 585–600.

6. Barbour, A. D. and Hall, P. (1984a). On the rate of Poisson convergence. *Mathematical Proceedings of the Cambridge Philosophical Society* **95**, 473–480.

7. Barbour, A. D. and Hall, P. (1984b). Stein's method and the Berry-Esseen theorem. *The Australian Journal Statistics* **26**, 8–15.

8. Barbour, A. D., Holst, L., and Janson, S. (1992). *Poisson Approximation.* Oxford Studies in Probability 2, Clarendon Press, Oxford.

9. Bikelis, A. (1966). Estimates of the remainder in the central limit

theorem. *Litovsk. Mat. Sb.* **6(3)**, 323–346 (in Russian).

10. Bolthausen, E. (1984). An estimate of the remainder in a combinatorial central limit theorem. *Zeitschrift Wahrscheinlichkeitstheorie und Verwandte Gebiete* **66**, 379–386.

11. Cacoullos, T., Papathanasiou, V. and Utev, S. A. (1994). Variational inequalities with examples and an application to the central limit theorem. *Annals of Probability* **22**, 1607–1618.

12. Chen, L. H. Y. (1975a). Poisson approximation for dependent trials. *Annals of Probability* **3**, 534–545.

13. Chen, L. H. Y. (1975b). An approximation theorem for convolutions of probability measures. *Annals of Probability* **3**, 992–999.

14. Chen, L. H. Y. (1986). The rate of convergence in a central limit theorem for dependent random variables with arbitrary index set. IMA Preprint Series #243, University of Minnesota.

15. Chen, L. H. Y. (1993). Extending the Poisson approximation. *Science* **262**, 379–380.

16. Chen, L. H. Y. (1998). Stein's method: some perspectives with applications. *Probability Towards 2000* (Eds., L. Accardi and C. Heyde), pp. 97–122. Lecture Notes in Statistics No. 128. Springer Verlag.

17. Chen, L. H. Y. and Choi, K. P. (1992). Some asymptotic and large deviation results in Poisson approximation. *Annals of Probability* **20**, 1867–1876.

18. Chen, L. H. Y. and Roos, M. (1995). Compound Poisson approximation for unbounded functions on a group, with application to large deviations. *Probability Theory and Related Fields* **103**, 515–528.

19. Chen, L. H. Y. and Shao, Q. M. (2000). A non-uniform Berry-Esseen bound via Stein's method. Preprint.

20. Chen, L. H. Y. and Soon, S. Y. T. (1994). On the number of ones in the binary expansion of a random integer. Unpublished manuscript.

21. Deheuvels, P. and Pfeifer, D. (1986). A semigroup approach to Poisson approximation. *Annals of Probability* **14**, 663–676.

22. Diaconis, P. (1977). The distribution of leading digits and uniform distribution mod 1. *Annals of Probability* **5**, 72–81.

23. Esseen, C.-G. (1945). Fourier analysis of distribution functions: a mathematical study of the Laplace-Gaussian law. *Acta Mathematica* **77** 1–125.

24. Ho, S. T. and Chen, L. H. Y. (1978). An L_p bound for the remainder in a combinatorial central limit theorem. *Annals of Probability* **6**, 231–249.

25. Loh, W. L. (1992). Stein's method and multinomial approximation. *Annals of Applied Probability* **2**, 536–554.

26. Michel, R. (1981). On the constant in the non-uniform version of the Berry-Esseen Theorem. *Zeitschrift Wahrscheinlichkeitstheorie und Verwandte Gebiete* **55**, 109–117.

27. Nagaev, S. V. (1965). Some limit theorems for large deviations. *Theory of Probability and its Applications* **10**, 214–235.

28. Paditz, L. (1977). Über die Annäherung der Verteilungsfunktionen von Summen unabhängiger Zufallsgröben gegen unberrenzt teilbare Verteilungsfunktionen unter besonderer berchtung der Verteilungsfunktion der standarddisierten Normalverteilung. *Dissertation*, A.TU Dresden.

29. Soon, S. Y. T. (1993). Some Problems in Binomial and Compound Poisson Approximations. *Ph.D. dissertation*, National University of Singapore.

30. Stein, C. (1972). A bound for the error in the normal approximation to the distribution of a sum of dependent random variables. *Proceedings of the Sixth Berkeley Symposium on Mathematics, Statistics and Probability* **2**, 583–602, University California Press. Berkeley, California.

31. Stein, C. (1986). *Approximation Computation of Expectations.* Lecture Notes 7, Institute of Mathematics and Statistics, Hayward, California.

32. Stroock, D. W. (1993). *Probability Theory: An Analytic View.* Cambridge University Press, Cambridge, U.K.

33. Utev, S. A. (1989). Probability problems connected with a certain integrodifferential inequality. *Siberian Mathematics Journal* **30**, 490–493.

34. Van Beek, P. (1972). An approximation of Fourier methods to the problem of sharpening the Berry-Esseen inequality. *Zeitschrift Wahrscheinlichkeitstheorie und Verwandte Gebiete* **23**, 187–196.

2

Probability Inequalities for Multivariate Distributions with Applications to Statistics

Joseph Glaz
University of Connecticut, Storrs, CT

ABSTRACT In this article we review positive and negative dependence structures for multivariate distributions. We present conditions under which product-type inequalities hold for the cumulative distribution function and the survival function of a sequence of random variables. When these conditions are not satisfied, we survey the Bonferroni-type inequalities that are often used. Applications to sequential inference, simultaneous inference, testing for randomness and outlier detection are discussed. Numerical results are presented to evaluate the performance of the inequalities discussed in this article.

Keywords and phrases Association, Bonferroni-type inequalities, moving sums, negative dependence, product-type inequalities, scan statistic

---

2.1 Introduction and Summary

In this article probability inequalities for multivariate distributions used in statistical inference are discussed. In Section 2.2 several positive dependence structures for multivariate distributions are defined and conditions are given for the validity of product-type inequalities of order $k \geq 1$ for the distribution and survival functions of a sequence of random variables.

In Section 2.3 only negative dependence structures most relevant to the development of product-type inequalities are mentioned.

In Section 2.4 a brief introduction to Bonferroni-type inequalities is presented. The emphasis is on the inequalities that are used in the applications discussed in Section 2.5. For more information on this subject the reader is referred to the book by Galambos and Simonelli (1996).

In Section 2.5 we discuss five applications that use the inequalities that have been presented in Sections 2.2–2.4. The first application deals with deriving an accurate upper bound for the tail probability of a stopping time associated with a sequential test. Based on this bound an accurate bound for the expected stopping time is obtained. Both bounds play an important role in sequential inference [Glaz and Kenyon (1995)].

A discrete scan statistic is the topic of the second application. Discrete scan statistics have been used extensively in various areas of science and technology to test the null hypotheses that the observations are identically distributed against a clustering alternative. For current research in this area and references for the applications see Glaz and Balakrishnan (1999). In this article numerical results are given to evaluate the inequalities developed in Glaz and Naus (1991) and Chen and Glaz (1995) for tail probabilities and expected values of a discrete scan statistic for 0-1 Bernoulli trials.

The third application discusses high order inequalities for the distribution function of a multinomial random vector. Both product-type and Bonferroni-type inequalities are used. These inequalities are specialized to approximate the distribution of the largest order statistic. Applications to testing for randomness and outlier detection are briefly discussed.

The fourth application is about approximating the distribution of a conditional discrete scan statistic for 0-1 Bernoulli trials. Bonferroni-type inequalities are used to accomplish this task. Based on these inequalities, quite accurate inequalities for the expected size of the scan statistic are obtained.

The last application presents simultaneous prediction intervals in time series models. To achieve a prescribed confidence level for data modeled by an AR(1) model both product-type and Bonferroni-type inequalities are used. Simultaneous confidence intervals for more general time series models are discussed in Glaz and Ravishanker (1991) and Ravishanker, Wu, and Glaz (1991).

2.2 Positive Dependence and Product-Type Inequalities

A sequence of random variables $X_1, ..., X_n$ or its distribution function is said to be *positive lower orthant dependent (PLOD)* if for all x_i

$$P(X_1 \leq x_1, ..., X_n \leq x_n) \geq \prod_{i=1}^{n} P(X_i \leq x_i). \qquad (2.2.1)$$

It is said to be *positive upper orthant dependent (PUOD)* if for all x_i

$$P(X_1 > x_1, ..., X_n > x_n) \geq \prod_{i=1}^{n} P(X_i > x_i). \qquad (2.2.2)$$

It is known that for $n = 2$ PLOD and PUOD are equivalent, while for $n \geq 3$ this fact is not true anymore [Ahmed *et al.* (1978) and Lehmann (1966)]. If $X_1, ..., X_n$ are both PLOD and PUOD we will say that they are *positive orthant dependent (POD)*. These inequalities have appeared first in the statistical literature in the context of simultaneous testing [Kimball (1951)] and simultaneous confidence interval estimation [Dunn (1959), Khatri (1967), Sidak (1967, 1971), and Scott (1967)].

In general it is quite difficult to verify the validity of PLOD or PUOD for a given sequence of random variables. To accomplish this task, Esary *et al.* (1967) introduced the following concept of positive dependence. A sequence of random variables $X_1, ..., X_n$ or its distribution function is said to be *associated,* if for all coordinatewise increasing real valued functions of n variables f and g such that $E[f(X_1, ..., X_n)]$ and $E[g(X_1, ..., X_n)]$ are finite

$$E[f(X_1, ..., X_n)g(X_1, ..., X_n)] \geq E[f(X_1, ..., X_n)]E[g(X_1, ..., X_n)].$$
$$(2.2.3)$$

The following theorem is of major importance.

THEOREM 2.2.1 *[Esary et al. (1967)]*
If $X_1, ..., X_n$ are associated then they are POD.

The class of distributions that are associated includes: multivariate normal distributions with nonnegative correlations [Pitt (1982) and Joag-Dev *et al.* (1983)], multivariate exponential distribution [Olkin and Tong (1994)], absolute value multivariate Cauchy distribution, negative multinomial, and multivariate logistic [Karlin and Rinott (1980a)].

In some applications the inequalities given in Equations (2.2.1) and (2.2.2) are not accurate, since they are based only on the univariate marginal distributions. This is especially true if the dependence between the random variables is strong and n is moderate or large [Glaz (1990, 1992) and Glaz and Johnson (1984, 1986)]. We introduce below concepts of positive dependence that lead to more accurate product-type inequalities that are based on k dimensional marginal distributions. We will say that these product-type inequalities are of order k.

A sequence of random variables $X_1, ..., X_n$ is said to be *sub-Markov* with respect to a sequence of intervals $I_1, ..., I_n$ if for any $1 \leq i \leq k \leq n$

$$P\left(X_k \in I_k | \bigcap_{j=1}^{k-1} (X_j \in I_j)\right) \geq P\left(X_k \in I_k | \bigcap_{j=i}^{k-1} (X_j \in I_j)\right), \quad (2.2.4)$$

where $\bigcap_{j=k}^{k-1}(X_j \in I_j)$ is to be interpreted as the entire space [Glaz (1990)]. It is easy to verify that if $X_1, ..., X_n$ are sub-Markov with respect to infinite intervals of the same type, i.e. $I_j = (-\infty, b_j]$, $1 \leq j \leq n$, or $I_j = [a_j, \infty)$, $1 \leq j \leq n$, (*SM with respect to ∞ intervals*), then they are associated. It turns out that the sub-Markov property is exactly what we need to establish the product-type inequalities presented below [Glaz (1990) and Glaz and Johnson (1984)]. Since it is quite difficult to establish the sub-Markov property for a sequence of random variables the following concepts of positive dependence play a major role.

A nonnegative real-valued function of two variables, $f(x, y)$, is *totally positive of order two*, TP_2, if for all $x_1 < x_2$ and $y_1 < y_2$

$$f(x_1, y_1)f(x_2, y_2) \geq f(x_1, y_2)f(x_2, y_1)$$

[Karlin (1968)]. A nonnegative real-valued function of n variables, $f(x_1, ..., x_n)$, is *multivariate totally positive of order two*, MTP_2, if for all $\mathbf{x} = (x_1, ..., x_n)$ and $\mathbf{y} = (y_1, ..., y_n)$

$$f(\mathbf{x} \vee \mathbf{y})f(\mathbf{x} \wedge \mathbf{y}) \geq f(\mathbf{x})f(\mathbf{y}),$$

where

$$\mathbf{x} \vee \mathbf{y} = (\max\{\mathbf{x}_1, \mathbf{y}_1\}, ..., \max\{\mathbf{x}_n, \mathbf{y}_n\})$$

and

$$\mathbf{x} \wedge \mathbf{y} = (\min\{\mathbf{x}_1, \mathbf{y}_1\}, ..., \min\{\mathbf{x}_n, \mathbf{y}_n\})$$

[Karlin and Rinott (1980a)]. A nonnegative real-valued function of n variables, $f(x_1, ..., x_n)$, is *totally positive of order two in pairs*, TP_2 *in pairs,* if for any pair of arguments x_i and x_j, the function f viewed as a function of x_i and x_j only, with the other arguments kept fixed, is TP_2 [Barlow and Proschan (1975)].

If f is MTP_2 then it is also TP_2 *in pairs.* But, if the support of f is a product space then f is MTP_2 if and only if it is TP_2 *in pairs* [Kemperman (1977) and Block and Ting (1981)]. We say that a sequence of random variables $X_1, ..., X_n$ is MTP_2 (resp. TP_2 *in pairs)* if its joint density function is MTP_2 (resp. TP_2 *in pairs).* We will assume that the support of the joint density functions is a product space.

The following results is often used to show that $X_1, ..., X_n$ are associated.

THEOREM 2.2.2 *[Sarkar (1969)]*
If $X_1, ..., X_n$ are MTP_2 then they are associated.

The above result is especially useful in determining when $|X_1|, ..., |X_n|$ are associated, as it is often easier to verify that $|X_1|, ..., |X_n|$ are MTP_2.

REMARK It is still an open problem to characterize when $|X_1|, ..., |X_n|$ are associated, if $X_1, ..., X_n$ have a multivariate normal distribution.

The following interesting examples of MTP_2 random variables are listed in Karlin and Rinott (1980a): multivariate normal with nonnegative partial correlations, absolute value multivariate normal with only nonnegative off-diagonal elements in $-D\Sigma^{-1}D$, where Σ is the covariance matrix and D is some diagonal matrix with elements ± 1, multivariate logistic, absolute value multivariate Cauchy, negative multinomial, certain classes of multivariate t, certain classes of multivariate gamma, certain classes of partial sums of i.i.d. random variables, and order statistics of i.i.d. random variables.

The following result gives sufficient conditions for the validity of product-type inequalities of order k.

THEOREM 2.2.3 *[Glaz and Johnson (1984)]*
Let $X_1, ..., X_n$ be MTP_2 random variables and $I_1, ..., I_n$ be infinite intervals, all of the same type. Then $X_1, ..., X_n$ are sub-Markov with respect to $I_1, ..., I_n$ and

$$P(X_1 \in I_1, ..., X_n \in I_n) \geq \gamma_k \geq \gamma_1$$

where

$$\gamma_k = P(X_1 \in I_1, ..., X_k \in I_k)$$

$$\times \prod_{i=k+1}^{n} \left[\frac{P(X_{i-k+1} \in I_{i-k+1}, ..., X_i \in I_i)}{P(X_{i-k+1} \in I_{i-k+1}, ..., X_{i-1} \in I_{i-1})} \right] \quad (2.2.5)$$

is an increasing function in k.

Note that γ_1 is the product-type inequality given by Equation (2.2.1) or (2.2.2). If the conditions of Theorem 2.3.1 are satisfied, γ_k, the kth order product-type inequality is tighter than γ_1. For $k = 2$ we get from Equation (2.2.5):

$$\gamma_2 = P(X_1 \in I_1, X_2 \in I_2) \prod_{i=3}^{n} \left[\frac{P(X_{i-1} \in I_{i-1}, X_i \in I_i)}{P(X_{i-1} \in I_{i-1})} \right]. \quad (2.2.6)$$

Also, if $X_1, ..., X_n$ are stationary and $I_i = I$ then

$$\gamma_k = P(X_1 \in I, ..., X_k \in I) \left[\frac{P(X_1 \in I, ..., X_k \in I)}{P(X_1 \in I, ..., X_{k-1} \in I)} \right]^{n-k}. \quad (2.2.7)$$

In Section 5 the performance of these product-type bounds will be evaluated for several applications.

Other concepts of positive dependence imply the validity of the product-type bounds given in Equation (2.2.5). We will mention here two such concepts.

A sequence of random variables $X_1, ..., X_n$ or its distribution function is said to be *right corner set increasing (RCSI)* or *left corner set decreasing (LCSD)* if for all $1 \le i \le n$ and x_i,

$$P(X_i > x_i | X_j > x_j, 1 \le j \le n, \ j \ne i)$$

is an increasing function in all the x_j's, or for all $1 \le i \le n$ and x_i, or

$$P(X_i \le x_i | X_j \le x_j, 1 \le j \le n, \ j \ne i)$$

is a decreasing function in all the x_js, respectively [Harris (1970)]. We will say that $X_1, ..., X_n$ are *positive corner set dependent (PCSD)* if they are both RCSI and LCSD.

A sequence of random variables $X_1, ..., X_n$ or its distribution function is said to be *right tail increasing in sequence (RTIS)* or *left tail decreasing in sequence (LTDS)* if for all $1 \le i \le n$ and x_i,

$$P(X_i > x_i | X_j > x_j, 1 \leq j \leq i-1)$$

is an increasing function in $x_1, ..., x_{i-1}$ or for all $1 \leq i \leq n$ and x_i, or

$$P(X_i \leq x_i | X_j \leq x_j, 1 \leq j \leq i-1)$$

is a decreasing function in $x_1, ..., x_{i-1}$, respectively [Esary and Proschan (1972)]. We will say that $X_1, ..., X_n$ are *positive tail sequence dependent (PTSD)* if they are both RTIS and LTDS. Properties and relationships among these and other related notions of positive dependence have been investigated be many authors, including: Alam and Wallenius (1976), Barlow and Proschan (1975), Block and Ting (1981), Brindley and Thompson (1972), and Glaz (1990).

The following relationships among the positive dependence structures presented above hold:

$$MTP_2 \implies PCSD \implies PTSD \implies SM \ \infty \ intervals$$

$$\implies Associated \implies POD.$$

It will be interesting to study in more detail the differences among these dependence structures and try to characterize them for certain classes of multivariate distributions. Investigations of these dependence structures for sequences of random vectors and stochastic processes are of interest as well.

Let $X_1, ..., X_n$ be a sequence of independent random variables. The sequence of moving sums $\{\sum_{j=i}^{m+i-1} X_j\}_{i=1}^{n-m+1}$ is *POD* and therefore the product-type inequalities given in Equations (2.2.1) and (2.2.2) hold. The moving sums of independent random variables are not MTP_2 and one can show that the product-type inequalities of order $k \geq 2$ given in Equations (2.2.5) do not hold in this case [Glaz and Johnson (1988)]. For a sequence of i.i.d. random variables Glaz and Naus (1991, 1996) have used a different approach to derive accurate product-type inequalities for a sequence of moving sums. We present here only one of the results for a sequence of i.i.d. integer valued random variables. For $2 \leq m \leq n_1 \leq n$, let

$$G(n_1) = P \left(\max_{1 \leq i \leq n_1 - m + 1} \sum_{j=i}^{m+i-1} X_j \leq k-1 \right). \qquad (2.2.8)$$

THEOREM 2.2.4   *[Glaz and Naus (1991)]*
Let $X_1, ..., X_n$ be a sequence of i.i.d. integer valued random variables. Then for $n \geq 3m$,

$$\frac{G(3m)}{\left[1 + \frac{G(2m) - G(2m-1)}{G(3m-1)}\right]^{n-3m}}$$

$$\leq G(n) \leq G(3m)\left[1 - G(3m) + G(3m-1)\right]^{n-3m}. \quad (2.2.9)$$

An application of these inequalities to a discrete scan statistic is presented in Section 2.5.

2.3   Negative Dependence and Product-Type Inequalities

In this section only the most relevant dependence structures for constructing high order product-type inequalities will be presented. In the statistical literature many interesting negative dependence structures for multivariate distributions have been discussed. In fact, most of these negative dependence structures are defined by reversing the inequalities given in Section 2.2 [Block *et al.* (1982), Ebrahimi and Ghosh (1981, 1984), Glaz (1990), Kim and Seo (1995) and Lee (1985)].

A nonnegative real-valued function of two variables, $f(x, y)$, is *reverse rule of order two, RR_2*, if for all $x_1 < x_2$ and $y_1 < y_2$

$$f(x_1, y_1)f(x_2, y_2) \leq f(x_1, y_2)f(x_2, y_1)$$

[Karlin (1968)]. A nonnegative real-valued function of n variables, $f(x_1, ..., x_n)$, is *multivariate reverse rule of order two, MRR_2*, if for all $\mathbf{x} = (x_1, ..., x_n)$ and $\mathbf{y} = (y_1, ..., y_n)$

$$f(\mathbf{x} \vee \mathbf{y})f(\mathbf{x} \wedge \mathbf{y}) \leq f(\mathbf{x})f(\mathbf{y}).$$

[Karlin and Rinott (1980b)]. If the domain of the function f is a product space then MRR_2 is equivalent to f being RR_2 in each pair of the arguments, while the rest of the arguments are kept fixed. An MRR_2 function has many nice properties, except that the MRR_2 property is not preserved under composition. Therefore if f is a density function of a random vector and it is MRR_2, the marginal densities are not necessarily MRR_2. In fact they can be even positively dependent [Karlin and Rinott (1980b, p. 510)]. To eliminate this defect, the following stronger concept of negative dependence is defined in Karlin and Rinott (1980b).

A density function f of a random vector $\mathbf{X} = (X_1, ..., X_n)$ is said to be *strongly* MRR_2, $S - MRR_2$, if it is MRR_2 and for any set of PF_2 functions $\{\varphi_m\}$ (a function φ defined on $(-\infty, \infty)$ is PF_2 if $\varphi(x - y)$ is TP_2 in the variables $-\infty < x, y < \infty$), the marginal density functions

$$g(x_{m_1}, ..., x_{m_k}) = \int \cdots \int f(x_1, ..., x_n) \prod_{i=1}^{n-k} \varphi_i(x_{j_i}) dx_{j_i} \qquad (2.3.1)$$

are MMR_2 in the variables

$$\{x_{m_1}, ..., x_{m_k}\} = \{x_1, ..., x_n\} \setminus \{x_{j_1}, ..., x_{j_{n-k}}\}.$$

The class of distributions with $S - MRR_2$ density functions includes: multinomial, multivariate hypergeometric, multivariate Hahn, Dirichlet family of densities on a simplex, multinormal with special covariance structure [Karlin and Rinott (1980b)].

A sequence of random variables $X_1, ..., X_n$ is said to be *super-Markov* with respect to a sequence of intervals $I_1, ..., I_n$ if for any $1 \leq i \leq k \leq n$

$$P\left(X_k \in I_k \Big| \bigcap_{j=1}^{k-1}(X_j \in I_j)\right) \leq P\left(X_k \in I_k \Big| \bigcap_{j=i}^{k-1}(X_j \in I_j)\right), \qquad (2.3.2)$$

where $\bigcap_{j=k}^{k-1}(X_j \in I_j)$ is to be interpreted as the entire space [Glaz (1990)]. For $k \geq 1$ let γ_k be defined in Equation (2.2.5). The following result establishes the product-type inequalities.

THEOREM 2.3.1 *[Glaz and Johnson (1984)]*
Let $X_1, ..., X_n$ be $S - MRR_2$ random variables and $I_1, ..., I_n$ be infinite intervals, all of the same type. Then $X_1, ..., X_n$ are super-Markov with respect to $I_1, ..., I_n$ and

$$P(X_1 \in I_1, ..., X_n \in I_n) \leq \gamma_k \leq \gamma_1,$$

where γ_k are decreasing in k.

In Section 2.5 the performance of these product-type bounds will be evaluated for several applications.

2.4 Bonferroni-Type Inequalities

The following classical Bonferroni inequalities for the probability of a union of n events have been derived in Bonferroni (1936). Let $A_1, ...,$

A_n be a sequence of events and let $A = \cup_{i=1}^n A_i$. Then for $2 \le k \le n$,

$$\sum_{j=1}^k (-1)^{j-1} U_j \le P(A) \le \sum_{j=1}^{k-1} (-1)^{j-1} U_j \qquad (2.4.1)$$

where k is an even integer and for $1 \le j \le n$,

$$U_j = \sum_{1 \le i_1 < \dots < i_j \le n} P\left(\bigcap_{m=1}^j A_{i_m}\right). \qquad (2.4.2)$$

The orders of these inequalities is given by the upper index of the sums in Equation (2.4.1). The first order Bonferroni upper bound is referred to in the statistical literature as Boole's inequality, derived in 1854 (Boole 1984). These classical lower (upper) Bonferroni inequalities have a shortcoming of not necessarily increasing (decreasing) with the increase of the order of the inequality. Moreover, these inequalities can be wide for a low order and the computational complexity increases rapidly with the increase of the order of the inequalities. Therefore attempts have been made to derive simpler inequalities that are more accurate than the classical Bonferroni inequalities. We refer to these inequalities as Bonferroni-type inequalities.

In this article we make use of the following class of upper Bonferroni-type inequalities proposed in Hunter (1976). The basic idea of this approach is to represent the event $A = \cup_{i=1}^n A_i$ as a union of disjoint event

$$A = A_1 \cup \left\{ \bigcup_{i=2}^n \left[A_i \cap \left(\bigcap_{j=1}^{i-1} A_j^c \right) \right] \right\}.$$

Therefore,

$$P(A) = P(A_1) + \sum_{i=2}^n P\left[A_i \cap \left(\bigcap_{j=1}^{i-1} A_j^c \right) \right].$$

For $2 \le k \le n-1$,

$$P(A) \le P(A_1) + \sum_{i=2}^k P\left[A_i \cap \left(\bigcap_{j=1}^{i-1} A_j^c \right) \right]$$

$$+ \sum_{i=k+1}^n P\left[A_i \cap \left(\bigcap_{j=i-k+1}^{i-1} A_j^c \right) \right].$$

This inequality can be rewritten for $k = 2$ as

$$P(A) \leq U_1 - \sum_{i=2}^{n} P(A_i \cap A_{i-1}) \qquad (2.4.3)$$

and for $k \geq 3$ as

$$P(A) \leq U_1 - \sum_{i=2}^{n} P(A_i \cap A_{i-1})$$

$$- \sum_{j=2}^{k-1} \sum_{i=1}^{n-j} P \left\{ A_i \cap \left[\left(\bigcap_{j=i+1}^{i+j-1} A_j^c \right) \cap A_{i+j} \right] \right\}. \qquad (2.4.4)$$

Note that these upper inequalities are of order k and they decrease as the order of the inequalities is increasing.

The inequality (2.4.3) is a member of a class of second order inequalities derived in Hunter (1976). The following concepts from graph theory have to be introduced. A *graph* $G(V, E)$, or briefly G, is a combinatorial structure comprised of a set of vertices V and a set of edges E. Each edge is associated with two vertices referred to as the end points of the edge. The graph G is called *undirected* if both endpoints of an edge have same relationship for each of the edges. A *path* is a sequence of edges such that two consecutive edges in it share a common end point. A *tree* is an undirected graph such that there is a unique path between every pair of its vertices. Let $v_1, \ldots, v_n$ be the vertices of a tree T, representing the events $A_1, \ldots, A_n$, respectively. The vertices v_i and v_j are joined by an edge e_{ij} if and only if $A_i \cap A_j \neq \phi$. The following inequality has been derived in Hunter (1976):

$$P(A) \leq S_1 - \sum_{\{(i,j);\ e_{ij} \in T\}} P(A_i \cap A_j). \qquad (2.4.5)$$

The optimal inequality in the class of inequalities (2.4.5) can be obtained via an algorithm of Kruskal [Worsley (1982)]. The inequality (2.4.3) is the most stringent within the class of inequalities (2.4.5) if the events $A_1, \ldots, A_n$ are exchangeable or if they are ordered in such a way that for $1 \leq i_1 < i_2 < n$, $P(A_{i_1} \cap A_{i_2})$ is maximized for $i_1 - i_2 = 1$ (Worsley 1982, Examples 3.1 and 3.2).

Inequality (2.4.4) is a member of the following class of inequalities investigated in Hoover (1990):

$$P(A) \leq P \left(\bigcup_{i=1}^{k} A_i \right) + \sum_{j=k+1}^{n} P \left\{ A_j \cap \left(\bigcap_{m \in T_j} A_m^c \right) \right\}, \qquad (2.4.6)$$

where T_j is a subset of $\{1, ..., j-1\}$ of size $k-1$ and $j \geq k+1$. For $k \geq 3$ there is no efficient algorithm to obtain the optimal inequality in the class of inequalities (2.4.6). If the events $A_1,...., A_n$ are naturally ordered in such a way that $P(\cap_{j=1}^{k-1} A_{i_j})$ is maximized for $i_j - i_{j-1} = 1$, $2 \leq j \leq k-1$ and $3 \leq k \leq n-1$, the natural ordering with $T_j = \{j-1, ..., j-k+1\}$ is recommended. In this case, inequalities (2.4.6) reduce to inequalities (2.4.4). If the events $A_1,...., A_n$ are exchangeable inequalities (2.4.4) have the following simplified form

$$P(A) \leq nP(A_1) - (n-1)P(A_1 \cap A_2)$$
$$- \sum_{j=2}^{k-1}(n-j)P\left\{A_1 \cap \left[\left(\bigcap_{i=2}^{j} A_i^c\right) \cap A_{j+1}\right]\right\}. \quad (2.4.7)$$

It is tedious but quite routine to verify that the Bonferroni-type inequalities (2.4.4) and (2.4.7) for $P(\cap_{i=1}^{n} A_i^c) = 1 - P(\cup_{i=1}^{n} A_i)$ are given by

$$P\left(\bigcap_{i=1}^{n} A_i^c\right) \geq \sum_{i=1}^{n-k+1} \alpha_{i,i+k-1} - \sum_{i=1}^{n-k} \alpha_{i+1,i+k-1} = \delta_k \quad (2.4.8)$$

and

$$P\left(\bigcap_{i=1}^{n} A_i^c\right) \geq (n+k-1)\alpha_{1,k} - (n-k)\alpha_{2,k}, \quad (2.4.9)$$

respectively, where

$$\alpha_{m,n} = P\left(\bigcap_{i=m}^{n} A_i^c\right).$$

In some examples it is easier to evaluate the terms $\alpha_{m,n}$, in which case inequalities (2.4.8) or (2.4.9) will be used. Many other interesting Bonferroni-type inequalities for the union of n events are discussed in a recent book by Galambos and Simonelli (1996).

The following lower Bonferroni-type inequality for $P(\cup_{i=1}^{n} A_i)$ had been derived in Dawson and Sankoff (1967). Let U_1 and U_2 be given in Equation (2.4.2) above. Then

$$P(A) \geq \frac{2U_1}{a} - \frac{2U_2}{a(a-1)}, \quad (2.4.10)$$

where a is the integer part of $2 + 2U_2/U_1$. Kwerel (1975), using a linear programming approach, has proved that inequality (2.4.10) is the optimal inequality in the class of all linear inequalities of the form

$$P(A) \geq b_1 U_1 + b_2 U_2.$$

In particular it outperforms the Bonferroni inequality $P(A) \geq S_1 - S_2$, whenever $a > 2$. Higher order lower Bonferroni-type inequalities for $P(\cup_{i=1}^{n} A_i)$ are discussed in Galambos and Simonelli (1996).

2.5 Applications

2.5.1 Sequential Analysis

Let $X_1, ..., X_n, ...$ be a sequence of independent and identically distributed $(i.i.d.)$ normal random variables with unknown mean θ and known variance σ^2. Without loss of generality we assume that $\sigma^2 = 1$. For $\theta_0 > 0$, we are interested in testing the following hypotheses

$$H_0 : \theta \leq \theta_0 \quad \text{vs} \quad H_1 : \theta > \theta_0.$$

Without loss of generality we assume $\theta_0 = 0$. The following stopping time is used in the sequential probability ratio test $(SPRT)$ for testing the above hypotheses:

$$\tau = \inf\{n \geq 1; T_n \notin I_n\},$$

where $T_n = \sum_{i=1}^{n} X_i$ is the sequence of partial sums of i.i.d. normal random variables, $I_n = (-a/\theta_1 + n\theta_1/2, a/\theta_1 + n\theta_1/2)$ is the continuation region, and a and θ_1 are the design parameters selected to control the type I and type II probability errors:

$$\beta(0) = \alpha \text{ and } \beta(\theta_1) = 1 - \beta_1,$$

and β is the power function of the sequential test. [Glaz and Kenyon (1995)]. It is convenient to transform the continuation region so that it will be symmetric about the n axis of the (n, T_n) plane. This is accomplished by the transformation $Y_i = X_i - \theta/2$ and defining

$$\tau = \inf\{n \geq 1; T_n^* \notin I_n^*\},$$

where $T_n^* = \sum_{i=1}^{n} Y_i$ and $I_n^* = (-a/\theta_1, a/\theta_1)$. The null hypothesis and the error rates requirements are transformed to $H_0 : \theta \leq -\theta^*$ and $\beta(-\theta^*) = \alpha$, $1 - \beta(\theta^*) = \beta_1$, respectively, where $\theta^* = \theta_1/2$ To satisfy these error rates requirements one uses $a = \ln(\gamma/\alpha) - 0.583$, where γ is the Laplace transform of the asymptotic distribution of the excess of the random walk $\theta_1 T_n^*$ over the boundary evaluated at the value 1 [Woodroofe (1982, Section 3.1)]. We present bounds for $P(\tau \geq n)$ and $E(\tau)$, which are of interest in evaluating the performance of sequential tests. For references and a more extensive discussion of sequential tests see Glaz and Johnson (1986) and Glaz and Kenyon (1995). The following result is central to deriving bounds for $P(\tau \geq n)$ and $E(\tau) = \sum_{i=1}^{\infty} P(\tau \geq n)$.

THEOREM 2.5.1   *[Glaz and Johnson (1986)]*
Let $X_1, ..., X_n, ...$ be a sequence of independent random variables from a density function $f(x)$, that has support on an interval. Then for any $n \geq 2$ the sequence of partial sums $T_i = \sum_{j=1}^{i} X_j$, $1 \leq i \leq n$, is MTP_2 (or TP_2 in pairs) if and only if

$$f(x) = e^{-\psi(x)}$$

where $\psi(x)$ is a convex function, or equivalently $f(x)$ is a log concave function.

It follows from this theorem that if $X_i, i \geq 1$, are i.i.d. normal random variables then the sequence of partial sums $\{T_i^*\}_{i=1}^{n}$ is MTP_2. Therefore the product-type bounds given in equation (2.2.5) are valid here. Glaz and Kenyon (1995) developed efficient algorithms to evaluate product-type inequalities for $P(\tau \geq n)$. In Tables 2.1 and 2.2, numerical results are presented for the $SPRT$ for normal data for selected values of the parameters.

There are many interesting open problems in this area of research. For example, development of probability inequalities for stopping times associated with sequential tests for a vector of parameters.

2.5.2   A Discrete Scan Statistic

Let $X_1, ..., X_n$ be a sequence of independent nonnegative integer valued random variables. A *discrete scan statistic*, denoted by S_n, is defined in terms of moving sums of length $2 \leq m \leq n$ as follows:

$$S_n = \max_{1 \leq i \leq n-m+1} \sum_{j=i}^{i+m-1} X_j. \qquad (2.5.1)$$

TABLE 2.1
Lower bounds and simulated values of $P(\tau > n)$
$\alpha = \beta_1 = .05$ **and** $\theta^* = .5$

n	γ_3	γ_7	$\widehat{P}(\tau > n)$
2	$.8273^*$	$.8273^*$	.8276
4	.5130	$.5133^*$	.5112
6	.3026	$.3057^*$	.3153
8	.1757	.1812	.1865
10	.1013	.1073	.1098
12	.0581	.0636	.0649
14	.0332	.0376	.0390
16	.0190	.0223	.0239
18	.0108	.0132	.0145

$\widehat{P}(\tau > n)$ is a simulated value of $P(\tau > n)$ based on $10,000$ trials. The starred values are exact.

TABLE 2.2
Lower bounds and simulated values of $E_\theta(\tau)$

$\alpha \backslash \theta$	$\theta^* = .25$			$\theta^* = .50$		
	-.25	-.125	0	-.50	-.25	0
.01	31.39	46.22	59.46	9.05	14.13	19.04
	34.55	53.87	72.04	9.28	15.21	21.16
	36.16	61.31	84.90	9.23	15.25	21.54
.05	19.51	25.77	29.53	5.62	7.70	8.99
	21.38	29.34	34.25	5.69	7.87	9.24
	22.32	30.58	36.20	5.72	7.99	9.25

Upper and middle values are lower bounds based on γ_3 and γ_7, respectively.
Lower value is simulated based on 10,000 trials.

TABLE 2.3
Product-type bounds for $P(S_n \geq k)$ Bernoulli model, $n = 500$

m	p	k	LB	UB	$\widehat{P}(k; m, n)$
10	.01	2	.3238	.3286	.3269
		3	.0159	.0159	.0141
		4	.0003	.0003	.0003
	.05	2	.9967	.9996	.9992
		3	.7274	.7479	.7328
		4	.1534	.1544	.1467
		5	.0134	.0134	.0141
20	.01	2	.5064	.5299	.5287
		3	.0635	.0639	.0624
		4	.0038	.0038	.0040
	.05	3	.9523	.9857	.9758
		4	.6263	.6509	.6476
		5	.1964	.1996	.1962
		6	.0352	.0353	.0315
		7	.0045	.0045	.0058

$\widehat{P}(k; m, n)$ is a simulated value of $P(k; m, n)$
based on 10,000 trials.

For the special case of i.i.d. $0 - 1$ Bernoulli trials this scan statistic generalizes the notion of the longest run of 1s that has been investigated extensively in the statistical literature. This discrete scan statistic is used for testing the null hypothesis of the observations being identically distributed against an alternative hypothesis of clustering, that specifies an increased occurrence of events in a consecutive subsequence of the observed data. For an extensive discussion about this scan statistic and the applications see Glaz and Naus (1991) and Chen and Glaz (1999). Based on Theorem 2.2.4, in Table 2.3 we present bounds for $P(S_n \geq k)$ for a sequence of i.i.d. $0 - 1$ Bernoulli trials.

2.5.3   An Approximation for a Multinomial Distribution

Let $\mathbf{X} = (X_1, ..., X_m)$ be a multinomial random vector with cell probabilities $p_1, ..., p_m$ and parameters $n_1, ..., n_m$, $0 < p_i < 1$, $\sum_{i=1}^{m} p_i = 1$ and $\sum_{i=1}^{m} n_i = N$. Since $\mathbf{X}$ is $S - MRR_2$, it follows from Theorem 6 that for $k \geq 1$,

$$P(X_1 \leq x_1, ..., X_m \leq x_m) \leq \gamma_k. \tag{2.5.2}$$

A Bonferroni-type lower bound for $P(X_1 \leq x_1, ..., X_m \leq x_m)$ is obtained from Equation (2.4.8):

$$P(X_1 \leq x_1, ..., X_m \leq x_m) \geq \delta_k. \tag{2.5.3}$$

TABLE 2.4
Bounds for P-values for a test of equal cell probabilities

n	5	6	7	8	9	10	11
γ_3	.9944	.8562	.4758	.1744	.0496	.0121	.0026
δ_3	1	1	.6200	.1894	.0507	.0121	.0026

For $1 \leq i \leq m$, $x_i = x$, inequalities (2.5.2) and (2.5.3) yield inequalities for the distribution function of $X_{(m)} = \max(X_1, ..., X_m)$. An approximation for the distribution of $X_{(1)} = \min(X_1, ..., X_m)$ is given in Glaz (1990). For the special case of $p_i = 1/m$, $1 \leq i \leq m$, inequalities (2.5.2) and (2.5.3) simplify to

$$(m + k - 1)\alpha_{1,k} - (m - k)\alpha_{1,k-1} \leq P(X_{(m)} \leq x)$$
$$\leq \frac{[\alpha_{1,k}]^{m-k+1}}{[\alpha_{1,k-1}]^{m-k}}, \qquad (2.5.4)$$

where for $j \geq 1$

$$\alpha_{1,j} = P(X_1 \leq x, ..., X_j \leq x).$$

The performance of inequalities (2.5.4) is evaluated for an example with a roulette with $m = 38$ numbers. Suppose we are interested to test the null hypotheses that $p_i = 1/38$, $1 \leq i \leq m$. Consider the test that rejects this null hypothesis for large values of $X_{(38)}$. Table 2.4 presents inequalities (2.5.4) for $k = 3$ for P-values of this test for $N = 100$. Inequalities for the distribution of $X_{(1)}$ and $X_{(m)}$ can be used to detect outliers in a specified multinomial model.

2.5.4 A Conditional Discrete Scan Statistic

Let $X_1, ..., X_n$ be i.i.d. $0 - 1$ Bernoulli trials with $P(X_i = 1) = p$, $0 < p < 1$. Suppose that we know that a successes (1s) and $n - a$ failures (0s) have been observed. Then

$$P\left(X_1 = x_1, ..., X_n = x_n \bigg| \sum_{i=1}^{n} X_i = a\right) = \frac{1}{\binom{n}{a}}.$$

In this case the joint distribution of the observed $0 - 1$ trials assigns equal probabilities to all the $\binom{n}{a}$ arrangements of a 1's and $n - a$ 0s. We are interested in deriving tight bounds for the upper tail probabilities for a conditional scan statistic denoted by

$$P(k; m, n, a) = P\left(S_m \geq k \mid \sum_{i=1}^{n} X_i = a\right),$$

where S_m is defined in Equation (2.5.1).

Let $Y_1, ..., Y_n$ be the random variables denoting one of the $\binom{n}{a}$ sequences of $0-1$ trials that contain a 1s and $n - a$ 0s. For $n = Lm$, $L \geq 4$ and $1 \leq i \leq L - 1$ define the events

$$D_i = \bigcap_{j=1}^{m+1} \left(Y_{(i-1)m+j} + ... + Y_{im+j-1} \leq k - 1\right).$$

It follows that

$$P(k; m, n, a) = P\left(\bigcup_{i=1}^{L-1} D_i^c\right).$$

Employing the second order Bonferroni-type inequality in Hunter (1976) and Worsley (1982) given in Equation (2.4.3) and the fact that D_i are stationary events we get

$$P(k; m, n, a) \leq \sum_{i=1}^{L-1} P\left(D_i^c\right) - \sum_{i=2}^{L-1} P\left(D_i^c \cap D_{i-1}^c\right)$$

$$= 1 + (L - 3)q_{2m}(a) - (L - 2)q_{3m}(a),$$

where for $1 \leq i \leq L - 1$, $q_{2m}(a) = P(D_i)$, $q_{3m}(a) = P(D_i \cap D_{i-1})$ and for $r = 2, 3$

$$q_{rm} = \sum_{j=0}^{\min(rk-r,a)} q(rm|j) \frac{\binom{rm}{j}\binom{n-rm}{a-j}}{\binom{n}{a}}$$

where $q(rm|j) = 1 - P(k; m, rm, j)$ are evaluated in Naus (1974, Theorem 1).

To derive a Bonferroni-type lower bound for $P(k; m, n, a)$ an inequality in Kwerel (1975) presented in Equation (2.4.10) gives:

$$P\left(\bigcup_{i=1}^{L-1} D_i^c\right) \geq \frac{2s_1}{b} - \frac{2s_2}{b(b-1)},$$

where $b = [2s_2/s_1 + 2]$,

$$s_1 = \sum_{i=1}^{L-1} P\left(D_i^c\right) = (L-1)(1 - q_{2m}(a))$$

and

$$
\begin{aligned}
s_2 &= \sum_{1 \leq i < j \leq L-1} P\left(D_i^c \cap D_j^c\right) \\
&= \sum_{i=1}^{L-2} P\left(D_i^c \cap D_{i+1}^c\right) + \sum_{1 \leq i < j-1 \leq L-2} P\left(D_i^c \cap D_j^c\right) \\
&= (L-2)[1 - 2q_{2m}(a) + q_{3m}(a)] \\
&\quad + .5(L-2)(L-3)[1 - 2q_{2m}(a) + q_{2m,2m}(a)] \\
&= (L-2)q_{3m}(a) + .5(L-2)(L-3)q_{2m,2m}(a) \\
&\quad + .5(L-1)(L-2)(1 - 2q_{2m}(a)),
\end{aligned}
$$

where

$$q_{2m,2m}(a)$$
$$= \sum_{j_1=0}^{\min(2k-2,a)} \sum_{j_2=0}^{\min(2k-2,a-j_1)} q(2m|j_1)q(2m|j_2) \frac{\binom{2m}{j_1}\binom{2m}{j_2}\binom{n-4m}{a-j_1-j_2}}{\binom{n}{a}}$$

and $q(rm|j)$, $r = 2, 3$ are given above. A different approach in deriving bounds for $P(k; m, n, a)$ is presented in Chen *et al.* (1999). In Tables 2.5 and 2.6 the bounds for $P(k; m, n, a)$ and

$$E\left(S_m \middle| \sum_{i=1}^{n} X_i = a\right) = \sum_{k=1}^{m} P(k; m, n, a)$$

are evaluated for selected values of the parameters.

Probability inequalities for a conditional scan statistic for a sequence of binomial or Poisson random variables has yet to be derived.

TABLE 2.5
Bonferroni-type bounds for $P(k; m, n, a)$

n	m	a	k	$BTLB$	$BTUB$	$\widehat{P}(k; m, n, a)$
100	10	5	3	.1665	.1665	.1678
			4	.0090	.0090	.0098
	20	10	5	.3505	.3513	.3465
			6	.0752	.0752	.0752
			7	.0096	.0096	.0102
500	10	50	4	.6850	1.000	.8523
			5	.2244	.2594	.2369
			6	.0271	.0274	.0265
	20	75	8	.3772	.4869	.4283
			9	.1223	.1276	.1254
			10	.0259	.0261	.0254

$\widehat{P}(k; m, n, a)$ is a simulated value based on 10,000 trials.

TABLE 2.6
Bonferroni-type bounds for $E\left(S_m \mid \sum_{i=1}^{n} X_i = a\right)$

n	m	a	LB	UB	$\widehat{E}\left(S_m \mid \sum_{i=1}^{n} X_i = a\right)$
100	10	5	1.95	2.18	2.08
		10	2.95	3.26	3.10
	20	10	4.16	4.33	4.30
		20	6.95	7.13	7.10
500	10	25	2.72	3.15	2.94
		50	3.87	4.29	4.13
	20	50	5.50	6.01	5.82
		75	7.09	7.65	7.46

$\widehat{E}\left(S_m \mid \sum_{i=1}^{n} X_i = a\right)$ is a simulated value based on 10,000 trials.

2.5.5 Simultaneous Prediction in Time Series Models

Let $X_1, ..., X_n$ be the observed time series data. We are interested in a rectangular simultaneous prediction region for the future observations $X_{n+1}, ..., X_{n+m}$. The following region has been investigated in Glaz and Ravishanker (1991) and Ravishanker, Wu and Glaz (1991):

$$X_{n+i}(i) \pm c\sqrt{Var(e_n(i))}, \ 1 \le i \le m,$$

where for $1 \le i \le m$, $X_{n+i}(i)$ are the forecasts for X_{n+i} and $e_n(i) = X_{n+i} - X_{n+i}(i)$ are the forecasts errors. The constant c is obtained from the following equation:

$$P\left(|U_n(i)| \le c; \ 1 \le i \le m\right) = 1 - \alpha,$$

where $U_n(i) = e_n(i)/\sqrt{Var(e_n(i))}$ are the standardized forecast errors and $1 - \alpha$ is the preassigned joint confidence level of the rectangular prediction region. Assume that the time series can be modeled by an $AR(1)$ process:

$$X_t = \phi X_{t-1} + \in_t, \ |\phi| < 1,$$

and $\{\in_t\}$ is a sequence of i.i.d. normal random variables with mean 0 and variance σ^2. Then $\mathbf{U}_n = (U_n(1), ..., U_n(m))$ has a multivariate normal distribution with mean vector $\mathbf{0}$ and correlation matrix Σ. It follows from Glaz and Ravishanker (1991) that $(|U_n(1)|, ..., |U_n(m)|)$ is MTP_2 and therefore, for $1 \le k \le 5$, product-type inequalities given in Equation (2.2.5) can used to evaluate the constant c via an algorithm in Schervish (1984). This algorithm can be used to evaluate the constant c from the Bonferroni-type inequalities given in Equation (2.4.8).

In Table 2.7 the constant c is evaluated for selected joint confidence levels for a data set Series D in Box and Jenkins (1976). The data set consists of $n = 310$ hourly readings of viscosity of a chemical process. An AR(1) model is used for this data set.

There are many interesting open problems in the area of simultaneous prediction regions in time series models. For example, simultaneous confidence regions for future observed random vectors in multivariate time series, using probability inequalities, have not been investigated yet.

TABLE 2.7
Evaluating the constant c using four methods

Method	k	Confidence level						
		.99	.95	.90	.88	.80	.70	.60
Product-type	1	3.29	2.80	2.56	2.41	2.29	2.11	1.96
	2	3.21	2.68	2.41	2.24	2.11	1.90	1.73
	3	3.20	2.66	2.39	2.21	2.08	1.87	1.70
	4	3.20	2.65	2.38	2.20	2.07	1.86	1.68
	5	3.20	2.65	2.38	3.20	2.07	1.85	1.68
Bonferroni-type	1	2.29	2.81	2.58	2.43	2.33	2.17	2.05
	2	3.21	2.69	2.43	3.26	2.14	1.95	1.81
	3	3.20	2.67	2.40	2.23	2.10	1.91	1.75
	4	3.20	2.66	2.39	2.22	2.09	1.89	1.73
	5	3.20	2.65	2.38	2.21	2.08	1.87	1.71
Simulation		3.19	2.64	2.37	2.19	2.06	1.85	1.68
Marginal		2.58	1.96	1.64	1.44	1.28	1.04	.84

The simulation method is based on 10,000 trials.
The marginal method is based on marginal confidence levels.

References

1. Ahmed, A. N., Langberg, N. A., Leon, R., and Proschan, F. (1978). Two concepts of positive dependence with applications in multivariate analysis. *Technical Report M486*, Department of Statistics, Florida State University.

2. Alam, K. and Wallenius, K. T. (1976). Positive dependence and monotonicity in conditional distributions. *Communications in Statistics — Theory and Methods* **5**, 525–534.

3. Barlow, R. E. and Proschan, F. (1975). *Statistical Theory of Reliability and Life Testing.* Holt, Reinhart, and Winston, Inc., New York.

4. Block, H. W., Savits, T. H., and Shaked, M. (1982). Some concepts of negative dependence. *Annals of Probability* **10**, 765–772.

5. Block, H. W. and Ting, M. L. (1981). Some concepts of multivariate dependence. *Communications in Statistics — Theory and Methods* **10**, 749–762.

6. Bonferroni, C. E. (1936). Teoria statistica delle classi e calcolo delle probabilita. *Pubblicazioni del R. Instituto Superiore di Scienze Economiche e Commerciali di Firenze* **8**, 1–62.

7. Boole, G. (1984). *Laws of Thought.* American reprint of 1854 ed., Dover, New York.

8. Box, G. E. P. and Jenkins, G. M. (1976). *Time Series Analysis: Forecasting and Control.* Holden Day, San Francisco.

9. Brindley, E. C. and Thompson, W. A. (1972). Dependence and aging aspects of multivariate survival. *Journal of the American Statistical Association* **67**, 822–830.

10. Chen, J. and Glaz, J. (1999). Approximations for the distribution and the moments of discrete scan statistics. In *Scan Statistics and Applications* (Eds., J. Glaz and N. Balakrishnan). Birkhäuser, Boston.

11. Chen, J., Glaz, J., Naus, J., and Wallenstein, S. (1999). Bonferroni-type inequalities for conditional scan statistics. *Technical Report*, Department of Statistics, University of Connecticut.

12. Dawson, D. A. and Sankoff, D. (1967). An inequality for probabilities. *Proceedings of the American Mathematical Society* **18**, 504–507.

13. Dunn, O. J. (1959). Confidence intervals for the means of dependent normally distributed variables. *Journal of the American*

Statistical Association **54**, 613–621.

14. Ebrahimi, N. and Ghosh, M. (1981). Multivariate negative dependence. *Communications in Statistics — Theory and Methods, Series A* **10**, 307–337. (Correction, 1984 **13**, p. 3251).

15. Esary, J. D. and Proschan, F. (1972). Relationships among some notions of bivariate dependence. *Annals of Mathematical Statistics* **43**, 651–655.

16. Esary, J. D., Proschan, F., and Walkup, D. W. (1967). Association of random variables with applications. *Annals of Mathematical Statistics* **38**, 1466–1474.

17. Galambos, J. and Simonelli, I. (1996). *Bonferroni-type Inequalities with Applications.* Springer, New-York.

18. Glaz, J. (1990). A comparison of Bonferroni-type and product-type inequalities in presence of dependence. *Topics in Statistical Dependence* (Eds., H. W. Block, A. R. Sampson, and T. H. Savits). IMS Lecture Notes – Monograph Series, Volume 16, pp. 223–235. IMS, Hayward, CA.

19. Glaz, J. (1992). Extreme order statistics for a sequence of dependent random variables. In *Stochastic Inequalities* (Eds., M. Shaked and Y. L. Tong). IMS Lecture Notes – Monograph Series, Volume 22, 100–115. IMS, Hayward, CA.

20. Glaz, J. and Balakrishnan, N. (Eds.) (1999). *Scan Statistics and Applications.* Birkhäuser, Boston.

21. Glaz, J. and Johnson, B. McK. (1984). Probability inequalities for multivariate distributions with dependence structures. *Journal of the American Statistical Association* **79**, 436–441.

22. Glaz, J. and Johnson, B. McK. (1986). Approximating boundary crossing probabilities with applications to sequential tests. *Sequential Analysis* **5**, 37–72.

23. Glaz, J. and Johnson, B. McK. (1988). Boundary crossing for moving sums. *Journal of Applied Probability* **25**, 81–88.

24. Glaz, J. and Kenyon, J. R. (1995). Approximating the characteristics of sequential tests. *Probability and Mathematical Statistics* **15**, 311–325.

25. Glaz, J. and Naus, J. (1991). Tight bounds and approximations for scan statistic probabilities for discrete data. *Annals of Applied Probability* **1**, 306–318.

26. Glaz, J. and Naus, J. (1996). Approximations and bounds for moving sums of iid continuous random variables. *Preliminary Report.*

27. Glaz, J. and Ravishanker, N. (1991). Simultaneus prediction intervals for multiple forecasts based on Bonferroni and product-type inequalities. *Statistics and Probability Letters* **12**, 57–63.

28. Harris, R. (1970). A multivariate definition for increasing hazard rate distributions. *Annals of Mathematical Statistics* **41**, 1456–1465.

29. Hoover, D. R. (1990). Subset complement addition upper bounds – an improved inclusion/exclusion method. *Journal of Statistical Planning and Inference* **24**, 195–202.

30. Hunter, D. (1976). An upper bound for the probability of a union. *Journal of Applied Probability* **13**, 597–603.

31. Joag-Dev, K., Perlman, M. D., and Pitt, L. D. (1983). Association of normal random variables and Slepian inequality. *Annals of Probability* **11**, 451–455.

32. Karlin, S. (1968). *Total Positivity*. Stanford University Press. Stanford, CA.

33. Karlin, S. and Rinott, Y. (1980a). Classes of orderings of measures and related correlations inequalities I: Multivariate totally positive distributions. *Journal of Multivariate Analysis* **10**, 467–498.

34. Karlin, S. and Rinott, Y. (1980b). Classes of orderings of measures and related correlations inequalities II: Multivariate reverse rule distributions. *Journal of Multivariate Analysis* **10**, 499–516.

35. Kemperman, J. H. B. (1977). On the FKG inequality measures on a partially ordered space. *Proc. Kon. Ned. Akad. Wet., Amsterdam, Series A* **80(4)**, 313–331.

36. Khatri, C. G. (1967). On certain inequalities for normal distributions and their applications to simultaneous confidence bounds. *Annals of Mathematical Statistics* **38**, 1853–1867.

37. Kim, T. S. and Seo, H. Y. (1995). A note on some negative dependence notions. *Communications in Statistics — Theory and Methods, Series A* **24**, 845-858.

38. Kimball, A. W. (1951). On dependent tests of significance in the analysis of variance. *Annals of Mathematical Statistics* **22**, 600–602.

39. Kwerel, S. M. (1975). Most stringent bounds on aggregated probabilities of partially specified probability systems. *Journal of the American Statistical Association* **70**, 472–479.

40. Lee, M. L. T. (1985). Dependence by reverse regular rule. *Annals of Probability* **13**, 583–591.

41. Lehmann, E. L. (1966). Some concepts of dependence. *Annals of Mathematical Statistics* **37**, 1137–1153.

42. Naus, J. I. (1974). Probabilities for a generalized birthday problem. *Journal of the American Statistical Association* **69**, 810–815.

43. Olkin, I. and Tong, Y. (1994). Positive dependence of a class of multivariate exponential distributions. *SIAM Journal of Computing* **32**, 965–974.

44. Pitt, L. D. (1982). Positively correlated normal random variables are associated. *Annals of Probability* **10**, 496–499.

45. Ravishanker, N., Wu, L. S. Y., and Glaz, J. (1991). Multiple prediction intervals for time series: comparison of simultaneous and marginal intervals. *Journal of Forecasting* **10**, 445–463.

46. Sarkar, T. K. (1969). Some lower bounds of reliability. *Technical Report No. 124*, Department of Operation Research and Statistics, Stanford University.

47. Schervish, M. K. (1984). Algorithm AS 195, multivariate normal probabilities with error bound. *Applied Statistics* **33**, 81-94.

48. Scott, A. (1967). A note on conservative confidence regions for the mean of a multivariate normal distribution. *Annals of Mathematical Statistics* **38**, 278–280. Correction: *Annals of Mathematical Statistics* **39**, 2161.

49. Sidak, Z. (1967). Rectangular confidence regions for means of multivariate normal distributions. *Journal of the American Statistical Association* **62**, 626–633.

50. Sidak, Z. (1971). On probabilities of rectangles in multivariate student distributions: their dependence on correlations. *Annals of Mathematical Statistics* **42**, 169–175.

51. Woodroofe, M. (1982). *Nonlinear Renewal Theory in Sequential Analysis*. CBMS-NSF Regional Conference Series in Applied Mathematics **39**. SIAM, Philadelphia.

52. Worsley, K. J. (1982). An improved Bonferroni inequality and applications. *Biometrika* **69**, 297–302.

3

Applications of Compound Poisson Approximation

A. D. Barbour, O. Chryssaphinou, and E. Vaggelatou
Universität Zürich, Switzerland
University of Athens, Greece
University of Athens, Greece

ABSTRACT Stein's method for compound Poisson approximation has not been as widely applied as in the Poisson context. One reason for this is that the bounds on the solutions of the Stein Equation are in general by no means as good as for Poisson approximation. Here, we illustrate how new bounds on the solutions have radically changed the situation, provided that the parameters of the approximating compound Poisson distribution satisfy certain conditions. The results obtained in this way are every bit as satisfactory as those obtained in the Poisson context.

Keywords and phrases Compound Poisson approximation, word counts, sequence comparison, success runs, Stein's method

3.1 Introduction

Stein's method has proved to be one of the most successful methods for determining the accuracy of Poisson approximation. Let W be any nonnegative integer valued random variable, with the property that

$$|\mathbb{E}\{\lambda g(W+1) - Wg(W)\}| \leq \varepsilon_0 M_0(g) + \varepsilon_1 M_1(g) \qquad (3.1.1)$$

for some $\lambda > 0$ and for all bounded $g : \mathbf{N} \to \mathbf{R}$, where $M_l(g) := \sup_{w \in \mathbf{N}} |\Delta^l g(w)|$, $l \in \mathbf{Z}_+$, and $\Delta g(w) := g(w+1) - g(w)$. Then, if the total variation distance between probability distributions on $\mathbf{Z}_+$ is defined by

$$d_{TV}(P,Q) := \sup_{A \subset \mathbf{Z}_+} |P(A) - Q(A)|,$$

it follows that

$$d_{TV}(\mathcal{L}(W), \mathrm{Po}(\lambda)) \leq \varepsilon_0 \sup_{A \subset \mathbf{Z}_+} M_0(g_A) + \varepsilon_1 \sup_{A \subset \mathbf{Z}_+} M_1(g_A), \qquad (3.1.2)$$

41

where g_A is the solution of the Poisson Stein Equation

$$\lambda g(j+1) - jg(j) = f(j) - \mathrm{Po}(\lambda)\{f\}, \qquad (3.1.3)$$

for $f = 1_A$. Bounds for the suprema are given in Barbour, Holst, and Janson (1992, Lemma 1.1.1 and Remark 1.1.2):

$$\sup_{A \subset \mathbf{Z}_+} M_0(g_A) \le \min\{1, \sqrt{2/e\lambda}\}; \qquad \sup_{A \subset \mathbf{Z}_+} M_1(g_A) \le \lambda^{-1}(1 - e^{-\lambda}).$$
$$(3.1.4)$$

The fact that both bounds decrease as λ increases is a very useful aspect of the error estimate (3.1.2), because it often ensures that increasing the size of the problem does not worsen the estimates. For instance, if $W = \sum_{i=1}^{n} I_i$ is a sum of independent Bernoulli Be (p_i) random variables, we can take $\lambda = \mathbb{E}W = \sum_{i=1}^{n} p_i$, $\varepsilon_0 = 0$ and $\varepsilon_1 = \sum_{i=1}^{n} p_i^2$, resulting in the simple bound [Barbour and Hall (1984); see also LeCam (1960)]

$$d_{TV}(\mathcal{L}(W), \mathrm{Po}(\lambda)) \le \lambda^{-1} \sum_{i=1}^{n} p_i^2,$$

a quantity which typically does not grow with n: if all the p_i are equal, the bound is exactly p for all n. Establishing (3.1.1) for small ε_0 and ε_1 is frequently a surprisingly tractable task; many examples are to be found in Arratia, Goldstein, and Gordon (1990) and in Barbour, Holst, and Janson (1992).

For compound Poisson approximation, there is an analogous approach. Let W be any nonnegative integer valued random variable, and let $\mu_i \ge 0$, $i \in \mathbf{N}$, satisfy $\sum_{i \ge 1} i\mu_i < \infty$ and $\sum_{i \ge 1} \mu_i = 1$. If it can be shown that

$$\left| \mathbb{E}\left\{ \sum_{i \ge 1} i\lambda\mu_i g(W + i) - Wg(W) \right\} \right| \le \varepsilon_0 M_0(g) + \varepsilon_1 M_1(g) \qquad (3.1.5)$$

for some $\lambda > 0$ and for all bounded $g : \mathbf{N} \to \mathbf{R}$, then it follows that

$$d_{\mathcal{F}}(\mathcal{L}(W), \mathrm{CP}(\lambda, \boldsymbol{\mu})) := \sup_{f \in \mathcal{F}} |\mathbb{E}f(W) - \mathrm{CP}(\lambda, \boldsymbol{\mu})\{f\}|$$

$$\le \varepsilon_0 \sup_{f \in \mathcal{F}} M_0(g_f) + \varepsilon_1 \sup_{f \in \mathcal{F}} M_1(g_f), \qquad (3.1.6)$$

for any set $\mathcal{F}$ of test functions, where g_f now solves the compound Poisson Stein Equation

$$\sum_{i \ge 1} i\lambda\mu_i g(j + i) - jg(j) = f(j) - \mathrm{CP}(\lambda, \boldsymbol{\mu})\{f\}, \quad j \ge 0. \qquad (3.1.7)$$

Here, CP $(\lambda, \boldsymbol{\mu})$ denotes the compound Poisson distribution of $\sum_{i \geq 1} i Z_i$, where the $Z_i \sim \text{Po}(\lambda \mu_i)$ are independent. As for (3.1.1), establishing (3.1.5) for small ε_0 and ε_1 may often be relatively simple. However, the resulting distance estimates (3.1.6) were not as powerful as they could have been, for lack of sharp bounds on the quantities $\sup_{f \in \mathcal{F}} M_l(g_f)$ for the commonest choices of test functions $\mathcal{F}$ and for most CP $(\lambda, \boldsymbol{\mu})$.

For the test functions $\mathcal{F}_{TV} = \{1_A, A \subset \mathbf{Z}_+\}$, appropriate to total variation approximation, the general bounds

$$H_l^{TV}(\lambda, \boldsymbol{\mu}) := \sup_{f \in \mathcal{F}_{TV}} M_l(g_f) \leq \min\{1, (\lambda \mu_1)^{-1}\} e^\lambda, \quad l = 0, 1, \quad (3.1.8)$$

were proved in Barbour, Chen, and Loh (1992); because of the exponential factor, these are only useful for small λ. However, under the additional condition on the μ_i that

$$\mu_1 \geq 2\mu_2 \geq 3\mu_3 \geq \dots; \quad (3.1.9)$$

they showed that

$$H_0^{TV}(\lambda, \boldsymbol{\mu}) \leq \min\left\{1, \frac{2}{\sqrt{\lambda \nu_1}}\right\};$$

$$H_1^{TV}(\lambda, \boldsymbol{\mu}) \leq \min\left\{1, \frac{1}{\lambda \nu_1}\left(\frac{1}{4\lambda \nu_1} + \log^+(2\lambda \nu_1)\right)\right\}, \quad (3.1.10)$$

where $\nu_1 = \mu_1 - 2\mu_2$. These bounds are much better, but that on M_1 is still weaker than could be hoped for, because of the logarithmic factor. This factor was removed in Barbour and Utev (1998, Theorem 3.1), in the setting of approximation with respect to the Kolmogorov distance d_K defined by

$$d_K(P, Q) := \sup_x |P[x, \infty) - Q[x, \infty)|,$$

by way of intricate Fourier arguments, but only at the expense of worsening the constant factors to a point at which they became unattractive for practical application.

In this paper, we illustrate the improvements in compound Poisson estimates that can be obtained by using two new bounds. The first of these, proved in Barbour and Xia (1999b), requires a different condition: if

$$\theta := \sum_{i \geq 1} i(i - 1)\mu_i \Big/ m_1 < 1/2, \quad (3.1.11)$$

where $m_1 := \sum_{i \geq 1} i \mu_i$, then

$$H_0^{TV}(\lambda, \boldsymbol{\mu}) \leq \frac{1}{(1 - 2\theta)\sqrt{\lambda m_1}}; \quad H_1^{TV}(\lambda, \boldsymbol{\mu}) \leq \frac{1}{(1 - 2\theta)\lambda m_1}. \quad (3.1.12)$$

These bounds are of exactly the same order as the Poisson bounds (3.1.4), and reduce to them if $CP(\lambda, \boldsymbol{\mu})$ is a Poisson distribution, so that $\mu_1 = 1$ and $\theta = 0$; they were established by using an analytic perturbation argument, starting from the Poisson solution.

The second bound is valid under the previous condition (3.1.9). It has been shown in Barbour and Xia (1999a) that, for the set of test functions $\mathcal{F}_K := \{f_k, \ k \in \mathbf{N} : \ f_k(x) = 1_{[k,\infty)}(x)\}$ appropriate to Kolmogorov distance,

$$H_0^K(\lambda, \boldsymbol{\mu}) \ := \sup_{f \in \mathcal{F}_K} M_0(g_f) \leq \min\left\{1, \sqrt{\frac{2}{e\lambda\mu_1}}\right\},$$

$$(3.1.13)$$

$$H_1^K(\lambda, \boldsymbol{\mu}) \ := \sup_{f \in \mathcal{F}_K} M_1(g_f) \leq \min\left\{\tfrac{1}{2}, \tfrac{1}{\lambda\mu_1 + 1}\right\}.$$

Here, the proof is based on probabilistic arguments, using the associated immigration–death process introduced in Barbour, Chen, and Loh (1992). Combining (3.1.12) and (3.1.13) with (3.1.6), we obtain the following proposition.

PROPOSITION 3.1.1
Let W be a nonnegative integer valued random variable, and suppose that (3.1.5) holds for all bounded $g : \mathbf{N} \to \mathbf{R}$, for some $\lambda > 0$ and $\mu_i \geq 0$, $i \in \mathbf{N}$, such that $\sum_{i \geq 1} i\mu_i < \infty$ and $\sum_{i \geq 1} \mu_i = 1$.

(A) *If Condition (3.1.11) is satisfied, then*

$$d_{TV}(\mathcal{L}(W), CP(\lambda, \boldsymbol{\mu})) \leq (1 - 2\theta)^{-1}\{(\lambda m_1)^{-1/2}\varepsilon_0 + (\lambda m_1)^{-1}\varepsilon_1\}.$$

(B) *If Condition (3.1.9) is satisfied, then*

$$d_K(\mathcal{L}(W), CP(\lambda, \boldsymbol{\mu}))$$

$$\leq \varepsilon_0 \min\left\{1, \sqrt{\frac{2}{e\lambda\mu_1}}\right\} + \varepsilon_1 \min\left\{\frac{1}{2}, \frac{1}{\lambda\mu_1 + 1}\right\}.$$

Neither pair of bounds has logarithmic factors, and both have small constant factors, making them well suited for practical application; we illustrate this in what follows, showing that a number of previously obtained compound Poisson error estimates can be substantially improved.

3.2 First Applications

3.2.1 Runs

As a first, frequently studied illustration, we consider compound Poisson approximation to the number W of k-runs of 1's in a sequence of independent Bernoulli Be(p) random variables X_i, $1 \leq i \leq n$: thus $W := \sum_{i=1}^{n-k+1} I_i$, where $I_i := \prod_{j=i}^{i+k-1} X_j$ and $\psi := \mathbb{E}I_i = p^k$. In order to simplify the definition of the approximating distribution, we start by adopting the 'circle convention', setting $X_{n+j} = X_j$ for all $j \in \mathbb{Z}$, and considering a new random variable $W' := \sum_{i=1}^{n} I_i$, which counts the number of consecutive k-runs of 1's when the X_i are arranged around a circle, rather than on the line. This is merely a way of avoiding edge effects, and is not essential to the method.

There have been a number of compound Poisson approximations to the distributions of W and W', with the precise choices of λ and $\boldsymbol{\mu}$ differing slightly from paper to paper. Arratia, Goldstein, and Gordon (1990) used Poisson process approximation to derive a total variation estimate of order $O(nk\psi^2(1-p))$ for the approximation error. Under the condition $p \leq \frac{1}{3}$, so that the bound (3.1.10) can be applied, Roos (1993) improved the error estimate to order $O(k\psi \log(n\psi))$; furthermore, Barbour and Utev (1999, Theorem 1.10) can be used for any value of p to give an error estimate of order $O(k\psi + \exp(-cn\psi))$ for some $c > 0$ (see Eichelsbacher and Roos (1998)), but with an uncomfortably large constant. Here, using the improved bounds on the Stein constants, we can establish substantially smaller error estimates, for Kolmogorov distance if $p \leq 1/2$ and for total variation distance if $p < 1/5$.

Roos (1993) addresses the problem using the 'local approach'. For W a sum of 0–1 random variables I_i, $1 \leq i \leq n$, this involves writing W in the form

$$W = I_i + U_i + Z_i + W_i \qquad (3.2.1)$$

for each $1 \leq i \leq n$, where U_i, Z_i and W_i are nonnegative, integer valued random variables, with U_i typically heavily dependent on I_i, and W_i almost independent of the pair (I_i, U_i). Then one can take

$$\varepsilon_0 = \sum_{i=1}^{n} \sum_{l \geq 1} \mathbb{E}|\mathbb{P}[I_i = 1, U_i = l - 1 \,|\, W_i] - \mathbb{P}[I_i = 1, U_i = l - 1]| \quad (3.2.2)$$

and

$$\varepsilon_1 = \sum_{i=1}^{n} \{\mathbb{E}I_i\, \mathbb{E}(I_i + U_i + Z_i) + \mathbb{E}(I_i Z_i)\} \qquad (3.2.3)$$

in (3.1.5), implying corresponding compound Poisson approximations by way of (3.1.6), if λ and μ have the canonical definitions

$$\lambda\mu_l := l^{-1}\sum_{i=1}^{n} \mathbb{E}\{I_i I[U_i = l - 1]\}, \quad l \geq 1; \qquad \lambda := \sum_{l \geq 1} \lambda\mu_l. \quad (3.2.4)$$

Note that then $\lambda \sum_{l \geq 1} l\mu_l = \mathbb{E}W$, so that W and the compound Poisson distribution CP (λ, μ) both have the same expectation.

For the example of runs and the random variable W', taking

$$U_i = \sum_{1 \leq |j-i| < k} I_j; \qquad Z_i = \sum_{k \leq |j-i| \leq 2(k-1)} I_j, \quad (3.2.5)$$

it follows that W_i is independent of (I_i, U_i), so that $\varepsilon_0 = 0$, and that I_i and Z_i are independent, which then easily gives $\varepsilon_1 = n\psi^2(6k - 5)$. The calculation of λ and μ is more complicated, eventually giving

$$\lambda\mu_l = \begin{cases} \mathbb{E}W' p^{l-1}(1 - p)^2 \\ \quad \text{for } l = 1, 2, ..., k - 1, \\[2mm] \mathbb{E}W' l^{-1}p^{l-1}[2(1 - p) + (2k - l - 2)(1 - p)^2] \\ \quad \text{for } l = k, ..., 2k - 2, \\[2mm] \mathbb{E}W' (2k - 1)^{-1}p^{2k-2} \\ \quad \text{for } l = 2k - 1, \end{cases} \quad (3.2.6)$$

with $\mathbb{E}W' = n\psi$. For these parameters, Condition (3.1.9) is satisfied for all k if $p \leq 1/3$ and for $k \geq 4$ if $p \leq 1/2$, and Proposition 3.1.1(B) then implies that

$$d_K(\mathcal{L}(W'), \text{CP}(\lambda, \mu)) \leq \min\left\{\frac{1}{2}, \frac{1}{\lambda\mu_1 + 1}\right\} n(6k - 5)\psi^2$$

$$< \begin{cases} p & \text{for } k = 1 \\ (6k - 5)\psi(1 - p)^{-2} & \text{for } k \geq 2. \end{cases} \quad (3.2.7)$$

For total variation distance, $\theta \leq 2p/(1 - p) < 1/2$ if $p < 1/5$, and Proposition 3.1.1(A) gives the error estimate

$$d_{TV}(\mathcal{L}(W'), \text{CP}(\lambda, \mu)) \leq (6k - 5)\psi(1 - p)/(1 - 5p). \quad (3.2.8)$$

Both of these estimates are simple, explicit, and of better order than the previous bounds in their domains of applicability.

The original runs problem is concerned with the distribution of $W := \sum_{i=1}^{n-k+1} I_i$, omitting the extra random variables I_i, $n - k + 2 \leq i \leq n$, whose definition involved the circle convention. The local approach described above can still be carried out in the same way, yielding a marginally smaller error bound; the drawback is that the canonical parameters obtained in this way are more complicated to specify, since they also require edge corrections. An alternative solution is to observe that W and W' differ only by the sum $\sum_{i=n-k+2}^{n} I_i$, so that

$$d_{TV}(\mathcal{L}(W), \mathcal{L}(W')) \le (k-1)\psi;$$

hence one can also approximate the distribution $\mathcal{L}(W)$ using the same compound Poisson distribution $\mathrm{CP}(\lambda, \boldsymbol{\mu})$, defined in (3.2.6), as was used for $\mathcal{L}(W')$, with an additional error of at most $(k-1)\psi$.

An even simpler approximating compound Poisson distribution is provided by the Pólya-Aeppli distribution $\mathrm{PA}(\lambda^*, p)$, defined as $\mathrm{CP}(\lambda^*, \boldsymbol{\mu}^*)$ with $\lambda^* := n\psi(1-p)$ and

$$\mu_l^* := p^{l-1}(1-p), \quad l \ge 1;$$

as can be seen from (3.2.6), $\lambda\mu_l = \lambda^*\mu_l^*$ for $1 \le l \le k-1$, and furthermore $\lambda\sum_{l\ge1} l\mu_l = n\psi = \lambda^*\sum_{l\ge1} l\mu_l^*$.

It then follows from Roos (1993, Lemma 2.4.2) that the distribution $\mathrm{CP}(\lambda, \boldsymbol{\mu})$ can be replaced by $\mathrm{PA}(\lambda^*, p)$ in the above approximations, provided that ε_1 is increased by an amount $\varepsilon^* := \sum_{l\ge1} l(l-1)|\lambda\mu_l - \lambda^*\mu_l^*|$. Using the definition of $\lambda\mu_l$ from (3.2.6), this gives

$$
\begin{aligned}
\varepsilon^* = n\psi &\left\{ \sum_{k\le l\le 2k-2} (l-1)p^{l-1}(1-p)|2(l-k)(1-p) - 2p| \right. \\
&\quad + (2k-2)p^{2k-2}|(2k-1)(1-p)^2 - 1| \\
&\quad \left. + \sum_{l\ge 2k} (l-1)lp^{l-1}(1-p)^2 \right\} \\
\le 2n\psi &\left\{ \sum_{l>k} (l-1)(l-k)p^{l-1}(1-p)^2 + (k-1)p^k(1-p) \right. \\
&\quad \left. + (k-1)p^{2k-2} \right\} \\
= 2n\psi^2 &\{k + 2p(1-p)^{-1} + (k-1)(1-p+p^{k-2})\} \\
\le 2n\psi^2 &\left\{ 2k + \frac{(3p-1)}{(1-p)} \right\}. \tag{3.2.9}
\end{aligned}
$$

The last inequality of (3.2.9) holds for $k \ge 3$ but it can be easily verified that it still remains a bound for ε^*, when $k = 2$ and $p \le 1/2$. Thus using instead the Pólya-Aeppli distribution PA $(n\psi(1-p), p)$ as approximation, requires the addition of at most $2(2k+1)\psi(1-p)^{-2}$ to the Kolmogorov estimate (3.2.7), and, what is more, the estimate is then valid irrespective of k for all $p \leq 1/2$, since Condition (3.1.9) holds for PA (λ^*, p) whenever $p \leq 1/2$. For the total variation estimate (3.2.8), valid in $p < 1/5$, one needs to add at most $4k\psi(1-p)/(1-5p)$.

The problem of k-runs can also be treated as one of counting the number of visits to a rare set in a Markov chain. From this standpoint, the general theorems of Erhardsson (1999) can be applied, in conjunction with the improved bounds on the Stein constants given by Proposition 3.1.1: see also Theorem 3.3.2 below. Using his approach, the approximating compound Poisson distribution is automatically the Pólya-Aeppli distribution PA (λ^*, p), and the error estimates obtained using the bounds given in (3.1.13) and (3.1.12) are somewhat smaller, being a little better than $4(k+1)\psi(1-p)^{-2}$ for Kolmogorov distance if $p \leq 1/2$, and a little better than $(4k+5)\psi/(1-5p)$ for total variation distance if $p < 1/5$.

3.2.2 Sequence Matching

Let $\xi_1, \ldots, \xi_m$ and $\eta_1, \ldots, \eta_n$ be two independent sequences of independently chosen letters from a finite alphabet $\mathcal{A}$, both uniformly distributed over $\mathcal{A}$. Fix k, and set

$$I_{ij} = I[\xi_i = \eta_j, \xi_{i+1} = \eta_{j+1}, \ldots, \xi_{i+k-1} = \eta_{j+k-1}],$$

so that

$$W := \sum_{i=1}^{m-k+1} \sum_{j=1}^{n-k+1} I_{ij}$$

counts the number of times that pairs of matching strings of length k can be found in the two sequences. In molecular sequence applications, an observed value of W higher than that expected according to the above model would indicate evolutionary relationship between the two sequences. Previous work (Arratia, Goldstein, and Gordon (1990), Neuhauser (1996)) has largely concentrated on approximating $\mathbb{P}[W = 0]$, which, by then varying k, translates into a statement about the length of the longest matching run; with this in mind, the strategy is typically to replace W by a random variable which counts distinct clumps of $k-$runs, and to approximate its distribution by a Poisson random variable. Here, as also in Månsson (1999), who explores a more general setting, we use compound Poisson approximation to treat the whole distribution of W, and to provide rather explicit bounds for the accuracy of the approximations obtained.

Once again, to simplify the canonical $CP(\lambda, \boldsymbol{\mu})$ approximation, we work instead with the random variable

$$W' := \sum_{i=1}^{m} \sum_{j=1}^{n} I_{ij},$$

adopting the 'torus convention' that $\xi_{m+i} = \xi_i$ and $\eta_{n+j} = \eta_j$ for all $i, j \in \mathbf{Z}$. The random variable W' then has expectation

$$\mathbb{E}W' = mn\psi,$$

where $\psi := \mathbb{E}I_{ij} = p^k$ and $p := 1/|\mathcal{A}|$, and we are typically interested in values of k less than, say, $2\log(mn)/\log(1/p)$, so that $\mathbb{E}W'$ is not extremely small. In order to use the local approach, we require a decomposition as in (3.2.1) of the form $W' = I_{ij} + U_{ij} + Z_{ij} + W_{ij}$, for each pair i, j. Noting that the indicators most strongly dependent on I_{ij} are those of the form $I_{i+l,j+l}$ with $|l| \leq k-1$, we take

$$U_{ij} = \sum_{1 \leq |l| \leq k-1} I_{i+l,j+l}$$

and

$$Z_{ij} = \left(\sum_{(r,s) \in N_{ij}} I_{rs} \right) - I_{ij} - U_{ij},$$

where

$$N_{ij} = \{(r,s); \ \min\{|r-i|, |s-j|\} \leq 2(k-1)\}.$$

This yields $W' = I_{ij} + U_{ij} + Z_{ij} + W_{ij}$, in such a way that W_{ij} is independent of the pair (I_{ij}, U_{ij}), so that $\varepsilon_0 = 0$.

The canonical parameters for the corresponding compound Poisson approximation are the same as those in (3.2.6) for success runs, but with the new definitions of p and $\mathbb{E}W'$, and, as before, Condition (3.1.9) is satisfied for all k if $p \leq 1/3$, and Condition (3.1.11) is satisfied with $(1-2\theta)^{-1} \leq (1-p)/(1-5p)$ if $p < 1/5$; a Pólya-Aeppli PA $(mn\psi(1-p), p)$ approximation would add only at most an extra $(4k+2)\psi(1-p)$ to Δ_1 in the estimates given below. In bounding the expression corresponding to (3.2.3) for ε_1, it is immediate that

$$\mathbb{E}I_{ij}\mathbb{E}(I_{ij} + U_{ij} + Z_{ij}) \leq (4k-3)(m+n)\psi^2. \tag{3.2.10}$$

For $\mathbb{E}(I_{ij}Z_{ij})$, note that I_{ij} is independent of each single I_{rs} in the sum defining Z_{ij}, so that we have

$$\mathbb{E}(I_{ij}Z_{ij}) \leq (4k-3)(m+n)\psi^2 \tag{3.2.11}$$

as well. Hence $\varepsilon_1 \leq 2mn(4k-3)(m+n)\psi^2$, and using Proposition 3.1.1 we deduce that

$$d_K(\mathcal{L}(W'), \mathrm{CP}\,(\lambda, \boldsymbol{\mu})) \quad \leq \{(1-p)^2 + (\mathbb{E}W')^{-1}\}^{-1}\,\Delta_1, \quad |\mathcal{A}| \geq 3;$$

$$d_{TV}(\mathcal{L}(W'), \mathrm{CP}\,(\lambda, \boldsymbol{\mu})) \quad \leq \{(1-p)/(1-5p)\}\,\Delta_1, \quad |\mathcal{A}| \geq 6,$$
$$\tag{3.2.12}$$

where λ and $\boldsymbol{\mu}$ are as given in (3.2.6), but now with $\mathbb{E}W' = mn\psi$, and

$$\Delta_1 := 2(4k-3)(m+n)\psi. \tag{3.2.13}$$

Conversion from W' to W is once again straightforward, since it is immediate that

$$d_{TV}(\mathcal{L}(W), \mathcal{L}(W')) \leq (k-1)(m+n)\psi.$$

Most emphasis has previously been placed on asymptotics in which m and n tend to infinity in such a way that both $\log m/\log n$ and $\lambda = \lambda_{mn}$ converge to finite, nonzero limits. Using (3.2.12) and (3.2.13), we can make more precise statements about how well the distribution of W' is being approximated, and under less restrictive conditions. For m and n given, $m \leq n$, set

$$k = k_{mn} := \log(mn)/\log(1/p) - c,$$

for any $c = c_{mn} \geq 0$, noting that then $\mathbb{E}W' = p^{-c}$. From (3.2.13), we immediately have

$$\Delta_1 \leq 8k(m+n)\psi \leq \left\{\frac{2\log(mn)}{\log(1/p)}\right\}(m^{-1} + n^{-1})p^{-c},$$

which is small so long as $\log n \ll m$ and c is not too large. In asymptotic terms, as $m, n \to \infty$ for fixed p, one would require that $\log n = o(m)$ and in addition that $c_{mn} \leq \delta \log\{m/\log n\}/\log(1/p)$ for some $\delta < 1$. Previous asymptotics have mostly assumed that c is bounded above, so that this last condition would then be automatically satisfied.

3.3   Word Counts

The next problem that we treat is counting the number of appearances W of a word $A = a_1 a_2 \ldots a_k$ of length k, in a sequence of letters $\xi_1, \ldots, \xi_n$ of length n from an infinite realization of an irreducible, aperiodic stationary Markov chain $(\xi_i, i \in \mathbf{Z})$ on a finite alphabet $\mathcal{A}$, having transition matrix Π and stationary distribution π. Thus $W := \sum_{i=k}^n I[B_k^i]$, where

$$B_j^i := \{\xi_{i-j+1} = a_1, \ldots, \xi_i = a_j\}, \quad 1 \leq j \leq k, \tag{3.3.1}$$

is the event that $\xi_{i-j+1} \dots \xi_i$ is a copy of the first j letters of A, and we set $\psi := \mathbb{E}I[B_k^i]$, noting that $\mathbb{E}W = (n - k + 1)\psi$. We discuss this problem in rather greater detail, making some numerical comparisons.

As a first approach, we illustrate how the 'declumping' procedure of Arratia, Goldstein, and Gordon (1990) can be used to derive accurate bounds, when applied in conjunction with the compound Poisson Stein equation, as in Barbour and Utev (1998, Section 5) and Barbour and Chryssaphinou (1999, Section 2.2). Schbath (1995) employed the Arratia, Goldstein, and Gordon (1990) procedure directly, establishing an estimate of order $O(n\psi^2)$ for the error in compound Poisson approximation to $\mathcal{L}(W)$, though no fully explicit formula for the estimate was given. As in the case of runs, this order of approximation is not accurate when $\mathbb{E}W$ is large. Here, we show how to modify her approach to obtain an explicit bound of the improved order $O(\psi \log n)$.

In what follows, we use the term "period" of a word as defined in Guibas and Odlyzko (1981). In particular, an integer p is a period of a word $A = a_1 a_2 \dots a_k$ if $a_i = a_{i+p}$ for all $i = 1, \dots, k - p$. Let also $\mathcal{P}'(A)$ denote the set of "principal" periods of A (Schbath (1995)), consisting of the shortest period p_0 of A together with those which are not multiples of p_0. For example, the word $A = a_1 a_1 a_2 a_1 a_1 a_2 a_1 a_1$ with $k = 8$ has periods $3, 6$ and 7, and $\mathcal{P}'(A) = \{3, 7\}$.

To introduce the declumping, let $I_{i,l}$ denote the indicator r.v. of the appearance of an l-clump at position i. More precisely, let $\mathcal{C}_l$ be the set of words consisting of exactly l overlapping appearances of A, $\mathcal{U}$ the set of $k - 1$ letter words that do not end with the p first letters of A for any $p \in \mathcal{P}'(A)$ and $\mathcal{V}$ the set of $k - 1$ letter words which do not start with the p last letters of A for any $p \in \mathcal{P}'(A)$. Then define

$$I_{i,l} = \sum_{U \in \mathcal{U}, C \in \mathcal{C}_l, V \in \mathcal{V}} (I_{i-k+1}(UCV)),$$

where $I_i(S)$ denotes the indicator of the event that a copy of the word S begins at index i, and note that

$$\mathbb{E}I_{i,l} = (1 - L)^2 L^{l-1} \psi, \tag{3.3.2}$$

where $L = \sum_{p \in \mathcal{P}'(A)} \prod_{t=1}^{p} \Pi(a_t a_{t+1})$.

The 'declumped' random variable $W' := \sum_{i=1}^{n-k+1} \sum_{l \geq 1} l I_{i,l}$ is easier to approximate than W, and $W = W'$ unless there is a copy of A which overlaps either the beginning or the end of the interval $\{1, 2, \dots, n\}$, so that

$$d_{TV}(\mathcal{L}(W), \mathcal{L}(W')) \leq 2(k - 1)\psi, \tag{3.3.3}$$

a quantity no larger than the other terms appearing in the approximation error. It follows from (3.3.2) that the Pólya-Aeppli distribution

PA $(\mathbb{E}W(1 - L), L)$ is the appropriate compound Poisson approximation, and Barbour and Chryssaphinou (1999, Section 2.2) shows that we
can take $\varepsilon_0 = b_3^*$ and $\varepsilon_1 = b_1^* + b_2^*$ in (3.1.6), where

$$b_1^* = \sum_{i=1}^{n-k+1} \sum_{l \geq 1} \sum_{(j,s) \in B_{(i,l)}} ls\mathbb{E}I_{i,l}\mathbb{E}I_{j,s};$$

$$b_2^* = \sum_{i=1}^{n-k+1} \sum_{l \geq 1} \sum_{(j,s) \in B_{(i,l)}, j \neq i} ls\mathbb{E}(I_{i,l}I_{j,s}); \qquad (3.3.4)$$

$$b_3^* = \sum_{i=1}^{n-k+1} \sum_{l \geq 1} l\mathbb{E}|\mathbb{E}\{I_{i,l} - \mathbb{E}I_{i,l} \mid \sigma(I_{j,s} : (j,s) \in B_{(i,l)}^c)\}|;$$

here, we specify

$$B_{(i,l)} = \{(j,s) : -(s+2)(k-1) - r + 1 \leq j - i$$
$$\leq (l+2)(k-1) + r - 1\}. \qquad (3.3.5)$$

This latter definition is appropriate, because the indicators $I_{i,l}$ and $I_{j,s}$
are measurable with respect to $\sigma(\xi_{i-k+1}, \ldots, \xi_{i+(l+1)(k-1)})$ and $\sigma(\xi_{j-k+1},$
$\ldots, \xi_{j+(s+1)(k-1)})$, respectively, so that $I_{i,l}$ and $I_{j,s}$ are only weakly dependent whenever $j - k + 1 \geq i + (l + 1)(k - 1) + r$ or $i - k + 1 \geq$
$j + (s + 1)(k - 1) + r$, for $r \geq 1$. This gives the conclusion of Theorem 3.3.1, for which we need some extra notation.

Let $\Pi^{(m)}(a, a')$ denote the m-step transition probabilities in the ξ-
chain, and $\Pi_R(a, a') := \Pi(a', a)\pi(a')/\pi(a)$ the transition probabilities in
the reversed ξ-chain. Define

$$\chi(A) := \pi(a_1)^{-1} \max_{m \geq 1} \Pi^{(m)}(a_k, a_1) \geq 1.$$

Then let $K \geq 1$ and $0 \leq \rho \leq 1$ be numbers such that

$$\max_{a \in \mathcal{A}} \max \left\{ \sum_{a' \in \mathcal{A}} |\Pi^{(m)}(a, a') - \pi(a')|, \sum_{a' \in \mathcal{A}} |\Pi_R^{(m)}(a, a') - \pi(a')| \right\} \leq K\rho^m$$
$$(3.3.6)$$

for all $m \geq 1$. It follows from an elementary coupling argument (see, for
example, Lindvall (1992), paragraph 2 on page 96, for a rather dismissive
description) that one can take $K = 2$ and $\rho = 1 - \min\{\delta(\Pi), \delta(\Pi_R)\}$,
where, for an $\mathcal{A} \times \mathcal{A}$ matrix P, $\delta(P) := \sum_{a' \in \mathcal{A}} \min_{a \in \mathcal{A}} P(a, a')$. There
are usually better choices, particularly if $\delta(\Pi) = 0$. Note also that (3.3.6)
is actually only needed for $m \geq r$, where $r \geq 1$ is as above, and may be
chosen to suit.

THEOREM 3.3.1
If $L \leq 1/2$, then

$$d_K\big(\mathcal{L}(W), \mathrm{PA}\,(\mathbb{E}W\,(1-L), L)\big)$$

$$\leq \Delta_1 \left(\frac{1}{(1-L)^2 + \{\mathbb{E}W\}^{-1}}\right) + \Delta_0 \left(\frac{2}{e\mathbb{E}W\,(1-L)^2}\right)^{1/2}$$

$$+ 2(k-1)\psi;$$

if $L < 1/5$, then

$$d_{TV}\big(\mathcal{L}(W), \mathrm{PA}\,(\mathbb{E}W\,(1-L), L)\big) \leq B_{3.3.1}$$

$$:= \{(1-L)/(1-5L)\}(\Delta_1 + \{\mathbb{E}W\}^{-1/2}\Delta_0) + 2(k-1)\psi,$$

where $\mathbb{E}W = (n-k+1)\psi$,

$$\Delta_0 := K\rho^r \mathbb{E}W \left\{2 + \rho^{3(k-1)+r} + K\rho^r\right\};$$

$$\Delta_1 := 2\psi\bigg\{3(k-1) + r + 2(k-1)\frac{L}{(1-L)}$$

$$+ \chi(A)\left(\frac{2L(k-1-p_0)}{(1-L)^3} + \frac{2(k-1)+r-1}{(1-L)^2}\right)\bigg\},$$

and $r \geq 1$ may be chosen to minimize the error estimates.

REMARK If the sequence ξ_i, $i \in \mathbf{Z}$, is independent, the estimates of Theorem 3.3.1 hold, with r and K replaced by zero. $\blacksquare$

PROOF In order to prove Theorem 3.3.1, we just need to bound the terms b_1^*, b_2^* and b_3^* of (3.3.4), and use (3.3.3). Then, as observed above, Proposition 3.1.1 can be applied with $\varepsilon_0 = b_3^*$ and $\varepsilon_1 = b_1^* + b_2^*$.

The computation of

$$b_1^* = \sum_{i=1}^{n-k+1} \sum_{l \geq 1} \sum_{s \geq 1} \sum_{j=i-(s+2)(k-1)-r+1}^{i+(l+2)(k-1)+r-1} ls\mathbb{E}I_{i,l}\mathbb{E}I_{j,s}$$

$$\leq 2 \sum_{i=1}^{n-k+1} \sum_{l \geq 1} \sum_{s \geq 1} \sum_{j=i}^{i+(l+2)(k-1)+r-1} ls\mathbb{E}I_{i,l}\mathbb{E}I_{j,s} \qquad (3.3.7)$$

is a simple matter, using (3.3.2), with (3.3.7) yielding exactly

$$b_1^* \leq 2\psi\mathbb{E}W \left\{ 3(k-1) + r + \frac{2(k-1)L}{1-L} \right\}. \qquad (3.3.8)$$

For the term b_2^*, we write

$$b_2^* = \sum_{i=1}^{n-k+1} \sum_{l\geq 1} \sum_{s\geq 1} \sum_{\substack{j=i-(s+2)(k-1)-r+1 \\ j\neq i}}^{i+(l+2)(k-1)+r-1} ls\mathbb{E}(I_{i,l}I_{j,s})$$

$$= 2 \sum_{i=1}^{n-k+1} \sum_{l\geq 1} \sum_{s\geq 1} ls \sum_{j=i+1}^{i+(l+2)(k-1)+r-1} \sum_{C\in\mathcal{C}_l} \mathbb{E}(I_{i,l}(C)I_{j,s}),$$

where

$$I_{i,l}(C) := \sum_{U\in\mathcal{U},V\in\mathcal{V}} I_{i-k+1}(UCV).$$

We now distinguish two cases, when bounding the expectations $\mathbb{E}(I_{i,l}(C)I_{j,s})$, $i+1 \leq j \leq i+(l+2)(k-1)+r-1$.

If $i+1 \leq j \leq i+|C|-1$, then $\mathbb{E}(I_{i,l}(C)I_{j,s}) = 0$, in view of the definition of $I_{j,s}$. If $i+|C| \leq j \leq i+(l+2)(k-1)+r-1$, observe that

$$\mathbb{E}(I_{i,l}(C)I_{j,s})$$
$$= \sum_{U\in\mathcal{U},V\in\mathcal{V}} \sum_{U'\in\mathcal{U},C'\in\mathcal{C}_s,V'\in\mathcal{V}} \mathbb{E}\{I_{i-k+1}(UCV)I_{j-k+1}(U'C'V')\}$$

$$\leq \sum_{U\in\mathcal{U}} \sum_{C'\in\mathcal{C}_s,V'\in\mathcal{V}} \mathbb{E}\{I_{i-k+1}(UC)I_j(C'V')\}, \qquad (3.3.9)$$

since

$$\sum_{V\in\mathcal{V}} I_{i-k+1}(UCV) \leq I_{i-k+1}(UC)$$

and

$$\sum_{U'\in\mathcal{U}} I_{j-k+1}(U'C'V') \leq I_j(C'V').$$

Moreover, the copies of the words UC and $C'V'$ appearing in (3.3.9) do not overlap, which implies that

$$\mathbb{E}(I_{i-k+1}(UC)I_j(C'V')) \leq \pi(UC)\Pi^{(j-i-|C|+1)}(a_k, a_1)\pi(C'V')/\pi(a_1)$$
$$\leq \chi(A)\pi(UC)\pi(C'V').$$

Hence, for $C \in \mathcal{C}_l$ and $i+|C| \leq j \leq i+(l+2)(k-1)+r-1$, we have

$$\mathbb{E}(I_{i,l}(C)I_{j,s}) \le \chi(A) \sum_{U \in \mathcal{U}} \sum_{C' \in \mathcal{C}_s, V' \in \mathcal{V}} \pi(UC)\pi(C'V')$$

$$\le \chi(A) \sum_{U \in \mathcal{U}} \pi(UC)(1-L)L^{s-1}\psi, \qquad (3.3.10)$$

and since $|C| \ge k+(l-1)p_0$ for all $C \in \mathcal{C}_l$, where p_0 is the shortest period of A, the number of indices j for which the bound (3.3.10) is needed is no greater than $(l-1)(k-1-p_0)+2(k-1)+(r-1)$. Combining these observations, we find that

$$b_2^* \le 2 \sum_{i=1}^{n-k+1} \sum_{l \ge 1} \sum_{s \ge 1} ls\{(l-1)(k-1-p_0)+2(k-1)+(r-1)\}$$

$$\sum_{C \in \mathcal{C}_l} \chi(A) \sum_{U \in \mathcal{U}} \pi(UC)(1-L)L^{s-1}\psi$$

$$= 2\chi(A)(n-k+1)\psi^2(1-L)^2$$

$$\sum_{l \ge 1} l\{(l-1)(k-1-p_0)+2(k-1)+(r-1)\}L^{l-1} \sum_{s \ge 1} sL^{s-1}$$

$$= 2\chi(A)\psi\mathbb{E}W \left(\frac{2L(k-1-p_0)}{(1-L)^3} + \frac{2(k-1)+r-1}{(1-L)^2} \right). \qquad (3.3.11)$$

It remains to examine the term b_3^*. Using the Markovian property and the fact that $\sigma(I_{j,s} : (j,s) \in B_{(i,l)}^c)$ is contained in $\sigma(\xi_1, \ldots, \xi_{i-k-r+1}, \xi_{i+(l+1)(k-1)+r}, \ldots, \xi_n)$, we obtain

$$b_3^* = \sum_{i=1}^{n-k+1} \sum_{l \ge 1} l\mathbb{E}|\mathbb{E}\{I_{i,l} - \mathbb{E}I_{i,l} \mid \sigma(I_{j,s} : (j,s) \in B_{(i,l)}^c)\}|$$

$$\le \sum_{i=1}^{n-k+1} \sum_{l \ge 1} l \sum_{U \in \mathcal{U}, C \in \mathcal{C}_l, V \in \mathcal{V}}$$

$$\mathbb{E}|\mathbb{E}\{I_{i-k+1}(UCV) - \mathbb{E}I_{i-k+1}(UCV) \mid \sigma(I_{j,s} : (j,s) \in B_{(i,l)}^c)\}|$$

$$\le \sum_{i=1}^{n-k+1} \sum_{l \ge 1} l \sum_{U \in \mathcal{U}, C \in \mathcal{C}_l, V \in \mathcal{V}} \sum_{x,y \in \mathcal{A}}$$

$$|\mathbb{E}\{I_{i-k+1}(UCV) \mid \xi_{i-k-r+1} = x, \xi_{i+(l+1)(k-1)+r} = y\} - \pi(UCV)|$$

$$\mathbb{P}(\xi_{i-k-r+1} = x, \xi_{i+(l+1)(k-1)+r} = y)$$

$$= \sum_{i=1}^{n-k+1} \sum_{l \geq 1} l \sum_{U \in \mathcal{U}, C \in \mathcal{C}_l, V \in \mathcal{V}} \sum_{x,y \in \mathcal{A}}$$

$$|\mathbb{P}(I_{i-k+1}(UCV) = 1, \xi_{i-k-r+1} = x, \xi_{i+(l+1)(k-1)+r} = y)$$

$$-\pi(UCV)\mathbb{P}(\xi_{i-k-r+1} = x, \xi_{i+(l+1)(k-1)+r} = y)|. \qquad (3.3.12)$$

Now

$$\mathbb{P}(I_{i-k+1}(UCV) = 1, \xi_{i-k-r+1} = x, \xi_{i+(l+1)(k-1)+r} = y)$$

$$= \pi(UCV)\Pi_R^{(r)}(u_1, x)\Pi^{(l(k-1)+r-|C|+1)}(v_{k-1}, y),$$

where u_1 and v_{k-1} denote the first and the last letters of the words U and V, respectively; furthermore,

$$\mathbb{P}(\xi_{i-k-r+1} = x, \xi_{i+(l+1)(k-1)+r} = y) = \pi(x)\Pi^{((l+2)(k-1)+2r)}(x, y).$$

Writing

$$|\Pi^{(s)}(a, a') - \pi(a')| := \eta^{(s)}(a, a'); \quad |\Pi_R^{(s)}(a, a') - \pi(a')| := \eta_R^{(s)}(a, a'),$$

where, from (3.3.6), for each $a \in \mathcal{A}$,

$$\sum_{a' \in \mathcal{A}} \eta^{(s)}(a, a') \leq K\rho^s; \quad \sum_{a' \in \mathcal{A}} \eta_R^{(s)}(a, a') \leq K\rho^s, \qquad (3.3.13)$$

we thus have

$$|\mathbb{P}(I_{i-k+1}(UCV) = 1, \xi_{i-k-r+1} = x, \xi_{i+(l+1)(k-1)+r} = y)$$

$$-\pi(UCV)\mathbb{P}(\xi_{i-k-r+1} = x, \xi_{i+(l+1)(k-1)+r} = y)|$$

$$\leq \pi(UCV) \left\{ \pi(y)\eta_R^{(r)}(u_1, x) + \pi(x)\eta^{(l(k-1)+r-|C|+1)}(v_{k-1}, y) \right.$$

$$+\eta_R^{(r)}(u_1, x)\eta^{(l(k-1)+r-|C|+1)}(v_{k-1}, y)$$

$$\left. +\pi(x)\eta^{((l+2)(k-1)+2r)}(x, y) \right\}.$$

Substituting this into (3.3.12), and using (3.3.13), it then follows that

$$b_3^* \leq (n - k + 1) \sum_{l \geq 1} l \sum_{U \in \mathcal{U}, C \in \mathcal{C}_l, V \in \mathcal{V}} \pi(UCV)$$

$$\times K \left\{ 2\rho^r + K\rho^{2r} + \rho^{3(k-1)+2r} \right\}$$

$$= K\rho^r \mathbb{E}W \left\{ 2 + K\rho^r + \rho^{3(k-1)+r} \right\}, \qquad (3.3.14)$$

since $\sum_{l\geq 1} l \sum_{U\in\mathcal{U}, C\in\mathcal{C}_l, V\in\mathcal{V}} \pi(UCV) = \psi$. The bounds (3.3.8), (3.3.11) and (3.3.14), used in the procedure given at the start of the proof, establish the theorem. ∎

The distribution of W can also be treated using the Markov chain approach. To do so, we introduce the achievement chain $X = (X_i, \ i \in \mathbf{Z})$, also Markovian, with state space $\mathcal{A}^* = ((\mathcal{A}\setminus\{a_1\})\times\{0\})\cup\{1,2,\ldots,k\}$, defined by

$$\{X_i = j\} := B_j^i \setminus \bigcup_{l=j+1}^{k} B_l^i, \quad j = 1,\ldots,k-1$$

$$\{X_i = k\} := B_k^i,$$

$$\{X_i = (a,0)\} := \{\xi_i = a\} \setminus \bigcup_{j=1}^{k} B_j^i, \qquad a \neq a_1, \qquad (3.3.15)$$

where the B_j^i are as in (3.3.1). Denote its transition matrix by P and its stationary distribution by ν. Then the number of visits of X_i to the state k between times k and n is also the number of copies of A in the sequence $\xi_1, \xi_2, \ldots, \xi_n$, and $\nu(k) = \psi$. We then use Erhardsson (1999, Theorem 4.3) and the improved bounds (3.1.13) and (3.1.12) to approximate $\mathcal{L}(W)$ by a compound Poisson distribution. Erhardsson also considers the more complicated problem of counting the number of times that a copy of any one of a set of words appears.

For $x \in \mathcal{A}^*$, let $\tau_x := \inf\{i \geq 0 : X_i = x\}$ and $\tau_x^R := \inf\{i \geq 0 : X_{-i} = x\}$ denote the first hitting times of the chain X and its time reversal X^R on the state x, and let $\tau_x^+ := \inf\{i > 0 : X_i = x\}$. Let $\mathbb{E}_k$ denote expectation conditional on $X_0 = k$ and $\mathbb{E}_\nu$ expectation conditional on $X_0 \sim \nu$. Then from Erhardsson (1999, Theorem 4.3 and Remark 4.2), and from the bounds (3.1.13) and (3.1.12), we have the following theorem.

THEOREM 3.3.2
Choose any $a_0 \neq a_1$, and define $L^* := \mathbb{P}_k(\tau_k^+ < \tau_{(a_0,0)})$. Then, if $L^* \leq 1/2$,

$$d_K(\mathcal{L}(W), \mathrm{PA}\,(\mathbb{E}W\,(1-L^*), L^*)) \leq \{(1-L^*)^2 + (\mathbb{E}W)^{-1}\}^{-1}\Delta_1' + \Delta_3;$$

if $L^* < 1/5$,

$$d_{TV}(\mathcal{L}(W), \mathrm{PA}\,(\mathbb{E}W\,(1-L^*), L^*))$$
$$\leq B_{3.3.2} := \{(1-L^*)/(1-5L^*)\}\Delta_1' + \Delta_3,$$

where $\mathbb{E}W = (n - k + 1)\psi$, $\Delta_3 := 2\psi\mathbb{E}_k(\tau^R_{(a_0,0)})$ *and*

$$\Delta_1' := 2\psi\{\mathbb{E}_k(\tau_{(a_0,0)}) + \mathbb{E}_k(\tau^R_{(a_0,0)}) + \mathbb{E}_\nu(\tau_{(a_0,0)})/\nu(a_0,0)\}.$$

The statement of this theorem is a little less explicit than that of Theorem 3.3.1, in particular because $L^* \neq L$ needs to be calculated. It is actually possible to prove rough bounds for the quantities $\mathbb{E}_k(\tau^R_{(a_0,0)})$ and $\mathbb{E}_z(\tau_{(a_0,0)})$, $z \in \mathcal{A}^*$, and for the difference between L^* and L, in terms of the elements of the transition matrix Π. However, all these quantities can be computed simply by solving small sets of linear equations if $|\mathcal{A}|$ is small, as is the case for the four letter DNA alphabet or the twenty letter amino acid alphabet, and this is the procedure that we adopt in the numerical comparisons. The bound in Theorem 3.3.2 is asymptotically better than that of Theorem 3.3.1, improving the order $O(\psi \log n)$ to $O(\psi)$. However, for moderate n, the logarithmic factor may not dominate the contributions from the constant elements in the bounds, so that numerical comparison for such n is also of interest.

3.4   Discussion and Numerical Examples

We compare the estimates $B_{3.3.1}$ of Theorem 3.3.1 and $B_{3.3.2}$ of Theorem 3.3.2 with the estimate B_S of Schbath (1995). Her estimate is not explicit, as far as the long range dependence is concerned, so we treat her b_3 much as we have treated our b_3^*, giving in total

$$d_{TV}(\mathcal{L}(W), \mathrm{CP}(\lambda, \mu)) \leq B_S$$

$$:= 2\psi\mathbb{E}W \left\{ (k-1)(1-L)(\frac{1}{\pi_{min}} + 1) \right.$$

$$\left. + (2k + r - 2 + \frac{r-k}{\pi_{min}})(1-L)^2 + \frac{2(k-1)}{\pi(a_1)} \right\}$$

$$+ K\rho^r \mathbb{E}W (1-L)\{2 + K\rho^r + \rho^{3(k-1)+r}\} + 2(k-1)\psi, \quad (3.4.1)$$

where $\pi_{min} := \min_{a \in \mathcal{A}} \pi(a)$.

In B_S and $B_{3.3.1}$, the value of r can be chosen to minimize the error estimate. Given the values of k, ψ, L and ρ, it only requires simple calculus to find the real value r which minimizes the expressions, and the best integer value of r can then be determined by experiment. In $B_{3.3.2}$, the choice of a_0 can also be made so as to obtain the best estimate. In the examples that we present, for the word ACGACG ($k = 6$) from the DNA alphabet $\mathcal{A} = \{A,C,G,T\}$ with three different matrices Π, we take $a_0 := T$. We tabulate the estimates for total variation distance; those for Kolmogorov distance are slightly smaller.

1. The following transition matrix corresponds to the *Lamda virus* (Reinert and Schbath (1998))

$$\Pi = \begin{pmatrix} 0.2994 & 0.2086 & 0.2215 & 0.2705 \\ 0.2830 & 0.2198 & 0.2740 & 0.2232 \\ 0.2540 & 0.2820 & 0.2480 & 0.2160 \\ 0.1813 & 0.2232 & 0.3164 & 0.2791 \end{pmatrix},$$

and has stationary distribution $(0.254374, 0.234228, 0.264254, 0.247145)$. The inequality (3.3.6) holds for $K = 2$ and $\rho = 0.1726$. The values of L and L^* are 0.0145177 and 0.0152569 respectively.

Word	n	r	$\mathbb{E}W$	$B_{3.3.2}$	$B_{3.3.1}$	B_S
	14500	6	3	0.0159630	0.0188487	0.1068091
	47500	6	10	0.0159630	0.0190079	0.3451702
ACGACG	95000	6	20	0.0159630	0.0191553	0.6882658
	142500	6	30	0.0159630	0.0192684	1.0313614
	237000	6	50	0.0159630	0.0194469	1.7139410
	474000	6	100	0.0159630	0.0197762	3.4258074

2. For the transition probability matrix

$$\Pi = \begin{pmatrix} 0.5 & 0.125 & 0.125 & 0.25 \\ 0.125 & 0.5 & 0.25 & 0.125 \\ 0.125 & 0.25 & 0.5 & 0.125 \\ 0.25 & 0.125 & 0.125 & 0.5 \end{pmatrix},$$

the stationary distribution is $(0.25, 0.25, 0.25, 0.25)$ and the inequality (3.3.6) holds for $K = 2$ and $\rho = 0.5$. The values of L and L^* are 0.0039063 and 0.0040525 respectively.

Word	n	r	$\mathbb{E}W$	$B_{3.3.2}$	$B_{3.3.1}$	B_S
	98500	15	3	0.0028436	0.0038838	0.0236823
	330000	16	10	0.0028436	0.0039900	0.0810643
ACGACG	655500	17	20	0.0028436	0.0040564	0.1661725
	985000	17	30	0.0028436	0.0040878	0.2495496
	1640000	17	50	0.0028436	0.0041371	0.4152917
	3280000	18	100	0.0028436	0.0041974	0.8590681

3. For the transition probability matrix

$$\Pi = \begin{pmatrix} 0.05 & 0.3 & 0.4 & 0.25 \\ 0.35 & 0.3 & 0.05 & 0.3 \\ 0.25 & 0.25 & 0.25 & 0.25 \\ 0.1 & 0.3 & 0.4 & 0.2 \end{pmatrix},$$

the stationary distribution is $(0.200778, 0.286976, 0,260486, 0.251761)$ and the inequality (3.3.6) holds for $K = 2$ and $\rho = 0.5426$. The values of L and L^* are 0.003750 and 0.003785 respectively.

Word	n	r	$\mathbb{E}W$	$B_{3.3.2}$	$B_{3.3.1}$	B_S
	267000	19	3	0.0007571	0.0017632	0.0119805
	890000	20	10	0.0007571	0.0018143	0.0408546
ACGACG	1771000	20	20	0.0007571	0.0018401	0.0811845
	2660000	20	30	0.0007571	0.0018602	0.1218807
	4430000	21	50	0.0007571	0.0018792	0.2091687
	8860000	21	100	0.0007571	0.0019108	0.4182248

We note that the bound $B_{3.3.2}$ remains stable as n increases, whereas $B_{3.3.1}$ increases slowly. In all the cases considered here, $B_{3.3.2}$ performs better. However, its advantage may be less in a larger alphabet, for moderate n, because the factor $1/\nu(a_0, 0)$ may then be expected to be bigger in comparison with $\log n$. For the calculations, we used Mathematica 2.2.

Acknowledgements. This work was supported by University of Athens Research Grant 70-4-2558 (A. D. Barbour and O. Chryssaphinou), by Swiss Nationalfonds Projekt Nr. 20-50686.97 (A. D. Barbour), by the grant Pened95 692 (O. Chryssaphinou and E. Vaggelatou), of the General Secretariat of Research and Technology of Greece and by the National Scholarship Foundation of Greece (E. Vaggelatou).

References

1. Arratia, R., Goldstein, L., and Gordon, L. (1990). Poisson approximation and the Chen-Stein method. *Statistical Science* **5**, 403-434.

2. Barbour, A. D. and Chryssaphinou, O. (1999). Compound Poisson approximation: a user's guide. preprint.

3. Barbour, A. D., Chen, L. H. Y., and Loh, W. (1992). Compound Poisson approximation for nonnegative random variables via Stein's method. *Annals of Probability* **20**, 1843-1866.

4. Barbour, A. D. and Hall, P. (1984). On the rate of Poisson convergence. *Mathematical Proceedings of the Cambridge Philosophical Society* **95**, 473-480.

5. Barbour, A. D., Holst, L., and Janson, S. (1992). *Poisson Approximation.*, Oxford University Press.

6. Barbour, A. D. and Utev, S. (1998). Solving the Stein equation in compound Poisson approximation. *Advances in Applied Probability* **30**, 449-475.

7. Barbour, A. D. and Utev, S. (1999). Compound Poisson approximation in total variation. *Stochastic Processes and Applications* **82**, 89-125.

8. Barbour, A. D. and Xia, A. (1999a). Estimating Stein's constants for compound Poisson approximation. *Bernoulli*, to appear.

9. Barbour, A. D. and Xia, A. (1999b). Poisson perturbations. *ESAIM Probability and Statistics* **3**, 131-150.

10. Chryssaphinou, O., Papastavridis, S., and Vaggelatou, E. (1999). On the number of appearances of a word in a sequence of i.i.d. trials. *Methodology and Computing in Applied Probability* **3**, 329-348.

11. Eichelsbacher, P. and Roos, M. (1998) Compound Poisson approximation for dissociated random variables via Stein's method. *Combinatorics, Probability and Computing* **8**, 335-346.

12. Erhardsson, T. (1999). Compound Poisson approximation for Markov chains using Stein's method. *Annals of Probability* **27**, 565-596.

13. Guibas, L. J. and Odlyzko, A. M. (1981). Periods in strings. *Journal of Combinatorial Theory* A **30**, 19-42.

14. LeCam, L. (1960). An approximation theorem for the Poisson binomial theorem. *Pacific Journal of Mathematics* **10**, 1181-1197.

15. Lindvall, T. (1992). *Lectures on the coupling method.* Wiley, New York.

16. Månsson, M. (1999). On compound Poisson approximation for sequence matching. preprint.

17. Neuhauser, C. (1996). A phase transition for the distribution of matching blocks. *Combinatorics, Probability and Computing* **5**, 139-159.

18. Reinert, G., Schbath, S. (1998). Compound Poisson and Poisson process approximations for occurrences of multiple words in Markov chains. *Journal of Computational Biology* **5**, 223-253.

19. Roos, M. (1993). Stein-Chen method for compound Poisson approximation. PhD thesis, University of Zürich, Switzerland.

20. Roos, M. (1994). Stein's method for compound Poisson approximation: the local approach. *Annals of Applied Probability* **4**, 1177-1187.

21. Schbath, S. (1995). Compound Poisson approximation of word counts in DNA sequences. *ESAIM Probability and Statistics* **1**, 1-16.

22. Wolfram, S. (1988). *Mathematica*, Addison-Wesley, Reading, MA.

4

Compound Poisson Approximation for Sums of Dependent Random Variables

Michael V. Boutsikas and Markos V. Koutras
University of Athens, Athens, Greece

ABSTRACT In the present article an upper bound is obtained for the Kolmogorov distance between a sum of (dependent) nonnegative r.v.s and an appropriate compound Poisson distribution. Two applications pertaining to moving sums distributions and success runs in a sequence of Markov dependent trials are considered and certain upper bounds for their distance to a compound Poisson distribution are established, along with asymptotic results (compound Poisson distribution convergence).

Keywords and phrases Compound Poisson approximation, sums of dependent variables, Kolmogorov distance, rate of convergence, moving sums, success runs in a Markov chain

4.1 Introduction

Let $X_1, X_2, ..., X_n$ be a sequence of real-valued, nonnegative r.v.s. The subject of the present paper is the approximation of the distribution of the sum of X_i when the masses of $\mathcal{L}(X_i), i = 1, 2, ..., n$ are concentrated around 0, and $X_i, i = 1, 2, ..., n$ are "weakly" dependent. Such problems arise quite naturally when we are dealing with a set of "rare" and "almost unrelated" events. The simplest case when X_i are binary i.i.d. r.v.s was initially treated in the fundamental work of Poisson as early as 1837 when he established the classical Poisson approximation of the binomial distribution. Since then a huge variety of generalizations and extensions of this result has appeared in the literature. In some of them the assumption of identical X_i was relaxed while in others the assumption of independence was replaced by a condition of "local" or "weak dependence". The latter generalization was thoroughly investigated after the introduction of the much acclaimed Stein-Chen method (c.f. Chen (1975); the definite reference for this method, is the elegant monograph by Barbour, Holst, and Janson (1992)). For more recent developments on Poisson or compound Poisson approximation using Stein-Chen method the reader may refer to Barbour, Chen, and Loh (1992), Roos (1994), Xia (1997), Barbour and Utev (1998) and Erhardsson (1999). A completely different approach of Compound Poisson approximations for sums of dependent r.v.s was taken by Vellaisamy and Chaudhuri (1999) (c.f. the references therein).

The Stein-Chen method provides quite tight error bounds for Poisson or compound Poisson approximations of sums of weakly dependent binary r.v.s. In a very recent work, Boutsikas and Koutras (2000a) replaced the assumption of binary r.v.s by that of integer valued ones and proceeded to the investigation of error bounds for compound Poisson approximations by taking a completely different approach than that of the Stein-Chen method. This was accomplished at the expense of restricting the nature of the dependence between X_i to that of associated or negatively associated (NA) r.v.s. More, specifically, the authors proved that if $X_1, X_2, \ldots, X_n$ are nonnegative, integer-valued, associated or NA r.v.'s with $\mathbb{E}(X_i), \mathbb{E}(X_i X_j) < \infty$ for $i, j = 1, 2, \ldots, n, i \neq j$, then

$$d_W(\mathcal{L}(\sum_{i=1}^{n} X_i), CP(\lambda, F)) \leq 2\left|\sum_{i<j} Cov(X_i, X_j)\right| + \sum_{i=1}^{n} \mathbb{E}(X_i)^2. \quad (4.1.1)$$

Here d_W stands for the Wasserstein distance, CP denotes a compound Poisson distribution, $\lambda = \sum_{i=1}^{n} \mathbb{P}(X_i > 0)$, and $F(x) = \frac{1}{\lambda}\sum_{i=1}^{n} \mathbb{P}(0 < X_i \leq x)$ (the latter can also be viewed as a mixture of the conditional distributions of X_i given that $X_i > 0$ with weights $\mathbb{P}(X_i > 0), i = 1, 2, \ldots, n$). Clearly, if the nonnegative, integer-valued, associated or NA r.v.s $X_i, i = 1, 2, \ldots, n$ are almost uncorrelated and their distributions are concentrated around zero, then the above result states that their sum can be approximated by an appropriate compound Poisson distribution. As a matter of fact, (4.1.1) is a by-product of a more general result which provides error estimates of the distance between $\sum_{i=1}^{n} X_i$ and a sum of independent r.v.s $\sum_{i=1}^{n} X_i'$ having the same marginals as the original ones ($\mathcal{L}(X_i) = \mathcal{L}(X_i')$).

It is worth stressing that, when X_i are binary associated or NA r.v.s, the bound provided by (4.1.1) has almost the same form as the respective Poisson approximation Stein-Chen bound (created by exploiting an appropriate coupling argument); the main difference in the final bounds is in the so-called "magic factor" which is known to improve substantially the error bound provided by the Stein-Chen method (especially when λ becomes large).

Should a compound Poisson approximation be pursued for sums of binary "weakly dependent" indicators, inequality (4.1.1) could be fruitfully applied to gain an error bound (e.g. by defining X_i as the size of clump clustered at the point where the i-th occurrence of the event we are interested in took place). In general this bound is simpler in form than the corresponding Stein-Chen bound which engages neighborhoods of strong/weak dependence for the binary indicators involved [see Arratia *et al.* (1989, 1990), Roos (1994), Barbour and Utev (1998)]. As demonstrated in many recent versions of the compound Poisson approximation Stein-Chen method, the incorporation of a factor similar to the ordinary Poisson approximation magic factor leads occasionally to tighter bounds than (4.1.1) especially when λ is large. Nevertheless, if λ is kept fixed (which is quite common in many asymptotic results), (4.1.1) is of the same order as the Stein-Chen bound.

When X_i are associated or *NA*, inequality (4.1.1) can be viewed as an extension of the Stein-Chen-type bound to integer-valued r.v.s . It is therefore reasonable to ask whether we could establish a similar extension, by rejecting the assumption of association or *NA* between the X_is.

The main purpose of the present paper is to provide a bound for the distance between $\mathcal{L}(\sum_{i=1}^n X_i)$, and an appropriate compound Poisson distribution where X_i are arbitrary nonnegative real-valued r.v.s. This is accomplished by using neighborhoods of strong and weak dependence similar in structure to the ones employed in Arratia's *et al.* (1989, 1990), ingenious approach. The core of our proof is very similar to that used in Boutsikas and Koutras (2000a), although the starting point of the two approaches is quite different. Note also that here the metric distance used for measuring the discrepancy between distributions and establishing asymptotic results is the Kolmogorov distance (instead of the Wasserstein or the total variation distance).

The new bound goes to 0 when the masses of $\mathcal{L}(X_i), i = 1, 2, ..., n$ are concentrated on 0, and $X_i, i = 1, 2, ..., n$ are "weakly" dependent. It is worth stressing though that, in contrast to inequality (4.1.1) the new result cannot be effectively used when we deal with "globally" dependent r.v.s [e.g. see Boutsikas and Koutras (2000b)].

The organization of the paper is as follows: In Section 4.2 we introduce the necessary notations and review the basic properties of the Kolmogorov distance. Section 4.3 provides our main result which offers an upper bound for the distance between the distribution of $\sum_{a=1}^n X_a$ and an appropriate compound Poisson distribution. Two interesting intermediate results, pertaining to the distance between a) a sum of dependent r.v.s $\sum_{a=1}^n X_a$ and the sum of their independent duplicates $\sum_{a=1}^n X_a'$ ($\mathcal{L}(X_a) = \mathcal{L}(X_a'), a = 1, 2, ..., n$), b) a sum of independent r.v.s and a compound Poisson distribution, are also presented. Finally, in

Section 4.4 two specific examples are considered: the first one deals with the distribution of the total exceedance amount (above a prespecified threshold) of moving sums in a sequence of i.i.d. r.v.s, and the second with the distribution of the number of overlapping success runs in a sequence of Markov dependent trials. In both cases, a bound for the distance between the distribution of interest and a proper compound Poisson distribution is obtained and an asymptotic result (compound Poisson convergence) is established.

4.2 Preliminaries and Notations

In the sequel $CP(\lambda, F)$ will denote the compound Poisson distribution with characteristic function (c.f.) $e^{-\lambda(1-\phi_F(z))}$, where $\phi_F(z)$ is the c.f. of the distribution F. Obviously, $CP(\lambda, F)$ coincides with the distribution of the sum $\sum_{i=1}^{N} Y_i$ with N being a Poisson r.v. with parameter λ and Y_i independent r.v.s with distribution function F. We shall also use the notation $Po(\lambda)$ for the ordinary Poisson distribution with parameter λ. Manifestly, $Po(\lambda) = CP(\lambda, \mathcal{L}(1))$.

In order to bound the error incurred when approximating the sum of X_i by an appropriate $CP(\lambda, F)$ we shall be using the Kolmogorov distance

$$d(\mathcal{L}(X), \mathcal{L}(Y)) = \sup_{w} |\mathbb{P}(X \leq w) - \mathbb{P}(Y \leq w)|.$$

For typographical convenience, we shall allow in the sequel an abuse of the notation and use $d(X, Y)$ instead of $d(\mathcal{L}(X), \mathcal{L}(Y))$. Manifestly, a sequence of random variables $\{X_n\}$ converges in distribution to Y if $d(X_n, Y)$ converges to 0. It can be easily verified that $d(aX, aY) = d(X, Y)$. Moreover, if Y is a r.v. independent of X, Z, we have

$$d(X + Y, Z + Y) = \sup_{w} \left| \int (\mathbb{P}(X \leq w - y) - \mathbb{P}(Z \leq w - y)) dF_Y(y) \right|$$

$$\leq d(X, Z) \tag{4.2.1}$$

Using the last inequality in conjunction with the triangle inequality we may easily verify that, if (X_1, Y_1) and (X_2, Y_2) are independent, then

$$d(X_1 + X_2, Y_1 + Y_2) \leq d(X_1 + X_2, Y_1 + X_2) + d(Y_1 + X_2, Y_1 + Y_2)$$

$$\leq d(X_1, Y_1) + d(X_2, Y_2) \tag{4.2.2}$$

A repeated application of (4.2.2) captures the subadditivity property of the Kolmogorov distance for independent summands, i.e. if (X_1, Y_1),

$(X_2, Y_2), ..., (X_n, Y_n)$ are independent, then

$$d(\sum_{i=1}^{n} X_i, \sum_{i=1}^{n} Y_i) \leq \sum_{i=1}^{n} d(X_i, Y_i). \tag{4.2.3}$$

Let now $X_1, X_2, ..., X_n$ be (independent) Bernoulli r.v.s such that $\mathbb{P}(X_i = 1) = p_i$. If $Y_1, Y_2, \ldots, Y_n$ are independent r.v.s such that $\mathcal{L}(Y_i) = Po(p_i)$, then, recalling (4.2.3), we get

$$d(\sum_{i=1}^{n} X_i, Po(\sum_{i=1}^{n} p_i)) = d(\sum_{i=1}^{n} X_i, \sum_{i=1}^{n} Y_i) \leq \sum_{i=1}^{n} d(X_i, Y_i)$$

$$= \sum_{i=1}^{n} \sup_{w=0,1,...} |\mathbb{P}(X_i > w) - \mathbb{P}(Y_i > w)|$$

$$= \sum_{i=1}^{n} \max \left\{ p_i - 1 + e^{-p_i}, 1 - e^{-p_i} - p_i e^{-p_i} \right\}$$

$$\leq \frac{1}{2} \sum_{i=1}^{n} p_i^2. \tag{4.2.4}$$

We finally mention that $d(X, Y) \leq \max\{\mathbb{P}(X < Y), \mathbb{P}(X > Y)\} \leq \mathbb{P}(X \neq Y)$ for every coupling (X, Y) of $\mathcal{L}(X), \mathcal{L}(Y)$. For an early review on some of the elementary properties of the Kolmogorov distance refer also to Serfling (1978).

4.3 Main Results

Consider a sequence of nonnegative r.v.s X_a, $a \in I_n = \{1, 2, ..., n\}$. For each $a \in I_n, a \neq 1$, we introduce a subset B_a of I_{a-1} so that X_a is independent or weakly dependent to $X_b, b \in B'_a = I_{a-1} \setminus B_a$. The set B_a will be referred in the sequel as the *left neighborhood of strong dependence of X_a*.

We shall now proceed to the statement of the main result of our paper which presents an upper bound for the Kolmogorov distance between the distribution of $\sum_{i=1}^{n} X_i$ and an appropriate compound Poisson distribution. In what follows, $X'_a, a = 1, 2, ..., n$ will denote independent r.v.'s having the same marginals as $X_a, a = 1, 2, ..., n$, i.e. $\mathcal{L}(X_a) = \mathcal{L}(X'_a)$; $X'_a, a = 1, 2, ..., n$ will also be assumed as being independent of $X_a, a = 1, 2, ..., n$.

THEOREM 4.3.1
If $X_a, a = 1, 2, ..., n$ is a sequence of nonnegative r.v.s, then

$$d\left(\sum_{a=1}^{n} X_a, CP(\lambda, F)\right)$$

$$\leq \sum_{a=2}^{n}(\mathbb{P}(X_a > 0, \sum_{b \in B_a} X_b > 0) + \mathbb{P}(X_a > 0)\mathbb{P}(\sum_{b \in B_a} X_b > 0))(4.3.1)$$

$$+\frac{1}{2}\sum_{i=1}^{n}\mathbb{P}(X_i > 0)^2 + \sum_{a=2}^{n} d(\sum_{b \in B'_a} X_b + X_a, \sum_{b \in B'_a} X_b + X'_a)$$

where $\lambda = \sum_{i=1}^{n} p_i$, and $F(x) = \frac{1}{\lambda}\sum_{i=1}^{n} p_i\mathbb{P}(X_i \leq x | X_i > 0), x \in \mathbf{R}$, $p_i = \mathbb{P}(X_i > 0)$.

   The proof is carried out in two steps: firstly we obtain an upper bound for the distance between $\sum_{a=1}^{n} X_a$ and $\sum_{a=1}^{n} X'_a$, and an upper bound for the distance between $\sum X'_a$ and $CP(\lambda, F)$. Secondly, the triangle inequality is exploited to arrive at a bound between $\sum X_a$ and $CP(\lambda, F)$. Due to its independent interest the development of the upper bounds for $d(\sum X_a, \sum X'_a)$ and $d(\sum X'_a, CP(\lambda, F))$ will be stated as separate theorems (see Theorems 4.3.3 and 4.3.4 below). Before proceeding to the statement and proof of the intermediate steps, let us discuss the applicability of it to some interesting special cases.
   Note first that an upper bound for the first term of the RHS of (4.3.1) is offered by the inequality

$$d\left(\sum_{a=1}^{n} X_a, CP(\lambda, F)\right)$$

$$\leq \sum_{a=2}^{n}\sum_{b \in B_a}(\mathbb{P}(X_a > 0, X_b > 0) + \mathbb{P}(X_a > 0)\mathbb{P}(X_b > 0)) \qquad (4.3.2)$$

$$+\frac{1}{2}\sum_{i=1}^{n}\mathbb{P}(X_i > 0)^2 + \sum_{a=2}^{n} d(\sum_{b \in B'_a} X_b + X_a, \sum_{b \in B'_a} X_b + X'_a)$$

In most applications the last bound is more convenient for performing the necessary calculations to establish upper bounds for $d(\sum X_a, CP(\lambda, F))$.
   It is clear that, should X_a be independent of X_b, $b \in B'_a$ for every $a \in I_n$ the last summand in the RHS of inequality (4.3.1) would vanish. The family of m-dependent r.vs offers a case where this last remark applies; under this condition the upper bound in (4.3.1) takes on a much

simpler form. A similar bound for the total variation distance has been mentioned by Barbour, Holst, and Janson (1992) (see Corollary 10.L.1, page 239).

If B'_a is chosen so that X_a is not independent but "weakly" dependent of X_b, $b \in B'_a$ we can ultimately bound the last term of (4.3.1) by an appropriate quantity. For example, consider a sequence of nonnegative r.v.s $X_1, X_2, ..., X_n$ taking values in $\{0, 1, 2, ..., r-1\}$ and assume that the sequence $\{X_n\}$ is a-mixing, i.e. there exist $a_m, m = 1, 2, ..$ such that $|\mathbb{P}(A \cap B) - \mathbb{P}(A)\mathbb{P}(B)| \leq a_m$ for every $A \in \sigma(X_1, ..., X_i)$, $B \in \sigma(X_{i+m}, X_{i+m+1}, ...), i \geq 1, m \geq 1$, and $a_m \to 0$ (see e.g. Billingsley (1986), p. 375). Then

$$d\left(\sum_{b=1}^{i-m} X_b + X_i, \sum_{b=1}^{i-m} X_b + X'_i\right)$$

$$= \sup_w \left|\mathbb{P}\left(\sum_{b=1}^{i-m} X_b + X_i \leq w\right) - \mathbb{P}\left(\sum_{b=1}^{i-m} X_b + X'_i \leq w\right)\right|$$

$$= \sup_w \left|\sum_{x=0}^{r-1} \left(\mathbb{P}\left(\sum_{b=1}^{i-m} X_b \leq w - x, X_i = x\right)\right.\right.$$

$$\left.\left. - \mathbb{P}\left(\sum_{b=1}^{i-m} X_b \leq w - x\right)\mathbb{P}(X_i = x)\right)\right|$$

$$\leq \sup_w \sum_{x=0}^{r-1} a_m = r a_m$$

and hence

$$d\left(\sum_{a=1}^{n} X_i, CP(\lambda, F)\right)$$

$$\leq \sum_{a=2}^{n} \sum_{b=\max\{1, i-m+1\}}^{i-1} (\mathbb{P}(X_a > 0, X_b > 0) + \mathbb{P}(X_a > 0)\mathbb{P}(X_b > 0))$$

$$+ \frac{1}{2}\sum_{i=1}^{n} \mathbb{P}(X_i > 0)^2 + n r a_m.$$

Therefore, if $\{X_n\}$ is a stationary Markov chain with state space $\{0, 1, ..., r-1\}$, then [see Billingsley (1986, p. 375, 128)] the above inequality is valid for $a_m = ar\rho^m$, for some $a \geq 0$ and $\rho \in (0, 1)$. Assuming furthermore that the Markov chain has positive transition probabilities p_{ij} then $a_m = r\rho^m$ where $\rho = 1 - r \min_{ij} p_{ij}$.

A final case where the last summand in (4.3.1) can be easily controlled (upper bounded) arises when the r.v.s $X_1, X_2, ..., X_n$ exhibit certain types of positive or negative dependence. Thus, when we deal with integer valued r.v.s and $X_a, X_b, b \in B'_a, a = 1, 2, ..., n$ are associated or negatively associated then one could exploit Theorem 1 in Boutsikas and Koutras (2000a) to deduce the upper bound

$$\sum_{a=2}^{n} d(\sum_{b \in B'_a} X_b + X_a, \sum_{b \in B'_a} X_b + X'_a) \leq \sum_{a=2}^{n} \left| Cov(\sum_{b \in B'_a} X_b, X_a) \right|$$

$$= \sum_{a=2}^{n} \left| \sum_{b \in B'_a} Cov(X_b, X_a) \right|.$$

It is worth mentioning that when the X_a are binary r.v.s, then the upper bound in (4.3.2) is of the form $(b_1 + b_2 + b_3^*)/2$ where b_1 and b_2 coincide to the quantities involved in the Chen-Stein upper bound, Arratia *et al.* (1989, 1990).

The following lemma can be considered as the main ingredient of the proofs to follow.

LEMMA 4.3.2
For any nonnegative random variables X, Y, Z, U we have

$$|d(X + Y + Z, X + Y + U) - d(X + Z, X + U)|$$

$$\leq \mathbb{P}(Y > 0, Z > 0) + \mathbb{P}(Y > 0, U > 0).$$

PROOF For every $w \geq 0$ it is obvious that the following equality holds true

$$\mathbb{P}(X + Y + Z > w) - \mathbb{P}(X + Z > w)$$
$$= \mathbb{P}(X + Y + Z > w, X + Z \leq w)$$
$$= \mathbb{P}(X + Z \leq w, X + Y > w)$$
$$\quad + \mathbb{P}(X + Y + Z > w, X + Z \leq w, X + Y \leq w)$$
$$= \mathbb{P}(X + Y > w) - \mathbb{P}(X + Z > w, X + Y > w, X \leq w) - \mathbb{P}(X > w)$$
$$\quad + \mathbb{P}(X + Y + Z > w, X + Z \leq w, X + Y \leq w).$$

Applying the above equality for U instead of Z we deduce that

$$(\mathbb{P}(X + Y + Z > w) - \mathbb{P}(X + Z > w))$$
$$- (\mathbb{P}(X + Y + U > w) - \mathbb{P}(X + U > w))$$
$$= -\mathbb{P}(X + Z > w, X + Y > w, X \leq w)$$
$$+ \mathbb{P}(X + Y + Z > w, X + Z \leq w, X + Y \leq w)$$
$$+ \mathbb{P}(X + U > w, X + Y > w, X \leq w)$$
$$- \mathbb{P}(X + Y + U > w, X + U \leq w, X + Y \leq w).$$

Hence

$$|\mathbb{P}(X + Y + Z > w) - \mathbb{P}(X + Y + U > w)|$$
$$\leq |\mathbb{P}(X + Z > w) - \mathbb{P}(X + U > w)|$$
$$+ \mathbb{P}(X + Z > w, X + Y > w, X \leq w)$$
$$+ \mathbb{P}(X + Y + Z > w, X + Z \leq w, X + Y \leq w)$$
$$+ \mathbb{P}(X + U > w, X + Y > w, X \leq w)$$
$$+ \mathbb{P}(X + Y + U > w, X + U \leq w, X + Y \leq w)$$
$$\leq |\mathbb{P}(X + Z > w) - \mathbb{P}(X + U > w)|$$
$$+ \mathbb{P}(Y > 0, Z > 0) + \mathbb{P}(Y > 0, U > 0)$$

and considering the supremum with respect to w in both sides of the inequality we deduce

$$d(X + Y + Z, X + Y + U) \leq d(X + Z, X + U) + \mathbb{P}(Y > 0, Z > 0)$$
$$+ \mathbb{P}(Y > 0, U > 0).$$

By the same token we may get,

$$d(X + Z, X + U) \leq d(X + Y + Z, X + Y + U) + \mathbb{P}(Y > 0, Z > 0)$$
$$+ \mathbb{P}(Y > 0, U > 0)$$

and the proof is complete. ∎

THEOREM 4.3.3
The distance between the sum of nonnegative dependent variables X_i and the sum of their independent duplicates X_i' ($\mathcal{L}(X_i) = \mathcal{L}(X_i')$, $i = 1, 2, ..., n$) is bounded from above by

$$d(\sum_{a \in I_n} X_a, \sum_{a \in I_n} X'_a) \le \sum_{a=2}^{n} (\mathbb{P}(X_a > 0, \sum_{b \in B_n} X_b > 0)$$

$$+ \mathbb{P}(X_a > 0)\mathbb{P}(\sum_{b \in B_n} X_b > 0))$$

$$+ \sum_{a=2}^{n} d(\sum_{b \in B'_n} X_b + X_a, \sum_{b \in B'_n} X_b + X'_a). \quad (4.3.3)$$

PROOF Applying Lemma 4.3.2 for the r.v.s $\sum_{b \in B'_n} X_b, \sum_{b \in B_n} X_b, X_n, X'_n$ we deduce

$$d(\sum_{b \in B'_n} X_b + \sum_{b \in B_n} X_b + X_n, \sum_{b \in B'_n} X_b + \sum_{b \in B_n} X_b + X'_n)$$

$$\le \mathbb{P}(\sum_{b \in B_n} X_b > 0, X_n > 0) + \mathbb{P}(\sum_{b \in B_n} X_b > 0)\mathbb{P}(X_n > 0)$$

$$+ d(\sum_{b \in B'_n} X_b + X_n, \sum_{b \in B'_n} X_b + X'_n) \quad (4.3.4)$$

Accordingly, for the r.v.s $\sum_{b \in B'_{n-1}} X_b + X'_n, \sum_{b \in B_{n-1}} X_b, X_{n-1}, X'_{n-1}$, we have that

$$d((\sum_{b \in B'_{n-1}} X_b + X'_n) + \sum_{b \in B_{n-1}} X_b + X_{n-1}, (\sum_{b \in B'_{n-1}} X_b + X'_n)$$

$$+ \sum_{b \in B_{n-1}} X_b + X'_{n-1})$$

$$\le \mathbb{P}(\sum_{b \in B_{n-1}} X_b > 0, X_{n-1} > 0) + \mathbb{P}(\sum_{b \in B_{n-1}} X_b > 0)\mathbb{P}(X_{n-1} > 0)$$

$$+ d(\sum_{b \in B'_{n-1}} X_b + X'_n + X_{n-1}, \sum_{b \in B'_{n-1}} X_b + X'_n + X'_{n-1}). \quad (4.3.5)$$

By virtue of inequalities (4.3.5) and (4.2.1) we may write

$$d(\sum_{b=1}^{n-1} X_b + X'_n, \sum_{b=1}^{n-2} X_b + \sum_{a=n-1}^{n} X'_a)$$

$$\le (\mathbb{P}(\sum_{b \in B_{n-1}} X_b > 0, X_{n-1} > 0)$$

$$+\mathbb{P}(\sum_{b\in B_{n-1}} X_b > 0)\mathbb{P}(X_{n-1} > 0))$$

$$+d(\sum_{b\in B'_{n-1}} X_b + X_{n-1}, \sum_{b\in B'_{n-1}} X_b + X'_{n-1}) \qquad (4.3.6)$$

and applying the triangle inequality (recall also relations (4.3.4),(4.3.6)) we obtain

$$d\left(\sum_{b=1}^{n} X_b, \sum_{b=1}^{n-2} X_b + \sum_{a=n-1}^{n} X'_a\right)$$

$$\leq \sum_{a=n-1}^{n} (\mathbb{P}(X_a > 0, \sum_{b\in B_a} X_b > 0) + \mathbb{P}(X_a > 0)\mathbb{P}(\sum_{b\in B_a} X_b > 0))$$

$$+ \sum_{a=n-1}^{n} d(\sum_{b\in B'_a} X_b + X_a, \sum_{b\in B'_a} X_b + X'_a).$$

The proof is easily completed on using induction ∎

Should we wish to secure that the upper bound in (4.3.3) converges to 0 and $\sum_{a\in I_n} X'_a < \infty$ a.s. as $n \to \infty$, it is necessary to impose the conditions that $\mathbb{P}(X_a > 0) \equiv p_a$ tends to 0 for every $a \in I_n$ and that $\sum_{a\in I_n} p_a \to \lambda \in (0,\infty)$ respectively. Under these conditions it is not hard to see that the distribution of $\sum_{a\in I_n} X'_a$ converges to a compound Poisson distribution. Wang (1989) proved an analogous result for integer-valued r.v.s (his approach is based on characteristic function), while Vellaisamy and Chaudhuri (1999) proposed a bound on the rate of convergence. Let us now proceed to the next theorem which provides a bound for the error (in terms of Kolmogorov distance) incurred when approximating the distribution of a sum of independent nonnegative r.v.s by a compound Poisson distribution.

THEOREM 4.3.4
If $X_a, a = 1, 2, ..., n$ is a sequence of independent nonnegative r.v.s, then

$$d\left(\sum_{a=1}^{n} X_a, CP(\lambda, F)\right) \leq \frac{1}{2}\sum_{i=1}^{n} p_i^2 \qquad (4.3.7)$$

where $p_i = \mathbb{P}(X_i > 0)$, $\lambda = \sum_{i=1}^{n} p_i$, and $F(x) = \frac{1}{\lambda}\sum_{i=1}^{n} p_i\mathbb{P}(X_i \leq x|X_i > 0), x \in \mathbf{R}$.

PROOF At first assume that X_i are simple r.v.s i.e. $X_i \in \{0, a_1, a_2, ...,$ $a_m\}$, $a_i > 0, a_i \neq a_j$ for $i \neq j$. The sum of X_i can be alternatively expressed as

$$\sum_{i=1}^{n} X_i = \sum_{i=1}^{n} \sum_{j=1}^{m} a_j I_{[X_i = a_j]} = \sum_{i=1}^{n} \sum_{j=1}^{m} a_j Y_{ji}$$

and employing Theorem 4.3.3 for the r.v.s $a_j Y_{ji}, j = 1, 2, ..., m, \quad i = 1, 2, ..., n$, and neighborhoods of strong dependence $B_{(j,i)} = \{(s,i) : s < j\}$ we deduce

$$d(\sum_{i=1}^{n} \sum_{j=1}^{m} a_j Y_{ji}, \sum_{i=1}^{n} \sum_{j=1}^{m} a_j Y'_{ji})$$

$$\leq \sum_{i=1}^{n} \sum_{j=2}^{m} \sum_{s=1}^{j-1} (\mathbb{P}(a_j I_{[X_i = a_j]} > 0, a_s I_{[X_i = a_s]} > 0)$$

$$+ \mathbb{P}(a_j I_{[X_i = a_j]} > 0)\mathbb{P}(a_s I_{[X_i = a_j]} > 0))$$

$$= \sum_{i=1}^{n} \sum_{j=2}^{m} \sum_{s=1}^{j-1} \mathbb{P}(X_i = a_j)\mathbb{P}(X_i = a_s)$$

$$= \frac{1}{2} \sum_{i=1}^{n} \mathbb{P}(X_i > 0)^2 - \frac{1}{2} \sum_{i=1}^{n} \sum_{j=1}^{m} \mathbb{P}(X_i = a_j)^2.$$

Moreover, $\sum_{i=1}^{n} \sum_{j=1}^{m} a_j Y'_{ji} = \sum_{j=1}^{m} a_j \sum_{i=1}^{n} Y'_{ji} = \sum_{j=1}^{m} a_j Z_j$, where $Z_j = \sum_{i=1}^{n} Y'_{ji}$ are independent r.v.s following a generalized binomial distribution with parameters $p_{aj} = \mathbb{P}(X_i = a_j), i = 1, 2, ..., n$. Therefore, (see also (4.2.1)–(4.2.4))

$$d(\sum_{j=1}^{m} a_j Z_j, \sum_{j=1}^{m} a_j W_j) \leq \sum_{j=1}^{m} d(a_j Z_j, a_j W_j)$$

$$= \sum_{j=1}^{m} d(Z_j, W_j) \leq \frac{1}{2} \sum_{j=1}^{m} \sum_{i=1}^{n} \mathbb{P}(X_i = a_j)^2$$

with $W_j, j = 1, 2, ..., m$ being independent Poisson r.v.s such that $\lambda_{a_j} = \mathbb{E}(W_j) = \sum_{i=1}^{n} \mathbb{P}(X_i = a_j)$. Combining the above inequalities we conclude that

$$d(\sum_{i=1}^{n} X_i, \sum_{j=1}^{m} a_j W_j) \leq \frac{1}{2} \sum_{i=1}^{n} \mathbb{P}(X_i > 0)^2$$

The proof for the case of simple r.v.s $X_i, i = 1, 2, ..., n$ is completed by observing that the r.v. $\sum_{j=1}^{m} a_j W_j$ follows a compound Poisson distribution. More specifically, the characteristic function of $\sum_{j=1}^{m} a_j W_j$ is

$$\mathbb{E}(e^{it \sum_{j=1}^{m} a_j W_j}) = \prod_{j=1}^{m} \mathbb{E}(e^{ita_j W_j}) = \prod_{j=1}^{m} e^{-\lambda_{a_j}(1 - e^{ita_j})}$$

$$= \exp\left(-\lambda(1 - \sum_{j=1}^{m} e^{ita_j} \frac{\lambda_{a_j}}{\lambda})\right)$$

$(\lambda = \sum_{j=1}^{m} \lambda_{a_j} = \sum_{i=1}^{n} \mathbb{P}(X_i > 0))$ which can be identified as the characteristic function of a compound Poisson distribution with parameter λ and compounding distribution

$$F(x) = \sum_{j: a_j \leq x} \frac{\lambda_{a_j}}{\lambda} = \frac{1}{\lambda} \sum_{j: a_j \leq x} \sum_{i=1}^{n} \mathbb{P}(X_i = a_j) = \frac{1}{\lambda} \sum_{i=1}^{n} \sum_{j: a_j \leq x} \mathbb{P}(X_i = a_j)$$

$$= \frac{1}{\lambda} \sum_{i=1}^{n} \mathbb{P}(0 < X_i \leq x) = \frac{1}{\sum_{i=1}^{n} p_i} \sum_{i=1}^{n} p_i \mathbb{P}(X_i \leq x | X_i > 0)$$

$(p_i = \mathbb{P}(X_i > 0))$. Note that according to the last expression $F(x)$ can be interpreted as a weighted mixture of the distributions of $X_i | X_i > 0$.

We shall now treat the general case, $X_i > 0$. Define

$$\Psi_s(x) = \begin{cases} (k-1)2^{-s}, & \text{if } (k-1)2^{-s} \leq x < k2^{-s}, 1 \leq k < s2^s, k \in \mathbf{Z} \\ s, & \text{if } x \geq s, \end{cases}$$

for $s = 1, 2, ...$. For each $i = 1, 2, ..., n$, the sequence of simple non-negative r.v.s $\Psi_s(X_i)$ converges (as $s \to \infty$) to X_i almost surely (a.s.) [e.g. see Billingsley (1986, pp. 185, 259)]. From the first part of the proof it follows that, for each $s = 1, 2, ...$

$$d\left(\sum_{a=1}^{n} \Psi_s(X_a), CP(\mu_s, F_s)\right) \leq \frac{1}{2} \sum_{i=1}^{n} \mathbb{P}\left(\Psi_s(X_i) > 0\right)^2$$

$$\leq \frac{1}{2} \sum_{i=1}^{n} \mathbb{P}\left(X_i > 0\right)^2 \qquad (4.3.8)$$

where $\mu_s = \sum_{i=1}^{n} \mathbb{P}(\Psi_s(X_i) > 0)$ and $F_s(x) = \frac{1}{\mu_s} \sum_{i=1}^{n} \mathbb{P}(0 < \Psi_s(X_i) \leq x), x \in \mathbf{R}$. It can be easily verified that

$$\mathbb{P}\left(X_i \leq w\right) \leq \mathbb{P}(\Psi_s(X_i) \leq w) \leq \mathbb{P}(X_i \leq w + \frac{1}{2^s}), \ 0 \leq w < s$$

and the right continuity of $\mathbb{P}(X_i \leq w)$ guarantee that, as $s \to \infty$,

$$\mathbb{P}(\Psi_s(X_i) \leq w) \to \mathbb{P}(X_i \leq w) \quad \text{for every } w \geq 0.$$

Thus, for every $w \geq 0$, we have $\mu_s \underset{s \to \infty}{\to} \lambda$ and

$$F_s(w) = \frac{1}{\mu_s} \sum_{i=1}^{n} \mathbb{P}(0 < \Psi_s(X_i) \leq w) \underset{s \to \infty}{\to} \frac{1}{\lambda} \sum_{i=1}^{n} \mathbb{P}(0 < X_i \leq w) = F(w)$$

which ascertain the convergence of the c.f. of $CP(\mu_s, F_s)$ to the c.f. of $CP(\lambda, F)$. Hence, by virtue of the continuity theorem,

$$CP(\mu_s, F_s)(-\infty, w] \to CP(\lambda, F)(-\infty, w]$$

at least for every continuity point w of $H(w) = CP(\lambda, F)(-\infty, w]$.

Since $\Psi_s(X_i) \underset{s \to \infty}{\to} X_i$ a.s., it follows that $\sum_{a=1}^{n} \Psi_s(X_a) \underset{s \to \infty}{\to} \sum_{a=1}^{n} X_a$ a.s. and thus it converges in distribution. Hence

$$\mathbb{P}(\sum_{a=1}^{n} \Psi_s(X_a) \leq w) \to \mathbb{P}(\sum_{a=1}^{n} X_a \leq w)$$

at least for every continuity point w of $G(w) = \mathbb{P}(\sum_{a=1}^{n} X_a \leq w)$. Combining the above results with (4.3.8) we conclude that

$$\left| \mathbb{P}(\sum_{a=1}^{n} X_a \leq w) - CP(\lambda, F)(-\infty, w] \right| \leq \frac{1}{2} \sum_{i=1}^{n} \mathbb{P}(X_i > 0)^2 \qquad (4.3.9)$$

for every $w \in \mathcal{C}_G \cap \mathcal{C}_H$. ($\mathcal{C}_F \subset \mathbf{R}$ denotes the set of continuity points of F, $\mathcal{C}'_F = \mathbf{R} \backslash \mathcal{C}_F$). Since H, G are nondecreasing, it follows that the sets $\mathcal{C}'_G, \mathcal{C}'_H$ are countable. Let now $w_0 \in (\mathcal{C}_G \cap \mathcal{C}_H)'$. Since $(\mathcal{C}_G \cap \mathcal{C}_H)' = \mathcal{C}'_G \cup \mathcal{C}'_H$ is countable, there exists a sequence $w_j \in \mathcal{C}_G \cap \mathcal{C}_H$, $j = 1, 2, ...$ such that $w_j \downarrow w_0$. Inequality (4.3.9) holds true for all w_j and due to the right-continuity of G, H it will hold true for w_0 as well. Hence (4.3.9) is valid for every $w \in \mathbf{R}$ and the proof is complete. ∎

4.4   Examples of Applications

4.4.1   A Compound Poisson Approximation for Truncated Moving Sums of i.i.d. r.v.s

Let $X_1, X_2, ..., X_n$ be a sequence of i.i.d. nonnegative unbounded random variables with a common distribution function F. The moving sums

$S_i = \sum_{j=i}^{i+r-1} X_j, i = 1, 2, ..., n - r + 1$ will be referred to as the r-scan process. Denote by F_m the distribution function of $\sum_{j=1}^{m} X_j$, the m-fold convolution of F. Fixing a threshold b, we can measure the exceeding amount of the moving sums above b, by

$$A_n^+(b) = \sum_{i=1}^{n-r+1} \max\{0, S_i - b\} = \sum_{i=1}^{n-r+1} I_{[S_i > b]}(S_i - b).$$

Rootzen, Leadbetter and De Haan (1998) studied the asymptotic distribution of tail array sums for strongly mixing stationary sequences. Their results pertain to compound Poisson and Normal convergence theorems for $A_n^+(b)$ for very high and moderate levels of b respectively. Note, though, that their approach does not provide any bounds or convergence rates and in the high b level case the parameters (Poisson intensity λ, compounding distribution G) of the limiting compound Poisson distribution $CP(\lambda, G)$ can not be explicitly described. The approach based on Theorem 4.3.1 offers an upper bound for the distance $d(A_n^+(b), CP(\lambda, G))$ and an explicit description of λ, G for the case of very high levels of the threshold b.

Dembo and Karlin (1992) studied the number of exceedances $N_n^+(b) = \sum_{i=1}^{n-r+1} I_{[S_i > b]}$ by the aid of the Stein-Chen method, and proved that, under certain conditions, $N_n^+(b)$ converges to a Poisson distribution. In order to exploit Theorem 4.3.1 to study the variable $A_n^+(b)$, let us introduce the nonnegative r.v.s $Y_i = I_{[S_i > b]}(S_i - b), i = 1, 2, ..., n - r + 1$ and choose B_i as $B_i = \{\max\{i - r + 1, 1\}, ..., i - 1\}$. Then Y_i are independent of $Y_j, j \in B_i'$ and a direct application of Theorem 4.3.1 in conjunction with inequality (4.3.2) yields (note that in this case the last sum in the RHS of (4.3.2) vanishes)

$$d\left(A_n^+(b), CP(\lambda, G)\right)$$

$$\leq \sum_{i=2}^{n-r+1} \sum_{j=\max\{i-r+1,1\}}^{i-1} \left(\mathbb{P}(Y_i > 0, Y_j > 0) + \mathbb{P}(Y_i > 0)^2\right)$$

$$+ \frac{1}{2} \sum_{i=1}^{n-r+1} \mathbb{P}(Y_i > 0)^2$$

$$\leq \sum_{i=2}^{n-r+1} \sum_{j=\max\{i-r+1,1\}}^{i-1} \mathbb{P}(S_i > b, S_j > b) + (n - r + 1)r\mathbb{P}(S_1 > b)^2$$

$$\leq \lambda \sum_{m=2}^{r} \mathbb{P}(S_m > b | S_1 > b) + \lambda r(1 - F_r(b)) := \lambda\epsilon(r, b) \qquad (4.4.1)$$

where $\lambda = \sum_{i=1}^{n-r+1} \mathbb{P}(S_i > b) = (n-r+1)(1-F_r(b))$. The compounding distribution G is given by the expression

$$G(x) = \mathbb{P}(S_i \le x + b | S_i > b) = 1 - \frac{1 - F_r(b+x)}{1 - F_r(b)}, x \ge 0.$$

Observe that the upper bound in (4.4.1) has almost the same form as the bound developed by Dembo and Karlin (1992) for the variable $N_n^+(b)$. Consequently, many of their results pertaining to limiting behavior of their upper bound can also be exploited here for the investigation on $A_n^+(b)$'s asymptotic distribution. Thus, resorting to Theorem 3 in Dembo and Karlin (1992) we can easily deduce the following lemma.

LEMMA 4.4.1
If for each constant $K > 0$,

$$\frac{1 - F(b - K)}{1 - F_2(b)} \to 0 \quad as\ b \to \infty, \tag{4.4.2}$$

then the term $\epsilon(r, b)$ appearing in the upper bound of (4.4.1) converges to 0 for any fixed r.

We can now state the next proposition which elucidates the asymptotic behavior of $A_n^+(b)$.

PROPOSITION 4.4.2
Let $X_1, X_2, \ldots$ be a sequence of i.i.d. nonnegative and unbounded r.v.s with common distribution function F satisfying condition (4.4.2). If r is fixed and $n, b \to \infty$ so that $n(1 - F_r(b)) \to \lambda \in (0, \infty)$ and

$$1 - \frac{1 - F_r(b+x)}{1 - F_r(b)} \underset{b \to \infty}{\to} G(x)$$

then

$$\mathcal{L}(A_n^+(b)) = \mathcal{L}(\sum_{i=1}^{n-r+1} \max\{0, S_i - b\}) \to CP(\lambda, G)$$

with the convergence rate given by (4.4.1).

According to Dembo and Karlin (1992), condition (4.4.2) holds true for any distribution function F which is a finite or infinite convolution of exponentials of any scale parameters or has a logconcave density.

As an example we shall examine the simplest case where X_i are exponentially distributed with parameter μ, i.e. $F(x) = 1 - e^{-\mu x}, x \ge 0$. Then

$$1 - F_m(x) = e^{-\mu x} \sum_{i=0}^{m-1} \frac{(\mu x)^i}{i!}, m = 1, 2, \ldots$$

and it is not difficult to verify that condition (4.4.2) holds true. Indeed,

$$\frac{1 - F(b - K)}{1 - F_2(b)} = \frac{e^{-\mu(b-K)}}{e^{-\mu b}(1 + \mu b)} = \frac{e^{\mu K}}{(1 + \mu b)} \xrightarrow[b \to \infty]{} 0$$

and applying Proposition 4.4.2 we conclude that if $n, b \to \infty$ so that

$$ne^{-\mu b} \sum_{i=0}^{r-1} \frac{(\mu b)^i}{i!} \sim ne^{-\mu b} \frac{(\mu b)^{r-1}}{(r-1)!} \to \lambda$$

(the symbol $\sim$ indicates that the ratio of the two sides tends to 1) then $A_n^+(b)$ converges to a compound Poisson distribution $CP(\lambda, G)$ with

$$G(x) = 1 - \lim_{b \to \infty} \frac{1 - F_r(b + x)}{1 - F_r(b)}$$

$$= 1 - \lim_{b \to \infty} \frac{e^{-\mu(b+x)} \sum_{i=0}^{r-1} \frac{(\mu(b+x))^i}{i!}}{e^{-\mu b} \sum_{i=0}^{r-1} \frac{(\mu b)^i}{i!}}$$

$$= 1 - e^{-\mu x}, x \geq 0$$

(such a result can be easily interpreted by recalling the lack of memory property of the exponential law). Should one be interested on an upper bound for $d(A_n^+(b), CP(\lambda, G))$ he could condition on $\sum_{i=m}^r X_i$ in the upper bound of (4.4.1) to deduce

$$(n - r + 1) \sum_{m=2}^{r} \left(\int_0^b (1 - F_{m-1}(b - x))^2 dF_{r-m+1}(x) + 1 - F_{r-m+1}(b) \right)$$

$$+ (n - r + 1)r(1 - F_r(b))^2$$

and apply Lemma 4.1 of Dembo and Karlin (1992) to get the upper bound

$$n(r - 1) \left(\int_0^b (1 - F(b - x))^2 dF_{r-1}(x) + 1 - F_{r-1}(b) \right) + nr(1 - F_r(b))^2$$

$$= n(r - 1) \left(\frac{\mu^{r-1} e^{-\mu b}}{(r - 2)!} \int_0^b c^{-\mu(b-x)} x^{r-2} dx + e^{-\mu b} \sum_{i=0}^{r-2} \frac{(\mu b)^i}{i!} \right)$$

$$+ nr \left(e^{-\mu b} \sum_{i=0}^{r-1} \frac{(\mu b)^i}{i!} \right)^2$$

$$\sim \frac{\lambda(r - 1)^2}{b^{r-1}} \int_0^b e^{-\mu(b-x)} x^{r-2} dx + \frac{\lambda(r - 1)^2}{\mu b} + \frac{r}{n} \lambda^2.$$

Since

$$\frac{1}{b^{r-2}} \int_0^b e^{-\mu(b-x)} x^{r-2} dx$$

$$= \sum_{i=0}^{r-2} \binom{r-2}{i} \frac{(-1)^i}{b^i} \int_0^b e^{-\mu y} y^i dy$$

$$= \sum_{i=0}^{r-2} \binom{r-2}{i} \frac{(-1)^i}{\mu^{i+1} b^i} \int_0^{\mu b} e^{-z} z^i dy \leq \sum_{i=0}^{r-2} \binom{r-2}{i} \frac{i!}{\mu^{i+1} b^i}$$

the upper bound in (4.4.1) tends to 0, its rate of convergence being at least

$$\frac{\lambda(r-1)^2}{b} \sum_{i=0}^{r-2} \binom{r-2}{i} \frac{i!}{\mu^{i+1} b^i} + \frac{\lambda(r-1)^2}{\mu b} + \frac{r}{n} \lambda^2$$

$$\sim \frac{2\lambda(r-1)^2}{\mu b} + \frac{r}{n} \lambda^2 = O(b^{-1}).$$

4.4.2   The Number of Overlapping Success Runs in a Stationary Two-State Markov Chain

Let $\{X_i\}_{i=0,1,\dots}$ be a stationary Markov chain with state space $\{0, 1\}$ and denote by W the number of overlapping success runs of length k (k consecutive 1's) in the sequence $X_1, X_2, \dots, X_n$, i.e.

$$W = \sum_{i=1}^{n-k+1} \prod_{j=1}^{k} X_{i+j-1}$$

The problem of approximating the distribution of W has been recently addressed by Geske *et al.* (1995). The approach considered there was based on the classical Stein-Chen method. We shall give here an alternative approach based on Theorem 4.3.1 which offers slightly better results. The one step transition matrix $\mathbf{P} = (\mathbb{P}(X_i = j | X_{i-1} = k))_{kj} = (p_{kj})_{kj}$ of the Markov chain equals

$$\mathbf{P} = \begin{pmatrix} p_{00} & p_{01} \\ p_{10} & p_{11} \end{pmatrix}, \quad p_{00} + p_{01} = p_{10} + p_{11} = 1, p_{ij} \in (0, 1)$$

with its i-step transition matrix given by

$$\mathbf{P}^{(i)} = \begin{pmatrix} p_{00}^{(i)} & p_{01}^{(i)} \\ p_{10}^{(i)} & p_{11}^{(i)} \end{pmatrix}$$

$$= \mathbf{P}^i = \frac{1}{p_{01} + p_{10}} \begin{pmatrix} p_{10} & p_{01} \\ p_{10} & p_{01} \end{pmatrix} + \frac{(p_{11} - p_{01})^i}{p_{01} + p_{10}} \begin{pmatrix} p_{01} & -p_{01} \\ -p_{10} & p_{10} \end{pmatrix}.$$

The stationary distribution of the Markov chain (which, due to the stationarity assumption, coincides with the initial distribution) reads

$$p_0 = \frac{p_{10}}{p_{01} + p_{10}}, \quad p_1 = \frac{p_{01}}{p_{01} + p_{10}}.$$

In order to exploit Theorem 4.3.1 let us denote by W_0 the sum of

$$Y_i = (1 - X_{i-1}) \sum_{r=0}^{k-1} \prod_{j=i}^{i+k+r-1} X_j, \quad i = 1, 2, ..., n-k+1$$

which represent the truncated number of success runs within a clump starting at position i. Manifestly, if there exist $j \geq k$ consecutive successes stretching from position i to position $i + j - 1$, bordered at each end by failures, then $Y_i = \min\{j - k + 1, k\}$.

The distance between W and an appropriate compound Poisson distribution will be assessed by establishing first upper bounds for $d(W, W_0)$, $d(W_0, CP)$ and applying next the triangle inequality

$$d(W, CP) \leq d(W, W_0) + d(W_0, CP) \tag{4.4.3}$$

An upper bound for $d(W, W_0)$ is offered by

$$d(W_0, W) \leq \mathbb{P}\left(\sum_{i=1}^{n-k+1} Y_i \neq W \right)$$

$$\leq \mathbb{P}\left(\bigcup_{i=1}^{n-2k+1} [X_{i-1} = 0, X_i = ... = X_{i+2k-1} = 1] \right)$$

$$+ \mathbb{P}(X_0 = ... = X_k = 1)$$

$$+ \mathbb{P}\left(\bigcup_{i=n-2k+3}^{n-k+1} [X_{i-1} = 0, X_i = ... = X_{n+1} = 1] \right)$$

$$\leq (n - 2k + 1) p_0 p_{01} p_{11}^{2k-1} + p_1 p_{11}^k + p_0 p_{01} p_{11}^k \frac{1 - p_{11}^{k-1}}{1 - p_{11}}$$

$$\leq (n - 2k + 1) p_0 p_{01} p_{11}^{2k-1} + 2 p_{11}^k p_1. \tag{4.4.4}$$

On the other hand, we could resort to Theorem 4.3.1 to establish an upper bound for $d(W_0, CP)$. Note first that in this case it is quite intricate to determine the left neighborhoods of strong dependence for $Y_i, i = 1, 2, ..., n - k + 1$ in an optimal way, i.e. so that the upper bound is minimized. Nonetheless, it seems plausible to consider $B_i = \{\max\{1, i - s\}, ..., i - 1\}$ for an arbitrary $s \geq 2k - 1$ and apply Theorem 4.3.1 to gain the inequality (recall also inequality (4.3.2))

$$d\left(W_0, CP(\lambda, F)\right)$$

$$\leq \sum_{i=2}^{n-k+1} \sum_{b=\max\{1,i-s\}}^{i-1} \left(\mathbb{P}(Y_i > 0, Y_b > 0) + \mathbb{P}(Y_i > 0)\mathbb{P}(Y_b > 0)\right)$$

$$+ \sum_{i=s+2}^{n-k+1} d\left(\sum_{b=1}^{i-s-1} Y_b + Y_i, \sum_{b=1}^{i-s-1} Y_b + Y_i'\right) + \sum_{i=1}^{n-k+1} \mathbb{P}\left(Y_i > 0\right)^2$$

$$(4.4.5)$$

where $\lambda = \sum_{i=1}^{n-k+1} \mathbb{P}(Y_i > 0)$, and $F(x) = \mathbb{P}(Y_i \leq x | Y_i > 0), x \in \mathbf{R}$. Now, it is clear that

$$\mathbb{P}(Y_i > 0) = \mathbb{P}(X_{i-1} = 0, \prod_{j=i}^{i+k-1} X_j = 1) = p_0 p_{01} p_{11}^{k-1},$$

$$i = 1, 2, ..., n - k + 1,$$

$$\mathbb{P}\left(Y_i > 0, Y_b > 0\right) = 0, \qquad b = i - k, ..., i - 1$$

$$\mathbb{P}\left(Y_i > 0, Y_b > 0\right) \leq \mathbb{P}(X_{b-1} = 0, \prod_{j=b}^{b+k-1} X_j = 1, X_{i-1} = 0, \prod_{j=i}^{i+k-1} X_j = 1)$$

$$= p_0 p_{01} p_{11}^{k-1} p_{10}^{(i-b-k)} p_{01} p_{11}^{k-1} \leq p_0 p_{01}^2 p_{11}^{2k-2},$$

$$b = i - s, ..., i - k - 1 \qquad (4.4.6)$$

and

$$\lambda = (n - k + 1) p_0 p_{01} p_{11}^{k-1}, \qquad F(x) = 1 - p_{11}^x, x = 1, 2, ..., k - 1. \quad (4.4.7)$$

The mid term of the RHS of inequality (4.4.5) could be bounded by a method similar to the ones described after the statement of Theorem 4.3.1. However, since in this example we are dealing with the simple case of a two-state Markov chain, we chose to use a direct approach that yields tighter bounds. Note first that

$$\left| \mathbb{P}\left(\sum_{b=1}^{i-s-1} Y_b + Y_i \le w \right) - \mathbb{P}\left(\sum_{b=1}^{i-s-1} Y_b + Y_i' \le w \right) \right|$$

$$\le \sum_{y=0}^{\min\{w,k\}} \left| \mathbb{P}\left(\sum_{b=1}^{i-s-1} Y_b \le w - y, Y_i = y \right) \right.$$

$$\left. - \mathbb{P}\left(\sum_{b=1}^{i-s-1} Y_b \le w - y \right) \mathbb{P}(Y_i' = y) \right|$$

$$\le \sum_{y=0}^{\min\{w,k\}} \sum_{x=0}^{1} \left| \mathbb{P}\left(\sum_{b=1}^{i-s-1} Y_b \le w - y, Y_i = y, X_{i-s+2k-3} = x \right) \right.$$

$$\left. - \mathbb{P}\left(\sum_{b=1}^{i-s-1} Y_b \le w - y, X_{i-s+2k-3} = x \right) \mathbb{P}(Y_i = y) \right|$$

and use the Markov property to write the RHS as

$$\sum_{y=0}^{\min\{w,k\}} \sum_{x=0}^{1} |\mathbb{P}(Y_i = y | X_{i-s+2k-3} = x) - \mathbb{P}(Y_i = y)|$$

$$\times \mathbb{P}\left(\sum_{b=1}^{i-s-1} Y_b \le w - y, X_{i-s+2k-3} = x \right)$$

$$\le \sum_{y=0}^{k} \sum_{x=0}^{1} |\mathbb{P}(Y_i = y | X_{i-s+2k-3} = x) - \mathbb{P}(Y_i = y)| \, \mathbb{P}(X_{i-s+2k-3} = x)$$

where

$$|\mathbb{P}(Y_i = 0 | X_{i-s+2k-3} = x) - \mathbb{P}(Y_i = 0)|$$

$$= |p_{x0}^{(s-2k+2)} - p_0| p_{01} p_{11}^{k-1}$$

$$= p_{1-x} \, p_{01} p_{11}^{k-1} |p_{11} - p_{01}|^{s-2k+2},$$

$$|\mathbb{P}(Y_i = y | X_{i-s+2k-3} = x) - \mathbb{P}(Y_i = y)|$$

$$= |p_{x0}^{(s-2k+2)} - p_0| p_{01} p_{11}^{k+y-2} p_{10}$$

$$= p_{1-x} \, p_{01} p_{11}^{k+y-2} p_{10} |p_{11} - p_{01}|^{s-2k+2},$$

$$|\mathbb{P}(Y_i = k | X_{i-s+2k-3} = x) - \mathbb{P}(Y_i = k)|$$

$$= |p_{x0}^{(s-2k+2)} - p_0| p_{01} p_{11}^{2k-2}$$

$$= p_{1-x}\, p_{01}\, p_{11}^{2k-2}\, |p_{11} - p_{01}|^{s-2k+2},$$

for $x = 0,1$ and $y = 1,...,k-1$. The validity of the next inequality is now evident

$$d\left(\sum_{b=1}^{i-s-1} Y_b + Y_i,\ \sum_{b=1}^{i-s-1} Y_b + Y_i' \right)$$

$$\leq \sum_{y=0}^{k} \sum_{x=0}^{1} |\mathbb{P}(Y_i = y | X_{i-s+2k-3} = x) - \mathbb{P}(Y_i = y)|\, p_x$$

$$= 2p_0 p_1 p_{01} p_{11}^{k-1}\, |p_{11} - p_{01}|^{s-2k+2} \left(1 + p_{10} \sum_{y=1}^{k-1} p_{11}^{y-1} + p_{11}^{k-1} \right)$$

$$= 4p_0 p_1 p_{01} p_{11}^{k-1}\, |p_{11} - p_{01}|^{s-2k+2}. \qquad (4.4.8)$$

A combined use of (4.4.3)-(4.4.8) produces the next inequality, for every $s \geq 2k - 1$,

$$d\,(W, CP(\lambda, F))$$

$$\leq (n-k)(s-k)p_0 p_{01}^2 p_{11}^{2k-2} + (n-k)s p_0^2 p_{01}^2 p_{11}^{2k-2}$$

$$+ (n-k-s)4p_0 p_1 p_{01} p_{11}^{k-1}\, |p_{11} - p_{01}|^{s-2k+2} + (n-k+1)p_0^2 p_{01}^2 p_{11}^{2k-2}$$

$$+ (n-2k+1)p_0 p_{01} p_{11}^{2k-1} + 2p_{11}^k p_1$$

$$\leq \lambda((s-k)p_{01} p_{11}^{k-1} + (s+1)p_0 p_{01} p_{11}^{k-1} + p_{11}^k + 4p_1\, |p_{11} - p_{01}|^{s-2k+2})$$

$$+ 2p_{11}^k p_1.$$

Several simpler upper bounds can be conferred by considering certain special values for s. For example if $s = 3k - 1$ we get

$$d\,(W, CP(\lambda, F)) \leq \lambda((2k-1)p_{01} p_{11}^{k-1} + 3k p_0 p_{01} p_{11}^{k-1} + p_{11}^k$$

$$+ 4p_1\, |p_{11} - p_{01}|^{k+1}) + 2p_{11}^k p_1$$

which offers an upper bound of the same order but of better quality than Geske's *et al.* (1995) corresponding bound (the latter performs better only when $p_{01} \to 0$). Obviously, if $n, k \to \infty$ so that $(n - k + 1)p_0 p_{01} p_{11}^{k-1} \to \lambda_0$ then the upper bound vanishes and W converges to a compound Poisson distribution at a convergence rate of order $O(kp_{11}^k)$. Since $F(x)$ is now describing a geometric distribution, W will follow (asymptotically) a Pólya-Aeppli distribution.

Acknowledgements   Research of the first author was supported by the National Scholarship Foundation of Greece.

References

1. Arratia, R., Goldstein, L., and Gordon, L. (1989). Two moments suffice for Poisson approximations: The Chen-Stein method. *Annals of Probability* **17**, 9–25.

2. Arratia, R., Goldstein, L., and Gordon, L. (1990). Poisson Approximation and the Chen-Stein method. *Statistical Science* **5**, 403–434.

3. Barbour, A. D., Chen, L. H. Y., and Loh, W. L. (1992). Compound Poisson approximation for nonnegative random variables via Stein's method. *The Annals of Probability* **20**, 1843–1866.

4. Barbour, A. D., Holst, L., and Janson, S. (1992). *Poisson Approximation*. Clarendon Press, Oxford.

5. Barbour, A. D. and Utev, S. (1998). Solving the Stein equation in compound Poisson approximation. *Advances in Applied Probability* **30**, 449–475.

6. Billingsley, P. (1986). *Probability and Measure*. John Wiley & Sons, New York .

7. Boutsikas, M. V. and Koutras, M. V. (2000a). A bound for the distribution of the sum of discrete associated or NA r.v.'s. *The Annals of Applied Probability* (to appear).

8. Boutsikas, M. V. and Koutras, M. V. (2000b). On the number of overflown urns and excess balls in an allocation model with limited urn capacity. *Journal of Statistical Planning and Inference* (to appear).

9. Chen, L. H. Y. (1975). Poisson approximation for dependent trials. *The Annals of Probability* **3**, 534–545.

10. Dembo, A. and Karlin, S. (1992). Poisson approximations for r-scan processes. *The Annals of Applied Probability* **2**, 329–357.

11. Erhardsson, T. (1999). Compound Poisson approximations for Markov chains using Steins method. *Annals of Probability* **27**, 565–595.

12. Geske, M. X., Godbole, A. P., Schaffner, A. A., Skolnick, A. M., and Wallstrom, G. L. (1995). Compound Poisson approximations for word patterns under Markovian hypotheses. *Journal of Applied*

Probability **32**, 877–892.

13. Roos, M. (1994). Stein's method for compound Poisson approximation: the local approach. *The Annals of Applied Probability* **4**, 1177–1187.

14. Rootzen, H., Leadbetter, M. R., and De Haan, L. (1998). On the distribution of tail array sums for strongly mixing stationary sequences. *Annals of Applied Probability* **8**, 868–885.

15. Serfling, R. J. (1978). Some elementary results on Poisson approximations in a sequence of Bernoulli trials. *SIAM Review* **20**, 567–579.

16. Vellaisamy, P. and Chaudhuri, B. (1999). On compound Poisson approximation for sums of random variables. *Statistics & Probability Letters* **41**, 179–189.

17. Wang, Y. H. (1989). From Poisson to compound Poisson approximations. *The Mathematical Scientist* **14**, 38–49.

18. Xia, A. (1997). On using the first difference in the Stein-Chen method. *The Annals of Applied Probability* **4**, 899–916.

5

Unified Variance Bounds and a Stein-Type Identity

N. Papadatos and V. Papathanasiou
University of Cyprus, Nicosia, Cyprus
University of Athens, Athens, Greece

ABSTRACT Let X be an absolutely continuous (a.c.) random variable (r.v.) with finite variance σ^2. Then, there exists a new r.v. X^* (which can be viewed as a transform on X) with a unimodal density, satisfying the extended Stein-type covariance identity

$$\operatorname{Cov}[X, g(X)] = \sigma^2 \, \mathbb{E}[g'(X^*)]$$

for any a.c. function g with derivative g', provided that $\mathbb{E}|g'(X^*)| < \infty$. Properties of X^* are discussed and, also, the corresponding unified upper and lower bounds for the variance of $g(X)$ are derived.

Keywords and phrases Stein-type identity, variance bounds, transformation of random variables

5.1 Introduction

The well-known Stein's identity, Stein (1972, 1981), for the standard normal r.v. Z is formulated as follows. For every absolutely continuous (a.c.) function g with derivative g' such that $\mathbb{E}|g'(Z)| < \infty$,

$$\mathbb{E}[Zg(Z)] = \mathbb{E}[g'(Z)] \tag{5.1.1}$$

(throughout this paper, the term 'a.c.'='absolutely continuous' will be used either to describe an r.v. having a density with respect to Lebesgue measure on $\mathbb{R}$, or to denote an ordinary a.c. function; in any case, the meaning will be clear from the context). This identity has had many important applications in several areas of Probability and Statistics; see for example Stein (1972), Hudson (1978) and Liu (1994). Several generalizations to other r.v.s can be found in Cacoullos and Papathanasiou (1989, 1995) and Hudson (1978).

Another interesting result from the point of view of upper variance bounds is the inequality of Chernoff, [Chernoff (1981)],

$$\mathrm{Var}\,[g(Z)] \leq \mathrm{E}\,[(g'(Z))^2] \tag{5.1.2}$$

with equality iff (if and only if) g is linear. This and its multivariate normal analogue, Chen (1982), motivated the generalizations to arbitrary discrete and continuous r.v.s as well as corresponding lower variance bounds, Cacoullos (1982) and Cacoullos and Papathanasiou (1985).

An important role of the general covariance identity given in Cacoullos and Papathanasiou (1989) (see (5.2.4), below) is in the derivation of simple and elementary proofs of the Central Limit Theorem [see Cacoullos, Papathanasiou and Utev (1992, 1994)]. Other applications concerning the rate of convergence in the Local Limit Theorem for sums of independent r.v.s are given in Cacoullos, Papadatos and Papathanasiou (1997).

However, all the preceding results [and, in particular, the validity of the generalized covariance identity given in Cacoullos and Papathanasiou (1989) and the variance bounds given in Cacoullos and Papathanasiou (1985, 1989)] require an interval support of the basic r.v. X, which is rather restrictive.

In the present paper we avoid the restriction of an interval support, introducing an appropriate (smooth) transformation X^*, which is in fact a new r.v. corresponding to X. The r.v. X^* is always uniquely defined (provided that X is a.c. with finite variance), having itself an absolutely continuous unimodal density. Thus, for any a.c. r.v. X with finite second moment, there exists a new smooth r.v. X^* with unimodal a.c. density satisfying the generalized Stein covariance identity. This transformation behaves well to convolutions of independent r.v.s. Moreover, it appears in the upper and lower bounds for the variance of any a.c. function g of X (and it is, in fact, the only r.v. with this property; Theorem 5.3.3).

It should be noted that in a recent paper, independently of our results, Goldstein and Reinert (1997) used a similar approach (the so-called *zero bias transformation*), which, in fact, is based on the same covariance identity when $\mathrm{E}[X] = 0$. They also fruitfully applied this identity to estimate the rate of convergence in the CLT, obtaining an $O(n^{-1})$ bound for smooth functions [see Goldstein and Reinert (1997, Corollary 3.1)]. Furthermore, they presented a nice application for dependent samples. However, except of the definition, the results of the present paper are completely different; our main interest is on unified variance bounds and their connection with Stein's identity. We are also interested on the behavior of the inverse transform $X^* \to X$; in fact we show that, under general conditions, the distribution of X^* uniquely determines that of X (Theorems 5.2.2 and 5.3.4). Finally, we also include some illustrative examples.

5.2 Properties of the Transformation

Let X be an a.c. r.v. with density f, mean μ, variance σ^2 and support $S(X)$ (for the sake of simplicity, we will always mean the support of an a.c. r.v. X with density f to be the set $S(X) = \{x : f(x) > 0\}$). We simply define X^* to be a random variable with density f^*, given by the relation

$$f^*(x) = \frac{1}{\sigma^2} \int_{-\infty}^{x} (\mu - t) f(t) \, dt = \frac{1}{\sigma^2} \int_{x}^{\infty} (t - \mu) f(t) \, dt. \qquad (5.2.1)$$

Obviously the right hand sides of (5.2.1) are equal and thus, f^* is non-negative. An application of Tonelli's Theorem shows that f^* integrates to 1, and therefore it is indeed a probability density [c.f. Cacoullos and Papathanasiou (1989) and Cacoullos and Papathanasiou (1989)]. The following Lemma summarizes the properties of X^* and shows the generalized Stein's identity (5.2.2).

LEMMA 5.2.1

(i) f^* is a unimodal a.c. density with mode μ and maximal value

$$f^*(\mu) = \frac{\mathbb{E}|X - \mu|}{2\sigma^2}.$$

Furthermore, the function $(f^*(x))'/(\mu - x)$ (defined almost everywhere) is nonnegative and integrable.

(ii)   For each a.c. function g with $\mathbb{E}|g'(X^*)| < \infty$,

$$\mathrm{Cov}[X, g(X)] = \sigma^2 \mathbb{E}[g'(X^*)]. \qquad (5.2.2)$$

(iii)   If the r.v. Y satisfies the identity

$$\mathrm{Cov}[X, g(X)] = \sigma^2 \mathbb{E}[g'(Y)]$$

for every a.c. g with $\mathbb{E}|g'(Y)| < \infty$, then $Y \stackrel{\mathrm{d}}{=} X^*$ (in the sense that Y and X^* have the same distribution).

(iv) $S(X^*) = (\operatorname{ess\,inf} S(X), \operatorname{ess\,sup} S(X))$.

(v)   For arbitrary scalars $a \neq 0$ and b,

$$(aX + b)^* \stackrel{\mathrm{d}}{=} aX^* + b.$$

(vi)   For independent a.c. r.v.s X_1, X_2 with means μ_1, μ_2, variances σ_1^2, σ_2^2 and arbitrary scalars a_1 and a_2 with $a_1 a_2 \neq 0$,

$$(a_1 X_1 + a_2 X_2)^* \stackrel{\mathrm{d}}{=} B(a_1 X_1^* + a_2 X_2) + (1 - B)(a_1 X_1 + a_2 X_2^*), \qquad (5.2.3)$$

where X_1, X_2, X_1^*, X_2^* and B are mutually independent with

$$\Pr[B = 1] = \frac{a_1^2 \sigma_1^2}{a_1^2 \sigma_1^2 + a_2^2 \sigma_2^2} = 1 - \Pr[B = 0].$$

PROOF (i), (iv) and (v) are obvious. (ii) follows from the definition of X^* using Fubini's theorem, since by the assumption that $\mathbb{E}|g'(X^*)| < \infty$, the nonnegative functions $g_1(x, t) = |g'(x)|(\mu - t)f(t)I(t < x)$ and $g_2(x, t) = |g'(x)|(t-\mu)f(t)I(t > x)$ are integrable over $(-\infty, \mu] \times (-\infty, \mu]$ and $[\mu, \infty) \times [\mu, \infty)$, respectively [see also Lemma 3.1 in Cacoullos and Papathanasiou (1989)]. (iii) follows from (ii) and the fact that

$$\mathbb{E}[G(X^*)] = \mathbb{E}[G(Y)]$$

for every bounded (measurable) function G. Regarding (vi), observe that for $S = a_1 X_1 + a_2 X_2$ and g' bounded, we have from (ii):

$$\text{Cov}[S, g(S)] = \sigma^2 \mathbb{E}[g'(S^*)],$$

where $\sigma^2 = a_1^2 \sigma_1^2 + a_2^2 \sigma_2^2$. On the other hand,

$$\text{Cov}[S, g(S)] = a_1 \text{Cov}[X_1, g(S)] + a_2 \text{Cov}[X_2, g(S)]$$

$$= a_1^2 \sigma_1^2 \mathbb{E}[g'(a_1 X_1^* + a_2 X_2)] + a_2^2 \sigma_2^2 \mathbb{E}[g'(a_1 X_1 + a_2 X_2^*)].$$

It follows that for any bounded function G,

$$\mathbb{E}[G(S^*)] = \frac{a_1^2 \sigma_1^2}{\sigma^2} \mathbb{E}[G(a_1 X_1^* + a_2 X_2)] + \frac{a_2^2 \sigma_2^2}{\sigma^2} \mathbb{E}[G(a_1 X_1 + a_2 X_2^*)]$$

$$= \mathbb{E}[G(B(a_1 X_1^* + a_2 X_2) + (1 - B)(a_1 X_1 + a_2 X_2^*))],$$

which completes the proof. ∎

It should be noted that in the previous Lemma, and elsewhere in this paper, the term 'unimodal' is reserved to denote a function $h : \mathbb{R} \to \mathbb{R}$ with the property that there exists some $m \in \mathbb{R}$ such that $h(x)$ is nondecreasing for $x \leq m$ and is nonincreasing for $x \geq m$; each m with this property is called a 'mode' of h. Of course, the assertion that h is a unimodal and a.c. function implies that the mode(s) of h form a compact interval $[a, b]$ with $-\infty < a \leq b < \infty$ (which, in the case $a = b$, reduces to single point).

The known identity of Cacoullos and Papathanasiou (1989), (5.2.4) below, requiring an interval support of X, follows immediately from (5.2.2).

Indeed, when $S(X)$ is a (finite or infinite) interval, $S(X)$ can always be taken to be open, and then $S(X) = S(X^*)$. Thus, for the nonnegative function $w(x) = f^*(x)/f(x)$ (defined on $S(X)$), (5.2.2) becomes

$$\mathrm{Cov}[X, g(X)] = \sigma^2 \, \mathbb{E}[w(X)g'(X)]. \tag{5.2.4}$$

The identity (5.2.2) remains valid for any nondecreasing (or nonincreasing) a.c. function g, even in the case where $\mathbb{E}[g'(X^*)] = \infty$ (or $-\infty$ if g is nonincreasing), as it follows by an application of Tonelli's (instead of Fubini's) theorem. In this case, $\mathbb{E}[(X - \mu)(g(X) - g(\mu))] = \infty$ (or $-\infty$). If, however, $\mathbb{E}|g'(X^*)| = \infty$ and g is arbitrary, it may happen that $\mathbb{E}|(X - \mu)(g(X) - g(\mu))| < \infty$, as the following example shows.

Example 1
Assume that X has density $f(x) = (3/8) \min\{1, x^{-4}\}$, $-\infty < x < \infty$ and

$$g(x) = \sum_{n=1}^{\infty} (2n - 1)(|x| - a_{2n-2}) I \left(a_{2n-2} \leq |x| < a_{2n-1}\right)$$

$$+ \sum_{n=1}^{\infty} 2n(a_{2n} - |x|) I \left(a_{2n-1} \leq |x| < a_{2n}\right),$$

where $a_0 = 0$ and $a_n = 1 + 1/2 + \cdots + 1/n$ for $n \geq 1$. It follows that $0 \leq g(x) \leq 1$ and $g(-x) = g(x)$ for all x. Moreover, g is a.c. with derivative $g'(x)$ (outside the set $\{0, \pm a_1, \pm a_2, \ldots\}$) satisfying $|g'(x)| \geq |x| \geq |g(x)| = g(x)$ for almost all x. Since $\mathbb{E}[X] = 0$, $\mathbb{E}|X| = 3/4$, $\mathrm{Var}[X] = 1$ and $S(X) = (-\infty, \infty)$, we conclude that X^* has the density $f^*(x) = (3/16)\left[(2 - x^2)I(|x| \leq 1) + x^{-2}I(|x| > 1)\right]$ supported by the entire real line, and $\mathbb{E}|g'(X^*)| \geq \mathbb{E}|X^*| = \infty$, while $\mathbb{E}|Xg(X)| \leq \mathbb{E}|X| = 3/4$. $\square$

We have shown in Lemma 5.2.1(iii) that X^* is the only r.v. satisfying the identity (5.2.2), and thus, an equivalent definition of X^* could be given via the covariance identity; the latter approach is due to Goldstein and Reinert [Goldstein and Reinert (1997, Definition 1.1)], who proved that such a transformation is uniquely defined by this identity. Moreover, their approach extends to r.v.s that are not necessarily a.c., e.g., for the symmetric Bernoulli r.v. X taking the values ± 1 with probability $1/2$, X^* is uniformly distributed over the interval $(-1, 1)$. Our approach, however, is restricted to a.c. r.v.s; for this reason, the analytic definition (2.1) is possible and, moreover, the density f^* itself turns out to be an a.c. unimodal function.

Our results also go to the opposite direction; the following Theorem shows that, in general, the distribution of X^* uniquely determines that of X.

THEOREM 5.2.2

Assume that the r.v. Y has a unimodal a.c. density h.

(i) *If the mode m of h is unique (i.e., $h(x) < h(m)$ for all $x \neq m$), there exists an r.v. X_m such that $X_m^* \overset{d}{=} Y$ iff the function $h'(x)/(m-x)$ is integrable. Moreover, $X_1^* \overset{d}{=} Y$ and $X_2^* \overset{d}{=} Y$ implies $X_1 \overset{d}{=} X_2$ (and thus, X_m is unique).*

(ii) *If $\{x : x \text{ is a mode of } h\} = [a,b]$ with $a < b$, then for each $\mu \in (a,b)$, there always exists a unique r.v. X_μ such that $\mathbb{E}[X_\mu] = \mu$ and $X_\mu^* \overset{d}{=} Y$. Moreover, for $\mu = a$ or $\mu = b$, there exists an r.v. X_μ such that $\mathbb{E}[X_\mu] = \mu$ and $X_\mu^* \overset{d}{=} Y$ iff the function $h'(x)/(\mu-x)$ is integrable. Finally, if $X_1^* \overset{d}{=} Y$, $X_2^* \overset{d}{=} Y$ and $\mathbb{E}[X_1] = \mathbb{E}[X_2]$, then $X_1 \overset{d}{=} X_2$.*

PROOF (i) If X is an r.v. with mean μ variance σ^2 and density f such that $X^* \overset{d}{=} Y$, it follows from Lemma 5.2.1(i) that μ must be a mode of $f^* = h$ and that $h'(x)/(\mu-x)$ is integrable. Thus, $\mu = m$ and $h'(x)/(m-x)$ is integrable. Assume now that the function $h'(x)/(m-x)$ (defined almost everywhere) is integrable. Observe that it is also nonnegative (because m is the mode of h) and define the r.v. X_m with density

$$f_m(x) = \frac{h'(x)}{c(m-x)}, \text{ where } c = \int_{-\infty}^{\infty} \frac{h'(x)}{m-x} \, dx > 0.$$

Since $\lim_{\pm\infty} h(x) = 0$ (because h is a unimodal density), we have $\mathbb{E}[m - X_m] = 0$. Applying Tonelli's Theorem we have

$$\int_{-\infty}^{m} (m-x)h'(x) \, dx = \int_{-\infty}^{m} \int_{x}^{m} h'(x) \, du \, dx = \int_{-\infty}^{m} h(u) \, du,$$

and similarly,

$$\int_{m}^{\infty} (x-m)(-h'(x)) \, dx = \int_{m}^{\infty} h(u) \, du,$$

yielding $\text{Var}[X_m] = 1/c$. Therefore, $X_m^* \overset{d}{=} Y$. We now show that X_m is unique. Indeed, if $X^* \overset{d}{=} Y$ for an r.v. X with mean μ, variance σ^2 and density f, it follows from Lemma 5.2.1(i) that μ must be a mode of

$f^* = h$ and therefore $\mu = m$. Hence,

$$f(x) = \frac{\sigma^2 h'(x)}{m - x} = c\sigma^2 f_m(x)$$

for almost all x, and thus $X \overset{d}{=} X_m$.

(ii) Since h is constant in $[a, b]$, $h' \equiv 0$ in (a, b). Hence, for all $\mu \in (a, b)$,

$$\int_{-\infty}^{\infty} \frac{h'(x)}{\mu - x} \, dx = \int_{-\infty}^{a} \frac{h'(x)}{\mu - x} \, dx + \int_{b}^{\infty} \frac{-h'(x)}{x - \mu} \, dx \le \frac{h(a)}{\mu - a} + \frac{h(b)}{b - \mu} < \infty.$$

Consider the r.v. X_μ with density

$$f_\mu(x) = \frac{h'(x)}{c_\mu(\mu - x)}, \quad \text{where } c_\mu = \int_{-\infty}^{\infty} \frac{h'(x)}{\mu - x} \, dx.$$

Then $\mathbb{E}[X_\mu] = \mu$, $\mathrm{Var}[X_\mu] = 1/c_\mu$, and therefore $X_\mu^* \overset{d}{=} Y$. It is easy to see that if $X^* \overset{d}{=} Y$ and $\mathbb{E}[X] = \mu$ for some r.v. X with density f and variance σ^2, then $X \overset{d}{=} X_\mu$. Indeed, it follows from (5.2.1) and Lemma 5.2.1(i) that

$$f(x) = \frac{\sigma^2 h'(x)}{\mu - x} = c_\mu \sigma^2 f_\mu(x)$$

for almost all x. Therefore, since the mean of any r.v. X satisfying $X^* \overset{d}{=} Y$ must be a mode of h, either $\mathbb{E}[X] = \mu \in (a, b)$ (and thus $X \overset{d}{=} X_\mu$) or $\mathbb{E}[X] = a$ (and $h'(x)/(a - x)$ is integrable) or $\mathbb{E}[X] = b$ (and $h'(x)/(b - x)$ is integrable). ∎

The following example shows that all the cases described by Theorem 5.2.2 are possible.

Example 2

(i) Assume that Y has the unimodal a.c. density

$$h(x) = \frac{1}{3} \left(\min \{2 - |x|, 1\} \right)^+.$$

with derivative $h'(x) = -(1/3)\mathrm{sign}(x)I(1 < |x| < 2)$ for almost all x. Then, for all $\mu \in (-1, 1)$, the r.v. X_μ with density

$$f_\mu(x) = (3c_\mu|x - \mu|)^{-1} I(1 < |x| < 2), \quad \text{where } c_\mu = \frac{1}{3} \log \left[\frac{4 - \mu^2}{1 - \mu^2} \right],$$

satisfies $\mathbb{E}[X_\mu] = \mu$ and $X_\mu^* \overset{d}{=} Y$. Moreover, $h'(x)/(\mu - x)$ is not integrable for $\mu = \pm 1$.

(ii) Assume that Y has the unimodal a.c. density

$$h(x) = \frac{6}{19} \left(\min\{x+2,1\} I(-2 < x < 1) + x(2-x)I(1 \leq x < 2) \right).$$

Then, for any $\mu \in (-1,1]$, the r.v. X_μ with density

$$f_\mu(x) = \frac{6}{19 c_\mu |x-\mu|} \left(I(-2 < x < -1) + 2(x-1)I(1 < x < 2) \right),$$

where

$$c_\mu = (6/19) \left(2 + \log\left[(\mu+2)/(\mu+1) \right] - 2(1-\mu)\log\left[(2-\mu)/(1-\mu) \right] \right),$$

satisfies $\mathbb{E}[X_\mu] = \mu$ and $X_\mu^* \overset{d}{=} Y$, while $h'(x)/(-1-x)$ is not integrable. Similarly, for the r.v. $W = -Y$ with density $h(-x)$, there exists an r.v. R_μ such that $\mathbb{E}[R_\mu] = \mu$ and $R_\mu^* \overset{d}{=} W$ iff $\mu \in [-1,1)$.

(iii) For the r.v. Y with density

$$h(x) = \frac{3}{10} \left(|x|(2-|x|)I(1 < |x| < 2) + I(|x| \leq 1) \right),$$

there exists an r.v. X_μ such that $\mathbb{E}[X_\mu] = \mu$ and $X_\mu^* \overset{d}{=} Y$ for any mode $\mu \in [-1,1]$. ☐

5.3   Application to Variance Bounds

Upper bounds for the variance of a function $g(X)$ of a normal r.v. X in terms of g' are known as *the inequality of Chernoff* [Chernoff (1981)] [see also Chen (1982) and Vitale (1989)]. Upper and lower variance bounds of $g(X)$ for an arbitrary r.v. X were considered in Cacoullos (1982) and Cacoullos and Papathanasiou (1985) [see also Cacoullos and Papathanasiou (1989), Cacoullos and Papathanasiou (1995) and references therein]. Both upper and lower variance bounds may be obtained as by-products of the Cauchy-Schwarz inequality. The following Lemma summarizes and unifies these bounds in terms of the r.v. X^*; in effect, (5.3.1) is a Chernoff-type, Chen (1982), upper bound; (5.3.2) is a Cacoullos-type, Papathanasiou (1990, 1993), lower bound as obtained in Cacoullos (1982) and Cacoullos and Papathanasiou (1985), in terms of a function w (see also (5.3.4) below).

LEMMA 5.3.1
Let X be an a.c. r.v. with mean μ and variance σ^2. Then, for every a.c. function g with derivative g', we have the following bounds.

(i)
$$\mathbb{E}[(g(X) - g(\mu))^2] \le \sigma^2 \mathbb{E}[(g'(X^*))^2], \qquad (5.3.1)$$

with equality iff either $\mathbb{E}[g^2(X)] = \infty$ or

$$g(x) - g(\mu) = \begin{cases} a_1(x - \mu) & \text{if } x \le \mu, \\ a_2(x - \mu) & \text{if } x \ge \mu, \end{cases}$$

for some constants a_1, a_2 and for all $x \in S(X^*)$.

(ii) If $\mathbb{E}|g'(X^*)| < \infty$,

$$\text{Var}[g(X)] \ge \sigma^2 \mathbb{E}^2[g'(X^*)], \qquad (5.3.2)$$

with equality iff $\Pr[g(X) = aX + b] = 1$ for some constants a and b.

PROOF (i) Let f be a density of X. We then have

$$\begin{aligned}
\sigma^2 \mathbb{E}[(g'(X^*))^2] &= \int_{-\infty}^{\mu} (g'(x))^2 \int_{-\infty}^{x} (\mu - t) f(t) \, dt \, dx \\
&\quad + \int_{\mu}^{\infty} (g'(x))^2 \int_{x}^{\infty} (t - \mu) f(t) \, dt \, dx \\
&= \int_{-\infty}^{\mu} f(t)(\mu - t) \int_{t}^{\mu} (g'(x))^2 \, dx \, dt \\
&\quad + \int_{\mu}^{\infty} f(t)(t - \mu) \int_{\mu}^{t} (g'(x))^2 \, dx \, dt \\
&\ge \int_{-\infty}^{\mu} f(t) \, (g(\mu) - g(t))^2 \, dt \\
&\quad + \int_{\mu}^{\infty} f(t) \, (g(t) - g(\mu))^2 \, dt \\
&= \mathbb{E}[(g(X) - g(\mu))^2],
\end{aligned}$$

from Tonelli's theorem and the Cauchy-Schwarz inequality for integrals. Observe that if $\mathbb{E}[g^2(X)] = \infty$, the equality holds in a trivial way ($\infty = \infty$); otherwise, the equality holds iff there exist constants a_1 and a_2 such that $g'(x) = a_1 + (a_2 - a_1)I(x \ge \mu)$ for almost all $x \in S(X^*)$, which completes the proof.

(ii) We have from (5.2.2),

$$\sigma^4 \mathbb{E}^2[g'(X^*)] = \text{Cov}^2[X, g(X)] \le \sigma^2 \text{Var}[g(X)]$$

by the Cauchy-Schwarz inequality for r.v.s, and the proof is complete. ∎

COROLLARY 5.3.2
For every a.c. function g,

$$\text{Var}\,[g(X)] \leq \sigma^2 \,\mathbb{E}\,[(g'(X^*))^2], \tag{5.3.3}$$

with equality iff either $\mathbb{E}\,[g^2(X)] = \infty$ *or g is linear on* $S(X^*)$.

Note that (5.3.2) continues to hold for any nondecreasing (nonincreasing) a.c. function g, even in the case where $\mathbb{E}\,[g'(X^*)] = \pm\infty$ (in this case, $\mathbb{E}\,[g^2(X)] = \infty$). Moreover, since $\Pr[X \in S(X^*)] = 1$ (because the measure produced by X is absolutely continuous with respect to that produced by X^*), equality in (5.3.3) implies the equality in (5.3.2). The converse is not always true, as the following example shows.

Example 3
Let X be uniformly distributed over $(-2, -1) \cup (1, 2)$ and

$$g(x) = xI(|x| > 1) + x^3 I(|x| \leq 1).$$

Then $\mu = 0$, $\sigma^2 = 7/3$ and X^* has the density

$$f^*(x) = (3/28)\left(\min\{4 - x^2, 3\}\right)^+.$$

It follows that $\text{Var}\,[g(X)] = \mathbb{E}\,[(g(X) - g(\mu))^2] = \sigma^2$ and $\sigma^2 \,\mathbb{E}^2[g'(X^*)] = \text{Var}\,[g(X)] < \sigma^2 \,\mathbb{E}\,[(g'(X^*))^2] = 53/15$. This shows that g is 'linear' with respect to the measure produced by X, and it is 'nonlinear' with respect to the measure produced by X^*. ☐

This example hinges on the fact that $S(X)$ fails to be an interval. If, however, $S(X)$ is a (finite or infinite) interval, the known upper and lower bounds for the variance of $g(X)$ take the form [see Cacoullos and Papathanasiou (1985, 1989)]

$$\sigma^2 \,\mathbb{E}^2[w(X)g'(X)] \leq \text{Var}\,[g(X)] \leq \sigma^2 \,\mathbb{E}\,[w(X)(g'(X))^2], \tag{5.3.4}$$

for some nonnegative function w defined on $S(X)$ (in fact, $w = f^*/f$), where both equalities hold iff g is linear on $S(X)$, provided that

$$\mathbb{E}\,[w(X)(g'(X))^2] < \infty.$$

The following result shows the equivalence between the variance bounds and the covariance identity.

THEOREM 5.3.3
Assume that for the a.c. r.v. X with finite variance σ^2, the r.v. Y satisfies one of the following.

(i) *For every a.c. function g with derivative g',*

$$\mathrm{Var}\,[g(X)] \leq \sigma^2\,\mathbb{E}\,[(g'(Y))^2].$$

(ii) *For every a.c. function g with derivative g' such that $\mathbb{E}\,|g'(Y)| < \infty$,*

$$\mathrm{Var}\,[g(X)] \geq \sigma^2\,\mathbb{E}^2[g'(Y)].$$

Then $Y \overset{\mathrm{d}}{=} X^*$.

PROOF Assume that (i) holds. Let h' be any (measurable) bounded function and consider the a.c. function $g(x) = x + \lambda h(x)$, where λ is an arbitrary constant and h an indefinite integral of h'. It follows that g is a.c. with bounded derivative $g' = 1 + \lambda h'$. Thus, $\mathrm{Var}\,[g(X)] < \infty$, $\mathrm{Var}\,[h(X)] < \infty$ and

$$\mathrm{Var}\,[g(X)] = \sigma^2 + \lambda^2\,\mathrm{Var}\,[h(X)] + 2\lambda\,\mathrm{Cov}\,[X, h(X)]$$

$$\leq \sigma^2\left(1 + \lambda^2\,\mathbb{E}\,[(h'(Y))^2] + 2\lambda\,\mathbb{E}\,[h'(Y)]\right).$$

Therefore, by using standard arguments [see Cacoullos and Papathanasiou (1989)], the quadratic $\theta\lambda^2 + 2\delta\lambda$ (where $\theta = \sigma^2\,\mathbb{E}\,[(h'(Y))^2] - \mathrm{Var}\,[h(X)] \geq 0$, $\delta = \sigma^2\,\mathbb{E}\,[h'(Y)] - \mathrm{Cov}\,[X, h(X)]$) is nonnegative for all λ, and thus $\delta = 0$. Hence, taking into account (5.2.2) we conclude that for any bounded function H,

$$\mathbb{E}\,[H(Y)] = \mathbb{E}\,[H(X^*)]$$

and the result follows. The same arguments apply to (ii). ∎

Finally, by using similar arguments and the results of Theorem 5.2.2, a converse of Theorem 5.3.3 and Lemma 5.2.1(iii) can be easily established:

THEOREM 5.3.4
Let Y be an arbitrary r.v. Then, Y has a unimodal a.c. density h such that the function $h'(x)/(m - x)$ is integrable for some mode m of h iff there exists some a.c. r.v. X with finite second moment such that any one of the following holds.

(i) *For every a.c. function g with derivative g' such that $\mathbb{E}\,|g'(Y)| < \infty$,*

$$\mathrm{Cov}\,[X, g(X)] = \mathrm{Var}\,[X]\,\mathbb{E}\,[g'(Y)].$$

(ii)   *For every a.c. function g with derivative g',*

$$\text{Var}\,[g(X)] \leq \text{Var}\,[X]\,\mathbb{E}\,[(g'(Y))^2].$$

(iii) *For every a.c. function g with derivative g' such that $\mathbb{E}\,|g'(Y)| < \infty$,*

$$\text{Var}\,[g(X)] \geq \text{Var}\,[X]\,\mathbb{E}^2[g'(Y)].$$

(iv) $X^* \overset{\mathrm{d}}{=} Y$.

Furthermore, X is unique iff the mode of h is unique.

Theorems 5.3.3 and 5.3.4 characterize the r.v.s X and Y which admit *Poincaré type inequalities* or *differential inequalities* [cf Chen (1988) and Vitale (1989)]. It should be noted that in the preceding inequalities the constant equals $\text{Var}\,[X]$, and this implies that the equality is attained for 'linear' g. There are, however, other kinds of Poincaré type inequalities where the constant does not equal $\text{Var}\,[X]$. The following example is relative to this subject.

Example 4

Let X be uniformly distributed over $(0,1)$ and Y have the density $h(x) = (3/4)(1-x^2)I(|x|<1)$. It follows that for every a.c. function g,

$$\text{Var}\,[g(X)] \leq \frac{1}{12}\mathbb{E}\,[(g'(X^*))^2] = \frac{1}{2}\int_0^1 x(1-x)(g'(x))^2\,dx$$

$$\leq c\mathbb{E}\,[(g'(Y))^2],$$

for some constant $c \leq 1/3$, since $(1/2)x(1-x)I(0 < x < 1) \leq (1/3)h(x)$ for $x \in (-1,1)$. On the other hand, an application to the function $g(x) = x^2$ shows that $c \geq 1/9 > 1/12 = \text{Var}\,[X]$.   ⬚

Acknowledgements Research partially supported by the Ministry of Industry, Energy and Technology of Greece under grant 1369. Part of this work was done when the second author was visiting the University of Bristol.

References

1. Cacoullos, T. (1982). On upper and lower bounds for the variance of a function of a random variable. *Annals of Probability* **10**, 799–809.

2. Cacoullos, T., Papadatos, N., and Papathanasiou, V. (1997). Variance inequalities for covariance kernels and applications to central limit theorems. *Theory of Probability and its Applications* **42**, 195–201.

3. Cacoullos, T. and Papathanasiou, V. (1985). On upper bounds for the variance of functions of random variables. *Statistics & Probability Letters* **3**, 175–184.

4. Cacoullos, T. and Papathanasiou, V. (1989). Characterizations of distributions by variance bounds. *Statistics & Probability Letters* **7**, 351–356.

5. Cacoullos, T. and Papathanasiou, V. (1995). A generalization of covariance identity and related characterizations. *Mathematical Methods of Statistics* **4**, 106–113.

6. Cacoullos, T., Papathanasiou, V., and Utev, S. (1992). Another characterization of the normal law and a proof of the central limit theorem connected with it. *Theory of Probability and its Applications* **37**, 648–657 (in Russian).

7. Cacoullos, T., Papathanasiou, V., and Utev, S. (1994). Variational inequalities with examples and an application to the central limit theorem. *Annals of Probability* **22**, 1607–1618.

8. Chen, L. H. Y. (1982). An inequality for the multivariate normal distribution. *Journal of Multivariate Analysis* **12**, 306–315.

9. Chen, L. H. Y. (1988). The central limit theorem and Poincaré type inequalities. *Annals of Probability* **16**, 300–304.

10. Chernoff, H. (1981). A note on an inequality involving the normal distribution. *Annals of Probability* **9**, 533–535.

11. Goldstein, L. and Reinert, G. (1997). Stein's method and the Zero Bias Transformation with application to simple random sampling. *Annals of Applied Probability* **7**, 935–952.

12. Hudson, H. M. (1978). A natural identity for exponential families with applications to multiparameter estimation. *Annals of Statistics* **6**, 473–484.

13. Liu, J. S. (1994). Siegel's formula via Stein's identities. *Statistics & Probability Letters* **21**, 247–251.

14. Papathanasiou, V. (1990). Characterizations of multidimensional exponential families by Cacoullos-type inequalities. *Journal of Multivariate Analysis* **35**, 102–107.

15. Papathanasiou, V. (1993). Some characteristic properties of the Fisher information matrix via Cacoullos-type inequalities. *Journal of Multivariate Analysis* **44**, 256–265.

16. Stein, C. (1972). A bound for the error in the normal approximation to the distribution of a sum of dependent random variables. In *Proceedings of the Sixth Berkeley Symposium on Mathematics, Statistics and Probability* **2**, 583–602. University California Press, Berkeley.

17. Stein, C. (1981). Estimation of the mean of a multivariate normal distribution. *Annals of Statistics* **9**, 1135–1151.

18. Vitale, R. A. (1989). A differential version of the Efron-Stein inequality: Bounding the variance of an infinitely divisible variable. *Statistics & Probability Letters* **7**, 105–112.

6

Probability Inequalities for U-statistics

Tasos C. Christofides
University of Cyprus, Nicosia, Cyprus

ABSTRACT Probability inequalities with exponential bounds are very central both in probability and statistics. In particular, such inequalities can be used in statistical (especially nonparametric) inference to provide rates of convergence for various estimates. In this paper, maximal inequalities and probability inequalities with exponential bounds are presented for various statistics including sums of i.i.d. random variables, sums of independent but not necessarily identically distributed random variables, and U-statistics. These inequalities generalize and extend results of Hoeffding (1963), Turner, Young, and Seaman (1992), Christofides (1994) and Qiying (1996).

Keywords and phrases U-statistics, maximal inequalities, exponential bounds

6.1 Introduction

Let $X_1, \ldots, X_n$ be independent observations of some space $\mathcal{X}$. For integer $m \geq 1$, let $h(x_1, \ldots, x_m)$ be a "kernel" mapping $\mathcal{X}^m$ to $\mathbf{R}$, and, without loss of generality, assume that h is symmetric in its arguments (i.e., invariant under permutations of arguments). To any such kernel h and sample $X_1, \ldots, X_n$ with $n \geq m$, we associate the average of the kernel over the sample observations taken m at a time, i.e., we define the corresponding U-*statistic*

$$U_n = \binom{n}{m}^{-1} \sum_c h(X_{i_1}, \ldots, X_{i_m}),$$

where $\sum_c$ denotes summation over the $\binom{n}{m}$ combinations of m distinct elements $\{i_1, \ldots, i_m\}$ from $\{1, \ldots, n\}$.

The class of U-statistics was first introduced and studied by Hoeffding (1948), as a generalization of the notion of sample mean. This class includes as special cases useful statistics such as the arithmetic mean and sample variance. They have a variety of applications including nonparametric estimation, hypothesis testing, and approximations to more complicated statistics. For an extensive treatment of U-statistics including examples, useful representations and applications see Serfling (1980) or Lee (1990).

A U-statistic by construction is an average of dependent observations except of course in the trivial case where $m = 1$. However, a U-statistic can be represented as an average of averages of independent (and identically distributed if $X_1, \ldots, X_n$ are i.i.d.) random variables. The representation, due to Hoeffding (1963) is the following: For $r = [n/m]$ where $[x]$ denotes the integer part of the real number x, put

$$W(x_1, \ldots, x_n) = \frac{1}{r}\{h(x_1, \ldots, x_m) + h(x_{m+1}, \ldots, x_{2m}) + \ldots$$
$$+ h(x_{rm-m+1}, \ldots, x_{rm})\}.$$

Clearly $W(X_1, \ldots, X_n)$ is an average of independent observations. One can easily verify that U_n can be expressed as

$$U_n = \frac{1}{n!} \sum_p W(X_{i_1}, \ldots, X_{i_n}),$$

where $\sum_p$ denotes summation taken over all permutations $(i_1, \ldots, i_n)$ of $\{1, \ldots, n\}$. The previous representation along with the results of Section 6.2 will be used in Section 6.3 to prove the main results of this paper.

Many U-statistics of interest are constructed based on bounded kernels. This class of U-statistics includes as a special case statistics constructed using variables which are indicator functions such as the empirical distribution function and the Wilcoxon one-sample statistic.

A very useful tool for various purposes is the following exponential probability inequality due to Hoeffding (1963).

THEOREM 6.1.1
For a U-statistic U_n based on the kernel $h(x_1, \ldots, x_m)$ with $a \leq h \leq b$ and $\epsilon > 0$ we have that

$$P(U_n - E(U_n) \geq \epsilon) \leq \exp(-2r\epsilon^2(b-a)^{-2}),$$

where $r = [n/m]$.

Hoeffding's inequality remains a powerful result having many applications not only in theoretical statistics but in other areas as well. See, for example, Clayton (1994) and Srivastar and Stangier (1996). Extensions of Hoeffding's result were given by various authors, including Roussas (1996) for negatively associated random variables and Devroye (1991) for kernel estimates of density functions.

The main objective of this paper is to establish maximal inequalities for U-statistics based on bounded kernels. Such inequalities can be used, among other things, to provide rates of convergence of various estimates. The results obtained in this paper improve and generalize the corresponding bounds of Hoeffding (1963), Turner *et al.* (1992), and Qiying (1996).

6.2 Preliminaries

For the development of the maximal inequalities in Section 6.3 we will make use of the following preliminary results.

LEMMA 6.2.1
Let U_n be a U-statistic and g a nondecreasing positive convex function. Then for $\epsilon > 0$ and $t > 0$

$$P(\sup_{k \geq n} U_k \geq \epsilon) \leq \{g(t\epsilon)\}^{-1} E\{g(tU_n)\}.$$

PROOF Let $\mathcal{F}_n = \sigma\{\mathbf{X}_{(n)}, X_{n+1}, X_{n+2}, \ldots\}$, where $\mathbf{X}_{(n)}$ denotes the vector of order statistics $(X_{n1}, \ldots, X_{nn})$. Then $\{U_k, \mathcal{F}_k, k \geq m\}$ is a reverse martingale. By the convexity of g, $\{g(tU_k), \mathcal{F}_k, k \geq m\}$ is a reverse submartingale. Therefore,

$$P(\sup_{k \geq n} U_k \geq \epsilon) = P(\sup_{k \geq n} g(tU_k) \geq g(t\epsilon))$$

$$\leq \{g(t\epsilon)\}^{-1} E\{g(tU_n)\}$$

where the inequality follows from a well-known maximal inequality for reverse submartingales [see, for example, Theorem 3.4.3 in Borovskikh and Korolyuk (1997) reference]. ∎

LEMMA 6.2.2

Let X be a random variable with $a \leq X \leq b$ and $E(X) = \mu$. If ϕ_X is the moment generating function of X then

$$\phi_X(t) \leq \frac{b - \mu}{b - a} e^{at} + \frac{\mu - a}{b - a} e^{bt}.$$

PROOF By the convexity of the exponential function

$$e^{tX} \leq \frac{b - X}{b - a} e^{at} + \frac{X - a}{b - a} e^{bt}$$

and by taking expectations the result follows. ∎

Two useful (determinist) inequalities are presented next. Lemma 6.2.3 combines results which can be found in Christofides (1991) and Christofides (1994).

LEMMA 6.2.3

Let $x, y \in \mathbf{R}$ such that $y = \alpha x$ for $\alpha \geq 0$.

(i) If $0 \leq \alpha < \frac{1}{4}$ then

$$x^{2p} - 2pxy^{2p-1} + (2p - 1)y^{2p} \geq (y - x)^{2p}, \quad p = 1, 2, \ldots$$

(ii) If $\alpha \geq \frac{1}{4}$ then

$$x^{2p} - 2pxy^{2p-1} + (2p - 1)y^{2p} \geq (2p - 1)(y - x)^{2p}, \quad p = 1, 2, \ldots$$

The next lemma provides lower bounds under different conditions for a quantity which appears very frequently in proofs of results concerning exponential inequalities.

LEMMA 6.2.4

Let

$$g(c, d) = (c + d) \ln(\frac{c + d}{d}) + (1 - c - d) \ln(\frac{1 - c - d}{1 - d})$$

for $c > 0$, $d > 0$ and $c + d < 1$.

(i) If $\frac{1}{2} < d < \frac{1}{2} + \frac{1}{3}c$ then

$$g(c, d) \geq (c + \frac{1}{2}) \ln(1 + 2c) + (\frac{1}{2} - c) \ln(1 - 2c).$$

(ii) If $c + d < \frac{1}{2}$ or $\frac{1}{2} + \frac{1}{3}c \leq d$ then

$$g(c, d) \geq -\frac{1}{2} \ln(1 - 4c^2).$$

(iii) If $\frac{1}{2} - c \le d \le \frac{1}{2}$ then

$$g(c, d) \ge 2c^2 + \frac{4}{3}c^4.$$

PROOF By Kambo and Kotz (1966) $g(c, d)$ can be expressed as

$$g(c, d) = \sum_{p=1}^{\infty} \frac{1}{2p(2p-1)} \{x^{2p} - 2pxy^{2p-1} + (2p-1)y^{2p}\}$$

where $x = 1 - 2c - 2d$ and $y = 1 - 2d$.

First assume that $\frac{1}{2} < d < \frac{1}{2} + \frac{1}{3}c$. Then $y = \alpha x$ for $0 < \alpha < \frac{1}{4}$. By Lemma 6.2.3

$$g(c, d) \ge \sum_{p=1}^{\infty} \frac{1}{2p(2p-1)} (y - x)^{2p}$$

$$= \sum_{p=1}^{\infty} \frac{1}{2p(2p-1)} (2c)^{2p}$$

$$= \sum_{p=1}^{\infty} \frac{1}{2p-1} (2c)^{2p} - \sum_{p=1}^{\infty} \frac{1}{2p} (2c)^{2p}$$

$$= 2c \sum_{p=1}^{\infty} \frac{1}{2p-1} (2c)^{2p-1} - \frac{1}{2} \sum_{p=1}^{\infty} \frac{1}{p} (4c^2)^p$$

$$= c \ln(\frac{1+2c}{1-2c}) + \frac{1}{2} \ln(1 - 4c^2)$$

$$= (c + \frac{1}{2}) \ln(1 + 2c) + (\frac{1}{2} - c) \ln(1 - 2c),$$

i.e., part (i) of the statement is established.

Now assume that $c + d < \frac{1}{2}$ or $\frac{1}{2} + \frac{1}{3}c \le d$. Then $y = \alpha x$ for some $\alpha \ge \frac{1}{4}$. By Lemma 6.2.3

$$g(c, d) \ge \sum_{p=1}^{\infty} \frac{1}{2p} (y - x)^{2p} = \frac{1}{2} \sum_{p=1}^{\infty} \frac{(4c^2)^p}{p} = -\frac{1}{2} \ln(1 - 4c^2),$$

and part (ii) is established as well. Part (iii) follows directly from Lemma 3 of Kambo and Kotz (1966). ∎

6.3   Probability Inequalities

In this section which is the main one of this paper we present maximal inequalities for U-statistics based on bounded kernels.

For the next theorem and its associated results we assume that the U-statistic U_n is constructed using i.i.d. random variables. However, with minor modifications in the proof the same results hold true for the case of random variables which are not necessarily identically distributed, as long as the kernel $h(X_{i_1}, \ldots, X_{i_m})$ is uniformly bounded below and above and has the same expectation for every choice of $\{i_1, \ldots, i_m\}$ from $\{1, \ldots, n\}$.

THEOREM 6.3.1
Let U_n be a U-statistic based on the kernel h with $a \leq h \leq b$. Let $E(h) = \mu$, $r = [n/m]$ and $\epsilon > 0$.

(i) If either $\epsilon + \mu < \frac{1}{2}(b-a)$ or $\epsilon \leq 3(\mu - \frac{1}{2}(b-a))$ then

$$P(\sup_{k \geq n}(U_k - \mu) \geq \epsilon) \leq (1 - 4\epsilon^2(b-a)^{-2})^{r/2}.$$

(ii) If $\frac{1}{2}(b-a) < \mu < \frac{1}{2}(b-a) + \frac{1}{3}\epsilon$ then

$$P(\sup_{k \geq n}(U_k - \mu) \geq \epsilon) \leq (1 - 2\epsilon(b-a)^{-1})^{r(\epsilon(b-a)^{-1} - \frac{1}{2})}$$
$$\times (1 + 2\epsilon(b-a)^{-1})^{-r(\epsilon(b-a)^{-1} + \frac{1}{2})}.$$

(iii) If $\frac{1}{2}(b-a) - \epsilon \leq \mu \leq \frac{1}{2}(b-a)$ then

$$P(\sup_{k \geq n}(U_k - \mu) \geq \epsilon) \leq \exp\{-r(2\epsilon^2(b-a)^{-2} + \frac{4}{3}\epsilon^4(b-a)^{-4})\}.$$

(iv) If $\epsilon + \mu = b$ then

$$P(\sup_{k \geq n}(U_k - \mu) \geq \epsilon) \leq (1 - \epsilon(b-a)^{-1})^r.$$

PROOF If $\epsilon + \mu > b$ then $P(\sup_{k \geq n}(U_k - \mu) \geq \epsilon) = 0$ and there is nothing to prove. First assume that $\epsilon + \mu < b$. Let $t > 0$. Using Hoeffding's representation of U_n as an average of averages of independent random variables and Lemma 6.2.1 we can write

$$P(\sup_{k \geq n}(U_k - \mu) \geq \epsilon) \leq e^{-t(\epsilon+\mu)}E(e^{tU_n})$$

$$= e^{-t(\epsilon+\mu)}E(e^{\frac{t}{n!}\sum_p W(\cdot)})$$

$$\leq e^{-t(\epsilon+\mu)}\frac{1}{n!}\sum_p E(e^{tW(\cdot)})$$

$$= e^{-t(\epsilon+\mu)}\frac{1}{n!}\sum_p E(e^{t/r\sum h(\cdot)})$$

$$= e^{-t(\epsilon+\mu)}\frac{1}{n!}\sum_p (\phi_h(t/r))^r$$

$$= e^{-t(\epsilon+\mu)}(\phi_h(t/r))^r$$

$$\leq e^{-t(\epsilon+\mu)}(\frac{b-\mu}{b-a}e^{at/r} + \frac{\mu-a}{b-a}e^{bt/r})^r$$

$$= e^{-f(t)} \tag{6.3.1}$$

where ϕ_h is the moment generating function of h, f is the function

$$f(t) = t(\epsilon+\mu) - r\log(\frac{b-\mu}{b-a}e^{at/r} + \frac{\mu-a}{b-a}e^{bt/r})$$

and the last inequality follows from Lemma 6.2.2. The function f is maximized at

$$t^\star = \frac{r}{b-a}\ln\{\frac{(\epsilon+\mu-a)(b-\mu)}{(b-\epsilon-\mu)(\mu-a)}\}$$

and

$$f(t^\star) = r\{\frac{\epsilon+\mu-a}{b-a}\ln(\frac{\epsilon+\mu-a}{\mu-a}) + \frac{b-\epsilon-\mu}{b-a}\ln(\frac{b-\epsilon-\mu}{b-\mu})\}.$$

Let $\epsilon^\star = \epsilon(b-a)^{-1}$ and $\mu^\star = (\mu-a)(b-a)^{-1}$. Then $f(t^\star) = rg(\epsilon^\star, \mu^\star)$ where

$$g(\epsilon^\star, \mu^\star) = (\epsilon^\star + \mu^\star)\ln(\frac{\epsilon^\star + \mu^\star}{\mu^\star}) + (1 - \epsilon^\star - \mu^\star)\ln(\frac{1 - \epsilon^\star - \mu^\star}{1 - \mu^\star}).$$

(i) If either $\epsilon+\mu < \frac{1}{2}(b-a)$ or $\epsilon \leq 3(\mu-\frac{1}{2}(b-a))$ then by the assumptions of the theorem $\epsilon^\star$ and $\mu^\star$ satisfy the conditions of Lemma 6.2.4 (ii) and thus $f(t^\star) \geq -\frac{r}{2}\ln(1 - 4\epsilon^{\star 2})$. Therefore by (6.3.1)

$$P(\sup_{k \geq n}(U_k - \mu) \geq \epsilon) \leq (1 - 4\epsilon^2(b-a)^{-2})^{r/2}.$$

(ii) If $\frac{1}{2}(b-a) < \mu < \frac{1}{2}(b-a) + \frac{1}{3}\epsilon$ then the conditions of Lemma 6.2.4 (i) are satisfied and thus

$$f(t^*) \geq r(\epsilon^* + \frac{1}{2})\ln(1 + 2\epsilon^*) + r(\frac{1}{2} - \epsilon^*)\ln(1 - 2\epsilon^*).$$

Therefore, again by (6.3.1)

$$P(\sup_{k \geq n}(U_k - \mu) \geq \epsilon) \leq (1 - 2\epsilon^*)^{r(\epsilon^* - \frac{1}{2})}(1 + 2\epsilon^*)^{-r(\epsilon^* + \frac{1}{2})}.$$

(iii) If $\frac{1}{2}(b-a) - \epsilon \leq \mu \leq \frac{1}{2}(b-a)$ then $\frac{1}{2} - \epsilon^* \leq \mu^* \leq \frac{1}{2}$ and by applying (iii) of Lemma 6.2.4 we have that

$$f(t^*) \geq r(2\epsilon^{*2} + \frac{4}{3}\epsilon^{*4}).$$

The result follows from (6.3.1).

(iv) If $\epsilon + \mu = b$ then

$$
\begin{aligned}
&P(\sup_{k \geq n}(U_k - \mu) \geq \epsilon) \\
&= P(\sup_{k \geq n} U_k = b) \\
&= P(h(X_{i_1}, \ldots, X_{i_m}) = b \ \forall \ (i_1, \ldots, i_m) \subset \{1, \ldots, n\}) \\
&\leq P(h(X_{jm+1}, \ldots, X_{jm+m}) = b, \ j = 0, \ldots, r - 1) \\
&= \prod_{j=0}^{r-1} P(h(X_{jm+1}, \ldots, X_{jm+m}) = b) \qquad\qquad (6.3.2) \\
&\leq \prod_{j=0}^{r-1} (\frac{E(h) - a}{b - a}) \\
&= (\frac{\mu - a}{b - a})^r \\
&= (1 - \epsilon^*)^r
\end{aligned}
$$

where (6.3.2) follows from independence. ∎

REMARK Simple algebraic manipulation shows that under its conditions Theorem 6.3.1 provides a better bound than the one of Theorem 6.1.1. In addition, part (i) of Theorem 6.3.1 generalizes Theorem 3.1 of Christofides (1994) in the sense that the result is established for a general bounded kernel h rather than a kernel which is a Bernoulli random variable. ∎

In many situations a two-sided version of the theorem might be more appropriate, if not necessary. Such a version is provided by the next corollary under appropriate conditions on the parameters.

COROLLARY 6.3.2
Let U_n be a U-statistic based on the kernel h with $a \leq h \leq b$. Let $E(h) = \mu$, $r = [n/m]$ and $\epsilon > 0$. Assume that $\mid \mu \mid + \epsilon < \frac{1}{2}(b - a)$ or $\epsilon + \mu < \frac{1}{2}(b - a) \leq -\mu - \frac{1}{3}\epsilon$ or $\epsilon - \mu < \frac{1}{2}(b - a) \leq \mu - \frac{1}{3}\epsilon$. Then

$$P(\sup_{k \geq n} \mid U_k - \mu \mid \geq \epsilon) \leq 2(1 - 4\epsilon^2(b - a)^{-2})^{r/2}.$$

PROOF We can write

$$P(\sup_{k \geq n} \mid U_k - \mu \mid \geq \epsilon) \leq P(\sup_{k \geq n}(U_k - \mu) \geq \epsilon)$$
$$+ P(\sup_{k \geq n}(-U_k + \mu) \geq \epsilon). \quad (6.3.3)$$

Observe that $-U_n$ is a U-statistic based on the kernel $-h$ with $E(-h) = -\mu$ and $-b \leq -h \leq -a$. The conditions on the parameters ϵ, μ, a and b are such that the assumptions of the theorem are satisfied for both U_n and $-U_n$. Thus, each term of the right hand side of (6.3.3) is bounded by $(1 - 4\epsilon^2(b - a)^{-2})^{r/2}$ and the result follows. ∎

REMARK Corollary 6.3.2 improves and generalizes Corollary 2 of Qiying (1996) in several ways. First, the result is stated for a U-statistic rather than the sample mean of i.i.d. random variables. In addition, in the special case where the U-statistic is the sample mean of i.i.d. random variables the bound provided is sharper than the one of Corollary 2 of Qiying (1996). To see this, just expand the logarithm of $(1 - 4\epsilon^2)$ to get

$$\ln(1 - 4\epsilon^2) = -4\epsilon^2 - 8\epsilon^4 - \frac{64}{3}\epsilon^6 \ldots$$

Clearly

$$\frac{n}{2}\ln(1 - 4\epsilon^2) < -2n\epsilon^2 - \frac{4}{3}n\epsilon^4,$$

i.e.,

$$(1 - 4\epsilon^2)^{n/2} < \exp(-2n\epsilon^2 - \frac{4}{3}n\epsilon^4),$$

where the right hand side of the previous inequality is the sharpest of the three upper bounds of Corollary 2 of Qiying (1996). ∎

Assume that $m = 1$, $h(x) = x$ and $X_1, \ldots, X_n$ are Bernoulli random variables with probability of success p, $0 < p < 1$. Then as a special case of Theorem 6.3.1 we have the following inequality for the Binomial distribution.

COROLLARY 6.3.3
Let Y_k be a binomial random variable with parameters k and p, $0 < p < 1$. Let $\epsilon > 0$.

(i) If either $\epsilon + p < \frac{1}{2}$ or $\epsilon \leq 3(p - \frac{1}{2})$ then

$$P(\sup_{k \geq n}(\frac{Y_k}{k} - p) \geq \epsilon) \leq (1 - 4\epsilon^2)^{\frac{n}{2}}.$$

(ii) If $\frac{1}{2} < p < \frac{1}{2} + \frac{1}{3}\epsilon$ then

$$P(\sup_{k \geq n}(\frac{Y_k}{k} - p) \geq \epsilon) \leq (1 - 2\epsilon^2)^{n(\epsilon - \frac{1}{2})}(1 + 2\epsilon^2)^{-n(\epsilon + \frac{1}{2})}.$$

(iii) If $\frac{1}{2} - \epsilon \leq p \leq \frac{1}{2}$ then

$$P(\sup_{k \geq n}(\frac{Y_k}{k} - p) \geq \epsilon) \leq \exp\{-n(2\epsilon^2 + \frac{4}{3}\epsilon^4)\}.$$

(iv) If $p + \epsilon < \frac{1}{2}$ or $0 < \epsilon - p < \frac{1}{2} \leq p - \frac{1}{3}\epsilon$ then

$$P(\sup_{k \geq n} | \frac{Y_k}{k} - p | \geq \epsilon) \leq 2(1 - 4\epsilon^2)^{\frac{n}{2}}.$$

REMARK Corollary 6.3.3 improves the result of Turner *et al.* (1992) who proved the inequality

$$P(\sup_{k \geq n}(\frac{Y_k}{k} - p) \geq \epsilon) \leq (1 - 2\epsilon^2)^{n(\epsilon - \frac{1}{2})}(1 + 2\epsilon^2)^{-n(\epsilon + \frac{1}{2})}$$

under the condition $p + \epsilon < \frac{1}{2}$ or $p \geq \frac{1}{2}$. As in the case of the previous remark to see this we can expand the logarithm of $(1 - 4\epsilon^2)$. Then simple algebraic manipulation shows that

$$(1 - 4\epsilon^2)^{\frac{n}{2}} < (1 - 2\epsilon^2)^{n(\epsilon - \frac{1}{2})}(1 + 2\epsilon^2)^{-n(\epsilon + \frac{1}{2})}.$$

∎

The next theorem is essentially a generalization of Theorem 6.1.1 to the case where the observations $X_1, \ldots, X_n$ are not necessarily identically distributed and the kernel h is not uniformly bounded.

THEOREM 6.3.4

Let $X_1, \ldots, X_n$ be independent random variables and U_n a U-statistic based on the kernel h. Assume that for $\{i_1, \ldots, i_m\} \subseteq \{1, \ldots, n\}$, $E\{h(X_{i_1}, \ldots, X_{i_m})\} = \theta_{i_1, \ldots, i_m}$ and $a_{i_1, \ldots, i_m} \leq h(X_{i_1}, \ldots, X_{i_m}) \leq b_{i_1, \ldots, i_m}$. Let $\bar{\theta}_n = \binom{n}{m}^{-1} \sum_c \theta_{i_1, \ldots, i_m}$, $r = [n/m]$ and $\epsilon > 0$. Then

$$P(\sup_{k \geq n} |U_k - \bar{\theta}_k| \geq \epsilon) \leq 2 \exp\{-2r^2\epsilon^2 / \max_{(i_1, \ldots, i_n)} D_{i_1, \ldots, i_n}\}$$

where

$$D_{i_1, \ldots, i_n} = \sum_{j=0}^{r-1} (b_{i_{jm+1}, \ldots, i_{(j+1)m}} - a_{i_{jm+1}, \ldots, i_{(j+1)m}})^2$$

and the maximum is taken over all permutations $(i_1, \ldots, i_n)$ of $\{1, \ldots, n\}$.

PROOF Let $t > 0$ and $\tilde{h}(X_{i_1}, \ldots, X_{i_m}) = h(X_{i_1}, \ldots, X_{i_m}) - \theta_{i_1, \ldots, i_m}$. For notational simplicity let $\mathbf{i}_{j,m} = (i_{jm+1}, \ldots, i_{(j+1)m})$. Following the proof of Theorem 6.3.1 we arrive at the following

$$P(\sup_{k \geq n}(U_k - \bar{\theta}_k) \geq \epsilon) \leq e^{-t\epsilon} \frac{1}{n!} \sum_p \prod_{j=0}^{r-1} E(e^{\frac{t}{r} \sum \tilde{h}(\cdot)})$$

$$\leq e^{-t\epsilon} \frac{1}{n!} \sum_p \prod_{j=0}^{r-1} G_{\mathbf{i}_{j,m}}(t), \qquad (6.3.4)$$

where

$$G_{\mathbf{i}_{j,m}}(t) = \exp(-\frac{t}{r}\theta_{\mathbf{i}_{j,m}}) \left\{ \frac{b_{\mathbf{i}_{j,m}} - \theta_{\mathbf{i}_{j,m}}}{b_{\mathbf{i}_{j,m}} - a_{\mathbf{i}_{j,m}}} \exp(\frac{t}{r} a_{\mathbf{i}_{j,m}}) \right.$$
$$\left. + \frac{\theta_{\mathbf{i}_{j,m}} - a_{\mathbf{i}_{j,m}}}{b_{\mathbf{i}_{j,m}} - a_{\mathbf{i}_{j,m}}} \exp(\frac{t}{r} b_{\mathbf{i}_{j,m}}) \right\},$$

and (6.3.4) is a consequence of Lemma 6.2.2. Let $u_{\mathbf{i}_{j,m}} = \dfrac{(\theta_{\mathbf{i}_{j,m}} - a_{\mathbf{i}_{j,m}})}{(b_{\mathbf{i}_{j,m}} - a_{\mathbf{i}_{j,m}})}$ and $s = t/r$. Then

$$G_{\mathbf{i}_{j,m}}(t) = \exp\{-s(\theta_{\mathbf{i}_{j,m}} - a_{\mathbf{i}_{j,m}})\}\{1 - u_{\mathbf{i}_{j,m}} + u_{\mathbf{i}_{j,m}} \exp\{s(b_{\mathbf{i}_{j,m}} - a_{\mathbf{i}_{j,m}})\}\}.$$

Following Hoeffding (1963) let

$$f_{\mathbf{i}_{j,m}}(s) = -s(\theta_{\mathbf{i}_{j,m}} - a_{\mathbf{i}_{j,m}}) + \ln\{1 - u_{\mathbf{i}_{j,m}} + u_{\mathbf{i}_{j,m}} \exp\{s(b_{\mathbf{i}_{j,m}} - a_{\mathbf{i}_{j,m}})\}\},$$

so that

$$G_{\mathbf{i}_{j,m}} = \exp\{f_{\mathbf{i}_{j,m}}(s)\}. \qquad (6.3.5)$$

Tasos C. Christofides

One can verify that $f_{i_{j,m}}(0) = 0$ and $f'_{i_{j,m}}(0) = 0$. In addition,

$$f''_{i_{j,m}}(s) = (b_{i_{j,m}} - a_{i_{j,m}})^2 g_{i_{j,m}}(s), \qquad (6.3.6)$$

where

$$g_{i_{j,m}}(s) = \frac{u_{i_{j,m}}(1 - u_{i_{j,m}}) \exp\{-s(b_{i_{j,m}} - a_{i_{j,m}})\}}{\{u_{i_{j,m}} + (1 - u_{i_{j,m}}) \exp\{-s(b_{i_{j,m}} - a_{i_{j,m}})\}\}^2}.$$

The function $g_{i_{j,m}}(s)$ is of the form $x(1-x)$ with $0 < x < 1$ and therefore $g_{i_{j,m}}(s) \leq \frac{1}{4}$. From (6.3.6)

$$f''_{i_{j,m}}(s) \leq \frac{1}{4}(b_{i_{j,m}} - a_{i_{j,m}})^2.$$

Using Taylor's formula up to terms involving the second derivative we have that

$$f_{i_{j,m}}(s) \leq \frac{1}{8}s^2(b_{i_{j,m}} - a_{i_{j,m}})^2.$$

Using the above inequality in (6.3.5) we have that

$$G_{i_{j,m}}(t) \leq \exp\{\frac{t^2}{8r^2}(b_{i_{j,m}} - a_{i_{j,m}})^2\}.$$

Then from (6.3.4) we obtain the inequalities

$$P(\sup_{k \geq n}(U_k - \bar{\theta}_k) \geq \epsilon) \leq e^{-t\epsilon} \frac{1}{n!} \sum_p \prod_{j=0}^{r-1} \exp\{\frac{t^2}{8r^2}(b_{i_{j,m}} - a_{i_{j,m}})^2\}$$

$$= e^{-t\epsilon} \frac{1}{n!} \sum_p \exp\{\frac{t^2}{8r^2} \sum_{j=0}^{r-1}(b_{i_{j,m}} - a_{i_{j,m}})^2\}$$

$$\leq \exp\{-t\epsilon + \frac{t^2}{8r^2} \max_{(i_1,\ldots,i_n)} D_{i_1,\ldots,i_n}\}.$$

Optimizing for t we get

$$P(\sup_{k \geq n}(U_k - \bar{\theta}_k) \geq \epsilon) \leq \exp\{-2r^2\epsilon^2/\max_{(i_1,\ldots,i_n)} D_{i_1,\ldots,i_n}\}.$$

The theorem follows using an analog of (6.3.3). ∎

References

1. Christofides, T. C. (1991). Probability inequalities with exponential bounds for U-statistics. *Statistics & Probability Letters* **12**, 257–261.

2. Christofides, T. C. (1994). A Kolmogorov inequality for U-statistics based on Bernoulli kernels. *Statistics & Probability Letters* **21**, 357–362.

3. Clayton, H. R. (1994). A combined bound for errors in auditing based on Hoeffding's inequality and the bootstrap. *Journal of Business and Economic Statistics* **12**, 437–448.

4. Devroye, L. (1991). Exponential inequalities in nonparametric estimation, *Nonparametric functional estimation and related topics (Spetses 1990)*, pp. 31–44, NATO Adv. Sci. Inst. Ser. C Math. Phys. Sci., 335, Kluwer Academic Publishers, Dordrecht.

5. Hoeffding, W. (1948). A class of statistics with asymptotically normal distributions. *Annals of Mathematical Statistics* **19**, 293–325.

6. Hoeffding, W. (1963). Probability inequalities for sums of bounded random variables. *Journal of the American Statistical Association* **58**, 13–30.

7. Kambo, N. S. and Kotz, S. (1966). On exponential bounds for binomial probabilities. *Annals of the Institute of Statistical Mathematics* **18**, 277–287.

8. Lee, A. J. (1990). *U-statistics: Theory and Practice*, Marcel Dekker, New York.

9. Qiying, W. (1996). On the maximal inequality. *Statistics & Probability Letters* **31**, 85–89.

10. Roussas, G. G. (1996). Exponential probability inequalities with some applications. *Statistics, Probability and Game Theory*, 303–319, IMS Lecture Notes Monogr. Ser., 30, *Inst. Math. Statist., Hayward, CA.*

11. Serfling, R. (1980). *Approximation Theorems of Mathematical Statistics.* John Wiley & Sons, New York.

12. Srivastar, A. and Stangier, P. (1996). Chernoff-Hoeffding inequalities in integer programming. *Random Structures Algorithms* **8**, 27–58.

13. Turner, D. W., Young, D. M., and Seaman, J. W. (1992). Improved Kolmogorov inequalities for the Binomial distribution. *Statistics & Probability Letters* **13** 223–227.

PART II

PROBABILITY AND STOCHASTIC PROCESSES

7

Theory and Applications of Decoupling

Victor H. de la Peña and Tze Leung Lai
Columbia University, New York, NY
Stanford University, Stanford, CA

ABSTRACT In this paper we provide a survey of the theory of decoupling, emphasizing its wide range of applications. Decoupling was born out of a need to extend the known martingale inequalities for real and Hilbert-space valued random variables to variables taking values in more general spaces like Banach spaces. Among its many uses is that it enables one to symmetrize highly dependent random variables. In fact, some of the basic results in decoupling theory were motivated by symmetrization techniques for U-processes which consist of U-statistics indexed by a class of functions. Other applications include sequential analysis, martingale theory, stochastic integration and weak convergence. We consider three types of decoupling. The first type, which we call complete decoupling, completely replaces the summands in a sum of dependent random variables by independent ones with the same marginal distributions. The second type of decoupling is decoupling of tangent sequences. In this approach, two sequences adapted to the same filtration and having the same conditional distributions (given the past) are compared. The effectiveness of this approach is realized through the use of a sequence with a more tractable dependence structure than the original one. The third type of decoupling is called "total decoupling of stopping times". In this approach, problems involving stopping times are handled by establishing inequalities that replace the original process by an independent copy which is therefore independent of the stopping time.

Keywords and phrases Conditional independence, decoupling inequalities, exponential bounds, martingales, stopping times, symmetrization, tangent sequences, U-statistics, weak convergence

7.1   Complete Decoupling of Marginal Laws and One-Sided Bounds

Let $(X_1, ..., X_n)$ be a vector of possibly dependent random variables. Let $(Y_1^{(i)}, ..., Y_n^{(i)})$, $i = 1, ..., n$, be n independent copies of the original random vector. To do this formally we might need to enlarge our probability space. Then all sequences have exactly the same joint distributions. Moreover, observe that $Y_i^{(i)}, 1 \leq i \leq n$, are independent random variables with $Y_i^{(i)}$ having the same distribution as X_i. With this construction we can now analyze the decoupling property obtained through the use of the linearity of expectation. That is,

$$E\left(\sum_{i=1}^{n} X_i\right) = E\left(\sum_{i=1}^{n} Y_i^{(i)}\right), \qquad (7.1.1)$$

where the first expectation involves a sum of dependent variables and the second a sum of independent random variables. Indeed, $E(\sum_{i=1}^{n} X_i) = \sum_{i=1}^{n} EX_i = \sum_{i=1}^{n} EY_i^{(i)} = E(\sum_{i=1}^{n} Y_i^{(i)})$. This elementary identity has been extended to the following one-sided bound by de la Peña (1990).

THEOREM 7.1.1
Let $\{X_i\}$ be a sequence of (arbitrarily dependent) random variables with $X_i \geq 0$ a.s. Let the sequence $\{Y_i^{(i)}\}$ be defined as above. Then, for all concave nondecreasing functions Ψ on $[0, \infty)$ with $\Psi(0) = 0$,

$$E\Psi\left(\sum_{i=1}^{n} X_i\right) \leq CE\Psi\left(\sum_{i=1}^{n} Y_i^{(i)}\right), \qquad (7.1.2)$$

for some universal constant C.

No reverse bound is possible in general for concave functions. However, the reverse bound holds for a class of convex functions. The following notation will be needed not only in the reverse bound but also in many subsequent inequalities. For $\alpha > 0$, let

$$\mathcal{A}_\alpha = \{\Phi : \ \Phi \text{ is a nondecreasing, continuous function on } [0, \infty) \text{ with}$$
$$\Phi(0) = 0 \text{ and } \Phi(cx) \leq c^\alpha \Phi(x) \text{ for all } c \geq 2 \text{ and } x \geq 0\}. \quad (7.1.3)$$

Then there exists a positive finite constant C_α depending only on α such that for all convex $\Phi \in \mathcal{A}_\alpha$,

$$E\Phi\left(\sum_{i=1}^{n} X_i\right) \geq C_\alpha E\Phi\left(\sum_{i=1}^{n} Y_i^{(i)}\right). \tag{7.1.4}$$

Results of this kind, which appear in de la Peña (1990), have obvious martingale analogues through the use of the square function inequalities of Burkholder, Davis and Gundy.

THEOREM 7.1.2
Let $\{d_i\}$ be a martingale difference sequence. Let $\{Y_i^{(i)}\}$ be a sequence of independent random variables with $Y_i^{(i)}$ having the same distribution as d_i (defined above). Let $\alpha > 0$ be fixed. Then, for all $\Phi \in \mathcal{A}_\alpha$ with $\Phi(x)$ convex and $\Phi(\sqrt{x})$ concave on $[0, \infty)$, there exists a positive finite constant C_α depending only on α such that

$$E\Phi\left(\max_{j \leq n} |\sum_{i=1}^{j} d_i|\right) \leq C_\alpha E\Phi\left(|\sum_{i=1}^{n} Y_i^{(i)}|\right). \tag{7.1.5}$$

Moreover, if both $\Phi \in \mathcal{A}_\alpha$ and $\Psi(\sqrt{x})$ are convex functions on $[0, \infty)$, then there exists a positive finite constant c_α depending only on α such that

$$E\Phi\left(\max_{j \leq n} |\sum_{i=1}^{j} d_i|\right) \geq c_\alpha E\Phi\left(|\sum_{i=1}^{n} Y_i^{(i)}|\right). \tag{7.1.6}$$

We can replace $\Sigma_{i=1}^{n}$ by $\max_{1 \leq i \leq n}$ in Theorem 7.1.1 in view of the following.

LEMMA 7.1.3
Let $\{U_i\}$ be a sequence of dependent sets. Let $\{V_i\}$ be a sequence of independent sets such that $P(U_i) \leq P(V_i)$ for all i. Then $P(\cup_{i=1}^{n} U_i) \leq 2P(\cup_{i=1}^{n} V_i)$.

To prove Lemma 7.1.3, first note that

$$P(\cup_{j=1}^{n} U_j) \leq P(\cup_{j=1}^{n} U_j \cap (\cup_{i=1}^{n} V_i)^c) + P(\cup_{j=1}^{n} V_j). \tag{7.1.7}$$

The first term in the right hand side of (7.1.7) is less than or equal to

$$\sum_{j=1}^{n} P(U_j \cap (\cap_{i=1}^{j-1} V_i^c)) = \sum_{j=1}^{n} P(U_j)P(\cap_{i=1}^{j-1} V_i^c) \leq \sum_{j=1}^{n} P(V_j)P(\cap_{i=1}^{j-1} V_i^c),$$

where the equality follows from independence. Note that the last sum above is equal to

$$P(\cup_{j=1}^{n}(V_j \cap (\cap_{i=1}^{j-1}V_i^c))) = P(\cup_{j=1}^{n}V_j).$$

From Lemma 7.1.3 it follows that

$$P(\max_{1\leq i\leq n} X_i > x) \leq 2P(\max_{1\leq i\leq n} Y_i^{(i)} > x) \quad \text{for all} \quad x. \tag{7.1.8}$$

Therefore, if $X_i \geq 0$ a.s. and Ψ is a nondecreasing, continuous function on $[0,\infty)$ with $\Psi(0) = 0$, then

$$E\Psi(\max_{1\leq i\leq n} X_i) \leq 2E\Psi(\max_{1\leq i\leq n} Y_i^{(i)}). \tag{7.1.9}$$

In particular, if the X_i have a common distribution function F, then the right hand side of (7.1.8) is equal to $2(1 - F^n(x))$, while the left hand side of (7.1.8) is majorized by $\min\{1, n(1 - F(x))\}$ and there exists a joint distribution of $X_1, \ldots, X_n$ with F as marginal such that

$$P(\max_{1\leq i\leq n} X_i > x) = \min\{1, n(1 - F(x))\} \quad \text{for all} \quad x, \tag{7.1.10}$$

cf. Lai and Robbins (1978) who call such random variables X_i "maximally dependent." Letting $p = 1 - F(x)$, we obtain from (7.1.8) and (7.1.10) the inequality $\min\{1, np\} \leq 2[1 - (1 - p)^n]$ for $0 \leq p \leq 1$.

We next mention two recent results of the type in Theorem 7.1.1 concerning associated variables and stochastic domination by independent random variables.

DEFINITION 7.1.4

A finite family of random variables $\{X_i, 1 \leq i \leq n\}$ is said to be negatively associated if for every pair of disjoint subsets A_1 and A_2 of $\{1, 2, ..., n\}$,

$$Cov\{f_1(X_i, i \in A_1), f_2(X_j, j \in A_2)\} \leq 0, \tag{7.1.11}$$

whenever f_1 and f_2 are coordinatewise nondecreasing and the covariance exists. The random variables X_i are said to be positively associated if the inequality in (1.11) is reversed.

As is well known, the following analogue of (7.1.8) holds for positively associated random variables:

$$P(\max_{1 \le i \le n} X_i > x) \le P(\max_{1 \le i \le n} Y_i^{(i)} > x) \quad \text{for all} \quad x. \qquad (7.1.12)$$

Examples of negatively associated random variables include the multinomial distribution, random sampling without replacement and the joint distribution of ranks; see Joag-Dev and Proschan (1983). Shao (1998) has obtained the following theorem, which generalizes Hoeffding's (1963) results for sampling without replacement.

THEOREM 7.1.5
Let $\{X_i, \ 1 \le i \le n\}$ be a negatively associated sequence, and let $\{Y_i^{(i)}\}$ be a sequence of independent random variables such that $Y_i^{(i)}$ has the same distribution as X_i for all $i = 1, ..., n$. Then

$$Ef(\sum_{i=1}^{n} X_i) \le Ef(\sum_{i=1}^{n} Y_i^{(i)}) \qquad (7.1.13)$$

for any convex function f, whenever the expectation on the right hand side of (7.1.13) exists. Moreover, if f is a nondecreasing convex function, then

$$Ef(\max_{1 \le k \le n} \sum_{i=1}^{k} X_i) \le Ef(\max_{1 \le k \le n} \sum_{i=1}^{k} Y_i^{(i)}) \qquad (7.1.14)$$

whenever the expectation on the right hand side of (7.1.14) exists.

THEOREM 7.1.6
Let $\{X_i\}$ be a sequence of random variables. Assume that for all n, $P(X_n > x|X_1, .., X_{n-1}) \le P(Y_n > x)$ for all real x, where $\{Y_i\}$ is a sequence of independent random variables. Then

$$P(\sum_{i=1}^{n} X_n > x) \le P(\sum_{i=1}^{n} Y_i > x), \qquad (7.1.15)$$

for all real x.

Theorem 7.1.6 was proved by Huang (1999) in her study of stochastic networks. It turns out to be a special case of a general domination inequality of Kwapien and Woyczynski (1992, Theorem 5.1.1).

THEOREM 7.1.7

On the probability space $(\Omega, \mathcal{F}, P)$ let $\{X_i\}$ and $\{Y_i\}$ be two sequences of random variables with values in a separable Banach space $(B, ||\cdot||)$, adapted to a filtration $\mathcal{F}_i$ increasing to $\mathcal{F}$ and such that Y_i is independent of $\mathcal{F}_{i-1}$ for all i. If $u : B \to [0, \infty)$ satisfies $E(u(x + X_i)|\mathcal{F}_{i-1}) \le Eu(x + Y_i)$ for all $x \in B$ and all i then,

$$Eu\left(x + \sum_{i=1}^{n} X_i\right) \le Eu\left(x + \sum_{i=1}^{n} Y_i\right),$$

for all $x \in B$.

A related result involving domination of variables from Kwapien and Woyczynski (1992, Theorem 5.2.1) helps to bridge the way between complete decoupling and decoupling of tangent sequences which will be introduced in the next two sections.

THEOREM 7.1.8

Let $\{X_i\}$ and $\{Y_i\}$ be two sequences of random variables adapted to $\mathcal{F}_i$, with $\{X_i\}$ taking values in a separable Banach space and $\{Y_i\}$ taking values in $[0, \infty)$. Assume that for all i and $x > 0$,

$$P(||X_i|| > x|\mathcal{F}_{i-1}) \le P(Y_i > x|\mathcal{F}_{i-1}). \tag{7.1.16}$$

Then for all $x > 0$ and concave $\Psi : [0, \infty) \to [0, \infty)$,

$$P(\max_{i \le i \le n} ||X_i|| > x) \le 2P(\max_{1 \le i \le n} Y_i > x), \tag{7.1.17}$$

$$E\Psi(\max_{1 \le j \le n} ||\sum_{i=1}^{j} X_i||) \le 3E\Psi\left(\max_{1 \le j \le n} \sum_{i=1}^{j} Y_i\right). \tag{7.1.18}$$

Moreover, for every continuous Ψ such that $\Psi(x + y) \le C(\Psi(x) + \Psi(y))$ for some $C > 0$ and all $x, y \in [0, \infty)$, there exists C' such that

$$E\Psi\left(\max_{1 \le j \le n} ||\sum_{i=1}^{j} X_i||\right) \le C'E\Psi\left(\max_{1 \le j \le n} \sum_{i=1}^{j} Y_i\right). \tag{7.1.19}$$

7.2 Tangent Sequences and Conditionally Independent Variables

DEFINITION 7.2.1
Let $\{d_i\}$ and $\{e_i\}$ be two sequences of random variables adapted to the same increasing sequence of σ-fields $\{\mathcal{F}_i\}$. Assume that the conditional distribution of d_i given $\mathcal{F}_{i-1}$ is the same as the conditional distribution of e_i given $\mathcal{F}_{i-1}$. Then the sequences $\{d_i\}$ and $\{e_i\}$ are said to be $\mathcal{F}_i$-tangent.

DEFINITION 7.2.2
Let $\{e_i\}$ be a sequence of variables adapted to an increasing sequence of σ-fields $\{\mathcal{F}_i\}$ contained in the σ-field $\mathcal{F}$. Then $\{e_i\}$ is said to satisfy the CI condition (conditional independence) if there exists a σ-field $\mathcal{G}$ contained in $\mathcal{F}$ such that $\{e_i\}$ is conditionally independent given $\mathcal{G}$ and the conditional distribution of e_i given $\mathcal{F}_{i-1}$ is the same as the conditional distribution of e_i given $\mathcal{G}$.

DEFINITION 7.2.3
A sequence $\{e_i\}$ which satisfies the CI condition and which is also tangent to d_i is said to be a decoupled tangent sequence to $\{d_i\}$.

Let T be a stopping time. It is easy to see that the sequences $\{X_i 1(T \geq i)\}$ and $\{\tilde{X}_i 1(T \geq i)\}$ are tangent with respect to $\mathcal{F}_i = \sigma(X_1, ..., X_i; \tilde{X}_1, ..., \tilde{X}_i)$. Moreover, $\{\tilde{X}_i 1(T \geq i)\}$ satisfies the CI condition given $\mathcal{G} = \sigma(X_1, X_2, ...)$. In the context of sampling schemes, the theory of decoupling provides a sampling scheme that lies between sampling with replacement and sampling without replacement. We call this sequence a CI (conditionally independent) sequence.

Example 1
Let $\{b_1, ..., b_N\}$ be a set of N balls inside a box. The sequence $\{d_i\}$ will represent a sample without replacement from the box and $\{e_i\}$ a CI (conditionally independent) sample, which can be generated as follows. At the ith stage, we first draw e_i, return the ball, and then draw d_i and put the ball aside. It is easy to see that the above procedure will make $\{e_i\}$ tangent to $\{d_i\}$ with $\mathcal{F}_i = \sigma(d_1, ..., d_i; e_1, ..., e_i)$. Moreover, $\{e_i\}$ satisfies the CI condition given $\mathcal{G} = \sigma(d_1, ..., d_N)$. $\quad\square$

A rather fortunate fact, which makes the theory of decoupling widely applicable, is closely linked to the above example. Broadly speaking, it states that decoupled tangent sequences always exist. This result is summarized in the following proposition.

PROPOSITION 7.2.4

For any sequence of random variables $\{d_i\}$ adapted to an increasing sequence of σ-fields $\{\mathcal{F}_i\}$, there always exists a decoupled sequence $\{e_i\}$ (on a possibly enlarged space) which is tangent to the original sequence and in addition satisfies the CI condition given a σ-field $\mathcal{G}$.

A constructive approach to obtaining the CI sequence proceeds by induction. First we take e_1 and d_1 to be two independent copies of the same random mechanism. Having constructed $d_1, \ldots, d_{i-1}, e_1, \ldots, e_{i-1}$, the ith pair of variables d_i and e_i comes from i.i.d. copies of the same random mechanism, given $d_1, \ldots, d_{i-1}$. It is easy to see that using this construction and taking

$$\mathcal{F}_i = \sigma(d_1, \ldots, d_i; e_1, \ldots, e_i),$$

the sequences $\{d_i\}$ and $\{e_i\}$ are $\mathcal{F}_i$-tangent and the sequence $\{e_i\}$ is conditionally independent given the σ-field $\mathcal{G}$ generated by $d_1, d_2, \ldots$ and is a decoupled version of $\{d_i\}$; see de la Peña (1994). This construction also provides the following refinement of Proposition 7.2.4 which will be used in Section 7.6.

PROPOSITION 7.2.5

For any sequence of random variables $\{d_i\}$ adapted to an increasing sequence of σ-fields $\{\mathcal{F}_i\}$ and for any stopping time T of $\{\mathcal{F}_i\}$, there exist a decoupled sequence $\{e_i\}$ and a σ-field $\mathcal{G}$ such that T is measurable with respect to $\mathcal{G}$, and $\{e_i\}$ is tangent to $\{d_i\}$ and is CI given $\mathcal{G}$.

7.3   Basic Decoupling Inequalities for Tangent Sequences

The following results constitute the backbone of the theory of decoupling for tangent sequences. We divide them into two classes. The first class concerns comparisons between two tangent sequences $\{d_i\}$ and $\{e_i\}$ and the second concerns inequalities between these two tangent sequences when one of them is decoupled. The first general result for tangent sequences involves the probabilities of unions, analogous to Lemma 7.1.3; see Kwapien and Woyczynski (1992) and Hitczenko (1988).

LEMMA 7.3.1
Let $\{D_i\}$ and $\{E_i\}$ be two sequences of events adapted to $\{\mathcal{F}_i\}$. Assume that these two sequences are $\mathcal{F}_i$-tangent (their indicator variables are tangent). Then for all $n \geq 1$,

$$P(\cup_{i=1}^n D_i) \leq 2P(\cup_{i=1}^n E_i). \tag{7.3.1}$$

Putting $D_i = 1(d_i > t)$ and $E_i = 1(e_i > t)$ in (7.3.1) yields

$$P(\max_{1 \leq i \leq n} d_i > t) \leq 2P(\max_{1 \leq i \leq n} e_i > t). \tag{7.3.2}$$

We remark that the above inequalities are two-sided as we can replace the role of the $D's$ and $E's$.

If the random variables are assumed to be either nonnegative or conditionally symmetric, Kwapien and Woyczynski (1992) give a detailed account of inequalities (7.3.3) and (7.3.6), while (7.3.5) and (7.3.8) (for variables in $\mathbf{R}^1$) are from Hitczenko (1988) and (7.3.4) and (7.3.7) come from de la Peña (1993); see also de la Peña and Giné (1999). The proof of (7.3.8) for Hilbert-space valued random variables (an easy extension) can be found in Kwapien and Woyczynski (1992) and de la Peña (1993). Two sequences of random variables $\{d_i\}$, $\{e_i\}$ adapted to $\mathcal{F}_i$ are said to be *conditionally symmetric* if $\mathcal{L}(d_i|\mathcal{F}_{i-1}) = \mathcal{L}(-d_i|\mathcal{F}_{i-1})$, i.e., if the conditional law of d_i is the same as that of $-d_i$. Define $\mathcal{A}_\alpha$ as in (7.1.3).

THEOREM 7.3.2
Let $\{d_i\}$, $\{e_i\}$ be two sequences of $\mathcal{F}_i$-adapted nonnegative random variables. Assume that the sequences are $\mathcal{F}_i$-tangent. Then for all $x, y > 0$,

$$P\left(\sum_{i=1}^n d_i > x\right) \leq \frac{y}{x} + 2P\left(\sum_{i=1}^n e_i > y\right), \tag{7.3.3}$$

$$P(\sum_{i=1}^j d_i > x) < 2\left\{\frac{E[(\sum_{i=1}^n e_i) \wedge y]}{x} + 2P(\sum_{i=1}^j e_i > y)\right\}. \tag{7.3.4}$$

Moreover, for every $\alpha > 0$, there exists C_α such that for all $\Phi \in \mathcal{A}_\alpha$,

$$E\Phi\left(\sum_{i=1}^n d_i\right) \leq C_\alpha E\Phi\left(\sum_{i=1}^n e_i\right). \tag{7.3.5}$$

THEOREM 7.3.3

Let $\{d_i\}$, $\{e_i\}$ be two sequences of $\mathcal{F}_i$-adapted Hilbert-space valued random variables. Assume that the sequences are $\mathcal{F}_i$-tangent and conditionally symmetric. Then for all $x, y > 0$

$$P\left\{\max_{j \leq n} \left\|\sum_{i=1}^{j} d_i\right\| > x\right\} \leq 3\left\{\frac{y}{x} + P\left(\max_{j \leq n} \left\|\sum_{i=1}^{j} e_i\right\| > y\right)\right\}, \quad (7.3.6)$$

$$P\left\{\max_{j \leq n} \left\|\sum_{i=1}^{j} d_i\right\|^2 > x\right\}$$

$$\leq 4\left\{6\frac{E(\max_{j \leq n} \left\|\sum_{i=1}^{j} e_i\right\|^2 \wedge y)}{x} + P(2\max_{j \leq n} \left\|\sum_{i=1}^{j} e_i\right\|^2 > y)\right\}.$$

$$(7.3.7)$$

Moreover, for every $\alpha > 0$, there exists C_α such that for all $\Phi \in \mathcal{A}_\alpha$,

$$E\Phi\left(\max_{j \leq n} \left\|\sum_{i=1}^{j} d_i\right\|\right) \leq C_\alpha E\Phi\left(\max_{j \leq n} \left\|\sum_{i=1}^{j} e_i\right\|\right). \quad (7.3.8)$$

Inequality (7.3.8) was extended to the case of UMD spaces by McConnell (1989). Assuming that one of the two tangent sequences satisfies the CI condition, de la Peña (1994) proved part (i) (with $g = 1$) while Hitczenko (1994) proved part (ii) of the following theorem. The case of (i) for general g comes from de la Peña (1999). See Section 6.2 of de la Peña and Giné (1999) for a detailed account of Theorem 7.3.4(i) and its corollaries.

THEOREM 7.3.4

Let $\{d_i\}$, $\{e_i\}$ be two sequences of $\mathcal{F}_i$-adapted random variables. Assume that the sequences are $\mathcal{F}_i$-tangent. Furthermore, assume that $\{e_i\}$ is conditionally independent given $\mathcal{G}$.

(i)   For all $\mathcal{G}$-measurable random variables $g \geq 0$,

$$Eg \exp\left\{\lambda \sum_{i=1}^{n} d_i\right\} \leq E^{1/2}\left\{g^2 \exp(2\lambda \sum_{i=1}^{n} e_i)\right\}. \quad (7.3.9)$$

(ii)   There exists a universal constant C such that for all $p \geq 1$,

$$E|\sum_{i=1}^{n} d_i|^p \leq C^p E|\sum_{i=1}^{n} e_i|^p.$$

COROLLARY 7.3.5
Let $\{d_i\}$ be a mean-zero martingale difference sequence. Then there exist a σ-field $\mathcal{G}$ and a $\mathcal{G}$-conditionally independent sequence $\{e_i\}$, tangent to $\{d_i\}$, such that for all λ,

$$Eg \exp\{\lambda \Sigma_{i=1}^{n} d_i\} \leq E^{1/2}\{g^2 \exp(4\lambda \Sigma_{i=1}^{n} e_i r_i)\}, \qquad (7.3.10)$$

where $\{r_i\}$ is a sequence of i.i.d. random variables independent of all the variables involved and with $P(r_i = 1) = P(r_i = -1) = \frac{1}{2}$.

COROLLARY 7.3.6
Let $\{D_i\}$, $\{E_i\}$ be two sequences of $\mathcal{F}_i$-adapted events. Assume that the sequences are $\mathcal{F}_i$-tangent (that is, their indicator variables are tangent). Furthermore, assume that $\{E_i\}$ is conditionally independent given $\mathcal{G}$. Then for all $\mathcal{G}$-measurable events G,

$$P(\cap_{i=1}^{n} D_i \cap G) \leq P(\cap_{i=1}^{n} E_i \cap G | \cap_{i=1}^{n} D_i \cap G). \qquad (7.3.11)$$

Taking $G = \Omega$ in (7.3.11) yields

$$P(\cap_{i=1}^{n} D_i) \leq P(\cap_{i=1}^{n} E_i | \cap_{i=1}^{n} D_i), \qquad (7.3.12)$$

and hence

$$P(\cap_{i=1}^{n} D_i) \leq P^{1/2}(\cap_{i=1}^{n} E_i). \qquad (7.3.13)$$

Setting $D_i = \{d_i < t\}$ and $E_i = \{e_i \leq t\}$, or alternatively $D_i = (d_i \geq t)$ and $E_i = (e_i \geq t)$, gives

$$P(\max_{1 \leq i \leq n} d_i \leq t) \leq P^{1/2}(\max_{1 \leq i \leq n} e_i \leq t), \qquad (7.3.14)$$

$$P(\min_{1 \leq i \leq n} d_i \geq t) \leq P^{1/2}(\min_{1 \leq i \leq n} e_i \geq t). \qquad (7.3.15)$$

From Theorem 7.3.4(ii) it follows that for all $p \geq 1$,

$$E \exp \left\{ |\sum_{i=1}^{n} d_i|^p \right\} \leq E \exp \left\{ C^p |\sum_{i=1}^{n} e_i|^p \right\}, \qquad (7.3.16)$$

in which C is a universal constant. Hitczenko and Montgomery-Smith (1996) have extended Theorem 7.3.4(ii) to functions $\Phi \in \mathcal{A}_\alpha$. Specifically, under the assumptions of the theorem, for all $\Phi \in \mathcal{A}_\alpha$,

$$E\Phi \left(\left| \sum_{i=1}^{n} d_i \right| \right) \leq C^{1+\alpha} E\Phi \left(\left| \sum_{i=1}^{n} e_i \right| \right). \qquad (7.3.17)$$

A generalization of (7.3.9) by de la Peña (1999) yields that for $p^{-1} + q^{-1} = 1 \ (p, q \geq 1)$,

$$E \exp \left(\lambda \sum_{i=1}^{n} d_i \right) \leq E^{1/p} \left\{ E^{p/q} \left[\exp \left(q\lambda \sum_{i=1}^{n} e_i \right) \mid \mathcal{G} \right] \right\}. \qquad (7.3.18)$$

In particular, consider the stochastic regression model $z_n = \beta^T u_n + \epsilon_n$, where $\epsilon_1, \epsilon_2 \ldots$ are independent random variables with zero means, representing unobservable disturbances, and the regressors u_n are $\mathcal{F}_{n-1}$-measurable random vectors. The sum $S_n = \sum_1^n u_i \epsilon_i$ plays a basic role in the analysis of the least squares estimate

$$\hat{\beta} = \left(\sum_1^n u_i u_i^T \right)^{-1} \sum_{i=1}^{n} u_i z_i = \beta + \left(\sum_1^n u_i u_i^T \right)^{-1} \sum_1^n u_i \epsilon_i.$$

Let $\psi_i(t) = \log(Ee^{t\epsilon_i})$ be the cumulant generating function of ϵ_i. Then (3.18) yields

$$E \exp \left(\lambda^T \sum_{i=1}^{n} u_i \epsilon_i \right) \leq E^{1/p} \left\{ \exp \left(pq^{-1} \sum_{i=1}^{n} \psi_i(q\lambda^T u_i) \right) \right\},$$

$$(7.3.19)$$

which can be used to derive tail probability bounds $\sum_1^n u_i \epsilon_i$.

7.4 Applications to Martingale Inequalities and Exponential Tail Probability Bounds

Let $\{d_i, \mathcal{F}_i, i \geq 1\}$ be a mean-zero martingale difference sequence. Let $\{e_i\}$ be CI (conditionally independent) given $\mathcal{G}$ and such that $\{d_i\}$ and

$\{e_i\}$ are tangent sequences. By Theorem 7.3.4(ii), for all $p \geq 1$,

$$\left\| \sum_{i=1}^{n} d_i \right\|_p \leq C \left\| \sum_{1}^{n} e_i \right\|_p$$

for some universal constant C. Since the e_i are independent zero-mean random variables given $\mathcal{G}$, it then follows from Rosenthal's inequality that

$$\left\{ E \left(\left| \sum_{i=1}^{n} e_i \right|^p \mid \mathcal{G} \right) \right\}^{1/p}$$

$$\leq C_p \left\{ \left[\sum_{i=1}^{n} E(e_i^2 | \mathcal{G}) \right]^{1/2} + [E(\sup_{k \leq n} |e_i|^p | \mathcal{G})]^{1/p} \right\}. \qquad (7.4.1)$$

Hence

$$\left\| \sum_{1}^{n} e_i \right\|_p = \left\| \left(E \left[\left| \sum_{1}^{n} e_i \right|^p \mid \mathcal{G} \right] \right)^{1/p} \right\|_p \leq B_p(\|s_n\|_p + \|e_n^*\|_p),$$

where $s_n^2 = \sum_{1}^{n} E(d_i^2 | \mathcal{F}_{i-1}) = \sum_{1}^{n} E(e_i^2 | \mathcal{F}_{i-1}) = \sum_{1}^{n} E(e_i^2 | \mathcal{G})$ and $e_n^* = \sup_{k \leq n} |e_k|$. Moreover, $\|e_n^*\|_p \leq 2^{1/p} \|d_n^*\|_p$ by (7.3.2). Hence, decoupling inequalities for tangent sequences and Rosenthal's inequality for sums of independent zero-mean random variables yield the martingale inequality

$$\left\| \sum_{i=1}^{n} d_i \right\|_p \leq B_p \{ \|s_n\|_p + \|d_n^*\|_p \}, \qquad (7.4.2)$$

which was derived by Burkholder (1973) using distribution function inequalities. Hitczenko (1990) used this approach to derive the "best possible" constant B_p, depending only on p, with the slowest growth rate $p/\log p$ as $p \to \infty$. In the case $p \geq 2$, Hitczenko (1994) also derived by a similar decoupling argument the following variant of (7.4.2):

$$\left\| \sum_{i=1}^{n} d_i \right\|_p \leq B \{ \sqrt{p} \, \|s_n\|_p + p \|d_n^*\|_p \} \qquad (7.4.3)$$

for some universal constant B.

Decoupling inequalities for moment generating functions and variants thereof (such as Theorem 7.3.4 and Corollary 7.3.5) are particularly useful for developing extensions of typical exponential inequalities for sums of zero-mean independent random variables to the case of the ratio of a martingale over its conditional variance. Let $\{d_i, \mathcal{F}_i, i \geq 1\}$ be a martingale difference sequence. Let $M_n = \sum_{i=1}^{n} d_i$ and $V_n^2 = \sum_{i=1}^{n} E(d_i^2 | \mathcal{F}_{i-1})$. Consider the problem of bounding the tail probability of $P(M_n / V_n^2 \geq x)$. Following the standard approach to such problems, one obtains

$$P\left(\sum_{i=1}^{n} d_i \geq xV_n^2\right) = P\left\{\exp(\lambda \sum_{i=1}^{n} d_i) \geq \exp(\lambda x V_n^2)\right\}$$
$$\leq \inf_{\lambda > 0} E\left\{\exp(-\lambda x V_n^2)\exp\left(\lambda \sum_{i=1}^{n} d_i\right)\right\}.$$

Corollary 7.3.5 can be used here (with $g = \exp(\lambda x V_n^2)$) to bound the last expression by

$$\inf_{\lambda > 0} E^{1/2}\left\{\exp(-2\lambda x V_n^2)\exp\left(2\lambda \sum_{i=1}^{n} e_i\right)\right\},$$

where $\{e_i\}$ is a decoupled version of $\{d_i\}$ and hence conditionally independent given $\mathcal{G} = \sigma(\{d_i\})$. Conditional on $\mathcal{G}$ one can proceed as if the variables were independent, and de la Peña (1999) used this method to derive new exponential tail probability bounds for M_n / V_n^2, including the following result extending sharply Prokhorov's inequality.

THEOREM 7.4.1
Let $\{d_i, \mathcal{F}_i\}$ be a martingale difference sequence. Let $M_n = \sum_{i=1}^{n} d_i$, $V_n^2 = \sum_{i=1}^{n} E(d_i^2 | \mathcal{F}_{i-1})$ or $V_n^2 = \|\sum_{i=1}^{n} E(d_i^2 | \mathcal{F}_{i-1})\|_\infty$. Assume that for some $c > 0$, $|d_j| \leq c$ a.s. for all j. Then for all $x \geq 0$

$$P\left(\frac{M_n}{V_n^2} > x, \frac{1}{V_n^2} \leq y \text{ for some } n\right)$$
$$\leq \exp\left(-\frac{cx}{2y}\text{arc sinh}\frac{cx}{2}\right). \tag{7.4.4}$$

7.5 Decoupling of Multilinear Forms, U-Statistics and U-Processes

In this section we let $X_1, X_2, \ldots$ be independent random variables taking values in some measurable space S. Motivated in part by the desire to extend the theory of multiple stochastic integration, the work of

McConnell and Taqqu (1986) provided decoupling inequalities for multilinear forms of independent symmetric variables. After their paper, a large amount of work has been done, providing extensions of this result in many directions, including Kwapien (1987), Bourgain and Tzafriri (1987), de Acosta (1987), Krakowiak and Szulga (1988), Kwapien and Woyczynski (1992), de la Peña (1992), de la Peña, Montgomery-Smith, and Szulga (1994), de la Peña and Montgomery-Smith (1994, 1995), and Szulga (1998). In particular, de la Peña (1992) provided an extension to a general class of statistics that include both U-statistics and multilinear forms, while de la Peña and Montgomery-Smith (1994, 1995) established the following tail probability comparison for such statistics.

THEOREM 7.5.1
Let $\{X_i\}$ be a sequence of independent random variables with values in a measurable space $(S, \mathcal{S})$. Let $\{X_i^{(1)}, X_i^{(2)}, ..., X_i^{(k)}\}$ be k independent copies of $\{X_i\}$. Let $f_{i_1,...,i_k} : S^k \to B$, where $(B, ||\cdot||)$ is a Banach space. Then for all $t \geq 0$,

$$P\left\{ \left\| \sum_{1 \leq i_1 \neq i_2 \neq .... \neq i_k \leq n} f_{i_1...i_k}(X_{i_1}, ..., X_{i_k}) \right\| \geq t \right\}$$

$$\leq C_k P\left\{ C_k \left\| \sum_{1 \leq i_1 \neq i_2 \neq .... \neq i_k \leq n} f_{i_1...i_k}(X_{i_1}^{(1)}, ..., X_{i_k}^{(k)}) \right\| \geq t \right\},$$

where $C_k > 0$ is a constant depending only on k and $1 \neq i_1 \neq i_2 \neq \cdots \neq i_k \leq n$ means $i_r \neq i_s$ for $r \neq s$.

The reverse inequality in Theorem 7.5.1 holds whenever

$$f_{i_1...i_k}(x_{i_1}, ..., x_{i_k}) = f_{i_1...i_k}(x_{i_{\sigma(1)}}, ..., x_{i_{\sigma(k)}})$$

for all permutations σ of $(1, 2, ..., k)$. To better understand this result, focus on the case $k = 2$. One can easily see that $f_{ij}(X_i, X_j) = a_{ij} X_i X_j$ reduces to quadratic forms while $f_{ij}(X_i, X_j) = f(X_i, X_j)/\binom{n}{2}$ corresponds to U-statistics. An important U-statistic is the sample variance

$$\frac{\sum_{i=1}^n (X_i - \bar{X})^2}{(n-1)} = \sum_{j=1}^n \sum_{i=1; i \neq j}^n \frac{(X_i - X_j)^2}{\binom{n}{2}},$$

whose tail probability is related by Theorem 7.5.1 to that of $\sum_{j=1}^n G_j(X_j^{(2)})$,

which is a sum of independent random variables when we condition on $\{X_i^{(1)}\}$, where $G_j(y) = \sum_{i=1; i \neq j}^n (X_i^{(1)} - y)^2 / \binom{n}{2}$.

There are potential applications of results like Theorem 7.5.1 to the study of random graphs. The following example, taken from Janson and Nowicki (1991) who study the properties of random graphs by introducing a generalized form of U-statistics, will help illustrate such applications,

Example 2

Let $\{X_i\}$ be a sequence of i.i.d. Bernoulli random variables with $P(X_i = 1) = p_1$, $P(X_i = 0) = 1 - p_1$. Let Y_{ij} be i.i.d. with $P(Y_{ij} = 1) = p_2$, $P(Y_{ij} = 0) = 1 - p_2$. In the context of random graphs, $Y_{ij} = 1$ indicates that there is an edge between vertices i and j, while $X_i = 1$ or 0 according as vertex i is colored red or blue. Letting $f(x_1, x_2, y) = y1(x_1 \neq x_2)$,

$$S_n(f) = \sum_{1 \leq i < j \leq n} f(X_i, X_j, Y_{ij})$$

is the number of edges with differently colored vertices. It is easy to see how to extend Theorem 7.5.1 to this type of statistics because in view of the assumed independence between $\{X_i\}$ and $\{Y_{ij}\}$, conditioning on $\{Y_{ij}\}$ yields a random variable to which Theorem 7.5.1can be applied. An application of this theorem therefore reduces the problem to one involving sums of conditionally independent random variables.   ☐

Decoupling inequalities like the one presented in Theorem 7.5.1 are also important in the study of multiple stochastic integrals of the form

$$\int \cdots \int_{0 \leq t_1 < \ldots < t_k \leq t} g(t_1, \ldots, t_k) dY_{t_1} \ldots dY_{t_k},$$

where Y_t is a process with independent increments. The reason for this is that such stochastic integrals are limits of the multilinear forms of the type

$$\sum_{1 \leq i_1 < \ldots < i_k \leq n} w_{i_1, \ldots, i_k} X_{i_1} \ldots X_{i_k},$$

where the X_i are independent random variables; see Kwapien and Woyczynski (1992).

An important application of Theorem 7.5.1 and its variants is to derive symmetrization inequalities by using the fact that the decoupled variables are conditionally independent. In particular, for U-statistics, Nolan and Pollard (1987) made use of such symmetrization inequalities to prove weak convergence of a functional version of U-statistics known as a U-process. Arcones and Giné (1993) extended this work to more general U-processes indexed by families of functions. A detailed account of these and subsequent developments and the role of symmetrization inequalities in the theory of U-processes can be found in Chapter 5 of de la Peña and Giné (1999).

In Theorem 7.5.4 below we present a key symmetrization inequality that can be derived by using decoupling and the following classical symmetrization inequality for sums of independent zero-mean random variables.

LEMMA 7.5.2

Let $\{X_i\}$ be a sequence of independent zero-mean random variables with values in a separable Banach space $(B, \|\cdot\|)$. Let $\{\varepsilon_i\}$ be a sequence of independent Bernoulli random variables, independent of $\{X_i\}$ and with $P(\varepsilon_i = 1) = P(\varepsilon_i = -1) = \frac{1}{2}$. Then for all convex increasing functions Φ,

$$E\Phi(\|\frac{1}{2}\sum_{i=1}^{n} X_i\varepsilon_i\|) \leq E\Phi(\|\sum_{i=1}^{n} X_i\|) \leq E\Phi(\|2\sum_{i=1}^{n} X_i\varepsilon_i\|).$$

DEFINITION 7.5.3

Let $\{X_i\}$ be a sequence of i.i.d. variables with values in a measurable space $(S, \mathcal{S})$. For $k \geq 1$ let $f : S^k \to \mathbf{R}$ be an integrable function such that $f(x_{\sigma(1)}, \ldots, x_{\sigma(k)}) = f(x_1, \ldots, x_k)$ for all permutations σ of $(1, \ldots, k)$. Then f is said to be P-degenerate of order $r - 1$ $(1 < r \leq k)$, with respect to the probability law P of X_1, if

$$E\{f(X_1, .., X_k)|X_1, ..., X_{r-1}\} = Ef(X_1, ..., X_k)$$

and $E(f(X_1, .., X_k)|X_1, ..., X_r)$ is not a constant function.

THEOREM 7.5.4

Let $1 < r \leq k$ and $p \geq 1$. Let $\{X_i\}$ be a sequence of i.i.d. random variables with values in a measurable space $(S, \mathcal{S})$. Let $f : S^k \to \mathbf{R}$ be a P-degenerate function of order $r - 1$ with $Ef(X_1, \ldots, X_k) = 0$ and $E|f(X_1, \ldots, X_k)|^p < \infty$. Then

$$E\left|\sum_{1 \leq i_1 \neq i_2 \neq \ldots \neq i_k \leq n}' f(X_{i_1}, \ldots, X_{i_k})\right|^p$$

$$\approx_p E\left|\sum_{1 \leq i_1 \neq i_2 \neq \ldots \neq i_k \leq n} f(X_{i_1}, \ldots, X_{i_k})\varepsilon_{i_1} \ldots \varepsilon_{i_r}\right|^p,$$

where $a \approx_p b$ denote two-sided inequalities between a and b possibly multiplied by constants that depend only on p.

In fact, due to the Banach-space nature of Theorem 7.5.1, the above result is still valid if one replaces f by a class of functions and the absolute value by the supremum this class. Moreover, Theorem 7.5.4 can also be stated for convex increasing functions instead of pth powers. Obviously, by conditioning on the X_i's, Theorem 7.5.4 permits the reduction of problems involving functions of the random variables X_i's to problems involving multilinear forms of the form $\sum_{1 \le i_1 \ne i_2 \ne \dots \ne i_k \le n} a_{i_1 \dots i_k} \varepsilon_{i_1} \dots \varepsilon_{i_r}$ for constants $a_{i_1 \dots i_k}$ and hence permits the use of Khintchine-type inequalities for such random variables. Another application of this symmetrization and decoupling inequalities is an extension of Bernstein's inequality for sums of independent random variables to the case of U-statistics, see Section 4.1 of de la Peña and Giné (1999).

7.6   Total Decoupling of Stopping Times

We begin by stating a decoupling reformulation of Wald's equation in sequential analysis. Let $X_1, X_2, \dots$ be i.i.d. random variables with $EX_i = \mu$. Then for any stopping time T with $ET < \infty$,

$$E\left(\sum_{i=1}^{T} X_i\right) = \mu ET = E\left(\sum_{i=1}^{T} \widetilde{X}_i\right), \qquad (7.6.1)$$

where $\{\widetilde{X}_i\}$ is an independent copy of $\{X_i\}$ and hence independent of T. Note that conditional on T, $\sum_{i=1}^{T} \widetilde{X}_i$ is a sum of independent random variables. Let $S_n = \sum_{i=1}^{n} X_i$, $\widetilde{S}_n = \sum_{i=1}^{n} \widetilde{X}_1$. The following result of Klass (1988, 1990) provides basic decoupling inequalities for randomly stopped sums.

THEOREM 7.6.1
Let $\{X_i\}$ be a sequence of independent random variables taking values in a Banach space $(B, || \cdot ||)$. Let T be a stopping time adapted to this sequence. Let $\{\widetilde{X}_i\}$ be an independent copy of $\{X_i\}$ and hence independent of T. Then, for all $\Phi \in \mathcal{A}_\alpha$ (see Definition 7.1.11), there exist constants $0 < b_\alpha, B_\alpha < \infty$ depending only on α for which

$$b_\alpha E \max_{1 \le n \le T} \Phi(||\widetilde{S}_n||) \le E \max_{1 \le n \le T} \Phi(||S_n||) \le B_\alpha E \max_{1 \le n \le T} \Phi(||\widetilde{S}_n||).$$

$$(7.6.2)$$

Making use of the decoupling inequality (7.6.2) with $\Phi(x) = x$ and the fact that $E(\max_{1 \leq k \leq n} \|S_k - ES_k\|) \leq 4E\|S_n - ES_n\|$, Klass (1988) established the following generalization of Wald's equation (7.6.1).

THEOREM 7.6.2
Let $\{X_i\}$ be a sequence of independent random variables with values in a Banach space $(B, \|\cdot\|)$ such that $EX_i = \mu_i$. Let T be a stopping time. Then $E(\sum_{i=1}^{T} X_i) = E(\sum_{i=1}^{T} \mu_i)$ whenever $Ea_T < \infty$, where $a_n = E\|\sum_{i=1}^{n} X_i\|$.

In particular, if the X_i are i.i.d. real-valued symmetric stable random variables with index $1 < q \leq 2$, then Theorem 7.6.2 yields that $E(\sum_{i=1}^{T} X_i) = 0$ whenever $ET^{1/q} < \infty$, since in this case

$$a_n = E|\sum_{i=1}^{n} X_i| = n^{1/q} E|X_1|.$$

By making use of tangent sequence decoupling, we can generalize this result far beyond the case of i.i.d. symmetric stable X_i. This is the content of the following theorem.

THEOREM 7.6.3
Let $\{X_n, \mathcal{F}_n, n \geq 1\}$ be a martingale difference sequence and let $S_n = \sum_{i=1}^{n} X_i$. Let $1 \leq p < q < 2$. Suppose there exist nonrandom constants κ and τ such that

$$P(|X_n| > t|\mathcal{F}_{n-1}) \leq \kappa t^{-q} \ a.s. \ for \ all \ n \geq 1 \ and \ t \geq \tau. \quad (7.6.3)$$

Then there exists a constant $C_{p,q,\kappa,\tau}$ depending only on p, q, κ, and τ such that for any stopping time T,

$$E(\max_{1 \leq n \leq T} |S_n|^p) \leq C_{p,q,\kappa,\tau} ET^{p/q}.$$

PROOF Apply (7.3.5) with $\Phi(x) = x^{p/2}$ to $d_i = X_i^2$ and e_i such that $\{d_i\}$ and $\{e_i\}$ are tangent sequences and $\{e_i\}$ is conditionally independent given $\mathcal{G}$ with respect to which T is measurable; see Proposition 7.2.5 for the existence of such $\mathcal{G}$ and e_i. Since $\{e_i 1\{T \geq i\}\}$ is tangent to $(d_i 1\{T \geq i\})$, it follows from (7.3.5) that

$$E\left(\sum_{i=1}^{T} X_i^2\right)^{p/2} \leq C_{p/2} E\left(\sum_{i=1}^{T} e_i\right)^{p/2}$$

$$= C_{p/2} E\left\{ E\left[\left(\sum_{i=1}^{T} e_i \right)^{p/2} | \mathcal{G} \right] \right\}. \qquad (7.6.4)$$

Recall that T is measurable with respect to $\mathcal{G}$ and that $\{e_n\}$ is conditionally independent given $\mathcal{G}$ with $\mathcal{L}(e_n|\mathcal{G}) = \mathcal{L}(X_n^2|\mathcal{F}_{n-1})$ a.s. Noting that $0 < \alpha := q/2 < 1$, let G_α be the distribution function of the positive stable random variable with Laplace transform $\exp(-\lambda^\alpha)$, $\lambda > 0$.

Take $b > 0$ such that $b^\alpha/\Gamma(1-\alpha) > \kappa$. Since $\lim_{x\to\infty} x^\alpha(1-G_\alpha(x)) = 1/\Gamma(1-\alpha)$ [cf. Feller (1971, p. 448)], (7.6.3) implies that there exists $\eta \geq \tau$ such that with probability 1,

$$P(e_n > x|\mathcal{G}) \leq \kappa x^{-q/2} < 1 - G_\alpha(x/b) \text{ for all } n \geq 1 \text{ and rational } x \geq \eta.$$
$$(7.6.5)$$

Let F_n be the conditional distribution of e_n given $\mathcal{G}$. In view of the right continuity of distribution functions, it follows from (7.6.5) that with probability 1, $1-F_n(x) \leq 1-G_\alpha(x/b)$ for all $n \geq 1$ and $x \geq \eta$. Therefore, letting $F^{-1}(u) = \sup\{x : F(x) \leq u\}$ for $0 \leq u < 1$ and $\theta = bG_\alpha^{-1}(\eta)$, we have $F_n^{-1}(u) \leq bG_\alpha^{-1}(u) + \theta$ for all $n \geq 1$ and $0 \leq u < 1$. Replacing u in the preceding inequality by i.i.d. uniform random variables $U_1, U_2, \ldots$ on $[0, 1)$ and noting that conditional on $\mathcal{G}$ (with respect to which T is measurable) the e_i are independent random variables with distribution functions F_i, we obtain that

$$P\left(\sum_{i=1}^{T} e_i > x|\mathcal{G} \right) \leq P\left\{ \sum_{i=1}^{T}(bY_i + \theta) > x \right\} \text{ a.s.,} \qquad (7.6.6)$$

where $Y_1, Y_2, \ldots$ are i.i.d. random variables with common distribution function G_α and independent of T. Since $\sum_{i=1}^{n} Y_i$ has the same distribution as $n^{1/\alpha}Y_1$, (7.6.6) implies that

$$E\left\{ E\left[\left(\sum_{i=1}^{T} e_i \right)^{p/2} | \mathcal{G} \right] \right\} \leq E(bY_1 T^{1/\alpha} + \theta T)^{p/2}. \qquad (7.6.7)$$

Noting that $T \leq T^{1/\alpha}$ and that $EY_1^{p/2} < \infty$ since $\alpha = q/2 > p/2$, the desired conclusion follows from (7.6.4), (7.6.7) and Burkholder's inequality

$$E(\max_{1\leq n\leq T} |S_n|^p) \leq B_p E\{(\sum_{i=1}^{T} X_i^2)^{p/2}\}. \qquad \blacksquare$$

One direction of extending total decoupling of a stopping time in Theorem 7.6.1 deals with randomly stopped multilinear forms, and more generally, de-normalized U-statistics of i.i.d. random variables $X_1, X_2, \ldots$ taking values in a measurable space $(S, \mathcal{S})$ such that $E|f(X_1, \ldots, X_k)|^p < \infty$ for some $f : S^k \to \mathbf{R}$. This has been undertaken by Chow, de la Peña and Teicher (1993), de la Peña and Lai (1997a,b) and Borovskikh and Weber (1998). Let $\{X_i^{(j)}\}, j = 1, \ldots, k$, be k independent copies of $\{X_i\}$. Then by combining the decoupling inequality (7.3.8) for tangent sequences with Burkholder's inequality for martingales, and using the bound $E|\sum_{i=1}^n Z_i|^p \le A_p n^{p/n} E|Z_1|^p$ for i.i.d. zero-mean random variables $Z_1, Z_2, \ldots$ together with Hölder's inequality to bound

$$
E\left\{ T^{p/2} \left| \sum_{i_{k-1}=k-1}^{T} \sum_{i_{k-2}=k-2}^{i_{k-1}-1} \cdots \sum_{i_1=1}^{i_2-1} f(X_{i_1}, \ldots, X_{i_{k-1}}, \tilde{X}_1^{(k)}) \right| \right\},
$$

de la Peña and Lai (1997b) proved the following decoupling inequality for randomly stopped de-normalized U-statistics in the case $p \ge 2$:

$$
E \max_{k \le n \le T} \left| \sum_{1 \le i_1 < \ldots < i_k \le n} f(X_{i_1}, \ldots, X_{i_k}) \right|^p
$$

$$
\le C_{k,p} (E T^{kp/2})(E|f(\tilde{X}_1^{(1)}, \ldots, \tilde{X}_1^{(k)})|^{kp})^{1/k}. \tag{7.6.8}
$$

Moreover, for the case $1 \le p < 2$, assuming that $f(x_{\sigma(1)}, \ldots, x_{\sigma(k)}) = f(x_1, \ldots, x_k)$ for any permutation of $(1, \ldots, k)$, de la Peña and Lai (1997b) also proved the decoupling inequality

$$
E \max_{k \le n \le T} \left| \sum_{1 \le i_1 < \ldots < i_k \le n} f(X_{i_1}, \ldots, X_{i_k}) \right|^p
$$

$$
\le C_{k,p} \sum_{h=0}^{k-1} \sum_{t=1}^{k-h} E \left| \sum_{i_{h+1}=1}^{T} \cdots \sum_{i_{h+t}=1}^{T} f_{T,k,h}(\tilde{X}_{i_{h+1}}^{(h+1)} \ldots, \right.
$$

$$
\left. \tilde{X}_{i_{h+1}}^{(h+t)}, \ldots, \tilde{X}_{i_{h+1}}^{(k)}) \right|^p, \tag{7.6.9}
$$

where $f_{T,k,0} = f$ and for $h \ge 1, f_{T,k,h} : S^{k-h} \to \mathbf{R}$ is given by

$$
f_{T,k,h}(y_{h+1}, \ldots, y_k) = \sum_{1 \le i_1 < \ldots < i_h \le T} f(X_{i_1}, \ldots, X_{i_h}, y_{h+1}, \ldots, y_k).
$$

Another direction of extending inequality (7.6.2) for total decoupling of a stopping time deals with continuous-time processes with independent increments. In particular, the following extension of Theorems 7.6.1

and 7.6.2 is given by de la Peña and Eisenbaum (1997) [see also de la Peña and Giné (1999)].

THEOREM 7.6.4
Let $\{N_t\}$ be a continuous-time process, taking values in a Banach space $(B, \|\cdot\|)$, with independent increments and right-continuous sample paths that have left-hand limits. Let T be a stopping time.

(i) *Let $\{\tilde{N}_t\}$ be an independent copy of $\{N_t\}$ and hence independent of T. Then, for all $\Phi \in \mathcal{A}_\alpha$, there exist constants $0 < b_\alpha, B_\alpha < \infty$ depending only on α for which*

$$b_\alpha E \sup_{0 \le s \le T} \Phi(\|\tilde{N}_s\|) \le E \sup_{0 \le s \le T} \Phi(\|N_s\|) \le B_\alpha E \max_{0 \le s \le T} \Phi(\|\tilde{N}_s\|).$$

(ii) *Let $m_t = EN_t$ and $b_t = E \sup_{s \le t} \|N_s - m_s\|$. Then $EN_T = Em_T$ whenever $Eb_T < \infty$.*

Concerning the decoupling of randomly stopped stochastic integrals, de la Peña and Eisenbaum (1994) introduced an approach based on the development of decoupling inequalities for randomly stopped local times. The fact that local times can be decoupled follows from the inequalities of Barlow and Yor (1981), leading to part (i) of the following theorem. Part (ii) is a further refinement that can be applied to prove the decoupling inequality in Theorem 7.6.6 for stochastic integrals; see de la Peña and Eisenbaum (1994) and Yang (1999) for a recent extension of Theorem 7.6.6.

THEOREM 7.6.5
Let $B_t, t \ge 0$ be a standard Brownian motion and let $\{L_t^x(B) : x \in \mathbf{R}, t \ge 0\}$ denote the family of local times of $\{B_t\}$. Let $\tilde{B}_t, t \ge 0$, be another standard Brownian motion independent of $B_t, t \ge 0$. Let T be a stopping time adapted to $\sigma(B_s, s \ge 0)$.

(i) *For all $p > 0$, there exist universal constants $0 < c_p, C_p < \infty$ depending only on p such that*

$$c_p E \sup_{x \in \mathbf{R}} [L_T^x(\tilde{B})]^p \le E \sup_{x \in \mathbf{R}} [L_T^x(B)]^p \le C_p E \sup_{x \in \mathbf{R}} [L_T^x(\tilde{B})]^p.$$

(ii) *For all $x \in \mathbf{R}$ and all $p \ge 1$, there exists a universal constant $C_p < \infty$ depending only on p such that*

$$E \sup_{y \in \mathbf{R}} [L_T^y(B)]^p \le C_p \{E[L_T^x(\tilde{B})]^p + |x|^p\}.$$

THEOREM 7.6.6

Let f be a Borel function such that $\int_0^\infty f^2(t)dt < \infty$. Let $B_t, t \geq 0$, be standard Brownian motion, and let $M_t = \int_0^t f(B_s)dB_s$. Let $\{\tilde{M}_t, t \geq 0\}$ be an independent copy of $\{M_t, t \geq 0\}$. Then for all $p \geq 1$, there exists a constant $C_p(f)$ such that for every stopping time T adapted to $\sigma(B_s, s \geq 0)$,

$$E \sup_{s \leq T} |M_s|^p \leq C_p(f)\{E \sup_{s \leq T} |\tilde{M}_s|^p + 1\}. \tag{7.6.10}$$

7.7 Principle of Conditioning in Weak Convergence

Decoupled tangent sequences are also useful for proving convergence in distribution of sums of dependent random variables. Jakubowski (1986) introduced a "principle of conditioning" for deriving such limit laws. This principle starts with a limit theorem for sums of independent random variables and extends it to sums of dependent random variables by replacing

(a) expectations of functions of summands by conditional expectations with respect to the past history,

(b) nonrandom number of summands by a random number, defined by a stopping time, of summands,

(c) convergence of nonrandom numbers by convergence in probability of the random variables.

Let $\{X_{nk}; 1 \leq k \leq \tau_n, n \geq 1\}$ be a double array of random variables adapted to a sequence of filtrations $\{\mathcal{F}_{nk}, k \geq 1\}$, of which τ_n is a stopping time. The corresponding independent random variables Y_{nk} in the principle of conditioning are such that (i) $Y_{nk}1\{\tau_n \geq k\}$ has the same conditional distribution given $\mathcal{F}_{n,k-1}$ as $X_{nk}1\{\tau_n \geq k\}$ and (ii) $Y_{n1}, Y_{n2}, \ldots$ are conditionally independent given the σ-field $\mathcal{G}_n = \sigma(\bigcup_{k=1}^\infty \mathcal{F}_{nk})$. The construction of such Y_{nk} has been discussed in Section 7.2. Jakubowski (1986) has proved the following result relating the limit law of $\sum_{k=1}^{\tau_n} X_{nk}$ to the conditional limit law of $\sum_{k=1}^{\tau_n} Y_{nk}$ given $\mathcal{G}_n$, justifying the preceding principle of conditioning.

THEOREM 7.7.1

Let $\mu_n(\tau_n)$ denote the conditional law of $\sum_{k=1}^{\tau_n} Y_{nk}$ given $\mathcal{G}_n$.

(i) Suppose $\mu_n(\tau_n)$ converges weakly in probability to some nonrandom

probability measure μ whose characteristic function is nowhere 0. Then $\sum_{k=1}^{\tau_n} X_{nk}$ converges weakly to μ as $n \to \infty$.

(ii) Suppose $\{\mu_n(\tau_n), n \geq 1\}$ is tight with probability 1. Then $\{\sum_{k=1}^{\tau_n} X_{nk}, n \geq 1\}$ is tight.

(iii) Suppose $\mu_n(\tau_n)$ converges weakly in probability to some nonrandom probability measure μ. Then $\{\sum_{k=1}^{\tau_n} X_{nk}, n \geq 1\}$ is tight. If furthermore the equation $\nu * \mu = \mu * \mu$ has a unique solution μ, then $\sum_{k=1}^{\tau_n} X_{nk}$ converges weakly to μ as $n \to \infty$.

(iv) In the case $X_{nk} = B_n^{-1} X_k$ and $\mathcal{F}_{nk} = \mathcal{F}_k$, where B_n are positive constants such that $\lim_{n \to \infty} B_n = \infty$, suppose $\mu_n(\tau_n)$ converges weakly in probability to a random measure μ such that $\hat{\mu}(t) \neq 0$ a.s. for t belonging to a dense subset of the real line, where $\hat{\mu}$ is the characteristic function of μ. Then $\sum_{k=1}^{\tau_n} X_{nk}$ converges weakly to $E\mu$ as $n \to \infty$.

Jakubowski (1986) also provides a functional version of Theorem 7.7.1 which is used to derive a functional central limit theorem for martingales. Theorem 7.7.1 can be used to derive nonnormal limiting distributions of martingales, as illustrated by the following.

Example 3

Let $\{X_{nk}, \mathcal{F}_{nk}, 1 \leq k \leq n\}$ be a martingale difference array (i.e.,

$$E(X_{nk}|\mathcal{F}_{n,k-1}) = 0).$$

Suppose $\max_{1 \leq k \leq n} E(X_{nk}^2 | \mathcal{F}_{n,k-1}) \xrightarrow{P} 0$ and there exists a bounded nondecreasing function H for which

$$\sum_{k=1}^{n} E\{X_{nk}^2 1(a < X_{nk} \leq b) | \mathcal{F}_{n,k-1}\} \xrightarrow{P} H(b) - H(a) \quad \text{as} \quad n \to \infty,$$

for all continuity points a, b of H. From Theorem 7.7.1 and the limiting infinitely divisible law of the sum $\sum_{k=1}^{n} Y_{nk}$ of independent random variables Y_{nk} (conditional on $\mathcal{G}_n$), it follows that $\sum_{k=1}^{n} X_{nk}$ converges weakly to an infinitely divisible limit law whose characteristic function $\phi(t)$ is given by

$$\log \phi(t) = \int_{-\infty}^{\infty} (e^{itx} - 1 - itx) x^{-2} dH(x).$$

This result was proved by Brown and Eagleson (1971) by a different argument which uses an elaborate analysis of conditional characteristic functions. ☐

Example 4

Let $\{X_k, \mathcal{F}_k, k \geq 1\}$ be a martingale difference sequence such that there exist nonrandom constants $B_n \to \infty$ for which $B_n^{-2} \sum_{k=1}^{n} E(X_k^2 | \mathcal{F}_{k-1})$ converges in probability to a random variable V with distribution function F. Assume also the conditional Lindeberg condition

$$B_n^{-2} \Sigma_{k=1}^{n} E\{X_k^2 1(|X_k| \geq \epsilon) | \mathcal{F}_{k-1}\} \xrightarrow{P} 0 \quad \text{for every} \quad \epsilon > 0.$$

Let $X_{nk} = B_n^{-1} X_k$ and apply Theorem 7.7.1(iv) in conjunction with Proposition 7.2.4. Conditional on $\mathcal{G}$, the Y_{nk} are independent zero-mean random variables satisfying the Lindeberg condition, so by the central limit theorem, $\sum_{k=1}^{n} Y_{nk}$ has a limiting normal distribution with mean 0 and variance V (which is $\mathcal{G}$-measurable). Hence by Theorem 7.7.1(iv), $\sum_{k=1}^{n} X_{nk}$ has a limiting distribution function of the form $\int_0^\infty \mathcal{N}(x/\sqrt{v}) dF(v)$, where $\mathcal{N}(x)$ denotes the standard normal distribution function. ▯

7.8 Conclusion

The basic idea behind decoupling is to reduce problems concerning dependent random variables to corresponding ones for independent random variables. It was introduced by Burkholder and McConnell to extend martingale inequalities to Banach-space valued martingales, for which difficulties in generalizing the concept of the square function have been circumvented by using a decoupled (conditionally independent) version that can be handled by symmetrization techniques; see Burkholder (1983), Zinn (1985), McConnell (1989). Since then it has found many different applications which have in turn inspired its extensive development. The relatively simple derivations of the new results in Theorem 7.6.3 and Example 4 illustrate the elegance and power of this idea, whose important applications to U-processes, multiple stochastic integration and randomly stopped sums or integrals are briefly reviewed in Sections 7.5 and 7.6. Section 7.2 describes the basic concept of decoupled tangent sequences, and the decoupling inequalities for tangent sequences in Sections 7.3 and 7.4 can be regarded as a "second generation" of martingale inequalities. This decoupling approach is particularly useful for the development of exponential inequalities for martingales, as illustrated in Theorems 7.3.4 and 7.4.1. Without appealing to conditioning arguments, Section 7.1 shows that it is also possible to obtain one-sided inequalities by using only the marginal distributions of the random variables to decouple.

Acknowledgement The research of the first author was partially supported by the National Science Foundation grant DMS-9626236 and DMS-99-72237, and that of the second author by the National Science Foundation grant DMS-9704324 and the National Security Agency grant MDA-904-98-1-0017.

References

1. Arcones, M. A. and Giné, E. (1993). Limit theorems for U-processes. *Annals of Probability* **21**, 1494–1542.

2. Barlow, M. T. and Yor, M. (1981). (Semi-)martingale inequalities and local time. *Zeitschrift Wahrscheinlichkeitstheorie und Verwandte Gebiete* **55**, 237–254.

3. Borovskikh, Y. V. and Weber, N. C. (1998). On Wald's equation for U-statistical sums. *Technical Report.* School of Mathematics and Statistics, University of Sydney.

4. Bourgain, J. and Tzafriri, L. (1987). Invertibility of "large" submatrices with applications to the geometry of Banach spaces and harmonic analysis. *Israel Journal of Mathematics* **57**, 137–224.

5. Brown, B. M. and Eagleson, G. K. (1971). Martingale convergence to infinitely divisible laws with finite variances. *Transactions of the American Mathematical Society* **162**, 449–453.

6. Burkholder, D. L. (1973). Distribution function inequalities for martingales. *Annals of Probability* **1**, 19–42.

7. Burkholder, D. L. (1983). A geometric condition that implies the existence of certain singular integrals of Banach-space-valued functions. In *Conference on Harmonic Analysis in Honor of Antoni Zygmund* (Eds., W. Beckner, A. P. Calderón, R. Feffcrman, and P. W. Jones). Wadsworth, Belmont, California.

8. Chow, Y. S., de la Peña, V. H., and Teicher, H. (1993). Wald's lemma for a class of de-normalized U-statistics. *Annals of Probability* **21**, 1151–1158.

9. de Acosta, A. (1987). A decoupling inequality for multilinear forms of stable vectors. *Probability and Mathematical Statistics* **8**, 71–76.

10. de la Peña, V. H. (1990). Bounds for the expectation of functions of martingales and sums of positive rv's in terms of norms of sums of independent random variables. *Proceedings of the American Mathematical Society* **108**, 233–239.

11. de la Peña, V. H. (1992). Decoupling and Khintchine's inequalities

for U-statistics. *Annals of Probability* **20**, 1877–1892.

12. de la Peña, V. H. (1993). Inequalities for tails of adapted processes with an application to Wald's lemma. *Journal of Theoretical Probability* **6**, 285–302.

13. de la Peña, V. H. (1994). A bound on the moment generating function of a sum of dependent variables with an application to simple random sampling without replacement. *Annals of the Institute of H. Poincaré. Probability and Statistics* **30**, 197–211. Correction note in **31**, 703–704.

14. de la Peña, V. H. (1996). From dependence to complete independence: The decoupling approach. *Proceedings of the IV Simposio de Probabilidad y Procesos Estocasticos*, Guanajuato, Mexico. Notas de Investigación No. 12, Sociedad Matemática Mexicana.

15. de la Peña, V. H. (1999). A general class of exponential inequalities for martingales and ratios. *Annals of Probability* **27**, 537–564.

16. de la Peña, V. H. and Eisenbaum, N. (1994). Decoupling inequalities for the local times of a linear Brownian motion. *Technical Report*, Department of Statistics, Columbia University.

17. de la Peña, V. H. and Eisenbaum, N. (1997). Exponential Burkholder-Davis-Gundy inequalities. *Bulletin of the London Mathematical Society* **29**, 239–242.

18. de la Peña, V. H. and Giné, E. (1999). *Decoupling: From Dependence to Independence*. Springer-Verlag, New York.

19. de la Peña, V. H. and Lai, T. L. (1997a) Wald's equation and asymptotic bias of randomly stopped U-statistics. *Proceedings of the American Mathematical Society* **125**, 917–925.

20. de la Peña, V. H. and Lai, T. L. (1997b) Moments of randomly stopped U-statistics. *Annals of Probability* **25**, 2055–2081.

21. de la Peña, V. H. and Montgomery-Smith, S. J. (1994). Bounds on the tail probability of U-statistics and quadratic forms. *Bulletin of the American Mathematical Society* **31**, 223–227.

22. de la Peña, V. H. and Montgomery-Smith S. J. (1995). Decoupling inequalities for the tail probabilities of multivariate U-statistics. *Annals of Probability* **23**, 806–816.

23. de la Peña, V. H., Montgomery-Smith S. J., and Szulga, J. (1994). Contraction and decoupling inequalities for multilinear forms and U-statistics. *Annals of Probability* **22**, 1745–1765.

24. Feller, W. (1971). *An Introduction to Probability Theory and Its Applications, Volume II*. Second Edition. John Wiley & Sons, New York.

25. Hitczenko, P. (1988). Comparison of moments for tangent sequences of random variables. *Probability Theory and Related Fields* **78**, 223–230.

26. Hitczenko, P. (1990). Best constants in martingale version of Rosenthal's inequality. *Annals of Probability* **18**, 1656–1668.

27. Hitczenko, P. (1994). On a domination of sums of random variables by sums of conditionally independent ones. *Annals of Probability* **22**, 453–468.

28. Hitczenko, P. and Montgomery-Smith, S. J. (1996). Tangent sequences in Orlicz and rearrangement invariant spaces. *Proceedings of the Cambridge Philosophical Society* **119**, 91–101.

29. Hoeffding, W. (1963). Probability inequalities for sums of bounded random variables. *Journal of the American Statistical Association* **58**, 13–30.

30. Huang, T. (1999). Steady-state asymptotics for queuing networks with heavy tailed service. *Ph.D. thesis.* Department of IEOR, Columbia University.

31. Jakubowski, A. (1986). Principle of conditioning in limit theorems for sums of random variables. *Annals of Probability* **14**, 902–915.

32. Janson, S. and Nowicki, K. (1991). The asymptotic distributions of generalized U-statistics with applications to random graphs. *Probability Theory and Related Fields* **90**, 341–375.

33. Joag-Dev, K. and Proschan, F. (1983). Negative association of random variables with applications. *Annals of Statistics* **11**, 268–295.

34. Klass, M. J. (1988). A best possible improvement of Wald's equation. *Annals of Probability* **16**, 840–863.

35. Klass, M. J. (1990). Uniform lower bounds for randomly stopped Banach space valued random sums. *Annals of Probability* **18**, 780–809.

36. Krakowiak, W. and Szulga, J. (1988). Hypercontraction principle and multilinear forms in Banach spaces. *Probability Theory and Related Fields* **77**, 325–342.

37. Kwapien, S. (1987). Decoupling for polynomial chaos. *Annals of Probability* **15**, 1062–1071.

38. Kwapien, S. and Woyczynski, W. (1992). *Random Series and Stochastic Integrals: Single and Multiple.* Birhäuser, New York.

39. Lai, T. L. and Robbins, H. (1978). A class of dependent variables and their maxima. *Zeitschrift Wahrscheinlichkeitstheorie und Verwandte Gebiete* **42**, 89–111.

40. McConnell, T. (1989). Decoupling and stochastic integration in UMD Banach spaces. *Probability and Mathematical Statistics* **10**, 283–295.

41. McConnell, T. and Taqqu, M. (1986). Decoupling inequalities for multilinear forms in independent symmetric Banach valued random variables. *Probability Theory and Related Fields* **75**, 499–507.

42. Nolan, D. and Pollard, D. (1987). U-processes: rates of convergence. *Annals of Statistics* **15**, 780–799.

43. Shao, Q. M. (1998). A comparison theorem on maximal inequalities between negatively associated and independent random variables. *Technical Report*, Department of Mathematics, University of Oregon.

44. Szulga, J. (1998). *Introduction to Random Chaos.* Chapman & Hall, New York.

45. Yang, M. (1999). On the order of ET_r^γ – the boundary crossing problem. *Ph.D. thesis*, Department of IEOR, Columbia University.

46. Zinn, J. (1985). Comparison of martingale difference sequences, In *Probability in Banach Spaces V*, 453–457. *Lecture Notes in Mathematics* **1153**. Springer-Verlag, New York.

8

A Note on the Probability of Rapid Extinction of Alleles in a Wright-Fisher Process

---

F. Papangelou
University of Manchester, Manchester, UK
University of Athens, Athens, Greece

ABSTRACT The standard formulation of a large deviation principle involves an upper bound for closed sets and a lower bound for open sets. Such a principle was established in Papangelou (2000) for a multiallele Wright-Fisher process. Here we extend under more restrictive assumptions the lower bound to boundary sets, thereby determining the exponential asymptotics of the probabilities of actually reaching the boundary within a short time.

Keywords and phrases Wright-Fisher process, Legendre transform, large deviation bounds, exponential asymptotics

---

8.1 Introduction

A multiallele Wright-Fisher process models the way in which the proportions of the various alleles of a gene in a population change from generation to generation. If the size of the population, N say, is large then the effects of random genetic drift are noticeable over a time span of N generations and if mutation and selection act on the same or a longer time scale then, according to a classical result, the continual changes of state of the suitably scaled process are approximated by a diffusion process in a sense that we will not make precise here. Over a much smaller number n of generations the changes are, with high probability, imperceptible but there is always a small probability that a large deviation may occur. It is shown in Papangelou (2000) that the appropriate space for studying the large deviations of the process is the d-sphere, if there are $d + 1$ alleles of the gene in the population. More precisely, if $p_1, \ldots, p_{d+1}$ are the proportions of the alleles, then $p_1 + \ldots + p_{d+1} = 1$

and therefore $(\sqrt{p_1}, \sqrt{p_2}, \ldots, \sqrt{p_{d+1}})$ is a point on the orthant of the d-sphere in $(d+1)$-dimensionsal Euclidean space. Now if there is a change from the point $(\sqrt{p_1}, \ldots, \sqrt{p_{d+1}})$ to the point $(\sqrt{q_1}, \ldots, \sqrt{q_{d+1}})$ over a span of n generations (where both n and $\dfrac{N}{n}$ are very large), then we can be nearly certain that the process has followed closely the arc of the great circle joining $(\sqrt{p_1}, \ldots, \sqrt{p_{d+1}})$ with $(\sqrt{q_1}, \ldots, \sqrt{q_{d+1}})$ on the sphere, at constant speed. The "probability" of such a deviation is of exponential order

$$\exp\left\{-\frac{4N}{n}\theta^2\right\}$$

where θ is the angle between $(\sqrt{p_1}, \ldots, \sqrt{p_{d+1}})$ and $(\sqrt{q_1}, \ldots, \sqrt{q_{d+1}})$ subtended at the centre of the sphere. See Theorems 1 and 2 in Papangelou (2000).

In order to make the above statements precise and explain the aim of the present note we need to give a full and exact statement of the large deviation result proved in Papangelou (2000). For a diploid population let $2N$ be the number of genes in the population and assume that the proportions of the $d+1$ alleles are $\dfrac{i_1}{2N}, \dfrac{i_2}{2N}, \ldots, \dfrac{i_{d+1}}{2N}$ where $i_1 + \ldots + i_{d+1} = 2N$. The state of our process is then the vector $y = (y_1, \ldots, y_d)$, where $y_k = \dfrac{i_k}{2N}, k = 1, 2, \ldots, d$. In the Wright-Fisher process the next generation of $2N$ genes is produced from the current generation "multinomially", i.e. the probability of a transition from state $y = (y_1, \ldots, y_d)$ to state $\tilde{y} = (\tilde{y}_1, \ldots, \tilde{y}_d) = \left(\dfrac{j_1}{2N}, \ldots, \dfrac{j_d}{2N}\right)$ is

$$[(2N)!]\left[\prod_{k=1}^{d+1}(j_k!)\right]^{-1} \pi_1^{j_1}\pi_2^{j_2}\ldots\pi_{d+1}^{j_{d+1}}$$

where π_k is y_k $(k = 1, \ldots, d)$ if there is no mutation or selection but in general π_k is y_k "perturbed" by mutation or selection effects, and

$$\pi_{d+1} = 1 - \sum_{k=1}^{d}\pi_k.$$

The large deviation result will be stated for a sequence $Y_t^{(n)}, t \geq 0; n = 1, 2, \ldots$ of scaled Wright-Fisher processes defined as follows. For each $n = 1, 2, \ldots$ let N_n be a positive integer such that $\dfrac{N_n}{n} \to \infty$ as $n \to \infty$. We stipulate that $Y_0^{(n)}, Y_{\frac{1}{n}}^{(n)}, Y_{\frac{2}{n}}^{(n)}, \ldots$ is a discrete Wright-Fisher process as defined above with

$$\pi_k = \pi_k(y, n) = y_k + \frac{g_k(y) + o_k(1)}{N_n}$$

where $g_1(y), \ldots, g_d(y)$ are continuous functions of the vector $y = (y_1, \ldots, y_d)$ on the simplex $\sum = \left\{ y : \sum_{k=1}^{d} y_k \leq 1, y_k \geq 0, k = 1, \ldots, d \right\}$ and $o_k(1) \to 0$ uniformly in y as $n \to \infty$. We further stipulate that $Y_t^{(n)}$ remains constant on $\dfrac{\nu}{n} \leq t < \dfrac{\nu+1}{n}$ for each $\nu = 0, 1, 2, \ldots$ and that $Y_0^{(n)} = p$ for all n, where p is an interior point of $\sum$. By Theorem 1 in Papangelou (2000), if F is a closed subset of $\sum$ and G a subset of $\sum$ open in the relative topology of $\sum$, then for any $T > 0$

$$\limsup_{n \to \infty} \frac{n}{2N_n} \log P(Y_T^{(n)} \in F) \leq - \inf_{q \in F} J_{p,T}(q) \tag{8.1.1}$$

$$\liminf_{n \to \infty} \frac{n}{2N_n} \log P(Y_T^{(n)} \in G) \geq - \inf_{q \in G} J_{p,T}(q) \tag{8.1.2}$$

where $\quad J_{p,T} = \dfrac{2}{T} \left[\cos^{-1} \left(\sum_{k=1}^{d+1} \sqrt{p_k q_k} \right) \right]^2 \quad (p_{d+1} = 1 - \sum_{k=1}^{d} p_k, q_{d+1} =$

$1 - \sum_{k=1}^{d} q_k)$. These two inequalities embody the large deviation property for the sequence $Y_T^{(n)}, n = 1, 2, \ldots$.

The situation is more complicated in the case of faster-acting mutation or selection, as shown for $d = 1$ in Papangelou (1998a,b).

The above large deviation principle raises an interesting problem, namely that of determining the rough asymptotics of the probabilities of actually *reaching* given subsets of the boundary of $\sum$ in the case of a large deviation. Such subsets are not open in the relative topology of $\sum$ and hence the large deviation lower bound can only tell us something about the probability of getting "close" to boundary sets. However, the event of actually reaching the boundary is of great importance. In the absence of mutation, for instance, the phenomenon of fixation whereby all but one of the alleles eventually become extinct is of primary interest.

In the present paper we consider the problem just mentioned for the case of a Wright-Fisher process subject only to random genetic drift, i.e. without mutation or selection. For such a process, the large deviation upper bound for boundary sets follows trivially from (8.1.1). Our purpose is to establish the lower bound for boundary sets under the assumption that $n^2 = o(N_n)$.

8.2   The Lower Bound for Boundary Sets

We base our calculations on the inequality stated in the lemma below, which is a variant of classical estimates: it is related for instance to one of the inequalities of Theorem 1.3.13 in Stroock (1993), which however we take a step further in our case by using the connections between the derivatives involved. More elaborate calculations based on an alternative approach lead to a weakening of the condition $n^2 = o(N_n)$ but the technicalities involved are out of proportion with the "improvement" obtained.

To state the lemma suppose $X_1, X_2, \ldots$ are independent, identically distributed random variables such that the cumulant generating function $G(z) = \log E e^{z X_1}$ is finite for all $z \in \mathbb{R}$ and let $H(u) = \sup_{z} \{zu - G(z)\}$, $u \in \mathbb{R}$, be the Legendre transform of G. The derivative $G'(z)$ exists for all z and is increasing, with $G'(0) = \mu$, where μ is the mean of X_1.

LEMMA 8.2.1
Suppose that $u > \mu$ and that u is in the interior of the range of G'. Then, for any $\delta > 0$ and any $N \geq 1$,

$$P(u - \delta < \frac{1}{N} \sum_{i=1}^{N} X_i < u + \delta)$$

$$\geq \exp\left[-NH(u+\delta)\right] \cdot \left\{1 - \left(NH''(u)\delta^2\right)^{-1}\right\}.$$

The proof is brief. By assumption there is a z_0 such that $G'(z_0) = u$. It is then clear that $H(u) = z_0 u - G(z_0) = z_0 G'(z_0) - G(z_0)$, hence $H'(u) = \dfrac{dH}{dz_0} \cdot \dfrac{dz_0}{du} = z_0 G''(z_0) \cdot [G''(z_0)]^{-1} = z_0$. If Q denotes the distribution of X_1, introduce the distribution

$$\tilde{Q}(dx) = \exp\{z_0 x - G(z_0)\} Q(dx)$$

and denote by $\tilde{Q}_N$ the distribution of $\dfrac{1}{N} \sum_{i=1}^{N} X_i$ under the convolution power $\tilde{Q}^N$. If $\delta > 0$, then [cf. Stroock (1993)]

$$P(u - \delta < \frac{1}{N} \sum_{i=1}^{N} X_i < u + \delta)$$

$$\geq \exp\left[-N(z_0(u+\delta) - G(z_0))\right] \cdot \tilde{Q}_N(u - \delta, u + \delta).$$

The right-hand side is greater than or equal to

$$\exp\left[-N(H(u) + H'(u)\delta)\right] \cdot \left\{1 - \frac{1}{N\delta^2} : \text{Variance} : (\tilde{Q})\right\}$$

by Chebyshev's inequality. Now $\text{Variance}(\tilde{Q}) = G''(z_0) = [H''(u)]^{-1}$ since the inverse of the function $u = G'(z_0)$ is the function $z_0 = H'(u)$. Also, $H(u) + H'(u)\delta \leq H(u + \delta)$ by the convexity of H. This proves the lemma.

Returning to our process, assume initially that $Y_t^{(n)}, t \geq 0$ is a scaled *one-dimensional* Wright-Fisher process ($d = 1$) as defined in the introduction, with no mutation or selection. For the sake of convenience we will write N for N_n. Suppose $T > 0$ and $\epsilon \in \left(0, \frac{1}{2}\right)$ and consider the curve $\phi(t) = 1 - \epsilon\left(1 - \frac{t}{T}\right)^2, 0 \leq t \leq T$. Then

$$P(Y_T^{(n)} = 1 \mid Y_0^{(n)} = 1 - \epsilon)$$

$$\geq \left[\prod_{k=0}^{nT-2} P\left(Y_{\frac{k+1}{n}}^{(n)} > \phi\left(\frac{k+1}{n}\right) \middle| Y_{\frac{k}{n}}^{(n)} = \phi\left(\frac{k}{n}\right)\right)\right]$$

$$\times P\left(Y_T^{(n)} = 1 \middle| Y_{T-\frac{1}{n}}^{(n)} = \phi\left(T - \frac{1}{n}\right)\right) \qquad (8.2.1)$$

assuming for convenience that T is an integral multiple of $\frac{1}{n}$. With a view to applying the lemma to each factor in the bracketed product we set, for $0 < y < 1$, $G(z) = G(y, z) = \log(1 + y(e^z - 1))$. Then $H(u) = H(y, u) = u \log \frac{u}{y} + (1 - u) \cdot \log \frac{1 - u}{1 - y}$ and hence $H''(u) = [u(1 - u)]^{-1}$. Note that if $u > y > \frac{1}{2}$ then

$$H(y, u) \leq u\left(\frac{u}{y} - 1\right) + (1 - u)\left(\frac{1 - u}{1 - y} - 1\right) = \frac{(u - y)^2}{y(1 - y)} \leq 2\frac{(u - y)^2}{1 - y}.$$

Taking

$$y = \phi\left(\frac{k}{n}\right), \quad u = \phi\left(\frac{k}{n}\right) + (1 + \beta)\left(\phi\left(\frac{k+1}{n}\right) - \phi\left(\frac{k}{n}\right)\right)$$

and

$$\delta = \beta\left(\phi\left(\frac{k+1}{n}\right) - \phi\left(\frac{k}{n}\right)\right),$$

where β is suitably small, we obtain from the lemma the following, in which $H''(y, u)$ denotes $\frac{\partial^2 H}{\partial u^2}$.

$$P\left(Y^{(n)}_{\frac{k+1}{n}} > \phi\left(\frac{k+1}{n}\right)\Big| Y^{(n)}_{\frac{k}{n}} = \phi\left(\frac{k}{n}\right)\right)$$

$$= P\left(Y^{(n)}_{\frac{k+1}{n}} > u - \delta \big| Y^{(n)}_{\frac{k}{n}} = y\right)$$

$$\geq \exp\left[-2NH(y, u+\delta)\right] \cdot \left\{1 - (2NH''(y,u)\delta^2)^{-1}\right\}$$

$$\geq \exp\left[-16(1+2\beta)^2 \frac{N\epsilon}{n^2T^2}\right] \cdot \left\{1 - \frac{n^2T^2}{8N\epsilon\beta^2}\right\}$$

since $H(y, u+\delta) \leq 2\dfrac{(u+\delta-y)^2}{1-y} \leq 8(1+2\beta)^2\dfrac{\epsilon}{n^2T^2}$ as can be checked, and

$$2NH''(y,u)\delta^2 = 2N\delta^2\left[u(1-u)\right]^{-1}$$

$$\geq 2N\delta^2\left[(u-\delta)(1-u+\delta)\right]^{-1}$$

$$\geq 2N\left(1 - \phi\left(\frac{k+1}{n}\right)\right)^{-1}\left[\frac{2\epsilon\beta}{nT}\left(1 - \frac{k+1}{nT}\right)\right]^2$$

$$= \frac{8N\epsilon\beta^2}{n^2T^2}.$$

Also,

$$P\left(Y^{(n)}_T = 1 \Big| Y^{(n)}_{T-\frac{1}{n}} = 1 - \frac{\epsilon}{n^2T^2}\right) = \left(1 - \frac{\epsilon}{n^2T^2}\right)^{2N}.$$

Now (8.2.1) implies that if $n^2 = o(N)$ as $n \to \infty$ then, for sufficiently large n,

$$\log P(Y^{(n)}_T = 1 \mid |Y^{(n)}_0 = 1 - \epsilon)$$

$$\geq (nT-1)\left[-16(1+2\beta)^2\frac{N\epsilon}{n^2T^2} + \log\left(1 - \frac{n^2T^2}{8N\epsilon\beta^2}\right)\right]$$

$$+2N\log\left(1 - \frac{\epsilon}{n^2T^2}\right)$$

$$\geq \frac{nT}{2}\left[-16(1+2\beta)^2\frac{N\epsilon}{n^2T^2} - \frac{n^2T^2}{4N\epsilon\beta^2}\right] - \frac{4N\epsilon}{n^2T^2}$$

$$= -\frac{8N}{n}\left[(1+2\beta)^2\frac{\epsilon}{T} + o(1)\right]$$

as $n \to \infty$. Hence

$$\liminf_{n\to\infty}\frac{n}{2N}\log P(Y^{(n)}_T = 1 \Big| Y^{(n)}_0 = 1 - \epsilon) \geq -4(1+2\beta)^2\frac{\epsilon}{T}.$$

In particular, if $T = \eta$ and $\epsilon = \eta^2$ where η is a small positive number, then

$$\liminf_{n\to\infty} \frac{n}{2N} \log P(Y_\eta^{(n)} = 1 \,\big|\, Y_0^{(n)} = 1 - \eta^2) \geq -4(1 + 2\beta)^2 \eta.$$

The inequality

$$P(Y_T^{(n)} = 1 \mid Y_0^{(n)} = p)$$
$$\geq P(Y_{T-\eta}^{(n)} > 1 - \eta^2 \mid Y_0^{(n)} = p) P(Y_T^{(n)} = 1 \mid Y_{T-\eta}^{(n)} = 1 - \eta^2)$$

combined with Theorem 8.2.2 implies

$$\liminf_{n\to\infty} \frac{n}{2N} \log P(Y_T^{(n)} = 1 \mid Y_0^{(n)} = p)$$
$$\geq -\frac{2}{T-\eta} \left[\cos^{-1} \left(\sqrt{p(1-\eta^2)} + \sqrt{(1-p)\eta^2} \right) \right]^2 - 4(1 + 2\beta)^2 \eta$$

and since this is true for all small η, we obtain

$$\liminf_{n\to\infty} \frac{n}{2N} \log P(Y_T^{(n)} = 1 \,\big|\, Y_0^{(n)} = p) \geq -\frac{2}{T} \left[\cos^{-1} \sqrt{p} \right]^2. \qquad (8.2.2)$$

We can now establish the following theorem, which was stated in Papangelou (2000) without proof. We drop the assumption that $Y_t^{(n)}$ is one-dimensional and return to the set-up of the introduction where $Y_t^{(n)}$ is d-dimensional.

THEOREM 8.2.2
Let $Y_t^{(n)}, 0 \leq t \leq T; n = 1, 2, \ldots$ be the sequence of d-dimensional processes defined in the introduction, with no mutation or selection effects. Set $\sum_m = \{q = (q_1, \ldots, q_d) \in \sum : q_1 = q_2 = \ldots = q_m = 0\}$ where $m \leq d$, and suppose that G is a subset of $\sum_m$ which is open in the relative topology of $\sum_m$. If $n^2 = o(N_n)$ as $n \to \infty$, then for $p (= Y_0^{(n)})$ in the interior of $\sum$,
$$\liminf_{n\to\infty} \frac{n}{2N_n} \log P(Y_T^{(n)} \in G) \geq -\inf_{q\in G} J_{p,T}(q).$$

We sketch the argument. Briefly, enlarge G to an open subset G^* of $\sum$ (in the relative topology) by including points $(q_1, \ldots, q_d)$ such that $q_1, \ldots, q_m$ are small rather than zero and $(0, \ldots, 0, q_{m+1}, \ldots, q_d) \in G$. The large deviation lower bound (8.1.2) implies that

$$P(Y_{T-\epsilon}^{(n)} \in G^*) \geq \exp \left\{ -\frac{2N}{n} \cdot \frac{2}{T-\epsilon} \inf_{q\in G^*} \left[\cos^{-1} \left(\sum_{k=1}^{d+1} \sqrt{p_k q_k} \right) + \beta \right]^2 \right\}$$

eventually ($\beta > 0$). Once $Y_{T-\epsilon}^{(n)}$ is in G^*, (8.2.2) applied to the one-dimensional process $1 - Z_t^{(n)}$ where $Z_t^{(n)}$ is the sum of the first m co-ordinates of $Y_t^{(n)}$, implies that the conditional probability of the event $\left\{Y_T^{(n)} \in G\right\}$ given $Y_{T-\epsilon}^{(n)} \in G^*$ is greater than or equal to

$$\exp\left\{-\frac{2N}{n} \cdot \frac{2}{\epsilon}\left[\cos^{-1}\sqrt{1-\eta}\right]^2\right\},$$

where η is small. The proof can be completed by letting $\beta \to 0$, $\eta \to 0$, $\epsilon \to 0$ in that order in the resulting lower bound of

$$\liminf_{n\to\infty} \frac{n}{2N} \log P(Y_T^{(n)} \in G).$$

The theorem naturally holds for the other components of the boundary of $\sum$ as well.

References

1. Papangelou, F. (1998a). Tracing the path of a Wright-Fisher process with one-way mutation in the case of a large deviation. In *Stochastic Processes and Related Topics – A Volume in Memory of Stamatis Cambanis* (Eds., I. Karatzas, B. Rajput, and M. S. Taqqu), pp. 315–330. Birkhäuser, Boston.

2. Papangelou, F. (1998b). Elliptic and other functions in the large deviations behavior of the Wright-Fisher process. *Annals of Applied Probability* **8**, 182–192.

3. Papangelou, F. (2000). The large deviations of a multi-allele Wright-Fisher process mapped on the sphere. *Annals of Applied Probability*, to appear.

4. Stroock, D. W. (1993). *Probability Theory, An Analytic View*. Cambridge University Press, Cambridge.

9

Stochastic Integral Functionals in an Asymptotic Split State Space

V. S. Korolyuk and N. Limnios
Ukrainian Academy of Science, Kiev, Ukraine
Université de Technologie de Compiègne, Compiègne, France

ABSTRACT Let $x^\varepsilon(t)$, $t \geq 0$, $\varepsilon > 0$ be a family of Markov jump processes on a measurable state space $(X, \mathcal{X})$, and consider a finite split of its state space. Under some additional conditions we obtain by a singular perturbation technique, averaging and diffusion approximation results in single and double merging scheme for the integral functional $\zeta^\varepsilon(t) := \int_0^t a(x^\varepsilon(s))ds$, where a is an $\mathcal{X}$-measurable function. We present here recent results given essentially in Korolyuk and Limnios (1998, 1999a,b), under a unified approach.

Keywords and phrases Markov process, Integral functional, averaging principle, diffusion approximation, merging state space, double merging

9.1 Introduction

Asymptotic behaviour of stochastic systems in series scheme and stability constitutes one of the major fields of investigation as well in theoretical as in applied fields of research [see, for example, Anisimov (1995) and Liptser (1984)]. In this paper we investigate the asymptotic behaviour of integral functionals of Markov jump processes with asymptotic state space merging.

In the study of complex systems we have very often to consider behaviour of different scales of time, one slow and one fast and in some cases even of more than two levels. On the other hand, the local characteristics of the systems (transition rates, probabilities, etc.) can be dependent on the current values of some other stochastic processes (natural environment, discrete event systems, etc.).

The study of complex systems with high dimensionality is not possible in most of the cases. Thus we need a simplification of the system. Many techniques were developed to face this problem. The most useful is the diffusion approximation and the asymptotic state space merging or consolidation. If the state space can be split in several regions and if the transition probabilities between them are small in some sense and if, in each region, states communicate asymptotically, then we can apply asymptotic consolidation, i.e., we simplify the state space of system by assuming that each region can be represented asymptotically by a single state.

In fact, in the merging scheme, we have three systems: the initial system, the supporting one and the merged one. The supporting (or intermediate) one is supposed to be ergodic in each region. In the merged system, each region is reduced to a single state, thus the number of the states is equal to the number of regions of the initial system. It is obvious that the study of the merged system is much more simple than the initial one. The initial system is considered as a perturbation of the supporting system. More precisely, we consider a sequence of systems including the initial system, converging to the supporting system. This constitutes the series scheme.

The underlying mathematical tools for results obtained here are based on the theory of singular perturbed reducible invertible operators [Korolyuk and Korolyuk (1999) and Korolyuk and Limnios (1999b)] and on the martingale characterization of Markov processes [Ethier and Kurtz (1986), Korolyuk and Korolyuk (1999) and Liptser and Shiryaev (1989)].

In Section 9.2, we give a more precise definition of the merging scheme of a Markov system and of the integral functional that will be studied. In Section 9.3, we give results for single and double merging of the Markov processes. In Section 9.4, we give results for the average convergence of the sequence of integral functionals. In Section 9.5, we give a diffusion approximation of it in single and double merging scheme. Finally, in Section 9.6, we give diffusion approximation results for a sequence of integral functionals with a perturbed kernel.

9.2 Preliminaries

Let $(X, \mathcal{X})$ be a measurable space with a countably generated σ-algebra $\mathcal{X}$. Consider now an ergodic irreducible Markov chain x_n, $n \geq 0$, with state space $(X, \mathcal{X})$ and transition probability function $P(x, B)$, $x \in X$, $B \in \mathcal{X}$, and stationary distribution $\rho(B)$, $B \in \mathcal{X}$, i.e., $\rho(B) = \int_X \rho(dx) P(x, B)$ and $\rho(X) = 1$.

Let $\mathbf{B}$ be a Banach space of measurable real-valued bounded functions defined on X, with the sup-norm $\|f\| := \sup_{x \in X} |f(x)|$. Let us denote by P too the operator of transition probabilities on $\mathbf{B}$ defined by : $Pf(x) = \int_X P(x, dy)f(y)$. Let us denote by $P^n(x, B)$ the n-step transition probability and P^n the corresponding operator.

The Markov chain x_n is called *uniformly ergodic* if $\sup_{\|f\| \le 1} \|(P^n - \Pi)f\|$ converges to zero as $n \to \infty$, where Π is the *stationary projector* in $\mathbf{B}$ defined from the stationary distribution $\rho(B)$ of the Markov chain x_n, as follows [Korolyuk and Korolyuk (1999)]

$$\Pi f(x) := \int_X \rho(dy)f(y)\mathbf{1}(x), \tag{9.2.1}$$

where $\mathbf{1}(x) = 1$ for all $x \in X$. Note that uniform ergodicity implies Harris recurrence.

For a uniformly ergodic Markov chain, the operator $Q := P - I$ (where I is the identity operator), is *reducible invertible*, i.e., $\mathbf{B} = N_Q \oplus R_Q$, where $N_Q = \{\varphi \in \mathbf{B} : Q\varphi = 0\}$ is the null-space of the operator Q, with $dim N_Q \ge 1$, and $R_Q = \{\psi \in \mathbf{B} : Qf = \psi\}$ is the space of values of the operator Q.

Let $x^\varepsilon(t)$, $t \ge 0$, $\varepsilon > 0$, be a family of Markov jump processes on a measurable state space $(X, \mathcal{X})$ defined by the generators

$$Q^\varepsilon \varphi(x) = q(x) \int_X P^\varepsilon(x, dy)[\varphi(y) - \varphi(x)]. \tag{9.2.2}$$

The *integral functional* in a merging and averaging scheme is considered in the following form:

$$\zeta^\varepsilon(t) = \int_0^t a(x^\varepsilon(s))ds \tag{9.2.3}$$

where a is an $\mathcal{X}$-measurable function, such that

$$\int_0^t |a(x^\varepsilon(s)| \, ds < \infty, \quad (a.s.), \ t > 0, \ \varepsilon > 0. \tag{9.2.4}$$

Concerning the above integral functional (9.2.3), we consider several averaging and diffusion approximation problems in combination with single and double asymptotic splitting with a singular perturbation approach.

Firstly, we consider the finite single splitting

$$X = \bigcup_{k=1}^N X_k, \quad X_k \bigcap X_{k'} = \phi, \quad k \ne k'. \tag{9.2.5}$$

The stochastic kernel $P^\varepsilon(x, dy)$ is then represented by

$$P^\varepsilon(x, dy) = P(x, dy) + \varepsilon P_1(x, dy), \qquad (9.2.6)$$

where the stochastic kernel $P(x, dy)$ defines the support imbedded Markov chain x_n, $n \geq 0$, uniformly ergodic in every class X_k, $1 \leq k \leq N$, with the stationary distributions $\rho_k(dx)$, $1 \leq k \leq N$. This kernel is related to the state space splitting (9.2.5), namely,

$$P(x, X_k) = 1_k(x) = \begin{cases} 1, & x \in X_k \\ 0, & x \notin X_k \end{cases}$$

$P_1(x, dy)$ is a perturbing kernel. Thus the support Markov process $x(t)$, $t \geq 0$, defined by the generator

$$Q\varphi(x) = q(x) \int_X P(x, dy)[\varphi(y) - \varphi(x)],$$

is uniformly ergodic too with the stationary distributions $\pi_k(dx)$, $1 \leq k \leq N$, represented as follows

$$\pi_k(dx)q(x) = q_k\rho_k(dx), \quad q_k = \int_{X_k} \pi_k(dx)q(x).$$

Secondly, we consider the following double finite splitting:

$$X = \bigcup_{k=1}^{N} X_k, \quad X_k = \bigcup_{r=1}^{N_k} X_k^r, \quad 1 \leq k \leq N,$$

$$X_k^r \bigcap X_{k'}^{r'} = \emptyset, \quad k \neq k' \text{ or } r \neq r'. \qquad (9.2.7)$$

The stochastic kernel $P^\varepsilon(x, dy)$, in this second case, has the following representation

$$P^\varepsilon(x, dy) = P(x, dy) + \varepsilon P_1(x, dy) + \varepsilon^2 P_2(x, dy). \qquad (9.2.8)$$

Here the stochastic kernel $P(x, dy)$ defines the support Markov chain x_n, $n \geq 0$, which is uniformly ergodic in every class X_k^r, $1 \leq r \leq N_k, 1 \leq k \leq N$, and $P_1(x, dy)$ and $P_2(x, dy)$ are perturbing kernels. The former concerns transitions between classes X_k^r and the latter between classes X_k. Of course, these kernels are not stochastic, in particular $P_1(x, X) = P_2(x, X) = 0$, for every $x \in X$. The supporting Markov process $x(t)$, $t \geq 0$, is uniformly ergodic too with generator Q.

9.3 Phase Merging Scheme for Markov Jump Processes

In the ergodic single split state space, define the merging function v as follows

$$v(x) = v, \quad \text{if } x \in X_k$$

Here we suppose that the perturbing kernel verifies the merging condition [Korolyuk and Limnios (1999a)], i.e.,

$$v(x^\varepsilon(t/\epsilon)) \Longrightarrow \hat{x}(t), \quad \text{as } \varepsilon \to 0, \tag{9.3.1}$$

where $\hat{x}(t), t \geq 0$, is the limit Markov process with state space $\hat{X} = \{1, ..., N\}$ and generator the contracted operator $\hat{Q}_1$, given by the following relation

$$\Pi Q_1 \Pi = \hat{Q}_1 \Pi, \quad Q_1 \varphi(x) = q(x) \int_X P_1(x, dy) \varphi(y). \tag{9.3.2}$$

The projector Π into the null space N_Q of the operator Q is defined by

$$\Pi \varphi(x) = \sum_{k=1}^{N} \hat{\varphi}_k 1_k(x), \quad \hat{\varphi}_k := \int_{X_k} \pi_k(dx) \varphi(x).$$

In the double split state space, let us assume that the perturbing kernels P_1 and P_2 satisfy the merging condition, that is, the following weak convergences take place:

$$\hat{v}(x^\varepsilon(t/\varepsilon)) \Longrightarrow \hat{x}(t) \tag{9.3.3}$$

$$\hat{\hat{v}}(x^\varepsilon(t/\varepsilon^2)) \Longrightarrow \hat{\hat{x}}(t) \tag{9.3.4}$$

where $\hat{v}$ and $\hat{\hat{v}}$ are the merging functions defined as follows

$$\hat{v}(x) = \hat{v}, \quad x \in X_k^r \quad \text{and} \quad \hat{\hat{v}}(x) = \hat{\hat{v}}, \quad x \in X_k.$$

The limit process $\hat{x}(t)$ has the state space $\hat{X} = \cup_{k=1}^{N} \hat{X}_k, \hat{X}_k = \{v_k^r : 1 \leq r \leq N_k\}$, and $\hat{\hat{x}}(t)$ has the state space $\hat{\hat{X}} = \{1, 2, ..., N\}$. The generators of processes $\hat{x}(t)$ and $\hat{\hat{x}}(t)$ are respectively $\hat{Q}_1$ and $\hat{Q}_2$, which are defined below.

The stationary distributions of the supporting uniformly ergodic Markov process $x(t)$, $t \geq 0$, of the merged Markov process $\hat{x}(t)$, $t \geq 0$, and of the twice merged Markov process $\hat{\hat{x}}(t)$, $t \geq 0$, are $(\pi_k^r(dx), 1 \leq r \leq N_k, 1 \leq k \leq N)$, $(\hat{\pi}_k^r, 1 \leq r \leq N_k, 1 \leq k \leq N)$ and $(\hat{\hat{\pi}}_k, 1 \leq k \leq N)$ respectively.

The contracted operator $\hat{Q}_1$ is given by (9.3.2). The double contracted operator $\hat{\hat{Q}}_2$ is defined as follows:

$$\Pi Q_2 \Pi = \hat{Q}_2 \Pi, \quad Q_2 \varphi(x) = q(x) \int_X P_2(x, dy) \varphi(y)$$

$$\hat{\Pi} \hat{Q}_2 \hat{\Pi} = \hat{\hat{Q}}_2 \hat{\Pi}.$$

The projectors Π and $\hat{\Pi}$ are defined as follows:

$$\Pi \varphi(x) = \sum_{k=1}^{N} \sum_{r=1}^{N_k} \int_{X_k^r} \pi_k^r(dy) \varphi(y) \mathbf{1}_k^r(x), \quad 1 \le r \le N_k, \ 1 \le k \le N$$

$$\hat{\Pi} \varphi(x) = \sum_{k=1}^{N} \sum_{r=1}^{N_k} \hat{\pi}_k^r \varphi_k^r \mathbf{1}_k^r(x)$$

with

$$\mathbf{1}_k^r(x) = \begin{cases} 1, & x \in X_k^r \\ 0, & x \notin X_k^r \end{cases}, \quad \mathbf{1}_k(x) = \begin{cases} 1, & x \in X_k \\ 0, & x \notin X_k \end{cases}$$

9.4   Average of Stochastic Integral Functional

Consider now the integral functional $\zeta^\varepsilon(t)$, $t \ge 0$, given by (9.2.3), in double merging and averaging scheme.

Let us define

$$\hat{a}(\hat{v}(x)) = \sum_{k=1}^{N} \sum_{r=1}^{N_k} \hat{a}_k^r \mathbf{1}_k^r(x),$$

$$\hat{\hat{a}}(\hat{\hat{v}}(x)) = \sum_{k=1}^{N} \hat{\hat{a}}(v_k) \mathbf{1}_k(x),$$

and

$$\hat{\hat{a}} = \sum_{k=1}^{N} \hat{\hat{a}}_k \hat{\hat{\pi}}_k$$

where

$$\hat{a}_k^r = \int_{X_k^r} \pi_k^r(dx) a(x), \quad 1 \le r \le N_k, \ 1 \le N \le N,$$

$$\hat{\hat{a}}_k = \sum_{r=1}^{N_k} \hat{\pi}_k^r \hat{a}_k^r, \quad 1 \le k \le N.$$

The following average limit results take place.

THEOREM 9.4.1
Under the above conditions, the following weak convergence holds, as
$\varepsilon \to 0$, *in the sense of the Skorokhod topology on the space* $D_{\mathbf{R}}[0,\infty)^1$:

1.
$$\varepsilon^3 \zeta^\varepsilon(t/\varepsilon^3) \Longrightarrow t\hat{\hat{a}},$$

2.
$$\varepsilon^2 \zeta^\varepsilon(t/\varepsilon^2) \Longrightarrow \hat{\zeta}(t) := \int_0^t \hat{a}(\hat{x}(s))ds.$$

3.
$$\varepsilon \zeta^\varepsilon(t/\varepsilon) \Longrightarrow \hat{\zeta}_k(t) := \int_0^t \hat{a}_k(\hat{x}(s))ds, \quad 1 \le k \le N,$$

with respect to $\mathbb{P}_x$, $x \in X_k$.

THEOREM 9.4.2
*Under the above merging conditions, the following weak convergence
holds, as* $\varepsilon \to 0$, *in the sense of the Skorokhod topology on the space*
$D_{\mathbf{R}}[0,\infty)$:
$$\varepsilon\hat{\zeta}(t/\varepsilon) \Longrightarrow t\hat{\hat{a}}$$

REMARK The following weak convergence, with respect to $\mathbb{P}_x$, $x \in X_k$,
$$\varepsilon\hat{\zeta}(t/\varepsilon) \Longrightarrow t\hat{\hat{a}}_k, \quad 1 \le k \le N$$

takes place. ∎

9.5 Diffusion Approximation of Stochastic Integral Functional

9.5.1 Single Splitting State Space

Let us define the potential matrix,
$$\hat{\mathbf{R}}_0 = [r_{kl}; 0 \le k, l \le N],$$

by the following relations [Korolyuk and Korolyuk (1999)]:

$$\hat{Q}\hat{\mathbf{R}}_0 = \hat{\mathbf{R}}_0\hat{Q} = \hat{\Pi} - I = [\pi_{lk} = \pi_k - \delta_{lk}; 0 \le l, k \le N].$$

---

[1] Continuity of sample paths of the limit process provides the weak convergence in
the space $C_{\mathbf{R}}[0,\infty)$.

The centering shift-coefficient $\hat{a}$ is defined by the relation

$$\hat{a} = \sum_{k=0}^{N} \pi_k a_k, \quad a_k := \int_{X_k} \pi_k(dx)a(x), \ 0 \le k \le N;$$

or, in an equivalent form, by

$$\hat{a} = q\sum_{k=0}^{N} \rho_k a_k, \quad a_k := \int_{X_k} \rho_k(dx)a(x), \ 0 \le k \le N,$$

where

$$q = \sum_{k=0}^{N} \pi_k q_k,$$

and

$$\rho_k = \pi_k q_k/q, \quad 0 \le k \le N,$$

is the stationary distribution of the imbedded Markov chain $\hat{x}_n, n \ge 0$. The vector $\tilde{a}$ is defined as $\tilde{a} = (\tilde{a}_k := a_k - \hat{a}, 0 \le k \le N)$.

THEOREM 9.5.1
Let the merging condition (9.3.1) be valid and the limit merged Markov process $\hat{x}(t)$, $t \ge 0$, be ergodic with the stationary distribution $\pi = (\pi_k, 1 \le k \le N)$. Then the normalized centered integral functional

$$\xi^\varepsilon(t) = \varepsilon^2 \zeta^\varepsilon(t/\varepsilon^3) - \varepsilon^{-1} t\hat{a}$$

converges weakly, as $\varepsilon \to 0$, to the diffusion process $\xi(t)$, $t \ge 0$, with zero mean and variance

$$\sigma^2 = 2\sum_{k,l=1}^{N} \pi_k \tilde{a}_k r_{kl} \tilde{a}_l = 2\sum_{i=1}^{N} \pi_i q_i b_i^2 - \sum_{i\ne j \ge 1}^{N} b_i b_j [\pi_i q_{ij} + \pi_j q_{ji}]. \quad (9.5.1)$$

REMARK The second equality in (9.5.1) is Liptser's formula [Liptser (1984) and Korolyuk and Limnios (1998)]. ∎

The generator of the coupled process $\xi^\varepsilon(t)$, $x^\varepsilon(t/\varepsilon^3)$, $t \ge 0$, has the following representation

$$L^\varepsilon = \varepsilon^{-3} Q + \varepsilon^{-2} Q_1 + \varepsilon^{-1} A(x)$$

with

$$Q_1\varphi(x) = q(x)\int_X P_1(x, dy)\varphi(y),$$

and

$$A(x)\varphi(u) = [a(x) - \hat{a}]\varphi'(u).$$

The proof of the above theorem is based on the following singular perturbation result.

PROPOSITION 9.5.2 *[Korolyuk (1998)]*
Let us assume that:

1. *the operator Q is reducible invertible;*

2. *the contracted (on N_Q) operator $\hat{Q}_1$ is reducible invertible with null-space $\hat{N}_{\hat{Q}_1} \subset \hat{N}_Q$;*

3. *the twice contracted operator $\hat{Q}_2$ which is defined by the relations*

$$\hat{Q}_2\hat{\Pi} = \hat{\Pi}\hat{Q}_2\hat{\Pi}, \quad \hat{Q}_2\Pi = \Pi Q_2\Pi$$

is a zero-operator, i.e., $\hat{Q}_2\varphi = 0$, for all $\varphi \in N_Q$.

Then the asymptotic representation

$$\left(\varepsilon^{-3}Q + \varepsilon^{-2}Q_1 + \varepsilon^{-1}Q_2 + Q_3\right)\left(\varphi + \varepsilon\varphi_1 + \varepsilon^2\varphi_2 + \varepsilon^3\varphi_3\right) = \psi + \theta^\varepsilon$$
(9.5.2)

can be realized by the following relations:

$$\hat{Q}_0\varphi = \hat{\psi}$$
(9.5.3)

$$\varphi_1 = -\hat{R}_0\hat{Q}_2\varphi$$
(9.5.4)

$$\varphi_2 = \hat{R}_0(\hat{\psi} - \hat{Q}_0\varphi)$$
(9.5.5)

$$\varphi_3 = R_0[\psi - Q_1\varphi_2 - Q_2\varphi_1 - Q_3\varphi]$$
(9.5.6)

where $\hat{Q}_0$ is the contraction of the operator $\hat{Q}_0 := \hat{Q}_3 - \hat{Q}_2\hat{R}_0\hat{Q}_2$ on the null-space of Q, and $\hat{R}_0$ is the potential operator of the contracted operator $\hat{Q}_1$.

From this proposition, the asymptotic representation of the above generator I^ε is realized in the following form :

$$L^\varepsilon\varphi^\varepsilon = L^0\varphi + \theta^\varepsilon,$$
(9.5.7)

where the limit operator L^0, by Proposition 9.5.2, is the twice contracted operator, defined by the following relations:

$$\hat{A}\Pi = \Pi A(x)\Pi$$
(9.5.8)

$$L^0\hat{\Pi} = \hat{\Pi}\hat{L}_0\hat{\Pi}, \tag{9.5.9}$$

where

$$\hat{L}_0 = \hat{A}\hat{R}_0\hat{A}. \tag{9.5.10}$$

Hence

$$L^0\varphi(u) = \frac{\sigma^2}{2}\varphi''(u), \tag{9.5.11}$$

is the operator of the limit diffusion process $\xi(t), t \geq 0$.

Now, the proof of weak convergence is based on the martingale characterization of the coupled Markov process $\xi^\varepsilon(t)$, $x^\varepsilon(t/\varepsilon^3)$, $t \geq 0$:

$$M^\varepsilon(t) = \varphi^\varepsilon(\xi^\varepsilon(t), x^\varepsilon(t/\varepsilon^3)) - \varphi^\varepsilon(u, x) - \int_0^t L^\varepsilon\varphi^\varepsilon(\xi^\varepsilon(s), x^\varepsilon(s/\varepsilon^3))ds. \tag{9.5.12}$$

By relation (9.5.7), the above martingale representation is transformed into the following form

$$M^\varepsilon(t) = \varphi(\xi^\varepsilon(t)) - \varphi^\varepsilon(t) - \int_0^t L^0\varphi^\varepsilon(\xi^\varepsilon(s))ds + \psi^\varepsilon(t) \tag{9.5.13}$$

where the negligible term $\psi^\varepsilon(t)$ satisfies assumptions of the pattern limit theorem [Korolyuk and Korolyuk (1999)].

9.5.2 Double Split State Space

Let $\hat{R}_0 = (\hat{r}_{k\ell}, 1 \leq k, \ell \leq N)$ be the potential matrix of operator $\hat{Q}_2$ defined by relations:

$$\hat{Q}_2\hat{R}_0 = \hat{R}_0\hat{Q}_2 = \hat{\Pi} - I = (\pi_{k\ell} = \pi_k - \delta_{k\ell}, 1 \leq k, \ell \leq N).$$

Note that $\hat{R}_0$ is the potential matrix of operator $\hat{Q}_1$.

Let $W(t)$, $t \geq 0$, be the standard Wiener process. Then the following result takes place.

THEOREM 9.5.3
If the merging condition (9.3.3) takes place and if the limit merged Markov process $\hat{x}(t)$, $t \geq 0$, has a stationary distribution, $\hat{\pi} = (\pi_k^r, 1 \leq$

$r \leq N_k$, $1 \leq k \leq N$) *say, then, under the balance condition* $\hat{\hat{A}} = 0$, *the following weak convergence holds,*

$$\alpha^\varepsilon(t) := \varepsilon^{-1} \int_0^t a(x^\varepsilon(s/\varepsilon^4))ds \Longrightarrow \sigma W(t), \quad as \ \varepsilon \to 0$$

in the sense of the Skorokhod topology on the space $D_{\mathbf{R}}[0, \infty)$, *with variance*

$$\sigma^2 = -2 \sum_{k=1}^{N} \sum_{\ell=1}^{N} \hat{\hat{a}}_k \hat{\hat{r}}_{k\ell} \hat{\hat{a}}_\ell, \tag{9.5.14}$$

where $\hat{\hat{a}}_k = \sum_{r=1}^{N_k} \hat{\pi}_k^r \int_{X_k^r} \pi_k^r(dx) a(x)$, $1 \leq k \leq N$.

REMARK For positiveness of the variance σ^2 defined by the above formula see Korolyuk and Limnios (1998, 1999a). ∎

THEOREM 9.5.4
If the merging condition (9.3.4) takes place and if the limit merged Markov process $\hat{\hat{x}}(t)$, $t \geq 0$, *has a stationary distribution,* $\hat{\hat{\pi}} = (\hat{\hat{\pi}}_k, 1 \leq k \leq N)$ *say, then, under the balance condition* $\hat{\hat{A}} = 0$, *the following weak convergence holds,*

$$\beta^\varepsilon(t) := \varepsilon^{-1} \int_0^t \hat{\hat{a}}(\hat{\hat{x}}(s/\varepsilon^2))ds \Longrightarrow \sigma W(t), \quad as \ \varepsilon \to 0$$

in the sense of the Skorokhod topology on the space $D_{\mathbf{R}}[0, \infty)$, *with variance* σ^2 *given by relation (9.5.14).*

THEOREM 9.5.5
If the merging condition (9.3.3) takes place and the support Markov process $x(t)$, $t \geq 0$, *has a stationary distribution,* $\pi(dx) = (\pi_k^r(dx), 1 \leq r \leq N_k, 1 \leq k \leq N)$ *say, then the following weak convergence holds,*

$$\eta^\varepsilon(t) := \varepsilon^3 \zeta^\varepsilon(t/\varepsilon^4) - \varepsilon^{-1} t \hat{\hat{a}} \Longrightarrow \sigma W(t), \quad as \ \varepsilon \to 0$$

in the sense of the Skorokhod topology on the space $D_{\mathbf{R}}[0, \infty)$, *with variance* σ^2 *given by*

$$\sigma^2 = -2 \sum_{k=1}^{N} \sum_{\ell=1}^{N} \tilde{\tilde{a}}_k \tilde{\tilde{r}}_{k\ell} \tilde{\tilde{a}}_\ell,$$

where $\tilde{\tilde{a}}_k = \hat{\hat{a}}_k - \hat{\hat{a}}$, $1 \leq k \leq N$.

9.6 Integral Functional with Perturbed Kernel

An analogous limit result can be obtained for the integral functional with a perturbed kernel $a^\varepsilon(x) = a(x) + \varepsilon a_1(x)$. In this case, the generator of the coupled process $\xi^\varepsilon(t), x^\varepsilon(t/\varepsilon^3),\ t \geq 0$, has the following representation:

$$L^\varepsilon = \varepsilon^{-3}Q + \varepsilon^{-2}Q_1 + \varepsilon^{-1}A(x) + A_1(x),$$

where the operator A_1 acts as follows: $A_1(x)\varphi(u) = a_1(x)\varphi'(u)$.

Consequently, we have the following singular perturbation problem:

$$[\varepsilon^{-3}Q + \varepsilon^{-2}Q_1 + \varepsilon^{-1}A(x) + A_1(x)](\varphi + \varepsilon\varphi_1 + \varepsilon^2\varphi_2 + \varepsilon^3\varphi_3) = \psi + \theta^\varepsilon,$$

where θ^ε is a negligible function, i.e., $\|\theta^\varepsilon\| \to 0$ as $\varepsilon \to 0$; ψ is given by limiting operator, i.e., $L^0\varphi = \psi$. From Proposition 9.5.2, this operator L^0 is a twice contracted operator defined by

$$L^0\hat\Pi = \hat\Pi \hat L_0 \hat\Pi.$$

Then

$$L^0\hat\Pi = \hat A_1 - \hat A \hat R_0 \hat A,$$

where $\hat\Pi$, $\hat A$, $\hat R_0$ are defined as in previous section and $\hat A_1(k)\varphi(u) = a_{1k}\varphi'(u),\ 0 \leq k \leq N$.

From this, we obtain the limit operator of the limit diffusion process.

$$L^0\varphi(u) = \hat a_1\varphi'(u) + \frac{\sigma^2}{2}\varphi''(u),$$

where the shift coefficient is twice averaged:

$$\hat a_1 = \sum_{k=0}^{N} \pi_k a_{1k}, \quad a_{1k} := \int_{X_k} \pi_k(dx)a_1(x),\ 1 \leq k \leq N.$$

Thus we have the following result.

THEOREM 9.6.1

Let the merging condition (9.3.3) be valid and let the limit merged Markov process $\hat x(t), t \geq 0$, be ergodic with the stationary distribution $\pi = (\pi_k, 1 \leq k \leq N)$. Then the normalized centered integral functional

$$\xi^\varepsilon(t) = \varepsilon^2\zeta^\varepsilon(t/\varepsilon^3) - \varepsilon^{-1}t\hat a$$

converges weakly, in the sense of the Skorokhod topology on the space
$D_{\mathbf{R}}[0, \infty)$, *as* $\varepsilon \to 0$, *to a diffusion process* $\xi(t)$, $t \geq 0$, *with drift* $\hat{a}_1$
and variance

$$\sigma^2 := 2 \sum_{k,l=0}^{N} \pi_k \tilde{a}_k r_{kl} \tilde{a}_l.$$

References

1. Anisimov, V. V. (1995). Switching processes: averaging principle, diffusion approximation and applications. *Acta Aplicandae Mathematica* **40**, 95–141.

2. Anisimov, V. V. (1999). Diffusion approximation for processes with semi-Markov switches, in *Semi-Markov Models and Applications* (Eds., J. Janssen and N. Limnios), pp. 77–101, Kluwer Academic Publishers, Dordrecht, The Netherlands.

3. Billingsley, P. (1968). *Convergence of Probability Measures.* John Wiley & Sons, New York.

4. Ethier, S. N. and Kurtz, T. G. (1986). *Markov Processes: Characterization and convergence*, John Wiley & Sons, New York.

5. Jacod, J. and Shiryaev, A. N. (1987). *Limit Theorems for Stochastic Processes.* Springer-Verlang, Berlin.

6. Korolyuk, V. S. and Korolyuk, V. V. (1999). *Stochastic Models of Systems.* Kluwer Academic Publishers, Dordrecht, The Netherlands.

7. Korolyuk, V. S. (1998). Stability of stochastic systems in diffusion approximation scheme. *Ukrainian Mathematics Journal* **50**, N 1, 36–47.

8. Korolyuk, V. S. and Limnios, N. (1998). A singular perturbation approach for Liptser's functional limit theorem and some extensions. *Theory of Probability and Mathematical Statistics* **58**, 76–80.

9. Korolyuk, V. S. and Limnios, N. (1999). Diffusion approximation of integral functionals in merging and averaging scheme. *Theory of Probability and Mathematical Statistics* **59**.

10. Korolyuk, V. S. and Limnios, N. (1999). Diffusion approximation of integral functionals in double merging and averaging scheme. *Theory of Probability and Mathematical Statistics* **60**.

11. Liptser, R. Sh. (1984). On a functional limit theorem for finite state space Markov processes, in *Steklov Seminar on Statistics and*

Control of Stochastic Processes, pp. 305–316, Optimization Software, Inc., NY.

12. Liptser, R. Sh. and Shiryaev, A. N. (1989). *Theory of Martingales*, Kluwer Academic Publishers, Dordrecht, The Netherlands.

10

Busy Periods for Some Queues with Deterministic Interarrival or Service Times

Claude Lefèvre and Philippe Picard
Université Libre de Bruxelles, Bruxelles, Belgique
Université de Lyon 1, Villeurbanne, France

ABSTRACT The queueing systems considered in this paper are the $D_g/M^{(Q)}/1$ queue with deterministic, not necessarily equidistant, interarrival times, and with exponential service times where customers are served in batches of random size, and its dual the $M^{(Q)}/D_g/1$ queue with Poisson arrival process where customers arrive in batches of random size, and with deterministic, not necessarily equidistant, service times. Our purpose is to determine the exact distribution of the statistic N_r that represents the number of customers served during some busy period initiated by r customers. The problem is analyzed as the first crossing of a compound Poisson trajectory with a fixed non-linear boundary, upper or lower respectively. For the $D_g/M^{(Q)}/1$ queue, a simple explicit formula is derived for the law of N_r that is expressed in terms of a generalization of Appell polynomials. For the $M^{(Q)}/D_g/1$ queue, the law of N_r is now written using a generalization of Abel-Gontcharoff polynomials.

Keywords and phrases Single server queue, busy period, deterministic interarrival or service times, bulk queue, first crossing problem, compound Poisson process, generalized Appell polynomials, generalized Abel Gontcharoff polynomials

10.1 Introduction

In queueing theory, the analysis of busy periods is an important problem, from a theoretical point of view and for practical applications. The literature on this subject, however, is relatively little abundant, especially in view of the large number of papers on the asymptotic behavior

of various queueing systems. A number of results can be found in the
books by Takács (1962), Bhat (1968) and Prabhu (1998), for instance.

Quite often studies rely on the use of Laplace transforms. Results
in distributional terms are generally more convenient and meaningful
but for many models, their derivation is quite difficult or even impossible. Nevertheless, there exist some well-known exceptions (see, e.g., the
references above).

Such exceptions, which have partly motivated our present research, are
the classical $M/D/1$ queue (Poisson arrival process with rate λ, service
times equal to a constant a) and its dual $D/M/1$ (interarrival times equal
to a constant a, exponential service times with rate λ). Our attention
will be mainly focused on the statistic N_r that represents the number of
customers served during the busy period initiated by r (≥ 1) customers;
obviously, $N_r \geq r$ a.s. For the $M/D/1$ queue, Borel (1942), and others
later, proved that the distribution of N_r is given by

$$P(N_r = n) = \frac{r}{n} e^{-\lambda an} \frac{(\lambda an)^{n-r}}{(n-r)!}, \quad n \geq r. \tag{10.1.1}$$

For the $D/M/1$ queue, Stadje (1995) showed (inter alia) that for $r = 1$,
the law of N_1 is provided by

$$P(N_1 \geq n) = e^{-\lambda a(n-1)} \sum_{k=0}^{n-2} \frac{[\lambda a(n-1)]^k}{k!} \frac{n-1-k}{n-1}, \quad n \geq 1. \tag{10.1.2}$$

Now the formulae (10.1.1) and (10.1.2), quite simple and explicit, have
been extended by Stadje (1995) to the cases where the service times or
the interarrival times are deterministic but not necessarily equidistant.
For clarity, we denote these models by $M/D_g/1$ and $D_g/M/1$, respectively. The approach followed by Stadje consists in investigating the
question as some first-passage problem. The formulae obtained, however, are rather intricate and of determinantal form.

In this paper, we are going to consider bulk queueing systems of both
kinds. Specifically, we will examine in Section 10.2 the queue, denoted by
$D_g/M^{(Q)}/1$, in which customers are served in batches of random size Q,
and in Section 10.3 its dual, denoted by $M^{(Q)}/D_g/1$, in which customers
arrive in batches of random size Q.

Our purpose is to derive for these two bulk queues a new and easily
tractable expression for the exact distribution of N_r ($r \geq 1$), the number
of customers served during the busy period initiated by r customers. It
is worth underlining here that the busy period is not defined in the same
way for both models. For the $D_g/M^{(Q)}/1$ queue, the busy periods stops
when the queue becomes empty, that is when all the customers, initial
and subsequent, have been served or at least begin to be served. For

the $M^{(Q)}/D_g/1$ queue, the busy period stops when the queueing system becomes empty, that is when the service of the last customer has been completed.

With respect to the work by Stadje (1995), a first difference is thus that we will discuss the case of bulk queues. A second difference, also important, is that the formulae obtained will be much simpler and with an enlightened algebraic structure.

The methodology followed is that developed in Picard and Lefèvre (1994, 1996, 1997) for the analysis of the first crossing of some counting processes in upper and lower nonlinear boundaries. For the $D_g/M^{(Q)}/1$ queue, we start by observing that N_r corresponds to the level of first crossing of a compound Poisson trajectory with random initial level through a given (increasing) upper nonlinear boundary. We then show that the distribution of N_r can be expressed in terms of a generalization of the classical Appell polynomials. For the $M^{(Q)}/D_g/1$ queue, we see that N_r corresponds, up to r, to the level of first crossing of a compound Poisson trajectory (starting at level 0) through a given (increasing) lower nonlinear boundary. We then establish that the distribution of N_r can be written in terms of a generalization of the less known Abel-Gontcharoff polynomials. The two families of polynomials used have remarkable mathematical structures which generate useful operational properties and make easy their numerical evaluation.

10.2   Preliminaries: A Basic Class of Polynomials

We start by introducing a basic class of polynomials $\{e_n, n \geq 0\}$, of degrees n, that will play a central role in the analysis of both queueing models. For further details, we refer the reader to Picard and Lefèvre (1997).

10.2.1   Construction of the Basic Polynomials

Let us consider a compound Poisson process, with Poisson parameter λ and jump sizes distributed as Q. For $t \geq 0$, let $N(t)$ be the number of jumps by time t, and let $S(t) = \sum_{k=1}^{N(t)} Q_k$ be the total height of all the jump sizes by time t. We have

$$\begin{cases} P[S(t) = 0] = e^{-\lambda t}, \\ P[S(t) = n] = e^{-\lambda t} \sum_{k=1}^{n} \frac{(\lambda t)^k}{k!} P(Q_1 + \ldots + Q_k = n), & n \geq 1. \end{cases}$$
$$(10.2.1)$$

Now let us put, for $t \geq 0$,

$$\begin{cases} e_0(t) = 1 \\ e_n(t) = \sum_{k=1}^{n} \frac{(\lambda t)^k}{k!} P(Q_1 + \ldots + Q_k = n), & n \geq 1, \end{cases} \qquad (10.2.2)$$

yielding

$$P[S(t) = n] = e^{-\lambda t} e_n(t), \quad n \geq 0. \qquad (10.2.3)$$

By construction, $e_n(t)$, $n \geq 0$, is a polynomial of degree n in t, $t \geq 0$. Note that $e_n(0) = \delta_{n,0}$, $n \geq 0$. Obviously, we may define the polynomials $e_n(t)$, $n \geq 0$, by (10.2.2) for any $t \in \mathbb{R}$, not necessarily positive.

This definition by (10.2.2), however, is not very convenient for the sequel. It will be often preferable to work with the generating function of the e_ns, which is given below.

DEFINITION 10.2.1
The polynomials $e_n(t)$, $n \geq 0$, are defined by their formal generating function

$$\sum_{n=0}^{\infty} e_n(t) s^n = e^{\lambda t q(s)}, \qquad (10.2.4)$$

where $q(s) = \displaystyle\sum_{j=1}^{\infty} q_j s^j$ is the probability generating function of Q.

REMARK The e_ns can be determined explicitly for various specific distributions of Q. For example, in the logarithmic case where $q_j = \rho \alpha^j / j$, $j \geq 1$, with $0 < \alpha < 1$ and $\rho = -1/\ln(1 - \alpha)$, then

$$e^{\lambda t q(s)} = (1 - \alpha s)^{-\lambda \rho t},$$

so that from (10.2.4),

$$e_n(t) = \binom{-\lambda \rho t}{n} (-\alpha)^n, \quad n \geq 0.$$

Other illustrations are given in Picard and Lefèvre (1997). ∎

A noteworthy particular situation is when the compound Poisson process reduces to a Poisson process, that is when $Q = 1$ a.s. From (10.2.1), we get the following simple expression for the e_ns.

SPECIAL CASE 10.2.2
In the Poisson case,

$$e_n(t) = (\lambda t)^n / n!, \quad n \geq 0. \tag{10.2.5}$$

10.2.2 A Generalized Appell Structure

With the class of polynomials $\{e_n, n \geq 0\}$, we now associate the operator Δ defined by

$$\begin{cases} \Delta e_0 = 0, \\ \Delta e_n = e_{n-1}, \quad n \geq 1, \end{cases} \tag{10.2.6}$$

the powers of Δ being built recursively from $\Delta^{i+1} = \Delta(\Delta^i)$, $i \geq 0$, with Δ^0 as the identity operator. In that way, $\{e_n, n \geq 0\}$ constitutes a family of generalized Appell polynomials (or Sheffer polynomials) with respect to the operator Δ (see, e.g., Sheffer (1937, 1939)).

The generalized Appell structure allows us to write a generalized Taylor expansion for any polynomial.

PROPERTY 10.2.3
Any polynomial $R(t)$ of degree k admits a generalized Taylor expansion with respect to the operator Δ and the class $\{e_n, n \geq 0\}$, namely for any real b,

$$R(t) = \sum_{i=0}^{k} [\Delta^i R(b)] e_i(t - b). \tag{10.2.7}$$

PROOF We may write

$$R(t) = \sum_{i=0}^{k} \alpha_i e_i(t - b),$$

for some coefficients α_i. By (10.2.6), we then get, for $0 \leq j \leq k$,

$$\Delta^j R(b) = \sum_{i=0}^{k} \alpha_i \Delta^j e_i(0) = \sum_{i=j}^{k} \alpha_i e_{i-j}(0)$$

$$= \sum_{i=j}^{k} \alpha_i \delta_{i,j} = \alpha_j,$$

hence (10.2.7). ∎

Let us go back to the particular situation where $Q = 1$ a.s. Inserting (10.2.5) in (10.2.6) and (10.2.7) yields the following corollary.

SPECIAL CASE 10.2.4
In the Poisson case, $\lambda\Delta$ corresponds to the usual differentiation operator D, and (10.2.7) reduces to a classical Taylor expansion.

The pair $(e_n, n \geq 0; \Delta)$ enjoys various other nice properties. Some of them are used in certain proofs that will not be developed hereafter (see Special cases 10.3.4 and 10.4.3).

10.3 The $D_g/M^{(Q)}/1$ Queue

10.3.1 Model and Notation

The *arrival process* is deterministic in the sense that the customers arrive at fixed points of time, not necessarily equidistant. Given r initial customers, we denote by $v_r, v_{r+1}, v_{r+2}, \ldots$ the arrival time of the 1st, 2nd, 3rd,... new customer; we put $v_0 = v_1 = \ldots = v_{r-1} = 0$ for the r initial customers.

The *service process* is a compound Poisson process. More precisely, the customers are served according to a Poisson process with parameter λ and in batches of random size Q (with $q_j \equiv P(Q = j)$, $j \geq 1$). If at some epoch the queue length is less than the service capacity, the server takes the available customers into service.

The system *stops being busy* at the instant T_r when the queue becomes empty, that is when all the customers, initial and subsequent, have been served or at least begin to be served (in other words, as soon as the server realizes that the service capacity is not being fully utilized). The total number of customers served during the time period $[0, T_r]$ is the statistic N_r under interest. Note that N_r includes also the number of customers in the last batch service, which often will not be filled up. We mention that in the literature, this last batch service is not always counted (see, e.g., Prabhu (1998)).

By definition of the busy period and since $Q \geq 1$, we have for $r = 1$ that $N_1 = 1$. In what follows, we will thus assume that $r \geq 2$.

In Figure 10.1 are drawn the deterministic curve F giving the cumulative number of customers arrived in the queue, including the r initial ones, and the trajectory τ giving the total number of services that are susceptible to have begun. Clearly, τ is the trajectory of the above compound Poisson process $S(t)$ but starting here at a random level distributed as Q, that is the trajectory of the process $Q + S(t)$ say.

Now, we observe that T_r and N_r correspond to the first crossing time and level of the random trajectory τ with the fixed boundary F. We

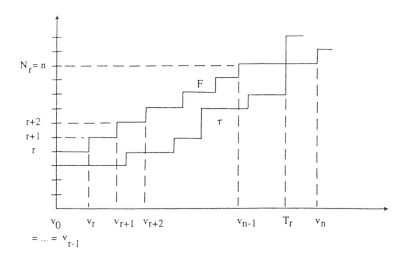

FIGURE 10.1
Arrivals (F) and services (τ) trajectories in the $D_g/M^{(Q)}/1$ queue

emphasize that F is an upper boundary for τ. Note that T_r is a continuous random variable valued in $\mathbb{R}^+$. Moreover, the distribution of N_r is connected with the law of T_r through the identity

$$P(N_r = n) = P(v_{n-1} \leq T_r < v_n), \quad n \geq r \qquad (10.3.1)$$

In order to derive the law of T_r, we are going to determine the probabilities

$$
\begin{aligned}
P_n(t) \quad &= P[\tau \text{ remains below } F \text{ during } (0,t), \text{ and its height at time } t \\
&\qquad\qquad\qquad\qquad\qquad\qquad\qquad\qquad \text{is equal to } n] \\
&= P[T_r > t \text{ and } Q + S(t) = n], \quad n \geq 1, t \in \mathbb{R}^+.
\end{aligned}
$$
$$(10.3.2)$$

10.3.2 Exact Distribution of N_r

When $1 \leq n \leq r - 1$, we have by (10.2.3) that

$$
\begin{aligned}
P_n(t) \quad &= P[Q + S(t) = n] \\
&= e^{-\lambda t} \sum_{j=1}^{n} q_j e_{n-j}(t) \\
&\equiv e^{-\lambda t} f_{n-1}(t), \quad \text{say}.
\end{aligned}
$$
$$(10.3.3)$$

Note that $f_{n-1}(t)$ is a polynomial of degree $n-1$ in t, with, in particular, $P_n(0) = f_{n-1}(0) = q_n$, as expected. Furthermore, for all $n \geq r$, by considering the instant v_n when F reaches the level $n+1$, we get that

$$\text{if } t \leq v_n, \quad P_n(t) = 0, \tag{10.3.4}$$

while

$$\text{if } t \geq v_n, \quad P_n(t) = \sum_{k=0}^{n-1} P_{n-k}(v_n)e^{-\lambda(t-v_n)}e_k(t-v_n). \tag{10.3.5}$$

Indeed, (10.3.4) is straightforward, and (10.3.5) means that at time v_n, $[T_r > v_n$ and $Q+S(v_n) = n-k]$ for some $0 \leq k \leq n-1$, and afterward k customers are served during $(v_n, t]$ (which occurs with probability given by (10.2.3)).

Let us look for an expression of $P_n(t)$ with the form

$$P_n(t) = e^{-\lambda t}H_{n-1}(t)I(t \geq v_n), \quad n \geq 1. \tag{10.3.6}$$

To satisfy (10.3.3) and (10.3.4), we need

$$H_{n-1}(t) = f_{n-1}(t), \quad 1 \leq n \leq r-1, \tag{10.3.7}$$

$$H_{n-1}(v_n) = 0, \quad n \geq r. \tag{10.3.8}$$

Concerning (10.3.5), when $t \geq v_n$ ($\geq v_{n-k}$), we should have

$$H_{n-1}(t) = \sum_{k=0}^{n-1} H_{n-1-k}(v_k)e_k(t-v_n), \quad n \geq r. \tag{10.3.9}$$

Now, we get that the three relations (10.3.7), (10.3.8) and (10.3.9) can be rewritten equivalently as

$$H_{n-1}(0) = q_n, \quad 1 \leq n \leq r-1, \tag{10.3.10}$$

$$H_{n-1}(v_n) = 0, \quad n \geq r, \tag{10.3.11}$$

$$H_{n-1}(t) = \sum_{k=0}^{n-1} H_{n-1-k}(v_n)e_k(t-v_n), \quad n \geq 1. \tag{10.3.12}$$

Indeed, for $1 \leq n \leq r-1$, we have $v_n = 0$ so that (10.3.12), using (10.3.10), reduces to (10.3.7). In particular, we then see that $H_{n-1}(t)$, $n \geq 1$, is a polynomial of degree $n-1$. Note that by (10.3.9), the relation (10.3.12) is restricted to $t \geq v_n$, but $H_n(t)$ being a polynomial, we may define it for any real t.

It remains to identify the class of polynomials $\{H_{n-1}, n \geq 1\}$. By Property 10.2.3, we observe that (10.3.12) will be satisfied if one has

$$\Delta H_{n-1}(t) = H_{n-2}(t), \quad n \geq 2, \tag{10.3.13}$$

since (10.3.12) will then be simply the generalized Taylor expansion of $H_n(t)$ with respect to the operator Δ and the class $\{e_n,\, n \geq 0\}$. As indicated before, the condition (10.3.13) is the characteristic property of a family of generalized Appell polynomials, $\{\bar{A}_{n-1},\, n \geq 1\}$ say. Obviously, (10.3.10) and (10.3.11) guarantee the unicity of the family.

Some useful complements on these polynomials $\bar{A}_{n-1}$ can be found in Picard and Lefèvre (1996, 1997). We recall, inter alia, that they can be expressed as

$$\bar{A}_{n-1}(t) = \sum_{k=0}^{n-1} b_k e_{n-1-k}(t), \quad n \geq 1, \tag{10.3.14}$$

for appropriate coefficients b_k that are independent of n. To sum up, we have proved the following result (see (10.3.6), (10.3.10), (10.3.11) and (10.3.14)).

PROPOSITION 10.3.1

$$P_n(t) = e^{-\lambda t} \bar{A}_{n-1}(t) I(t \geq v_n), \quad n \geq 1, \tag{10.3.15}$$

where the generalized Appell polynomials $\bar{A}_{n-1}$ are evaluated by (10.3.14), the polynomials e_n being given by (10.2.4) and the coefficients b_k being computed recursively from

$$\sum_{k=0}^{n-1} b_k e_{n-1-k}(0) = q_n, \quad 1 \leq n \leq r-1, \tag{10.3.16}$$

$$\sum_{k=0}^{n-1} b_k e_{n-1-k}(v_n) = 0, \quad n \geq r. \tag{10.3.17}$$

This result, combined with (10.3.1), yields directly the distributions of T_r and N_r.

PROPOSITION 10.3.2

$$P(T_r > t) = e^{-\lambda t} \sum_{n=1}^{\infty} \bar{A}_{n-1}(t) I(t \geq v_n), \tag{10.3.18}$$

that is, when $v_j \leq t < v_{j+1},\, j \geq r-1$,

$$P(T_r > t) = e^{-\lambda t} \sum_{n=1}^{j} \bar{A}_{n-1}(t). \tag{10.3.19}$$

Moreover,

$$P(N_r = n) = e^{-\lambda v_{n-1}} \sum_{k=1}^{n-1} \bar{A}_{k-1}(v_{n-1}) - e^{-\lambda v_n} \sum_{k=1}^{n-1} \bar{A}_{k-1}(v_n), \quad n \geq r.$$

$$(10.3.20)$$

Let us examine the particular situation when $Q = 1$ a.s. Using the conclusions of Special case 10.2.4, we deduce from (10.3.13) that the polynomials $\bar{A}_n(t)$ are here given as follows

SPECIAL CASE 10.3.3
In the Poisson case,

$$\bar{A}_{n-1}(t) = \lambda^{n-1} A_{n-1}(t), \quad n \geq 1, \qquad (10.3.21)$$

where $\{A_{n-1}, n \geq 1\}$ is a classical family of Appell polynomials.

Now, for illustration, let us assume that the interarrival times are equal to a constant a, i.e. $v_r - v_{r-1} = v_{r+1} - v_r = \ldots = a$. This is the situation which is usually discussed in the literature. We underline that the boundary F becomes here a straight line. A more explicit formula for the $\bar{A}_{n-1}$'s is available; the proof is omitted for brevity reasons.

SPECIAL CASE 10.3.4
When the interarrival times are equal to a,

$$\bar{A}_{n-1}(t) = f_{n-1}(t), \quad for\ 1 \leq n \leq r - 1, \qquad (10.3.22)$$

$$= \sum_{j=0}^{r-2} \frac{t - a(n - r + 1)}{t - a(j - r + 2)} f_j[a(j - r + 2)] e_{n-1-j}[t - a(j - r + 2)],$$

$$for\ n \geq r. \qquad (10.3.23)$$

In the Poisson case,

$$A_{n-1}(t) = t^{n-1}/(n-1)!, \quad for\ 1 \leq n \leq r - 1, \qquad (10.3.24)$$

$$= \frac{[t - a(n - r + 1)]}{(n-1)!} \sum_{j=0}^{r-2} \binom{n-1}{j} [a(j - r + 2)]^j$$

$$\times\ [t - a(j - r + 2)]^{n-j-2}, \quad for\ n \geq r. \qquad (10.3.25)$$

REMARK In the Poisson case, (10.3.24) and (10.3.25) give, for example, when $r = 2$,

$$A_{n-1}(t) = [t - a(n-1)]t^{n-2}/(n-1)!, \quad n \geq 1, \qquad (10.3.26)$$

and when $r = 3$,

$$A_{n-1}(t) = [t - a(n-2)](t+a)^{n-2}/(n-1)!, \quad n \geq 1.$$

Now, in this case, τ is the trajectory of the compound Poisson process $S(t)$ starting at level 1, and which represents the total number of services that can begin. On the other hand, the total number of services that can be completed yields the trajectory $\tilde{\tau}$ of the same compound Poisson process, but starting now at level 0. Let $\tilde{T}_r$ and $\tilde{N}_r$ be the first crossing time and level of the trajectory $\tilde{\tau}$ through the boundary F. In particular, for $r = 1$, $\tilde{N}_1$ is the variable examined by Stadje (1995) and referred to in (10.1.2) above. Now, we observe that by construction,

$$T_r = \tilde{T}_{r-1} \quad \text{and} \quad N_r = 1 + \tilde{N}_{r-1}. \qquad (10.3.27)$$

For $r = 2$, we have shown that the distribution of N_2 is provided explicitly from (10.3.20), (10.3.21) and (10.3.26). By (10.3.27), the distribution of $\tilde{N}_1$ then follows. It is easily checked that it corresponds precisely to the formula (10.1.2) obtained by Stadje (1995). ∎

10.4 The $M^{(Q)}/D_g/1$ Queue

10.4.1 Model and Notation

The *arrival process* is a compound Poisson process. More precisely, the customers arrive according to a Poisson process with parameter λ and in batches of random size Q (with $q_j \equiv P(Q = j)$, $j \geq 1$).

The *service process* is deterministic in the sense that the customers are served at fixed points of time, not necessarily equidistant. Given r initial customers, we denote by $-u_0$ the service time of the r initial customers, and by $u_0 - u_1, u_1 - u_2, u_2 - u_3 \ldots$ the service times of the 1st, 2nd, 3rd ... new customer; obviously, $0 < -u_0 < -u_1 < -u_2 < \cdots$ This unusual notation will be justified later.

It is convenient to assume for a while that the arrival process starts only at time $-x$, with $-x \leq -u_0$. Ultimately, we will take $x = 0$.

The system *stops being busy* at the instant T_r when the queueing system becomes empty, that is when the service of the last customer has been completed. We are interested in the statistic N_r that counts the number of customers served during $[0, T_r]$.

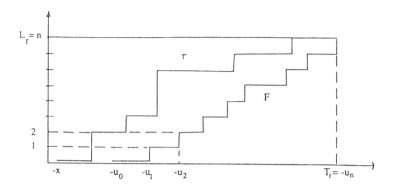

FIGURE 10.2
Arrivals (τ) and services (F) trajectories in the $M^{(Q)}/D_g/1$ queue

In Figure 10.2 are drawn the trajectory τ giving the cumulative number of new arrivals, and the deterministic curve F giving the total number of services of new customers that are susceptible to be completed (after the r initial ones). Clearly, τ is the trajectory of the above compound Poisson process $S(t)$, but starting now at level 0.

We observe here that T_r corresponds to the first crossing time of the random trajectory τ with the fixed boundary F. The associated first crossing level, denoted by L_r, represents the number of new customers served during $[0, T_r]$; thus, $N_r = r + L_r$, and $L_r \geq 0$ a.s. We point out that this time, F is a lower boundary for τ. Note that T_r is a discrete random variable valued in $\{-u_i, i \geq 0\}$. Moreover, L_r is connected with T_r through the relation

$$T_r = -u_{L_r}. \tag{10.4.1}$$

The law of N_r (or L_r) depends, of course, on the initial value x and the family $U \equiv \{u_i, i \geq 0\}$. To stress this dependence, we will use the notation $P(L_r = n | x, U)$, $n \geq 0$, when necessary.

10.4.2   Exact Distribution of N_r

We have

$$P(L_r = 0 | x, U) = P[S(-u_0) = 0] = e^{-\lambda(-u_0 + x)}. \tag{10.4.2}$$

Furthermore, for all $n \geq 1$, by considering the instant $-u_0$ when the service of the r initial customers has been completed, we get that for $-x = -u_0$,

$$P(L_r = n | u_0, U) = 0, \tag{10.4.3}$$

and for $-x < -u_0$, by (10.2.3),

$$P(L_r = n|x, U) = \sum_{k=0}^{n} e^{-\lambda(-u_0 + x)} e_k(-u_0 + x) P(L_r = n - k|u_0, E^k U),$$

$$(10.4.4)$$

where $E^k U$, $k \geq 0$, is the shifted family

$$E^k U \equiv \{u_{k+i}, i \geq 0\}. \tag{10.4.5}$$

Indeed, (10.4.3) is obvious, and (10.4.4) means that k customers arrived during $(-x, -u_0]$, for some $0 \leq k \leq n$, and the first crossing will then occur at level $n - k$, for the trajectory τ starting now at $-u_0$ (instead of $-x$) through the boundary F starting now at $-u_k$ (instead of $-u_0$).

Let us look for an expression of $P(L_r = n|x, U)$ with the form

$$P(L_r = n|x, U) = e^{-\lambda(-u_n + x)} K_n(x|U), \quad n \geq 0. \tag{10.4.6}$$

The relations (10.4.2), (10.4.3) and (10.4.4) require that for $-x \leq -u_0$,

$$K_0(x|U) = 1, \tag{10.4.7}$$

$$K_n(u_0|U) = 0, \quad n \geq 1, \tag{10.4.8}$$

$$K_n(x|U) = \sum_{k=0}^{n} K_{n-k}(u_0|E^k U) e_k(x - u_0), \quad n \geq 1. \tag{10.4.9}$$

Clearly, these three conditions can be rewritten equivalently as

$$K_n(u_0|U) = \delta_{n,0}, \tag{10.4.10}$$

$$K_n(x|U) = \sum_{k=0}^{n} K_{n-k}(u_0|E^k U) e_k(x - u_0), \quad n \geq 0. \tag{10.4.11}$$

In particular, we see that $K_n(x|U)$, $n \geq 0$, is a polynomial of degree n. Note that (10.4.11) may be defined for any real x, so that the restriction $-x \leq -u_0$ is superfluous.

Now, by Property 10.2.3, we observe that (10.4.11) will be satisfied if

$$\Delta K_n(x|U) = K_{n-1}(x|EU), \quad n \geq 1, \tag{10.4.12}$$

since (10.4.11) will then be the generalized Taylor expansion of $K_n(x|U)$ with respect to the operator Δ and the class $\{e_n, n \geq 0\}$. As developed in Picard and Lefèvre (1996), the condition (10.4.12) is the characteristic property of a family of generalized Abel-Gontcharoff polynomials, $\{\bar{G}_n, n \geq 0\}$ say. The unicity of the family is guarateed by (10.4.10).

We recall that these polynomials $\bar{G}_n$ can be determined recursively by

$$\begin{cases} \bar{G}_0(x|U) = 1, \\ \bar{G}_n(x|U) = e_n(x) - \displaystyle\sum_{k=0}^{n-1} e_{n-k}(u_k)\bar{G}_k(x|U), \quad n \geq 1. \end{cases} \quad (10.4.13)$$

Other useful properties are given in Picard and Lefèvre (1996). To obtain the law of L_r, it suffices to take $x = 0$ as announced, which yields the following result (see (10.4.6) and (10.4.13)).

PROPOSITION 10.4.1

$$P(N_r = r + n) = e^{\lambda u_n} \bar{G}_n(0|U), \quad n \geq 0, \quad (10.4.14)$$

where the generalized Abel-Gontcharoff polynomials $\bar{G}_n$ are evaluated by (10.4.13), the polynomials e_n being given by (10.2.4).

In the particular situation when $Q = 1$ a.s., we obtain from (10.4.12) that the polynomials $\bar{G}_n(x|U)$ reduce, up to λ^n, to the more standard Abel-Gontcharoff polynomials (see, e.g., Lefèvre and Picard (1990)).

SPECIAL CASE 10.4.2
In the Poisson case,

$$\bar{G}_n(x|U) = \lambda^n G_n(x|U), \quad n \geq 0, \quad (10.4.15)$$

where $\{G_n(x|U), n \geq 0\}$ is a classical family of Abel-Gontcharoff polynomials.

Finally, for illustration, let us assume, as generally made so far, that the service times are equal to a constant a, i.e. $-u_0 = ra$, $-u_1 = (r+1)a, \ldots$ Here thus, the boundary F is a straight line. The $\bar{G}_n$'s can then be determined quite explicitly; the proof is omitted.

SPECIAL CASE 10.4.3

$$\bar{G}_n(x|U) = \frac{x - u_0}{x - u_n} e_n(x - u_n), \quad n \geq 0. \quad (10.4.16)$$

In the Poisson case,

$$G_n(x|U) = (x - u_0)(x - u_n)^{n-1}/n!, \quad n \geq 0, \quad (10.4.17)$$

that is, the G_n's become the classical Abel polynomials.

REMARK For the Poisson case, inserting (10.4.15) and (10.4.17) in (10.4.14) yields the formula (10.1.1) derived by Borel (1942). ∎

References

1. Bhat, U. N. (1968). *A Study of Queueing Systems M/G/1 and GI/M/1*. Lectures Notes, Springer-Verlag, NewYork.

2. Borel, E. (1942). Sur l'emploi du théorème de Bernoulli pour faciliter le calcul d'une infinité de coefficients. Application au problème de l'attente à un guichet. *Comptes Rendus de l'Académie des Sciences, Paris* **214**, 452–456.

3. Kaz'min, Y. A. (1988). Appell polynomials. In *Encyclopedia of Mathematics* **1**, Reidel Kluwer, Dordrecht, 209–210.

4. Lefèvre, C. and Picard, P. (1990). A non standard family of polynomials and the final size distribution of Reed-Frost epidemic processes. *Advances in Probability* **22**, 25–48.

5. Picard, P. and Lefèvre, C. (1994). On the first crossing of the surplus process with a given upper barrier. *Insurance: Mathematics and Economics* **14**, 163–179.

6. Picard, P. and Lefèvre, C. (1996). First crossing of basic counting processes with lower non-linear boundaries: a unified approach through pseudopolynomials (I). *Advances in Applied Probability* **28**, 853–876.

7. Picard, P. and Lefèvre, C. (1997). The probability of ruin in infinite time with discrete claim size distribution. *Scandinavuan Actuarial Journal* **1**, 58–69.

8. Prabhu, N. U. (1998). *Stochastic Storage Processes: Queues, Insurance Risk, Dams and Data Communication*. Springer-Verlag, New York (second edition).

9. Sheffer, I. M. (1937). Concerning Appell sets and associated linear functional equations. *Duke Mathematics Journal* **3**, 593–609.

10. Sheffer, I. M. (1939). Some properties of polynomial sets of type zero. *Duke Mathematics Journal* **5**, 590–622.

11. Stadje, W. (1995). The busy periods of some queueing systems. *Stochastic Processes and their Applications* **55**, 159–167.

12. Takács, L. (1962) *Introduction to the Theory of Queues*. Oxford University Press, New York.

11

The Evolution of Population Structure of the Perturbed Non-Homogeneous Semi-Markov System

P.-C. G. Vassiliou and H. Tsakiridou
Aristotle University of Thessaloniki, Thessaloniki, Greece

ABSTRACT In the present we study the evolutions of various population structures in a Perturbed Non-Homogeneous Semi-Markov System (P-NHSMS) with different goals in mind. Firstly we start with the expected population structure with respect to the first passage time probabilities and we follow with the study of the expected population structures with respect to the duration of a membership in a state, the state occupancy of a membership and the counting transition probabilities.

Keywords and phrases Semi-Markov, population models, non-homogeneous Markov system

11.1 Introduction

The concept of the Perturbed Non-Homogeneous Semi-Markov System (P-NHSMS) was introduced and defined for the first time in Vassiliou and Tsakiridou (1998). A general theory for semi-Markov process is included in the book by Howard (1971). The theory of semi-Markov models for the evolution of population structures derives its motives from the use of semi-Markov models in manpower systems. In this respect a basic reference useful also for practical purposes is the book by Bartholomew, Forbes and McClean (1991). Elegant theory for semi-Markov models in manpower systems is included in Bartholomew (1982). Moreover an interesting account of theoretical results and important applications of semi-Markov models can be found in Bartholomew (1986), McClean (1980, 1986, 1993), McClean, Montgomery, and Ugwuowo (1997). The evolution of the theory of non-homogeneous Markov systems is described in Vassiliou (1997). The concept of a non-homogeneous semi-Markov system (NHSMS) was introduced and defined for the first time in Vassiliou and Papadopoulou (1992). This provided also a general framework for a number of semi-Markov chain models in manpower systems and a great variety of applied probability models. The asymptotic behavior of the NHSMS was found in analytic form by Papadopoulou and Vassiliou (1994) and Papadopoulou (1997) studied the concepts of counting transitions and entrance probabilities. The idea of a perturbation of a stochastic matrix was introduced in Meyer (1975) and an interesting general theory is included in Campell and Meyer (1979). The concept of a perturbed non-homogeneous Markov system was introduced in Vassiliou and Symeonaki (1997, 1999). The non-homogeneous semi-Markov system in a stochastic environment was studied in Vassiliou (1996). In Section 11.2 of the present we introduce the concept of a perturbed non-homogeneous semi-Markov system (P-NHSMS). In Section 11.3 we introduce for a P-NHSMS the first passage times probabilities and we find them in closed analytic form. Then we establish the expected population structure in relation with the first passage time from a state. In Section 11.4 we introduce for a P-NHSMS the concept of the duration of a membership in a state. In what follows in a matrix form the probabilities of a membership of the system which entered the system at time s to remain in the same state up to time n having made ν transitions are calculated in closed analytic form. This form is useful for many purposes and as an example we could refer to the asymptotic behaviour of the system. The section is closed with the study of the evolution of the expected population structure in the system with respect to the duration of a membership in a state. In Section 11.5 we introduce the concept of the state occupancy of a membership. In this respect the probabilities of a membership of the system which entered in a state at time s that will enter the same state on ν occasions in n-steps are calculated in closed analytic form useful for many purposes. The section terminates with the study of the evolution of the expected population structure in the system with respect to the state occupancies of a membership. Finally in Section 11.6 we study counting transition probabilities for a P-NHSMS a concept first introduced for a NHSMS by Papadopoulou (1998) and calculated with the use of geometric transformations. In the present we study the counting transition probabilities with pure probabilistic methods and we conclude with finding the evolution of the population structure with respect to the counting transition probabilities. The study of the evolutions of the various population structures in a P-NHSMS and finding them in closed analytic form apart from providing the only predictive tools and much deeper understanding of the population also provides the tools for controlling the system locally and asymptotically in time and means for the study of the asymptotic behavior of a P-NHSMS.

11.2 The Perturbed Non-Homogeneous Semi-Markov System

Consider a population (system) which is stratified into classes (states) according to various characteristics. The members of the system could be sections of human societies, animals, biological microorganisms, particles in a physical phenomenon, various types of machines etc. Assume that the sequence $\{T(t)\}_{t=0}^{\infty}$ of all members of the system is known or that it is a known realization of a known stochastic process. Let $S = \{1, 2, ..., k\}$ be the set of states that are assumed to be exclusive so that each member of the system may be in one and only one state at any given time. We consider that initially there are $T(0)$ members in the system and a member entering the system holds a particular membership which moves within the states with the member. When the system is expanding $(\Delta T(t) = T(t) - T(t-1) > 0)$ then new memberships are created in the system which behave like the initial ones. Let $\mathbf{N}(t) = [N_1(t), N_2(t), ..., N_k(t)]$ where $N_i(t)$ is the number of members of the system in the i-th state at time t.

Now let $f_{ij}(t) = \Pr\{$a member of the system who entered state i at time t to choose to move to state j at its next transition$\}$. Let $\mathbf{F}(t) = \{f_{ij}(t)\}_{i,j \in S}$ and assume that $\mathbf{F}(t)$ is of the form

$$\mathbf{F}(t) = \mathbf{F} - \mathcal{E}_f(t) \quad \text{for } t = 1, 2, ...$$

with $\mathbf{F}(t)\mathbf{1}' \leq \mathbf{1}'$, $\mathbf{F}\mathbf{1}' \leq \mathbf{1}'$ and $\mathbf{F}(t) \geq \mathbf{0}$, $\mathbf{F} \geq \mathbf{0}$ for $t = 1, 2, ...$

$$(11.2.1)$$

where $\mathbf{1}' = [1, 1, ..., 1]'$.

Moreover assume that the matrix $\mathcal{E}_f(t)$ is chosen randomly from a finite set $\mathcal{E}_f = \{\mathcal{E}_f(1), \mathcal{E}_f(2), ..., \mathcal{E}_f(\nu)\}$ with

$$\Pr\{\mathcal{E}_f(t) = \mathcal{E}_f(i)\} = c_i(t) > 0 \text{ for } i = 1, 2, ..., \nu. \qquad (11.2.2)$$

Let $p_{i,k+1}(t) = \Pr\{$a member of the system who entered state i at time t to choose to leave the system at its next transition$\}$; then if we define by $\mathbf{p}'_{k+1}(t) = [\, p_{1,k+1}(t), p_{2,k+1}(t), ..., p_{k,k+1}(t)]'$ we have

$$\mathbf{p}'_{k+1}(t) = [\mathbf{I} - \mathbf{F} + \mathcal{E}_f(t)\,]\mathbf{1}'. \qquad (11.2.3)$$

From (11.2.1),(11.2.2),(11.2.3) we get that

$$\mathbf{p}'_{k+1}(t) = \mathbf{p}'_{k+1} - \varepsilon'_{k+1}(t) \tag{11.2.4}$$

where

$$\mathbf{p}'_{k+1} = [\mathbf{I} - \mathbf{F}]\mathbf{1}' \quad \text{and} \quad \varepsilon'_{k+1}(\mathbf{t}) = -\mathcal{E}_{\mathbf{f}}(\mathbf{t})\,\mathbf{1}' \tag{11.2.5}$$

and consequently $\varepsilon'_{k+1}(t)$ is randomly chosen with probabilities

$$\Pr\{\varepsilon'_{k+1}(t) = \varepsilon'_{k+1}(i)\} = c_i(t) > 0 \quad \text{for} \quad t = 1, 2, \dots \quad i = 1, 2, \dots, \nu. \tag{11.2.6}$$

Let $p_{0i}(t) = \Pr\{$a new member who enters the system in state i as a replacement of a member who entered its last state at time $t\}$ and $\mathbf{p_0}(\mathbf{t}) = [p_{01}(t),\ p_{02}(t), \dots,\ p_{0k}(t)]$. Assume that for every t the vector $\mathbf{p}_0(t)$ is of the form

$$\mathbf{p_0}(t) = \mathbf{p_0} - \varepsilon_0(t) \text{ with } \mathbf{p_0}(t)\mathbf{1}' = \mathbf{1}' \text{ and } \mathbf{p_0}\mathbf{1}' = \mathbf{1}' \text{ for every } \mathbf{t}. \tag{11.2.7}$$

Let $\mathbb{M}_{n,m}$ be the set of all $n \times m$ matrices with elements from $\mathbb{R}$. Also let $S\mathbb{M}_n$ be the set of $n \times n$ stochastic matrices. If $\mathbf{p} \in S\mathbb{M}_{1,n}$ then the vector $\varepsilon \in \mathbb{M}_{1,n}$ is a *perturbation vector* for $\mathbf{p}$ if the vector $\tilde{\mathbf{p}} = p - \varepsilon$ is a stochastic vector. Assume that in (11.2.7) $\varepsilon_0(t)$ is a perturbation vector for $\mathbf{p_0}$ and is randomly chosen from a finite set $\mathcal{E}_0 = \{\varepsilon_0(1),\ \varepsilon_0(2), \dots,\ \varepsilon_0(\mu)\}$ with $\Pr\{\varepsilon_0(t) = \varepsilon_0(i)\} = c_{0i}(t) > 0$ for every t and $i=1, 2, \dots, \mu$.

Assume that the stochastic evolutions of selecting $\mathcal{E}_f(t)$ and $\varepsilon_0(t)$ are independent.

Now let $p_{ij}(t) = \Pr\{$a membership of the system which entered state i at time t to choose to move in state j at its next transition$\} = f_{ij}(t) + p_{i,k+1}(t)p_{0j}(t)$.

Let $\mathbf{P}(t) = \{\ p_{ij}(t)\}_{i,j \in S}$; then

$$\mathbf{P}(t) = [\mathbf{F} - \mathcal{E}_f(t)] + [\mathbf{p}_{k+1} - \varepsilon_{k+1}(t)\ ]'[\mathbf{p}_0 - \varepsilon_0(t)\ ] = \mathbf{P} - \mathcal{E}_p(t) \tag{11.2.8}$$

with

$$\mathcal{E}_p(t) = \mathcal{E}_f(t) + \mathbf{p}'_{k+1}\varepsilon_0(t) + \varepsilon'_{k+1}(t)\,\mathbf{p}_0 - \varepsilon'_{k+1}(t)\varepsilon_0(t) \quad \text{for} \quad t = 0, 1, 2, \dots$$

and let

$$\Pr\{\mathcal{E}_p(t) = \mathcal{E}_p(i)\} = c_{pi}(t) > 0 \quad \text{for} \quad t = 1, 2, \dots, \quad i = 1, 2, \dots, \mu\nu. \tag{11.2.9}$$

From Meyer (1975) we get that if $\mathbf{Q} \in S\mathbb{M}_n$ and fully regular, then the matrix $\mathcal{E} \in \mathbb{M}_n$ is a *perturbation matrix* for $\mathbf{Q}$ if the matrix $\tilde{\mathbf{Q}} = \mathbf{Q} - \mathcal{E}$ is also fully regular. In (11.2.8) we assume that $\mathcal{E}_p(t)$ is a perturbation matrix for $\mathbf{P}$.

Define by τ_{ij} $(i,j = 1,2,...,k)$ to be the time that a membership "holds" in state i after j has been selected before the actual transition to j takes place. The holding times τ_{ij} are positive integer-valued random variables with probability mass function $h_{ij}(m,t) = \Pr\{\tau_{ij} = m$ / the membership entered state i at time t and state j has been selected$\}$; where $h_{ij}(0,t) = 0$ for every i,j,t. Now let $\mathbf{H}(m,t) = \{\ h_{ij}(m,t)\}_{i,j\in S}$ and we assume that it is of the form $\mathbf{H}(m,t) = \mathbf{H}(m) - \mathcal{E}_H(m,t)$ where $\mathcal{E}_H(m,t)$ is a *perturbation matrix* for $\mathbf{H}(m,t)$ i.e. the following properties hold:

$$\mathbf{H}(m,t) \geq \mathbf{0} \quad \text{and} \quad \sum_{m=0}^{\infty} \mathbf{H}(\mathbf{m},\mathbf{t}) = \mathbf{U} \quad \text{with}$$

$$\mathbf{H}(m) \geq \mathbf{0} \quad \text{and} \quad \sum_{m=0}^{\infty} \mathbf{H}(m) = \mathbf{U} \tag{11.2.10}$$

where $\mathbf{U}$ is a matrix of 1's. Moreover we assume that $\mathcal{E}_H(m,t)$ is randomly selected from the set of perturbation matrices

$$\mathcal{E}_H = \{\mathcal{E}_H(m,1), \mathcal{E}_H(m,2), ..., \mathcal{E}_H(m,l)\}$$

with probability

$$\Pr\{\mathcal{E}_H(m,t) = \mathcal{E}_H(m,j)\} = c_{Hj}(t) > 0, \quad j = 1,2,...,l. \tag{11.2.11}$$

We assume that the stochastic evolutions of selecting $\mathcal{E}_p(t)$, $\varepsilon_0(t)$ and $\mathcal{E}_H(m,t)$ are independent.

For the new memberships created by expansion at time t let $r_{0i}(t) = \Pr\{$a new membership to enter state i given that it enters the system at time $t\}$ and $\mathbf{r}_0(t) = [r_{01}(t), r_{02}(t), ..., r_{0k}(t)]$. Assume that for every t the vector $\mathbf{r}_0(t)$ is of the form

$$\mathbf{r}_0(t) = \mathbf{r}_0 - \varepsilon_r(t) \quad \text{with} \quad \mathbf{r}_0(t)\mathbf{1}' = \mathbf{1}' \quad \text{and} \quad \mathbf{r}_0\mathbf{1}' = \mathbf{1}'. \tag{11.2.12}$$

The vector $\varepsilon_r(t)$ is assumed to be a perturbation vector for $\mathbf{r}_0$ and is randomly selected from a finite set $\mathcal{E}_r = \{\varepsilon_r(1), \varepsilon_r(2), ..., \varepsilon_r(\theta)\}$ with probabilities $\Pr\{\varepsilon_r(t) = \varepsilon_r(i)\} = c_{ri}(t) > 0$. We assume that the stochastic evolution of selecting $\varepsilon_r(t)$ is independent of the selection of $\mathcal{E}_p(t)\varepsilon_0(t)$ and $\mathcal{E}_H(m,t)$ which in fact seems quite realistic and is a well established inherent assumption (see Figure 11.1).

A population whose evolution is adequately described by a model as above is called *a perturbed non-homogeneous semi-Markov system*.

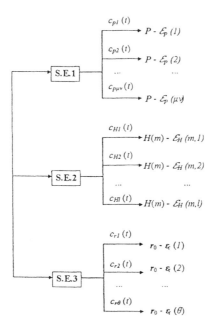

FIGURE 11.1
Stochastic evolutions at time t

11.3   The Expected Population Structure with Respect to the First Passage Time Probabilities

In a P-NHSMS the sequences $\{\mathbf{P}(t)\}_{t=0}^{\infty}$ and $\{\mathbf{H}(m,t)\}_{t=0}^{\infty}$ define uniquely the *inherent perturbed semi-Markov process*. First passage time probabilities for semi-Markov process play the same important role that they played for the basic Markov process [Howard (1971)]. The perspective has obvious practical implications for P-NHSMS especially when the members are human beings and the system is a manpower system. Thus, let $\{\mathcal{F}(\nu,s,n)\}_{\nu,s,n=0}^{\infty}$ be the sequence of matrices the (i,j)-element of which is the probability

$$f_{ij}(\nu,s,n)$$

$$= \Pr\{\text{a membership of the system which entered state } i \text{ at time } s$$

to make $\nu - 1$ transitions in the time interval $(s, s + n)$ and

enters at time $s + n$ for the first time in state j}

It is necessary to find $\{\mathcal{F}(\nu, s, n)\}_{\nu, s, n=0}^{\infty}$ as a function of the basic parameters of the system which for the inherent perturbed semi-Markov process are the sequences $\{\mathbf{P(t)}\}_{t=0}^{\infty}$, $\{\mathbf{H(m, t)}\}_{t=0}^{\infty}$. When a membership enters a state at anytime, let us say t for example, then as described in the previous section four independent stochastic evolutions are taken place for the selection of $\mathbf{P}(t) = \mathbf{P} - \mathcal{E}_p(t)$, $\mathbf{H}(m, t) = \mathbf{H}(m) - \mathcal{E}_H(m, t)$ and $\mathbf{r}_0(t) = \mathbf{r}_0 - \varepsilon_r(t)$ respectively. This fact constitutes the elements of $\{\mathbf{P}(t)\}_{t=0}^{\infty}$, $\{\mathbf{H}(m, t)\}_{t=0}^{\infty}$ and $\{\mathcal{F}(\nu, s, n)\}_{\nu, s, n=0}^{\infty}$ random variables and thus for many reasons it is natural to seek for $\{E[\mathcal{F}(\nu, s, n)]\}_{\nu, s, n=0}^{\infty}$ as a function of $\{E[\mathbf{P}(t)]\}_{t=0}^{\infty}$ and $\{E[\mathbf{H}(m, t)]\}_{t=0}^{\infty}$. It could be proved that

$$E[\mathbf{P}(t)] = \mathbf{P} - \hat{\mathbf{E}}_p(t) \tag{11.3.1}$$

where

$$\hat{\mathbf{E}}_p(t) = E[\mathcal{E}_p(t)] = \sum_{h=1}^{\nu} \mathcal{E}_p(h)c_h(t) + \{\sum_{h=1}^{\nu} \varepsilon'_{k+1}(h)c_h(t)\}\mathbf{p}_0$$

$$+ \mathbf{p}'_{k+1} \sum_{b=1}^{\mu} \varepsilon_0(b)c_{0b}(t) - \sum_{h=1}^{\nu}\sum_{b=1}^{\mu} \varepsilon'_{k+1}(h)\varepsilon_0(b)c_h(t)c_{0b}(t)$$

$$\tag{11.3.2}$$

$$E[\mathbf{H}(m, t)] = \mathbf{H}(m) - \sum_{\alpha=1}^{l} \mathcal{E}_H(m, \alpha)c_{H\alpha}(t) = \mathbf{H}(m) - \hat{\mathbf{E}}_H(t) \tag{11.3.3}$$

$$E[\mathbf{r}_0(t)] = \mathbf{r}_0 - \sum_{h=1}^{\theta} \varepsilon_r(h)c_{rh}(t) = \mathbf{r}_0 - \hat{\mathbf{E}}_r(t). \tag{11.3.4}$$

Let $\mathbf{A} = \{a_{ij}\}$ and $\mathbf{B} = \{b_{ij}\} \in \mathbf{M}_{n,m}$ then the Hadamard product of these matrices is the matrix $\mathbf{A} \diamond \mathbf{B} = \{a_{ij}b_{ij} \in \mathbf{M}_{n,m}$.

We will now firstly find $E[\mathcal{F}(1, s, n)]$. It can be proved that

$$E[\mathcal{F}(1, s, n)] = E[\mathbf{P}(s) \diamond \mathbf{H}(n, s)] = [\mathbf{P} - \hat{\mathbf{E}}_p(s)] \diamond [\mathbf{H}(n) - \hat{\mathbf{E}}_H(s)] \tag{11.3.5}$$

where $\mathbf{P}(s) \diamond \mathbf{H}(n, s)$ is the Hadamard product of the two matrices.

Now let us find $E[\mathcal{F}(2, s, n)]$. Consider the probabilities $f_{ij}(2, s, n)$ for $i, j \in S$ then (see Figure 11.2) taking all the mutually exclusive events where the membership will move at time $s + m_1$ to a state x where $x \neq j$ and then from x to state j at time $s + n$ we can prove that

$$f_{ij}(2, s, n) = \sum_{m_1=1}^{n-1}\sum_{x \neq j} p_{ix}(s)h_{ix}(m_1, s)p_{xj}(s + m_1)h_{xj}(n - m_1, s + m_1).$$

$$\tag{11.3.6}$$

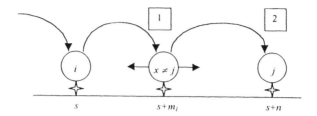

FIGURE 11.2
**Entrance for the first time at state j at time $s + n$, after one
transition**

Taking expectations in both sides due to the independency of the stochastic evolutions we get

$$E[f_{ij}(2, s, n)]$$

$$= \sum_{m_1=1}^{n-1} \sum_{x \neq j} E[p_{ix}(s)] E[h_{ix}(m_1, s)] E[p_{xj}(s + m_1)] E[h_{xj}(n - m_1, s + m_1)].$$

$$(11.3.7)$$

Define the matrix $\hat{\mathbf{H}}$ to be a $k \times k$ matrix of the form

$$\hat{\mathbf{H}} = \begin{pmatrix} 0 & 1 & ... & 1 \\ 1 & 0 & ... & 1 \\ ... & ... & ... & ... \\ 1 & 1 & ... & 0 \end{pmatrix}.$$

Then from (11.3.1) and (11.3.3) equation (11.3.7) it can be proved that takes the form

$$E[\mathcal{F}(2, s, n)] = \sum_{m_1=1}^{n-1} \{[\mathbf{P} - \hat{\mathbf{E}}_p(s)] \diamond [\mathbf{H}(m_1) - \hat{\mathbf{E}}_H(s)]\}$$

$$\times \{([\mathbf{P} - \hat{\mathbf{E}}_p(s + m_1)] \diamond [\mathbf{H}(n - m_1) - \hat{\mathbf{E}}_H(s + m_1)]) \diamond \hat{\mathbf{H}}\}.$$

$$(11.3.8)$$

We will now find $E[\mathcal{F}(3, s, n)]$ as a function of the basic parameters of the system. Following the steps with which we arrived in equation (11.3.8) we arrive at the following (see also Figure 11.3)

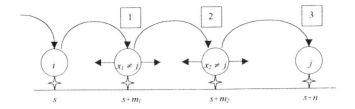

FIGURE 11.3
Entrance for the first time in state j at time $s + n$, after two
transitions

$$\mathcal{F}(3, s, n) = \sum_{m_1=1}^{n-2} \{\mathbf{P}(s) \diamond \mathbf{H}(m_1, s)\}\{\mathcal{F}(2, s + m_1, n) \diamond \hat{\mathbf{H}}\}. \quad (11.3.9)$$

From the equation (11.3.8) we can prove that

$$E[\mathcal{F}(3, s, n)]$$

$$= \sum_{m_1=1}^{n-2} \sum_{m_2=1+m_1}^{n-1} [(\mathbf{P} - \hat{\mathbf{E}}_p(s)) \diamond (\mathbf{H}(m_1) - \hat{\mathbf{E}}_H(s))]$$

$$\times \{[(\mathbf{P} - \hat{\mathbf{E}}_p(s + m_1)) \diamond (\mathbf{H}(m_2 - m_1) - \hat{\mathbf{E}}_H(s + m_1))]$$

$$\times \{[(\mathbf{P} - \hat{\mathbf{E}}_p(s + m_2)) \diamond (\mathbf{H}(n - m_2) - \hat{\mathbf{E}}_H(s + m_2))] \diamond \hat{\mathbf{H}}\} \diamond \hat{\mathbf{H}}\}.$$

$$(11.3.10)$$

Consider the probability which is the (i, j) element of the matrix

$$[\mathbf{P}(s) \diamond \mathbf{H}(m_1 - m_0, s)]\{[\mathbf{P}(s + m_1) \diamond \mathbf{H}(m_2 - m_1, s + m_1)]\{[\mathbf{P}(s + m_2)$$

$$\diamond \mathbf{H}(m_3 - m_2, s + m_2)]\{...\{[\mathbf{P}(s + m_{\nu-2})$$

$$\diamond \mathbf{H}(m_{\nu-1} - m_{\nu-2}, s + m_{\nu-2})]$$

$$\times \{[\mathbf{P}(s + m_{\nu-1}) \diamond \mathbf{H}(m_\nu - m_{\nu-1}, s + m_{\nu-1})] \diamond \hat{\mathbf{H}}\} \diamond \hat{\mathbf{H}}\}... \diamond \hat{\mathbf{H}}\} \diamond \hat{\mathbf{H}}\}$$

$$(11.3.11)$$

with $m_0 = 0$ and $m_\nu = n$.

Then this is (see Figure 11.4) Pr{a membership of the system which entered state i at time s makes $\nu - 1$ transitions in the time interval

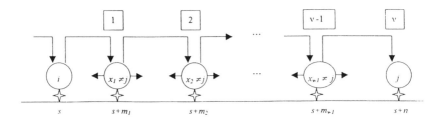

FIGURE 11.4
Entrance for the first time in state j at time $s + n$, after $\nu - 1$ transitions

$(s, s + n)$ at exactly the times $s + m_1, s + m_2, ..., s + m_{\nu-1}$ and enters state j for the first time at time $s + n\}$.

We can write the above product as

$$\prod_{r=0}^{\nu-2} {}^{\{\hat{\mathbf{H}}\}}[\mathbf{P}(s + m_r) \diamond \mathbf{H}(m_{r+1} - m_r, s + m_r)]$$

$$\times \{[\mathbf{P}(s + m_{r+1}) \diamond \mathbf{H}(m_{r+2} - m_{r+1}, s + m_{r+1})] \diamond \hat{\mathbf{H}}\}$$

$$(11.3.12)$$

with $m_0 = 0$ and $m_\nu = n$.

Now if we sum the above product for all possible values of $m_1, m_2, ..., m_{\nu-1}$ the holding times we get

$$\mathcal{F}(\nu, s, n)$$

$$= \sum_{m_1=1}^{n-\nu+1} \{ \sum_{m_2=1+m_1}^{n-\nu+2} \{...\{ \sum_{m_{\nu-1}=1+m_{\nu-2}}^{n-1} \prod_{r=0}^{\nu-2} {}^{\{\hat{\mathbf{H}}\}}[\mathbf{P}(s + m_r)$$

$$\diamond \mathbf{H}(m_{r+1} - m_r, s + m_r)]$$

$$\times \{[\mathbf{P}(s + m_{r+1}) \diamond \mathbf{H}(m_{r+2} - m_{r+1}, s + m_{r+1})] \diamond \hat{\mathbf{H}}\}\}...\}\}$$

$$\text{for } \nu = 4, 5, ..., n. \quad (11.3.13)$$

Using the fact that $E[E[X/Y, Z]] = E[X]$ and the independence of the various stochastic evolutions we get that the expected values of the above probabilities are

$$E[\mathcal{F}(\nu, s, n)]$$

$$= \sum_{m_1=1}^{n-\nu+1} \left\{ \sum_{m_2=1+m_1}^{n-\nu+2} \left\{ ... \left\{ \sum_{m_{\nu-1}=1+m_{\nu-2}}^{n-1} \prod_{r=0}^{\nu-2} \{\hat{\mathbf{H}}\} \right. \right. \right.$$

$$\times \left[(\mathbf{P} - \hat{\mathbf{E}}_p(s + m_r)) \diamond (\mathbf{H}(m_{r+1} - m_r) - \hat{\mathbf{E}}_H(s + m_r)) \right]$$

$$\times \left\{ [(\mathbf{P} - \hat{\mathbf{E}}_p(s + m_{r+1})) \diamond (\mathbf{H}(m_{r+2} - m_{r+1}) - \hat{\mathbf{E}}_H(s + m_{r+1}))] \right.$$

$$\left. \diamond \hat{\mathbf{H}} \} ... \} \right\}. \qquad (11.3.14)$$

Now let us define by

$$N_i^{(f)}(\nu, t)$$

$$= \Pr\{\text{the number of memberships which enter state } i \text{ for the first}$$

$$\text{time at time } t \text{ after } \nu \text{ transitions in the system}\}.$$

Let $\mathbf{N}^{(f)}(\nu, \mathbf{t}) = [\mathbf{N}_1^{(\mathbf{f})}(\nu, \mathbf{t}), \mathbf{N}_2^{(\mathbf{f})}(\nu, \mathbf{t}), ..., \mathbf{N}_\mathbf{k}^{(\mathbf{f})}(\nu, \mathbf{t})]$ then a classical problem in population system theory [Bartholomew (1982) and Bartholomew, Forbes and McClean (1991)] is to find $E[\mathbf{N}^{(f)}(\nu, t)]$ as a function of the basic parameters of the P-NHSMS i.e. $\{\mathbf{P} - \mathcal{E}_p(t)\}_{t=0}^\infty$, $\{\mathbf{H}(m) - \mathcal{E}_H(m, t)\}_{t=0}^\infty$, $\{\mathbf{r}_0 - \varepsilon_r(t)\}_{t=0}^\infty$ and $\{T(t)\}_{t=0}^\infty$.

We consider the population divided into two classes. In the first belong the initial memberships that survived and they enter for the first time in state j at time t. In the second we have all the new memberships that are created due to expansion of the population after $t = 0$ and which enter at time t for the first time in state j. With careful probabilistic analysis we can prove that

$$E[N_j^{(f)}(\nu, t)] = \sum_{i=1}^k E[f_{ij}(\nu, 0, t)] N_i(0)$$

$$+ \sum_{m=1}^t \sum_{i=1}^k \Delta T(m) E[r_{0i}(m)] E[f_{ij}(\nu, m, t - m)]$$

$$(11.3.15)$$

for $t = 1, 2, ...,$ $\nu = 0, 1, 2, ..$ and $f_{ij}(0, m, 0) = 1$ for every i, j and $m = 0, 1, 2, ...$

The above equation in matrix notation takes the form

$$E[\mathbf{N}^{(f)}(\nu, t)] = \mathbf{N}(0) E[\mathcal{F}(\nu, 0, t)]$$

$$+ \sum_{m=1}^{t} \Delta T(m)[\mathbf{r}_0 - \hat{\mathbf{E}}_r(t)] E[\mathcal{F}(\nu, m, t - m)].$$

(11.3.16)

Where the expected value of the matrices $E[\mathcal{F}(\nu, 0, t)]$, $E[\mathcal{F}(\nu, m, t - m)]$ are given in (11.3.14). The relation apart from being the only one to predict the memberships that visit for the first time a state is important in solving a large number of problems in the theory of the P-NHSMS.

11.4 The Expected Population Structure with Respect to the Duration of a Membership in a State

In a P-NHSMS a membership is allowed to make a transition in the same state. As a distinction between real transitions, which require an actual change of state indices as a result of a transition we call *virtual transitions* the situation where the state indices could be the same after the transition. The distinction between real and virtual transitions is useful in the theory of the semi-Markov process and as a consequence to the P-NHSMS. Some physical processes require that only real transitions be allowed, in other processes the virtual transitions are most important. For example when the state of the process represents the last brand purchased by the customer in a marketing model, a virtual transition represents a repeat purchase of a brand, an event of frequent importance to the analyst. As another example in a manpower system, a virtual transition represents a consideration for promotion for a member which eventually fails to be promoted. Thus there are important reasons to preserve our ability to speak of virtual transitions. We are now in a position to introduce the aspect of the duration in a state for a membership in a P-NHSMS. In this respect let us define by

$d_i(\nu, s, n)$

$= \mathrm{Pr}\{$a membership which entered state i at time s remains in

state i up to time n having made ν transitions$\}$

We define by $\mathcal{D}(\nu, \int, \backslash)$ the diagonal matrix with $d_i(\nu, s, n)$ in the (i, i)-position. Obviously due to the stochastic evolutions taken place at any time in a P-NHSMS the probabilities $\mathcal{D}(\nu, s, n)$ are random variables and thus we seek $E[\mathcal{D}(\nu, s, n)]$.

Now define

$$w_i(n, s)$$

$$= \Pr\{\text{a membership of the system which entered state } i \text{ at time } s$$

$$\text{to stay } n \text{ time units in state } i \text{ before its next transition}\}$$

Let $\mathbf{w}(n, s)$ the $k \times k$ matrix which has zeros everywhere apart from the diagonal which has in position i the element $w_i(n, s)$.

It can be proved that the matrix which in the (i, i)-position contains the probabilities that a membership which entered state i at time s stays n time units in state i before making a real transition is given by

$$\mathbf{w}(n, s) - \mathbf{P}(s) \diamond \mathbf{H}(n, s) \diamond \mathbf{I} = \mathbf{I} \diamond \{[\mathbf{P}(s) \diamond \mathbf{H}(n, s)]\mathbf{U}\} - [\mathbf{P}(s) \diamond \mathbf{H}(n, s) \diamond \mathbf{I}]$$

For $\nu = 2, 3, \ldots$ consider the probability which is the (i, j)-element of the matrix

$$[\mathbf{P}(s) \diamond \mathbf{H}(m_1, s) \diamond \mathbf{I}][\mathbf{P}(s + m_1) \diamond \mathbf{H}(m_2 - m_1, s + m_1) \diamond \mathbf{I}\ldots]$$

$$\ldots[\mathbf{P}(s + m_{\nu-2}) \diamond \mathbf{H}(m_{\nu-1} - m_{\nu-2}, s + m_{\nu-2}) \diamond \mathbf{I}]$$

$$\times [\mathbf{w}(n - m_{\nu-1}, s + m_{\nu-1}) - \mathbf{P}(s + m_{\nu-1})$$

$$\diamond \mathbf{H}(n - m_{\nu-1}, s + m_{\nu-1}) \diamond \mathbf{I}] \tag{11.4.1}$$

then it can be proved that this is the probability (see Figure 11.5) $\Pr\{$a membership of the system which entered state i at time s makes $\nu - 1$ virtual transitions in the time interval $(s, s + n)$ at exactly the times $s + m_1, \ s + m_2, \ \ldots, s + m_{\nu-1}$ and makes a real transition at time $s + n\}$

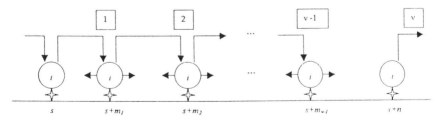

FIGURE 11.5
Duration in state i

We can write the above product as

$$\{\prod_{r=1}^{\nu-2}[\mathbf{P}(s+m_r)\diamond\mathbf{H}(m_{r+1}-m_r,s+m_r)\diamond\mathbf{I}]\}$$

$$\times[\mathbf{w}(n-m_{\nu-1},s+m_{\nu-1})-\mathbf{P}(s+m_{\nu-1})\diamond\mathbf{H}(n-m_{\nu-1},s+m_{\nu-1})\diamond\mathbf{I}].$$

$$(11.4.2)$$

Define by

$$\delta(\nu)=\begin{cases}1,&\text{if }\nu=0\\0,&\text{otherwise}\end{cases}.$$

Then if among other steps we sum the above product for all possible values of the holding times $m_1,m_2,...,m_{\nu-1}$ then we can prove that

$$\mathcal{D}(\nu,s,n)$$

$$=\sum_{m_1=1}^{n-\nu+1}\{\sum_{m_2=1+m_1}^{n-\nu+2}\{...\{\sum_{m_{\nu-1}=1+m_{\nu-2}}^{n-1}\{\prod_{r=1}^{\nu-2}[\mathbf{P}(s+m_r)$$

$$\diamond\mathbf{H}(m_{r+1}-m_r,s+m_r)\diamond\mathbf{I}]\}$$

$$\times[\mathbf{w}(n-m_{\nu-1},s+m_{\nu-1})-\mathbf{P}(s+m_{\nu-1})$$

$$\diamond\mathbf{H}(n-m_{\nu-1},s+m_{\nu-1})\diamond\mathbf{I}]\}...\}\}$$

$$+\delta(\nu-1)[\mathbf{w}(n,s)-\mathbf{P}(s)\diamond\mathbf{H}(n,s)\diamond\mathbf{I}]\}\qquad(11.4.3)$$

where obviously the first part in the second side of the above equation is zero for $\nu=1$.

We can prove that

$$E[\mathbf{w}(n,s)]=\mathbf{I}\diamond\{[(\mathbf{P}-\hat{\mathbf{E}}_p(s))\diamond(\mathbf{H}(m)-\hat{\mathbf{E}}_H(s))]\mathbf{U}\}.\qquad(11.4.4)$$

Using the fact that $E[E[X/Y,Z]]=E[X]$ and the independence of the various stochastic evolutions of selections we can prove that the expected values of the above probabilities are

$$E[\mathcal{D}(\nu,s,n)]$$

$$=\sum_{m_1=1}^{n-\nu+1}\{\sum_{m_2=1+m_1}^{n-\nu+2}\{...\{\sum_{m_{\nu-1}=1+m_{\nu-2}}^{n-1}\{\prod_{r=1}^{\nu-2}\{[\mathbf{P}-\hat{\mathbf{E}}_p(s+m_r)]$$

$$\diamond[\mathbf{H}(m_{r+1}-m_r)-\hat{\mathbf{E}}_H(s+m_r)]\diamond\mathbf{I}\}\}$$

$$\times\{\mathbf{I}\diamond\{[(\mathbf{P}-\hat{\mathbf{E}}_p(s+m_{\nu-1}))\diamond(\mathbf{H}(n-m_{\nu-1})-\hat{\mathbf{E}}_H(s+m_{\nu-1}))]\mathbf{U}\}$$

$$- \left[(\mathbf{P} - \hat{\mathbf{E}}_p(s + m_{\nu-1})) \right] \diamond (\mathbf{H}(n - m_{\nu-1}) - \hat{\mathbf{E}}_H(s + m_{\nu-1})) \right] \diamond \mathbf{I} \} \}...$$

$$+ \delta(\nu - 1) \{ \mathbf{I} \diamond \{ [(\mathbf{P} - \hat{\mathbf{E}}_p(s)) \diamond (\mathbf{H}(m) - \hat{\mathbf{E}}_H(s))] \mathbf{U} \}$$

$$- [\mathbf{P} - \hat{\mathbf{E}}_p(s)] \diamond [\mathbf{H}(m) - \hat{\mathbf{E}}_\mathbf{H}(s)] \diamond \mathbf{I} \}. \tag{11.4.5}$$

Now let us define by $N_i^{(d)}(\nu, t) = \{$the number of memberships which entered in state i in the system and remain in i up to time t having made ν transitions$\}$ and $\mathbf{N}^{(d)}(\nu, t) = [N_1^{(d)}(\nu, t), N_2^{(d)}(\nu, t), ..., N_k^{(d)}(\nu, t)]$. With probabilistic argument we can prove that

$$E[\mathbf{N}^{(d)}(\nu, t)] = \mathbf{N}(0) E[\mathcal{D}(\nu, \prime, \sqcup)]$$

$$+ \sum_{m=1}^{t} \Delta T(m)[\mathbf{r}_0 - \hat{\mathbf{E}}_r(t)] E[\mathcal{D}(\nu, m, t - m)] \tag{11.4.6}$$

Relations (11.4.5), (11.4.6) provide a complete description of the evolution of the expected population structure of the system in relation with their duration in the state.

11.5 The Expected Population Structure with Respect to the State Occupancy of a Membership

In this section we will study the number of times a membership occupies a specific state in a time interval. In many applications this aspect is important in practice. In this respect let us define by

$\omega_{ij}(\nu, s, n)$

= Pr{a membership of the system which entered in state i at time

  s will enter state j on ν occasions in the time period $(s, s + n)$}

Let $\otimes(\nu, \int, \backslash)$ be the matrix the (i, j) element of which is the above probability.
Now define by

$f_{ij}(s, n)$

= Pr{a membership of the system which entered in state i at time s

  will enter state j for the first time at time n}

Let $\mathcal{F}(s,n) = \{f_{ij}(s,n)\}_{i,j\in S}$ then it is clear that

$$E[\otimes(\prime,\int,\backslash)] = \sum_{\langle=\backslash+\infty}^{\infty} \mathcal{E}[\mathcal{F}(\int,\langle)] = \mathbf{U} - \sum_{h=0}^{n} \mathbf{E}[\mathcal{F}(\int,\langle)] = \mathcal{E}[\mathcal{F}^{*}(\int,\langle)].$$

(11.5.1)

Consider the expected value of the probability (see Figure 11.6) Pr{a membership of the system which entered in state i at time s will enter for the first time in state j at time $s+m_1$, then will enter again for the first time in state j at time $s+m_2$, ..., will enter again for the first time since $s+m_{\nu-1}$ in state j at time $s+m_{\nu}$ which is the ν-th time it enters since time s and then will not enter again up to time $s+n$}.

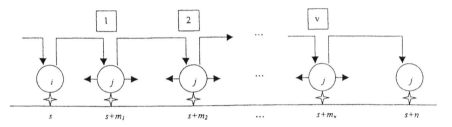

FIGURE 11.6
Occupancy of state j for ν times in the time interval $(s, s+n)$

It can be proved that it is the (i,j)-element of the matrix

$$E[\mathcal{F}(s,m_1)]\{E[\mathcal{F}(s+m_1, m_2-m_1)]\{E[\mathcal{F}(s+m_2, m_3-m_2)]\{...$$
$$...\{E[\mathcal{F}(s+m_{\nu-1}, m_\nu - m_{\nu-1})][E[\mathcal{F}^*(s+m_\nu, n-m_\nu)] \diamond \mathbf{I}]$$
$$\diamond\mathbf{I}\}... \diamond \mathbf{I}\} \diamond \mathbf{I}\}$$
$$= E[\prod(\mathcal{F}(s, m_1, m_2, ..., m_\nu))].$$

(11.5.2)

Now if we sum the above product for all possible values of the holding times then we get

$$E[\otimes(\nu,s,n)] = \sum_{m_1=1}^{n-\nu+1} \sum_{m_2=1+m_1}^{n-\nu+2} \cdots \sum_{m_\nu=1+m_{\nu-1}}^{n} E[\prod(\mathcal{F}(s, m_1, m_2, ..., m_\nu))].$$

(11.5.3)

In order to complete the calculation of $E[\otimes(\nu,s,n)]$ we have to find $E[\mathcal{F}(s,n)]$ for any value of n and s. However it can be easily seen that

$$E[\mathcal{F}(s,n)] = \sum_{\nu=1}^{n} E[\mathcal{F}(\nu,s,n)]$$

(11.5.4)

and the probabilities $E[\mathcal{F}(\nu, s, n)]$ are calculated in Section 11.3.

In order to find the population structure with respect the number of times a membership has visited its present state we define $N_j^{(\omega)}(\nu, t) =$ Pr{the number of memberships which entered state j on ν occasions in the time period $(0, t)$}.

Let $\mathbf{N}^{(\omega)}(\nu, t) = [N_1^{(\omega)}(\nu, t), N_2^{(\omega)}(\nu, t), ..., N_k^{(\omega)}(\nu, t)]$ then with probabilistic arguments we can prove that

$$E[\mathbf{N}^{(\omega)}(\nu, t)] = \mathbf{N}(0)E[\otimes(\prime, \int, \backslash)]$$

$$+ \sum_{m=1}^{t} \Delta T(m)[\mathbf{r}_0 - \hat{\mathbf{E}}_r(m)]E[\otimes(\nu, m, t - m)] \quad (11.5.5)$$

where the expected values of the matrices $E[\otimes(0, s, n)]$, $E[\otimes(\nu, m, t - m)]$ and $E[\mathcal{F}(\nu, s, n)]$, are given through relations (11.5.1), (11.5.3) and (11.3.14), which completes the study of the evolution of the expected structure of the system in relation with the number of times a membership visited a specific state.

11.6 The Expected Population Structure with Respect to the Counting Transition Probabilities

The semi-Markov processes allows a distinction between the number of time units that have passed and the number of transitions that have occurred. Thus it is important to study not only the probability of a membership being in each state at time t but also the probability distribution of the number of transitions made by that time. In this respect let us define by

$\phi_{ij}(\nu, s, n)$

$= \text{Pr}\{\text{a membership which entered state } i \text{ at time } s \text{ will make } \nu$

$\text{transitions in the time interval } (s, s + n) \text{and be in state } j$

$\text{at time } s + n\}$

Now we define by $\mathbf{\Phi}(\nu, s, n)$ to be the matrix the (i, j)-element of which is the probability $\phi_{ij}(\nu, s, n)$. We call the probabilities $\mathbf{\Phi}(\nu, s, n)$ the *counting transition probabilities* of the P-NHSMS. Counting transition probabilities for NHSMS were studied by Papadopoulou (1997) using geometric transformations. In the present we will be interested in finding the expected values $E[\mathbf{\Phi}(\nu, s, n)]$ which are necessary in finding the expected population structure in respect with counting transitions.

Consider the probability $\Pr\{$a membership of the system which entered in state i at time s will make ν transitions at times $s+m_1, s+m_2, ..., s+n$ and enter in state j at time $s + n\}$.

Then it can be proved that the expected value of the above probability is the (i, j)-element of the matrix

$$
\{[\mathbf{P} - \hat{\mathbf{E}}_p(s)] \diamond [\mathbf{H}(m_1) - \hat{\mathbf{E}}_H(s)]\}
$$
$$
\times \{[\mathbf{P} - \hat{\mathbf{E}}_p(s + m_1)] \diamond [\mathbf{H}(m_2 - m_1) - \hat{\mathbf{E}}_H(s + m_1)]\}...
$$
$$
...\{[\mathbf{P} - \hat{\mathbf{E}}_p(s + m_{\nu-1})] \diamond [\mathbf{H}(m_\nu - m_{\nu-1}) - \hat{\mathbf{E}}_H(s + m_{\nu-1})]\}
$$
$$
= \prod_{r=1}^{\nu} \{[\mathbf{P} - \hat{\mathbf{E}}_p(s + m_{r-1})] \diamond [\mathbf{H}(m_r - m_{r-1}) - \hat{\mathbf{E}}_H(s + m_{r-1})]\}
$$

$$(11.6.1)$$

with $m_0 = 0$ and $m_\nu = n$.

With the above result as basis of our analysis of events we can prove that

$$
E[\mathbf{\Phi}(\nu, s, n)]
$$
$$
= \sum_{m_1=1}^{n-\nu+1} \{ \sum_{m_2=1+m_1}^{n-\nu+2} \{...\{ \sum_{m_{\nu-1}=1+m_{\nu-2}}^{n-1} \{\prod_{r=1}^{\nu} \{[\mathbf{P} - \hat{\mathbf{E}}_p(s + m_{r-1})]
$$
$$
\diamond [\mathbf{H}(m_r - m_{r-1})
$$
$$
- \hat{\mathbf{E}}_H(s + m_{r-1})]\} \cdot E[\mathbf{W}(s + m_r, n - m_r)]
$$
$$
+ \sum_{m_1=1}^{n-\nu+1} \{ \sum_{m_2=1+m_1}^{n-\nu+2} \{...\{ \sum_{m_{\nu-1}=1+m_{\nu-2}}^{n-1}
$$
$$
\times \{\prod_{r=1}^{\nu} \{[\mathbf{P} - \hat{\mathbf{E}}_p(s + m_{r-1})] \diamond [\mathbf{H}(m_r - m_{r-1}) - \hat{\mathbf{E}}_H(s + m_{r-1})]\}
$$

$$(11.6.2)$$

where $m_0 = 0$ and $m_\nu = n$, and from Vassiliou and Tsakiridou (1999) we get that

$$E[\mathbf{W}(n,s)] = \sum_{m=n+1}^{\infty} \mathbf{I} \diamond \{E[\mathbf{P}(s) \diamond \mathbf{H}(m,s)]\mathbf{U}\}$$

$$= \mathbf{I} \diamond \{([\mathbf{P} - \hat{\mathbf{E}}_p(s)] \diamond \sum_{m=n+1}^{\infty} [\mathbf{H}(m) - \hat{\mathbf{E}}_H(s)])\mathbf{U}\}$$

Now define by $N_i^{(\phi)}(\nu,t) = \{$the number of memberships of the P-NHSMS that are in state i at time t having made ν transitions from their entrance into the system$\}$ and $\mathbf{N}^{(\phi)}(\nu,t) = [N_1^{(\phi)}(\nu,t), N_2^{(\phi)}(\nu,t), ...,$ $N_k^{(\phi)}(\nu,t)]$. Then the expected population structure with respect the number of transitions made up to time t it can be proved that it is given by

$$E[\mathbf{N}^{(\phi)}(\nu,t)] = \mathbf{N}(0)E[\mathbf{\Phi}(\nu,0,n)]$$

$$+ \sum_{m=1}^{t} \Delta T(m)[\mathbf{r}_0 - \hat{E}_r(m)]E[\mathbf{\Phi}(\nu,m,t-m)]$$

$$(11.6.3)$$

where the expected value of the matrices $E[\mathbf{\Phi}(\nu,0,n)]$, $E[\mathbf{\Phi}(\nu,m,t-m)]$ are given in (11.6.2). The relation apart from being the only one to predict the memberships in each state in relation with the number of transitions made up to time t is important in solving a large number of problems in the theory of the P-NHSMS.

It is easy to see that

$$E[\mathbf{N}(t)] = \sum_{\nu=0}^{t} E[\mathbf{N}^{(\phi)}(\nu,t)].$$

References

1. Bartholomew, D. J. (1982). *Stochastic models for social processes*, Third Edition. John Wiley & Sons, Chichester.

2. Bartholomew, D. J. (1986). Social applications of semi-Markov processes. In *Semi-Markov Models: Theory and Applications* (Ed., J. Janssen). Plenum Press, New York.

3. Bartholomew, D. J., Forbes, A. F., and McClean, S. I. (1991). *Statistical Techniques for Manpower Planning*. John Wiley & Sons, Chichester.

4. Campbell, S. L. and Meyer, C. D. (1979). *Generalized inverses of Linear Transformations.* Pitman, London.

5. Howard, R. A. (1971). *Dynamic Probabilistic Systems*, Vol. II. John Wiley & Sons, Chichester.

6. Janssen, J. (1986) (Ed.). *Semi-Markov Models: Theory and Applications.* Plenum Press, New York.

7. McClean, S. I. (1980) A semi-Markov model for a multigrade population with Poisson recruitment. *Journal of Applied Probability* **17**, 846–852.

8. McClean, S. I. (1986). Semi-Markov models for manpower planning. In *Semi-Markov Models: Theory and Applications* (Ed., J. Janssen). Plenum Press, New York.

9. McClean, S. I. (1993). Semi-Markov models for human resource modelling. *IMA Journal of Mathematics Applied in Business and Industry* **4**, 307–315.

10. McClean, S. I., Montgomery, E., and Ugwuowo, F. (1997). Non-homogeneous continuous time and semi-Markov manpower models. *Applied Stochastic Models and Data Analysis* **13**, 191–198.

11. Mehlman, A. (1979). Semi-Markovian manpower models in continuous time. *Journal of Applied Probability* **6**, 416–422.

12. Meyer, C. D. (1975). The role of the group generalized inverse in the theory of finite Markov chains. *SIAM Review* **17**, 443–464.

13. Papadopoulou, A. A. (1997). Counting transitions-entrance probabilities in non-homogeneous semi-Markov systems. *Applied Stochastic Models and Data Analysis* **13**, 199–206.

14. Papadopoulou, A. A. and Vassiliou, P.-C. G. (1994). Asymptotic behavior of non-homogeneous semi-Markov systems. *Linear Algebra and its Applications* **210**, 153–198.

15. Vassiliou, P.-C. G. (1996). The non-homogeneous semi-Markov system in a stochastic environment. In *Applied Probability in Honor of J. M. Gani* (Eds., C. C. Heyde, Yu. V. Parohorov, R. Pyke, and S. T. Rachev). Springer-Verlag.

16. Vassiliou, P.-C. G. (1998). The evolution of the theory of Non-homogeneous Markov systems. *Applied Stochastic Models and Data Analysis* **13**, 159–176.

17. Vassiliou, P.-C. G. and Papadopoulou, A. A. (1992). Non-homogeneous semi-Markov systems and maintainability of the state sizes. *Journal of Applied Probability* **29**, 519–534.

18. Vassiliou, P.-C. G. and Symeonaki, M. A. (1998). The perturbed non-homogeneous Markov System in continuous time. *Applied*

Stochastic Models and Data Analysis **13**, 207–216.

19. Vassiliou, P.-C. G. and Symeonaki, M. A. (1999). The perturbed non-homogeneous Markov system. *Linear Algebra and its Applications* **289**, 319–332.

20. Vassiliou, P.-C. G. and Tsakiridou, H. (1999). The perturbed non-homogeneous semi-Markov system. To appear.

PART III

DISTRIBUTIONS, CHARACTERIZATIONS, AND APPLICATIONS

12

Characterizations of Some Exponential Families Based on Survival Distributions and Moments

M. Albassam, C. R. Rao, and D. N. Shanbhag
University of Sheffield, Sheffield, UK
The Pennsylvania State University, University Park, PA
University of Sheffield, Sheffield, UK

ABSTRACT Shanbhag (1972b) characterized a family of exponential distributions via unimodality. With a minor modification in Shanbhag's argument, one can arrive at a characterization of a family of gamma distributions, considering α-unimodality in place of unimodality. A discrete version of the latter result has recently been obtained by Sapatinas (1993, 1999). In the present paper, we give a unified approach to arrive at the aforementioned results and certain of their variations. In the process of doing this, we also show as to how these results are linked with the results of Laha and Lukacs (1960), Shanbhag (1979), and Morris (1982), as well as with a damage model introduced by Rao (1965).

Keywords and phrases Damage model, unimodality, α-unimodality, α-monotonicity, Laha-Lukacs theorem, Shanbhag-Morris theorem, Rao-Rubin theorem, Poisson, negative binomial distributions, power series distribution, exponential family

12.1 Introduction

Khintcthine (1938) [also see Lukacs (1970, p. 92)] showed that a distribution function F on $\mathbb{R}$ is unimodal with vertex 0 if and only if a random variable X with distribution function F satisfies the relation

$$X \stackrel{d}{=} U Z , \tag{12.1.1}$$

where U and Z are independent random variables such that U is uniformly distributed on $(0, 1)$. Olshen and Savage (1970) gave an extended version of the characterization showing that (in the notation as above) F

is α-unimodal, with $\alpha > 0$, if and only if (12.1.1) with $U^{1/\alpha}$ in place of U and U and Z satisfying the stated conditions holds. A discrete analogue relative to distributions concentrated on $\{0, 1, \ldots\}$ of α-unimodality has recently been introduced by Steutel (1988); he defined an $\mathbb{N}_\nvDash$-valued r.v. X to be α-montone (discrete α-unimodal) where $\alpha > 0$, if

$$X \stackrel{d}{=} U^{1/\alpha} \circ Z ,$$

where U is as defined above, Z is a nonnegative integer-valued r.v. independent of U, and the operator $\circ$ is in the sense of Steutel and van Harn (1979). The latter definition can easily be seen to be linked with a damage model introduced by Rao (1965). Indeed, it follows that a distribution concentrated on $\{0, 1, \ldots\}$ is α-unimodal in the sense of Steutel if and only if it denotes, in a damage model, the distribution of the resulting random variable when the original random variable (which is nonnegative integer-valued) is subjected to a destructive process according to the survival distribution

$$S(x \mid z) = \frac{\binom{-\alpha}{x}\binom{-1}{z-x}}{\binom{-\alpha-1}{z}} , \quad x = 0, 1, 2, \ldots, z \; ; z = 0, 1, \ldots \quad . \qquad (12.1.2)$$

[Refer to Rao (1965) and M. B. Rao and Shanbhag (1982) for a precise definition of a damage model]. The survival distribution in (12.1.2) is a specialized version of the survival distribution considered by Shanbhag (1977) to give a generalization of the celebrated Rao-Rubin theorem (1964).

Shanbhag (1972b) showed, among other things, that for all θ (in an open interval)

$$X_\theta \stackrel{d}{=} U \, Z_\theta , \qquad (12.1.3)$$

with U as uniformly distributed on $(0, 1)$ independently of Z_θ and the distributions of X_θ and Z_θ forming respectively certain exponential families, if and only if X_θ or $-X_\theta$ is exponentially distributed, with mean of a specific form, for each θ. With a minor modification in the argument of Shanbhag, it follows that replacing, in (12.1.3), U by $U^{1/\alpha}$ with α positive and independent of θ, one gets a characterization of a gamma distribution with index α in place of that of an exponential distribution. Essentially a discrete analogue of the latter result, following implicitly Shanbhag's idea, has recently been given by Sapatinas (1993, 1999). [Incidentally, Alamastaz (1985) has extended yet another result in Shanbhag (1972b), which is not directly linked with the results in our paper; for a bibliography of the literature relevant to the Alamastaz result, see Pakes (1992, 1994) and Rao and Shanbhag (1994, Chapter 6).]

The purpose of the present paper is to unify and extend the aforementioned results of Shanbhag and Sapatinas, and show that these results are linked with certain results of Laha and Lukacs (1960), Shanbhag (1972a, 1979), and Morris (1982). In view of the observation that we have made above concerning the α-unimodality in the discrete case, it follows that the result of Sapatinas referred to is linked with a certain result on damage models; we also arrive at variations of this latter result of the type in Shanbhag and Clark (1972) and Patil and Ratnaparkhi (1975, 1977).

12.2   An Auxiliary Lemma

The following lemma plays a crucial role in the present investigation.

LEMMA 12.2.1
Let $\{(\alpha(\theta), \beta(\theta)) : \theta \in \Theta\}$ *be a collection of 2-component real-vectors such that there exist at least three points,* θ_0, θ_1 *and* θ_2, *in* Θ *such that* $\beta(\theta_0), \beta(\theta_1)$ *and* $\beta(\theta_2)$ *are distinct. Also, let* $c_{i1}, c_{i2}, c_{i3}, i = 1, 2$, *be real numbers with* c_{11} *and* c_{21} *as positive and distinct. Then*

$$\exp\{c_{i1}\alpha(\theta)\} = c_{i2} + c_{i3}\beta(\theta) , \quad i = 1, 2; \ \theta \in \Theta \tag{12.2.1}$$

if and only if $c_{13} = c_{23} = 0$, $\alpha(\theta) \equiv \alpha_0$ *on* Θ *for some number* α_0, *and* $\exp\{c_{i1}\alpha_0\} = c_{i2}, i = 1, 2$.

PROOF The "if" part of the lemma is obvious. To prove the "only if" part of the lemma, it is sufficient if we show that under the assumptions in the lemma, (12.2.1) holds only if $c_{13} = c_{23} = 0$. We do this, by showing that if we assume that at least one of c_{i3}s is nonzero, then we are led to a contradiction. Indeed, If we have the assumption, then (12.2.1) implies that $\alpha(\theta)$ has distinct values at $\theta = \theta_0, \theta_1$ and θ_2, where θ_0, θ_1 and θ_2 as in the statement of the lemma, and hence that both c_{13} and c_{23} are nonzero. Assume then this is so. Without loss of generality, we can assume that $c_{21} < c_{11}$, $c_{13} > 0$, and that $\beta(\theta_0)$ is the smallest of $\beta(\theta_0), \beta(\theta_1)$ and $\beta(\theta_2)$. Then (12.2.1) implies that

$$\exp\{c_{i1}(\alpha(\theta) - \alpha(\theta_0))\} - 1 = c_{i3}\left(\beta(\theta) - \beta(\theta_0)\right)\exp\{-c_{i1}\alpha(\theta_0)\} ,$$
$$i = 1, 2; \theta_0, \theta_1, \theta_2. \tag{12.2.2}$$

Eq. (12.2.2), in turn, implies that $\alpha(\theta_0)$ is the smallest of $\alpha(\theta_0), \alpha(\theta_1)$ and $\alpha(\theta_2)$ and that

$$\exp\{c_{21}(\alpha(\theta) - \alpha(\theta_0))\} - 1$$

$$= \frac{c_{23}}{c_{13}} \exp\{-(c_{21} - c_{11})\alpha(\theta_0)\} \left(\exp\{c_{11}(\alpha(\theta) - \alpha(\theta_0))\} - 1\right) \, , \theta$$

$$= \theta_0, \theta_1, \theta_2. \tag{12.2.3}$$

The function f defined by

$$f(x) = \frac{(1+x)^{c_{21}/c_{11}} - 1}{x} \, , \quad x > 0 \, ,$$

is strictly decreasing under our assumption that $c_{21} < c_{11}$. (One of the approaches to see this is that based on the binomial theorem because for each x,

$$f(x) = \left\{ (1+x) \left(1 - \frac{x}{1+x}\right)^{1 - \frac{c_{21}}{c_{11}}} - 1 \right\} / x. \;)$$

As (12.2.3) implies that $f(x)$ at $x = \exp\{c_{11}(\alpha(\theta_i) - \alpha(\theta_0))\} - 1$, $i = 1, 2$, are identical, we have a contradiction. Hence the lemma follows. ∎

12.3 Characterizations Based on Survival Distributions

Let $\{(X_\theta, Z_\theta) : \theta \in \Theta\}$ be a family of random vectors with Θ having at least three points, such that for each θ, X_θ and Z_θ have respectively the distributions of the form

$$F_\theta(x) \propto \int_{(-\infty, x]} e^{\lambda_1(\theta)y} \, \mu_1(dy) \, , \quad x \in \mathbb{R} \, ,$$

and

$$G_\theta(z) \propto \int_{(-\infty, z]} e^{\lambda_2(\theta)y} \, \mu_2(dy) \, , \quad z \in \mathbb{R} \, ,$$

where μ_1 and μ_2 are σ-finite measures defined on the Borel σ-field of $\mathbb{R}$ and λ_1 and λ_2 are real functions on Θ with λ_1 also as a one-to-one function. Motivated by the link between the α-unimodality property and the survival distribution of a certain type in a damage model, we now establish the following theorems. These theorems extend the existing characterizations of exponential families based on α-unimodality.

THEOREM 12.3.1
Let $\{(X_\theta, Z_\theta) : \theta \in \Theta\}$ be a family of random vectors with nonnegative integer-valued components such that both μ_1 and μ_2 are concentrated on $\{0, 1, ...\}$ with $\mu_2(\{0, 1\}^c) > 0$. Also, let $\{a_x : x = 0, 1, ...\}$ be a sequence of positive real numbers. Then for all z with $\mu_2(\{z\}) > 0$ and $\theta \in \Theta$,

$$P_\theta\{X_\theta = x \mid Z_\theta = z\} = \frac{a_x}{A_z} \,, x = 0, 1, ..., z, \qquad (12.3.1)$$

where $A_z = \sum_{i=0}^{z} a_i$, $z = 0, 1, ...$, if and only if, for all $\theta \in \Theta$, $P_\theta\{X_\theta \leq Z_\theta\} = 1$, and, for all $\theta \in \Theta$ and some $\beta > 0$,

$$P_\theta\{X_\theta = x, Z_\theta = z\} \propto a_x \left(e^{\lambda_1(\theta)}\beta\right)^z , \quad x = 0, 1, ..., z; \; z = 0, 1, ... \; . \qquad (12.3.2)$$

PROOF The "if" part follows easily. To prove the "only if" part, assume that for all $\theta \in \Theta$ the conditional distribution of X_θ given Z_θ is as in the statement of the theorem. Then there exists a function $C : \Theta \longrightarrow (0, \infty)$ such that

$$\sum_{z=x}^{\infty} e^{\lambda_2(\theta)z} \frac{\mu_2(\{z\})}{A_z} = C(\theta) \, e^{\lambda_1(\theta)x} \frac{\mu_1(\{x\})}{a_x} \,, x = 0, 1, ..., \qquad (12.3.3)$$

where $A_z = \sum_{x=0}^{z} a_x$. From (12.3.3), we get that

$$e^{\lambda_2(\theta)x} \frac{\mu_2(\{x\})}{A_x} = C(\theta) \, e^{\lambda_1(\theta)x} \left(\frac{\mu_1(\{x\})}{a_x} - e^{\lambda_1(\theta)} \frac{\mu_1(\{x+1\})}{a_{x+1}}\right) ,$$
$$x = 0, 1, ... \quad ,$$

and hence that

$$e^{(\lambda_2(\theta)-\lambda_1(\theta))x} \frac{\mu_2(\{x\})}{A_x} = C(\theta) \left(\frac{\mu_1(\{x\})}{a_x} - e^{\lambda_1(\theta)} \frac{\mu_1(\{x+1\})}{a_{x+1}}\right) ,$$
$$x = 0, 1, ... \quad . \qquad (12.3.4)$$

Also, in view of the assumptions in the theorem including especially that $\mu_2(\{0, 1\}^c) > 0$, it follows from (12.3.3) that $\mu_1(\{x\}) > 0$ at least for $x = 0, 1, 2$. (12.3.4), for $x = 0$, then implies that $\frac{\mu_1(\{0\})}{a_0} - e^{\lambda_1(\theta)} \frac{\mu_1(\{1\})}{a_1} > 0$ on Θ and $\mu_2(\{0\}) > 0$. In view of this, it follows from (12.3.4) that

$$e^{(\lambda_2(\theta) - \lambda_1(\theta))x} \frac{\mu_2(\{x\})}{A_x}$$

$$= \left(\frac{\mu_2(\{0\})}{a_0}\right) \left(\frac{\mu_1(\{x\})}{a_x} - e^{\lambda_1(\theta)} \frac{\mu_1(\{x+1\})}{a_{x+1}}\right)$$

$$\times \left(\frac{\mu_1(\{0\})}{a_0} - e^{\lambda_1(\theta)} \frac{\mu_1(\{1\})}{a_1}\right)^{-1} , \quad x = 0, 1, ...; \ \theta \in \Theta. \ (12.3.5)$$

For $x = 1, 2$, and inductively for $x = k, k + 1$ with k as a positive integer, (12.3.5) can be reduced to the form of (12.2.1) with $\alpha(\theta) = \lambda_2(\theta) - \lambda_1(\theta)$ and $\beta(\theta) = \left(\frac{\mu_1(\{0\})}{a_0} - e^{\lambda_1(\theta)} \frac{\mu_1(\{1\})}{a_1}\right)^{-1}$. Hence it follows from the lemma that $\lambda_2(\theta) = \lambda_1(\theta) + c$, $\theta \in \Theta$, for some constant c, and that

$$e^{cx} \frac{\mu_2(\{x\})}{A_x} = \frac{\mu_2(\{0\})}{a_x \mu_1(\{0\})} \mu_1(\{x\}) , x = 0, 1, ... \quad (12.3.6)$$

with $\{\frac{\mu_1(\{x\})}{a_x} : x = 0, 1, ...\}$ as a geometric sequence. From (12.3.1) and (12.3.6), it is immediate that, for each $\theta \in \Theta$, the joint distribution of X_θ and Z_θ is of the form in (12.3.2) with $P_\theta(X_\theta \leq Z_\theta) = 1$. ∎

THEOREM 12.3.2
Let $\{(X_\theta, Z_\theta) : \theta \in \Theta\}$ *be such that μ_2 has at least one positive support point and μ_1 and μ_2 are concentrated on $\mathbb{R}_+$, and let a be a positive real-valued continuous function on $(0, \infty)$ such that it is integrable with respect to Lebesgue measure on $(0, x)$ for some, and hence all, $x \in (0, \infty)$. For all $\theta \in \Theta$, there exist versions of the conditional distributions of X_θ given Z_θ satisfying*

$$P_\theta\{X_\theta \leq x \mid Z_\theta = z\} = \begin{cases} 1 & \text{for all } x \in \mathbb{R}_+ \text{ if } z = 0 \\ \frac{A(x)}{A(z)} & \text{for all } x \in (0, z] \text{ if } z > 0, \end{cases} \quad (12.3.7)$$

where $A(x) = \int_0^x a(y)dy$ (and $A(z)$ is defined in obvious way) if and only if, for all $\theta \in \Theta$, $P_\theta(0 < X_\theta < Z_\theta) = 1$, and, for some positive number β and all $\theta \in \Theta$, (X_θ, Z_θ) has an absolutely continuous distribution with probability density function of the following form (with respect to Lebesgue measure on $\mathbb{R}_+^2$)

$$f_\theta(x, z) \propto a(x) \left(e^{\lambda_1(\theta)} \beta\right)^z , \quad x \in (0, z], \ z \in (0, \infty).$$

PROOF The "if" part of the assertion is obvious and to prove the "only if" part of the assertion proceed as follows. (12.3.7) implies that the restriction to $(0, \infty)$ of μ_1 is absolutely continuous with respect to Lebesgue measure with left continuous Radon-Nikodym derivative m_1 such that, for each $\theta \in \Theta$,

$$m_1(x) = (K(\theta))^{-1} \left\{ a(x) \frac{\int_{[x,\infty)} (A(y))^{-1} e^{\lambda_2(\theta)y} \mu_2(dy)}{e^{\lambda_1(\theta)x}} \right\},$$

$$x \in (0,\infty). \tag{12.3.8}$$

for some positive real-valued function K on Θ. (12.3.8) implies that for each $x_0 > 0$ such that $\mu_2((x_0,\infty)) > 0$, the hazard measure relative to the survival function

$$\left(\int_{[x,\infty)} (A(y))^{-1} e^{\lambda_2(\theta)y} \mu_2(dy) \right)$$

$$\times \left(\int_{[x_0,\infty)} (A(y))^{-1} e^{\lambda_2(\theta)y} \mu_2(dy) \right)^{-1}, \quad x \in (x_0,\infty),$$

where the measure is as defined in Kotz and Shanbhag (1980), is such that it is not independent of θ while its value for each Borel set B of Lebesgue measure zero is, for $\theta \in \Theta$. This is impossible unless the restriction to $(0,\gamma)$, where $\gamma = \sup\{x : x \in supp[\mu_2]\}$ (in standard notation), of μ_2 is absolutely continuous with respect to Lebesgue measure, with its Radon-Nikodym derivative, m_2, satisfying, for each $\theta \in \Theta$, for almost all $x \in (0,\gamma)$,

$$-(A(x))^{-1} e^{\lambda_2(\theta)x} m_2(x)$$

$$= \left(\int_{[x,\infty)} (A(y))^{-1} e^{\lambda_2(\theta)y} \mu_2(dy) \right) (\lambda_1(\theta) + g(x)) \tag{12.3.9}$$

for some Borel measurable function g that is independent of θ. Appealing to (12.3.8) once more, one can see that, for each $\theta \in \Theta$, for almost all $x \in (0,\gamma)$,

$$-(A(x))^{-1} e^{(\lambda_2(\theta)-\lambda_1(\theta))x} m_2(x) = K(\theta) m_1(x) (a(x))^{-1} (\lambda_1(\theta) + g(x)). \tag{12.3.10}$$

(In the last two statements by "almost all", we mean "almost all with respect to Lebesgue measure".)

Consequently, it follows that given $\theta_0, \theta_1 \in \Theta$ with $\theta_0 \neq \theta_1$, we can take m_2 in (12.3.9) and (12.3.10) to be such that

$$m_2(x) \, (A(x))^{-1} \left((K(\theta_1))^{-1} \, e^{(\lambda_2(\theta_1) - \lambda_1(\theta_1))x} \right.$$

$$\left. - (K(\theta_0))^{-1} \, e^{(\lambda_2(\theta_0) - \lambda_1(\theta_0))x} \right)$$

$$= (\lambda_1(\theta_0) - \lambda_1(\theta_1)) \, m_1(x)(a(x))^{-1}.$$

This, in turn, implies that we can choose m_2 and g in (12.3.9) and (12.3.10) to be left continuous, and that, with this choice, the equations referred to hold for all $\theta \in \Theta$ and $x \in (0, \gamma)$. Understand by (12.3.9) and (12.3.10) henceforth their new versions with g and m_2 left continuous. Given any $x \in (0, \gamma)$, we can find x_0, x_1 such that $0 < x_0 < x_1 < x$ and obtain from (12.3.10), the expressions for $\exp\{(\lambda_2(\theta) - \lambda_1(\theta))(x - x_0)\}$ and $\exp\{(\lambda_2(\theta) - \lambda_1(\theta))(x_1 - x_0)\}$ as linear functions of $(\lambda_1(\theta) + g(x_0))^{-1}$ (on noting that $\lambda_1(\theta) + g(x_0) < 0$) on Θ. In view of the lemma, we get then that $\lambda_2(\theta) = \lambda_1(\theta) + c$ for all $\theta \in \Theta$ and some constant c. This, in turn, implies, because of the validity of (12.3.10) for all $\theta \in \Theta$ and $x \in (0, \gamma)$, that, for some constant ξ, $K(\theta)(\lambda_1(\theta) + \xi)$ is independent of θ on Θ and $g(x) = \xi$ for all x in $(0, \gamma)$. We have hence that (12.3.9) with $g \equiv \xi$ holds for all $x \in (0, \gamma)$ and $\theta \in \Theta$; this implies that $\gamma = \infty$ and

$$(A(x))^{-1} \, e^{\lambda_2(\theta)x} \, m_2(x) = \eta(\theta) \, e^{(\lambda_1(\theta) + \xi)x} \,, \quad x \in (0, \gamma), \theta \in \Theta,$$
(12.3.11)

for some function η on Θ. (The relation between λ_i's implies then that η is independent of θ on Θ). Also, as K in (12.3.8) is not identically equal to a constant on Θ, it is clear that the ratio of the normalizing functions of the two exponential families cannot be identically equal to a constant on Θ; this implies that $\mu_1(\{0\}) = \mu_2(\{0\}) = 0$. In view of what we have proved, we can then claim that for each $\theta \in \Theta$, (X_θ, Z_θ) has an absolutely continuous distribution (w.r.t. Lebesgue measure) on $\mathbb{R}_+^2$ such that $P_\theta(0 < X_\theta < Z_\theta) = 1$ and it has a probability density function of the form

$$f_\theta(x, z) \propto a(x) \left(e^{\lambda_1(\theta)} \beta \right)^z \,, \quad x \in (0, z], \, z \in (0, \infty),$$

with β as a positive constant. (β here equals $\exp\{\xi\}$ where ξ is as in (12.3.11).) Hence, we have the theorem. $\blacksquare$

REMARK If we take in Theorem 12.3.1, $a_x = \binom{-\alpha}{x} (-1)^x$, $x = 0, 1, ...,$ then the result reduces to that corresponding to α-unimodal distributions, extending the result of Sapatinas (1993, 1999). We can obtain characterizations of several other discrete distribution families via the property in the theorem, taking specific $\{a_x\}$. In particular, if we take

$a_x = \frac{1}{x!}, x = 0, 1, ...$, we get the result with

$$P_\theta\{X_\theta = x, Z_\theta = x\} \propto \frac{(e^{\lambda_1(\theta)} \beta)^z}{x!} , \quad x = 0, 1, ..., z; z = 0, 1, ... \quad .$$

Note that in this latter case, the marginal distribution of X_θ is $P_0(e^{\lambda_1(\theta)} \beta)$ and that of Z_θ is such that

$$P_\theta\{Z_\theta = z\} \propto \left(\sum_{x=0}^{z} \frac{1}{x!}\right) \left(e^{\lambda_1(\theta)} \beta\right)^z , \quad z = 0, 1, ...$$

∎

REMARK In the case of $a(x) = x^{\alpha-1}, x \in (0, \infty)$ with, $\alpha > 0$, Theorem 12.3.2 provides us with a characterization of a gamma distribution family based on α-unimodality. In the notation used in the theorem, the family that is characterized in this case has $X_\theta \sim Ga(|\lambda_1(\theta) + \log \beta|, \alpha)$ and $Z_\theta \sim Ga(|\lambda_1(\theta) + \log \beta|, \alpha + 1)$ where $\lambda_1(\theta) + \log \beta < 0$. We can obviously use other forms of the function $a(\cdot)$ to characterize various other distribution families with X_θ having a well known distribution on $\mathbb{R}_+$. ∎

REMARK In view of Theorems 12.3.1 and 12.3.2, it follows that it is not possible to have $\{(X_\theta, Z_\theta) : \theta \in \Theta\}$ with $\mu_i(\mathbb{R}_+) > \mathcal{K}, \mu_{\exists}(-\mathbb{R}_+) > \mathcal{K}$ for $i = 1, 2$ and the families of the conditional distributions $\{P_\theta\{X_\theta \leq x, Z_\theta \leq z \mid X_\theta \geq 0, Z_\theta \geq 0\}, x, z \in \mathbb{R}_+ : \theta \in \mathcal{\nleq}\}$ and of the conditional distributions $\{P_\theta\{-X_\theta \leq x, -Z_\theta \leq z \mid X_\theta \leq 0, Z_\theta \leq 0\}, x, z \in \mathbb{R}_+ : \theta \in \mathcal{\nleq}\}$ are both of the form characterized in Theorems 12.3.1 or Theorems 12.3.2. In particular, this implies that the characterization of gamma distributions based on α-unimodality, mentioned in the introduction follows essentially as a corollary to Theorem 12.3.2. ∎

REMARK The characterization of gamma distributions referred to in the introduction does not hold if (in the notation of the present section) we take $\lambda_i(\theta)|y|$ in place of $\lambda_i(\theta)y$ for each of $i = 1, 2$ in the definitions of F_θ and G_θ. This is illustrated by the following example:

EXAMPLE
Let $\{(X_\theta, Z_\theta) : \theta \in (0, \infty)\}$ be a family of random vectors with absolutely continuous distributions (w.r.t. Lebesgue measure) on $\mathbb{R}^{\mathcal{K}}$ such that for each θ, the corresponding density is given by

$$f_\theta(x, z) = \begin{cases} \frac{\theta^{\alpha+1}}{2\,\Gamma(\alpha)} |x|^{\alpha-1} e^{-\theta\,|z|} & \text{if } 0 < x < z \text{ or } z < x < 0 \\ 0 & \text{otherwise,} \end{cases}$$

with $\alpha > 0$ (and fixed). Note that we have here for each $\theta \in (0, \infty)$, a random variable U_θ independent of Z_θ such that it is uniformly distributed on $(0, 1)$ and

$$X_\theta \overset{d}{=} U_\theta^{1/\alpha} Z_\theta.$$

Moreover, we have in the present case, for each $\theta \in (0, \infty)$, with the notation used above,

$$F_\theta(x) = \int_{-\infty}^{x} \frac{\theta^\alpha}{2\,\Gamma(\alpha)} \, |y|^{\alpha-1} \, e^{-\theta\,|y|} \, dy, \quad x \in \mathbb{R}$$

and

$$G_\theta(z) = \int_{-\infty}^{z} \frac{\theta^{\alpha+1}}{2\,\Gamma(\alpha+1)} \, |y|^\alpha \, e^{-\theta\,|y|} \, dy, \quad z \in \mathbb{R}.$$

∎

REMARK The counter example given in the paper of Sapatinas (1999) (i.e. in his Remark 2) is not valid. This is so because (in his notation) for the P of the example, we have $\rho = \infty$, contradicting the parameter space of G that is given in the paper. ∎

REMARK Instead of stating Sapatinas's (1999) result in terms of power series and modified power series families, one can restate it, with reparameterization, in terms of exponential families concentrated on $\{0, 1, \ldots\}$. (Note that the point $'0'$ may be excluded from the parameter space, as it leads only to a degenerate member.) Also, the way Sapatinas's theorem is stated, it holds with "$b_i > 0$ for some $i \geq 2$" deleted provided we understand that the case of $P_\theta\{X_\theta = 0\} \equiv 1$ is excluded from consideration. ∎

REMARK Both Theorem 12.3.1 and 12.3.2 hold even when in place of the condition that λ_1 is a one-to-one function, we take only that it has three distinct values. However, this does not apply to the results given in the next section (i.e. in Section 12.4). ∎

12.4 Characterizations Based on Moments

The relations of the type

$$X_\theta \overset{d}{=} V \, Z_\theta, \quad \theta \in \Theta, \tag{12.4.1}$$

and

$$X_\theta \overset{d}{=} V \circ Z_\theta, \quad \theta \in \Theta, \tag{12.4.2}$$

where "∘" is the operator referred to in the introduction, with $V \in (0,1)$ and as a random variable independent of Z_θ, lead us to some versions of the problems addressed by the next two theorems. Note that (12.4.2) is equivalent to stating that for each θ, (X_θ, Z_θ) is as in Rao's (1965) damage model with the survival distribution as mixed binomial with the mixing distribution fixed. These results are in the spirit of those given by Morris (1982) and Shanbhag (1979), and by Shanbhag and Clark (1972), Patil and Ratnaparkhi (1975, 1977) and Sapatinas and Aly (1994) respectively.

THEOREM 12.4.1
Let $\{(X_\theta, Z_\theta) : \theta \in \Theta\}$ be as mentioned in Section 12.3 before the statement of Theorems 12.3.1, but, with at least one μ_i as nondegenerate, Θ as an open interval, λ_1 as continuous and λ_2 such that

$$\lambda_2(\theta) = \lambda_1(\theta) + c , \quad \theta \in \Theta$$

with c as a constant. Further let $0 < c_2 < c_1$. Then for all $\theta \in \Theta$,

$$E_\theta(X_\theta) = c_1 \ E_\theta(Z_\theta) \tag{12.4.3}$$

and

$$E_\theta(X_\theta^2) = c_2 \ E_\theta(Z_\theta^2) , \tag{12.4.4}$$

if and only if $c_1^2 < c_2$, and, for a given $\theta_0 \in \Theta$, $(X_{\theta_0}, Z_{\theta_0})$ or $(-X_{\theta_0}, -Z_{\theta_0})$ has its components to be gamma distributed with the same scale parameter and the index parameters as $\frac{c_1(c_1-c_2)}{c_2-c_1^2}$ and $\frac{(c_1-c_2)}{c_2-c_1^2}$ respectively.

PROOF The "if" part follows easily because of the form of the distributions of X_θ and Z_θ for each $\theta \in \Theta$. To prove the "only if" part of the theorem, assume that (12.4.3) and (12.4.4) are valid for all $\theta \in \Theta$. There is no loss of generality in assuming that $\lambda_1(\theta) \equiv \theta$. Define now

$$\beta_i(\theta) = \int_{\mathbb{R}} \exp\{\theta x\} \ \mu_i(dx) , i = 1, 2, \quad \theta \in \Theta.$$

In view of the validity of (12.4.3) and (12.4.4) for all $\theta \in \Theta$, it follows (in the standard notation for the first and second derivatives) that

$$\frac{\beta_1'(\theta)}{\beta_1(\theta)} = c_1 \frac{\beta_2'(\theta)}{\beta_2(\theta)} , \quad \theta \in \Theta \tag{12.4.5}$$

and that

$$\frac{\beta_1''(\theta)}{\beta_1(\theta)} = c_2 \frac{\beta_2''(\theta)}{\beta_2(\theta)} , \quad \theta \in \Theta . \tag{12.4.6}$$

Eq. (12.4.5) implies, in view of (12.4.6), that

$$\frac{\beta_1''(\theta)}{\beta_1(\theta)} - \left(\frac{\beta_1'(\theta)}{\beta_1(\theta)}\right)^2 = c_1 \left[\frac{\beta_2''(\theta)}{\beta_2(\theta)} - \left(\frac{\beta_2'(\theta)}{\beta_2(\theta)}\right)^2\right]$$

$$= c_1 \left[\frac{1}{c_2}\frac{\beta_1''(\theta)}{\beta_1(\theta)} - \frac{1}{c_1^2}\left(\frac{\beta_1'(\theta)}{\beta_1(\theta)}\right)^2\right], \theta \in \Theta.$$

$$(12.4.7)$$

On simplifying (12.4.7) gives that

$$Var_\theta(X_\theta) = \frac{\beta_1''(\theta)}{\beta_1(\theta)} - \left(\frac{\beta_1'(\theta)}{\beta_1(\theta)}\right)^2 = \frac{c_2 - c_1^2}{c_1(c_1 - c_2)}\left(\frac{\beta_1'(\theta)}{\beta_1(\theta)}\right)^2$$

$$\left(= \frac{c_2 - c_1^2}{c_1(c_1 - c_2)}\left[(E_\theta(X_\theta))^2\right]\right), \quad \theta \in \Theta. \qquad (12.4.8)$$

Eq. (12.4.5) implies also that

$$\beta_1(\theta) \propto (\beta_2(\theta))^{c_1}, \quad \theta \in \Theta. \qquad (12.4.9)$$

Essentially, in view of Morris (1982) or Rao and Shanbhag (1994; Corollary 9.2.4) and the fact that one of the μ_i's is nondegenerate, the assertion follows from (12.4.8) and (12.4.9). ▌

THEOREM 12.4.2
Let $\{(X_\theta, Z_\theta) : \theta \in \Theta\}$, c_1 and c_2 be as in Theorems 12.4.1. Then for all $\theta \in \Theta$,

$$E_\theta(X_\theta) = c_1 E_\theta(Z_\theta)$$

$$E_\theta(X_\theta(X_\theta - 1)) = c_2 E_\theta(Z_\theta(Z_\theta - 1))$$

if and only if one of the following conditions holds:

(i) $c_2 > c_1^2$ and, for a given $\theta_0 \in \Theta$, $(X_{\theta_0}, Z_{\theta_0})$ or

$$\left(-X_\theta - \frac{c_1(c_1 - c_2)}{c_2 - c_1^2}, -Z_\theta - \frac{c_1 - c_2}{c_2 - c_1^2}\right)$$

has its components to be negative binomially distributed with parameter vectors $\left(\frac{c_1(c_1 - c_2)}{c_2 - c_1^2}, p\right)$ and $\left(\frac{c_1 - c_2}{c_2 - c_1^2}, p\right)$ respectively, for some $p \in (0, 1)$.

(ii) $c_2 < c_1^2$ *with* $\frac{c_1(c_1-c_2)}{c_1^2-c_2}$ *and* $\frac{c_1-c_2}{c_1^2-c_2}$ *as integers, and, for a given* $\theta_0 \in \Theta$, X_{θ_0} *and* Z_{θ_0} *are binomially distributed with parameter vectors* $\left(\frac{c_1(c_1-c_2)}{c_1^2-c_2}, p\right)$ *and* $\left(\frac{c_1-c_2}{c_1^2-c_2}, p\right)$ *for some* $p \in (0,1)$.

(iii) $c_2 = c_1^2$ *and, for a given* $\theta_0 \in \Theta$, X_{θ_0} *and* Z_{θ_0} *are Poisson random variables with expected values* α *and* α/c_1 *respectively for some* $\alpha > 0$.

PROOF The result follows essentially via the argument used to prove Theorem 12.4.1. In this case, in place of (12.4.6) in the proof of the previous theorem, we get

$$\frac{\beta_1''(\theta)}{\beta_1(\theta)} - \frac{\beta_1'(\theta)}{\beta_1(\theta)} = c_2 \left(\frac{\beta_2''(\theta)}{\beta_2(\theta)} - \frac{\beta_2'(\theta)}{\beta_2(\theta)}\right), \quad \theta \in \Theta.$$

Essentially the argument that led to (12.4.8), leads us in the present case to

$$Var_\theta(X_\theta) = \frac{c_2 - c_1^2}{c_1(c_1 - c_2)} \left(\frac{\beta_1'(\theta)}{\beta_1(\theta)}\right)^2 + \frac{\beta_1'(\theta)}{\beta_1(\theta)}, \quad \theta \in \Theta.$$

The result then follows using the remainder of the argument of the proof of the previous theorem. ∎

REMARK That the Morris (1982) result follows via the Laha-Lukacs (1960) result is clear essentially from what appears in Shanbhag (1979). ∎

REMARK In the light of Morris (1982) and Shanbhag (1972a, 1979), some of the steps in the proof of Theorem 12.4.1 are routine. However, for the sake of clarity, we have reproduced these here. Also, now it is an exercise to identify the exponential families for which, for all $\theta \in \Theta$, (12.4.3) in conjunction with

$$E_\theta(X_\theta^2) = c_2 E_\theta(Z_\theta^2) + c_3 E_\theta(Z_\theta) + c_4,$$

where c_2, c_3 and c_4 are real numbers, holds; one is then essentially led to the identification of the six exponential families (but for scale and location changes) met in the characterizations given by Morris (1982) and Shanbhag (1972a, 1979) respectively. ∎

REMARK In the case where X_θ is a weighted random variable (i.e. a random variable with a weighted distribution) relative to Z_θ for each $\theta \in$

Θ, with weight function independent of θ, Theorems 12.4.1 and 12.4.2 imply certain characterization results involving weighted distributions; these results subsume some of the variations of the Rao-Rubin theorem (1964) appearing in Shanbhag and Clark (1972) and Patil and Ratnaparkhi (1975, 1977). ▌

References

1. Alamatsaz, M. H. (1985). A note on an article by Artikis. *Acta. Math. Acad. Sci. Hungar.* **45**, 159–162.

2. Khintchine, A. Y. (1938). On unimodal distributions. *Izv. Nauchno-Issled. Inst. Mat. Mech. Tonsk. Gos. Univ.* **2**, 1–7 (in Russian).

3. Kotz, S. and Shanbhag, D. N. (1980). Some new approaches to probability distributions. *Advances in Applied Probability* **12**, 903–921.

4. Laha, R. and Lukacs, E. (1960). On a problem connected with quadratic regression. *Biometrika* **47**, 335–343.

5. Lukacs, E. (1970). *Characterstic Functions*, Second Edition. Griffin, London.

6. Morris, C. (1982). Natural exponential families with quadratic variance functions. *Annals of Statistics* **10**, 65–80.

7. Olshen, R. A. and Savage, L. J. (1970). A generalized unimodality. *Journal of Applied Probability* **7**, 21–34.

8. Pakes, A. G. (1992). A characterization of gamma mixtures of stable laws motivated by limit theorems. *Statistica Neerlandica* **46**, 209-218.

9. Pakes, A. G. (1994). Necessary conditions for the characterization of laws via mixed sums. *Annals of the Institute of Statistical Mathematics* **46**, 797-802.

10. Patil, G. P. and Ratnaparkhi, M. V. (1975). Problems of damaged random variables and related characterizations. In *Statistical Distributions in Scientific Work*, Vol. 3 (Eds., G. P. Patil, S. Kotz, and J. K. Ord), pp. 255–270. Dordrecht, Reidel.

11. Patil, G. P. and Ratnaparkhi, M. V. (1977). Characterizations of certain statistical distributions based on additive damage models involving Rao-Rubin condition and some of its variants. *Sankhyā, Series B* **39**, 65–75.

12. Rao, C. R. (1965). On discrete distributions arising out of methods

of ascertainment. *Sankhyā, Series A* **27**, 311–324 .

13. Rao, C. R. and Rubin, H. (1964). On a characterization of the Poisson distribution. *Sankhyā, Series A* **26**, 295–298.

14. Rao, C. R. and Shanbhag, D. N. (1994). *Choquet-Deny Type Functional Equations with Applications to Stochastic Models*. John Wiley & Sons, New York.

15. Rao, M. B. and Shanbhag, D. N. (1982). Damage models. In *Encyclopedia of Statistical Science*, Vol.2 (Eds., N. L. Johnson and S. Kotz), pp. 262-265, John Wiley & Sons, New York.

16. Sapatinas, T. (1993). Characterization and identifibility results in some stochastic models. Ph.D. thesis, University of Sheffield.

17. Sapatinas, T. (1999). A characterization of the negative binomial distribution via α-monotonicity. *Statistics & Probability Letters* **45**, 49–53.

18. Sapatinas, T. and Aly, M. A. H. (1994). Characterizations of some well-known discrete distributions based on variants of the Rao-Rubin condition. *Sankhyā, Series A* **56**, 335–346.

19. Shanbhag, D. N. (1972a), Some characterizations based on the Bhattacharyya matrix. *Journal of Applied Probability* **9**, 580–587.

20. Shanbhag, D. N. (1972b), Characterizations under unimodality for exponential distributions. Research Report 99/DNS3, Sheffield University.

21. Shanbhag, D. N. (1977). An extension of the Rao-Rubin characterization of the Poisson distribution. *Journal of Applied Probability* **14**, 640–646.

22. Shanbhag, D. N. (1979), Diagonality of the Bhattacharyya matrix as a characterization. *Theory of Probability and its Applications* **24**, 430–433.

23. Shanbhag, D. N. and Clark, R. M. (1972). Some characterizations of the Poisson distribution starting with a power series distribution. *Proceedings of the Cambridge Philosophical Soceity* **71**, 517–522.

24. Steutel, F. W. (1988). Note on discrete α-unimodality. *Statistica Neerlandica* **48**, 137–140.

25. Steutel, F. W. and van Harn, K. (1979). Discrete analogues of self-decomposability and stability. *Annals of Probability* **7**, 893–899.

13

Bivariate Distributions Compatible or Nearly Compatible with Given Conditional Information

Barry C. Arnold, Enrique Castillo, and José María Sarabia
University of California, Riverside, CA
University of Cantabria, Santander, Spain
University of Cantabria, Santander, Spain

ABSTRACT Let (X, Y) denote a two dimensional discrete random variable. Assume that exact or approximate statements are provided with regard to the conditional distributions of X given Y and of Y given X. This list of statements may be incompatible in the sense that no joint distribution for (X, Y) will satisfy all the statements in the list. Measurement of the degree of incompatibility and identification of most nearly compatible distributions is discussed in the present paper.

Keywords and phrases Conditional specification, imprecise probabilities, ϵ-compatibility, Kullback-Leibler information, Markov chains

13.1   Introduction

Consider a two dimensional discrete random variable (X, Y) with possible values $x_1, x_2, \ldots, x_I$ and $y_1, y_2, \ldots, y_J$ for X and Y respectively. First, suppose that we are given two putative families of conditional distributions: One of X given Y, namely

$$P(X = x_i | Y = y_j) = a_{ij}, \quad i = 1, 2, \ldots, I, \quad j = 1, 2, \ldots, J \quad (13.1.1)$$

and one for Y given X, namely

$$P(Y = y_j | X = x_i) = b_{ij}, \quad i = 1, 2, \ldots, I, \quad j = 1, 2, \ldots, J. \quad (13.1.2)$$

The matrices $A = (a_{ij})$ and $B = (b_{ij})$ will be said to be compatible if there exists a joint distribution $P = (p_{ij})$ for (X, Y) whose corresponding conditional distributions are given by A and B [i.e., by (13.1.1) and (13.1.2)]. In the absence of compatibility, we wish to measure the degree of incompatibility and to produce an appropriate most-nearly compatible distribution.

More generally we might be provided with imprecise specifications of the conditional distributions of the form

$$a'_{ij} \leq P(X = x_i | Y = y_j) \leq a''_{ij}, \ i = 1, 2, \dots, I, \ j = 1, 2, \dots, J \quad (13.1.3)$$

and

$$b'_{ij} \leq P(Y = y_j | X = x_i) \leq b''_{ij}, \ i = 1, 2, \dots, I, \ j = 1, 2, \dots, J \ . \quad (13.1.4)$$

Here again we will ask if the information given, i.e., (13.1.3) and (13.1.4), is compatible. Specifically, does there exist a joint distribution $P = (p_{ij})$ for (X, Y) satisfying (13.1.3) and (13.1.4)? Again, in the absence of compatibility we will endeavor to measure the degree of incompatibility and identify a suitable nearly compatible distribution.

Problems of this nature can be expected to arise from attempts to elicit subjective probabilities by asking a series of questions about acceptability of bets. A fertile source of such examples is the arena of prior identification in Bayesian analysis.

The precise specification problem, (13.1.1) and (13.1.2), has received most attention in the literature. A broad spectrum of possible solutions are available. Some will extend readily to cover the imprecise specification scenario, others will not. We will briefly outline the nature of the imprecise specification problem and a variety of possible approaches to solution in Section 13.2. Section 13.3 will focus in more detail on the precise specification case with appropriate indications regarding whether and how the techniques can be modified to account for imprecise information. In Section 13.4, a simple precise specification example is analysed in detail to provide a sense of the different results that can be expected from different approaches to the problem.

13.2 Imprecise Specification

Suppose that $P = (p_{ij})$ is the joint distribution matrix for the random variable (X, Y). Each constraint in (13.1.3) and (13.1.4) can be rewritten as a linear constraint involving the elements of P. For example the constraint $a'_{12} \leq P(X = x_1 | Y = y_2)$ is equivalent to:

$$a_{12} \sum_{i=1}^{I} p_{i2} - p_{12} \leq 0 \ . \quad (13.2.1)$$

Let us arrange the elements of P in a long column vector (by stacking the columns of P) denoted by $\boldsymbol{p}$. Rewriting the $4 \times I \times J$ constraints in a form analogous to (13.2.1), we seek a vector $\boldsymbol{p}$ such that

$$A\boldsymbol{p} \leq 0 \qquad (13.2.2)$$

where A is a known matrix of dimension $(4 \times I \times J)$ by $(I \times J)$. In addition to (13.2.2) we require that $\boldsymbol{p}$ have nonnegative elements which sum to 1. We do not include the conditions $\boldsymbol{p} \geq 0$ and $\boldsymbol{p}'\boldsymbol{1} = 1$ in our list (13.2.2) because these conditions will be insisted on. We will however, in many cases, have to be content with finding a vector $\boldsymbol{p}$ which almost satisfies (13.2.2), to be called an almost compatible distribution.

The search for an almost compatible distribution can be viewed as a problem of selecting $\boldsymbol{p}$ to minimize some measure of the degree to which the constraints (13.2.2) are violated. Candidate objective functions include

$$D_1(\boldsymbol{p}) = \sum_{i=1}^{4 \times I \times J} (A_i\boldsymbol{p})^+, \qquad (13.2.3)$$

$$D_2(\boldsymbol{p}) = \max_i (A_i\boldsymbol{p})^+, \qquad (13.2.4)$$

and

$$D_3(\boldsymbol{p}) = \sum_{i=1}^{4 \times I \times J} [(A_i\boldsymbol{p})^+]^2, \qquad (13.2.5)$$

where A_i denotes the i-th row of the coefficient matrix A. The attained value of the objective function can, for each case, be viewed as a measure of incompatibility of the given information.

Another approach to this problem involves the concept of ϵ-compatibility introduced in Arnold, Castillo, and Sarabia (1999) in the context of precise specification. The concept remains viable in the current context and provides an alternative interpretation of the discrepancy measure defined in terms of (13.2.4).

We say that information of the form $A\boldsymbol{p} \leq 0$ is ϵ^*-compatible if ϵ^* is the smallest value of ϵ such that the system

$$A\boldsymbol{p} - \epsilon\boldsymbol{1} \leq 0 \qquad (13.2.6)$$

has a solution $\boldsymbol{p}$ with $\boldsymbol{p} \geq 0$ and $\boldsymbol{p}'\boldsymbol{1} = 1$. In this formulation the determination of ϵ^* can be viewed as a linear programming problem with ϵ as the objective function and linear constraints (13.2.6) plus $\boldsymbol{p} \geq 0$ and $\boldsymbol{p}'\boldsymbol{1} = 1$. Of course, if ϵ^* turns out to be 0, the information $A\boldsymbol{p} \leq 0$ is actually compatible.

Arnold, Castillo, and Sarabia (1999) also discussed a concept of weighted ϵ-compatibility which might be used if violations of the inequalities in (13.2.2) are judged to be of unequal importance. In this case our constraints (13.2.6) will be replaced by

$$Ap - \epsilon w'1 \leq 0, \tag{13.2.7}$$

where w is a known vector of suitable weights. Again, a simple linear programming problem is to be solved. In practice, specification of an appropriate weight vector may be too difficult and one would expect to usually resort to the use of the unweighted constraints (13.2.6).

13.3   Precise Specification

First we remark that the material in Section 13.2 can be readily adjusted to handle the precise specification case by merely assuming that $a'_{ij} = a''_{ij} = a_{ij}$ and $b'_{ij} = b''_{ij} = b_{ij}$, $\forall i, j$. However in the special case of precise specification, a variety of compatibility approaches have been discussed in the literature. There are thus many competitors to the distance based approaches (13.2.3), (13.2.4), (13.2.5) and the ϵ-compatibility approach, (13.2.6). Which compatibility measure is to be appropriately used clearly depends on the proposed use of the most compatible distribution so determined. Another criterion which deserves attention is ease of interpretability. Many of the competitors will be seen to be mathematically attractive but not generally easy to interpret.

Some review of background material about compatible conditional distributions will be helpful.

Conditions for compatibility were provided in Arnold and Press (1989). To be compatible, A and B, defined in (13.1.1) and (13.1.2) must have the same incidence matrix, i.e., we must have $a_{ij} \neq 0$ if and only if $b_{ij} \neq 0$. A and B are then compatible if and only if there exist vectors τ (of dimension I) and η (of dimension J) with nonnegative elements such that

$$\tau_i b_{ij} = \eta_j a_{ij}, \quad \forall i, j . \tag{13.3.1}$$

If τ and η are both normalized to have their coordinates sum to 1, then we can identify them as representing the marginal distributions of X and Y respectively, i.e., $\tau_i = P(X = i)$ and $\eta_j = P(Y = j)$. If we consider $A'B$ as the transition matrix of a Markov chain, a sufficient condition for uniqueness of the solution to (13.3.1) is that the Markov chain with transition matrix $A'B$ be irreducible. Arnold and Gokhale (1998b) discuss Markovian measures of incompatibility in some detail. They observe that, whether A and B are compatible or not, two simple Markov chains can be defined in terms of them. First an I-state chain with transition matrix BA' and second a J-state chain with transition matrix $A'B$. Denote the long run distribution of BA' by μ and the long run distribution of $A'B$ by ν. The vectors μ and ν then satisfy

$$\mu = \mu BA' \tag{13.3.2}$$

and

$$\nu = \nu A'B . \tag{13.3.3}$$

If A and B are compatible then the solutions to (13.3.2) and (13.3.3) will satisfy

$$\mu_i b_{ij} = \nu_j a_{ij} \ \forall i, j , \tag{13.3.4}$$

i.e., they will provide solutions to (13.3.1). If A and B are not compatible, the difference between the left-hand and right-hand sides of the equation (13.3.4), can be used to develop a measure of incompatibility. Observe that the solutions to (13.3.2) and (13.3.3) are intimately related. It is always true that if μ and ν satisfy (13.3.2) and (13.3.3) then $\mu = \nu A'$, so that only one of the systems needs to be solved. Arnold and Gokhale (1998b) suggest that a most nearly compatible distribution corresponding to A and B be defined as one with elements $\frac{1}{2}(\mu_i b_{ij} + \nu_j a_{ij})$, and that the degree of incompatibility of A and B be measured by a measure such as:

$$M_1 = \frac{1}{IJ} \sum_{i=1}^{I} \sum_{j=1}^{J} (\mu_i b_{ij} - \nu_j a_{ij})^2 \tag{13.3.5}$$

$$M_2 = \frac{1}{IJ} \sum_{i=1}^{I} \sum_{j=1}^{J} |\mu_i b_{ij} - \nu_j a_{ij}| \tag{13.3.6}$$

$$M_3 = \sqrt{\sum_{i=1}^{I} \sum_{j=1}^{J} (\mu_i b_{ij} - \nu_j a_{ij})^2} . \tag{13.3.7}$$

Not included in their list, but worthy of consideration is

$$M_4 = \max_{i,j} |\mu_i b_{ij} - \nu_j a_{ij}| . \tag{13.3.8}$$

If we define $P_\mu = (\mu_i b_{ij})$ and $P_\nu = (\nu_j a_{ij})$, then there is an attractive alternative to $\frac{1}{2}(P_\mu + P_\nu)$ as a most-nearly compatible distribution. It can be verified that P_μ and P_ν have identical marginals and we can search for a matrix P^* with those marginals such that the Kullback-Leibler information distance between P^* and P_μ equals that between P^* and P_ν. An iterative procedure for obtaining such a P^* may be found in Gokhale and Kullback (1978, p. 182).

Several other measures of incompatibility, not involving the Markov chains BA' and $A'B$, were suggested by Arnold and Gokhale (1998a). The following quadratic incompatibility measure was suggested.

$$Q = \min_{P} \sum_{i=1}^{I} \sum_{j=1}^{J} [(p_{ij} - a_{ij}p_{\cdot j})^2 + (p_{ij} - b_{ij}p_{i\cdot})^2] \ . \tag{13.3.9}$$

The corresponding most nearly compatible matrix P would be P_Q, the choice of P for which the expression in (13.3.9) is minimized. The three other measures suggested in the Arnold and Gokhale (1998a) paper were

$$Q' = \min_{P} \sum_{i=1}^{I} \sum_{j=1}^{J} [|p_{ij} - a_{ij}p_{\cdot j}| + |p_{ij} - b_{ij}p_{i\cdot}|] \tag{13.3.10}$$

$$Q'' = \min_{P} \sum_{i=1}^{I} \sum_{j=1}^{J} [(a_{ij} - \frac{p_{ij}}{p_{\cdot j}})^2 + (b_{ij} - \frac{p_{ij}}{p_{i\cdot}})^2] \tag{13.3.11}$$

and

$$Q''' = \min_{P} \sum_{i=1}^{I} \sum_{j=1}^{J} [|a_{ij} - \frac{p_{ij}}{p_{\cdot j}}| + |b_{ij} - \frac{p_{ij}}{p_{i\cdot}}|] \ . \tag{13.3.12}$$

Corresponding most nearly compatible distributions $P_{Q'}, P_{Q''}$ and $P_{Q'''}$ are defined as values of P that minimize the respective right-hand sides of (13.3.10)–(13.3.12).

A plausible approach would involve selection of a distribution P whose conditional distributions are minimally distant from the distributions given by A and B using some convenient measure of distance between distributions. Criteria (13.3.10) and (13.3.11) are actually of this form. An attractive alternative is to use the Kullback-Leibler information to measure distance between distributions. From this viewpoint Arnold and Gokhale (1998a) suggest choosing P to minimize the following objective function

$$\sum_{i=1}^{I} \sum_{j=1}^{J} [a_{ij} \log(\frac{a_{ij}p_{\cdot j}}{p_{ij}}) + b_{ij} \log(\frac{b_{ij}p_{i\cdot}}{p_{ij}})] \ . \tag{13.3.13}$$

The distribution P^* which minimizes (13.3.13) can be obtained by an application of the Darroch-Ratcliff (1972) iterative scaling algorithm.

As remarked earlier, the concept of ϵ-compatibility can be used in the precise specification context. A clear advantage of ϵ-compatibility is the fact that it is readily adapted to handle imprecise formulations of A and B, such is not the case for many of the other measures discussed in the current section.

In fact, in the precise specification context, Arnold, Castillo and Sarabia (1999) suggested 3 variant concepts of ϵ-compatibility of putative conditional distribution matrices A and B. They described weighted ϵ-compatibility, but for simplicity we only consider unweighted versions here. The 3 options are:

Option 1: Minimize ϵ subject to

$$|p_{ij} - a_{ij}p_{\cdot j}| \leq \epsilon \ \forall i,j$$
$$|p_{ij} - b_{ij}p_{i\cdot}| \leq \epsilon \ \forall i,j$$
$$p_{ij} \geq 0 \qquad \forall i,j \qquad (13.3.14)$$

and $\sum_i \sum_j p_{ij} = 1$.

Option 2: Minimize ϵ subject to

$$|a_{ij}\eta_j - b_{ij}\tau_i| \leq \epsilon \ \forall i,j$$
$$\eta_j \geq 0 \qquad \forall j$$
$$\tau_i \geq 0 \qquad \forall i \qquad (13.3.15)$$
$$\sum_j \eta_j = 1$$

and $\quad \sum_i \tau_i = 1$.

Option 3: Minimize ϵ subject to

$$\left| a_{ij} \sum_{k=1}^{I} b_{kj}\tau_k - b_{ij}\tau_i \right| \leq \epsilon \ \ \forall i,j$$
$$\tau_i \geq 0 \ \ \forall i \qquad (13.3.16)$$
$$\text{and} \ \sum_i \tau_i = 1 .$$

Option 1 will be recognizable as the version of ϵ-compatibility introduced in Section 13.2 of the present paper. Using Option 1, if the objective function ϵ is minimized by the choice (P_1, ϵ_1^*) then we will say that A and B are ϵ_1^*-compatible and a most nearly compatible matrix is given by P_1. Using Option 2, if the objective function ϵ is minimized by the choice $(\boldsymbol{\tau}^{(2)}, \boldsymbol{\eta}^{(2)}, \epsilon_2^*)$ then we will say that A and B are ϵ_2^*-compatible and a most nearly compatible matrix is given by P_2 with elements equal to $\frac{1}{2}(a_{ij}\eta_j^{(2)} + b_{ij}\tau_i^{(2)})$. Finally, using Option 3, if the objective function ϵ is minimized by the choice $(\boldsymbol{\tau}^{(3)}, \epsilon_3^*)$ then we will say that A and B are ϵ_3^*-compatible and a most nearly compatible matrix is given by P_3 with elements equal to $\frac{1}{2}(a_{ij}\sum_{k=1}^{I} b_{kj}\tau_k^{(3)} + b_{ij}\tau_i^{(3)})$.

All of the measures of compatibility thus far discussed may be used to quantify the compatibility of any A (with columns summing to 1) and any B (with rows summing to 1). If every element of A and every element of B is assumed to be strictly positive (the "positivity" assumption), then several more measures of compatibility merit consideration. A and B will be compatible (under a positivity assumption) if and only if their uniform marginal representations [Mosteller (1968)], $UMR(A)$ and $UMR(B)$, are the same. Consequently $d(UMR(A), UMR(B))$ for any distance function d can be used as a compatibility measure. Under the positivity assumption, A and B will also be compatible if and only if their cross product ratios are equal. Following Arnold and Gokhale (1998b) we can define a cross product ratio matrix corresponding to A to be a matrix with elements

$$a_{ij}a_{IJ}/a_{iJ}a_{Ij} \qquad (13.3.17)$$

Analogously we may define $CPR(B)$ and a measure of compatibility is provided by $d(CPR(A), CPR(B))$ for any distance function d.

The final criterion for measuring compatibility between $A > 0$ and $B > 0$ is based on the compatibility condition introduced by Arnold and Press (1989) [equation (13.3.1)]. If we define a matrix C_1 to have elements b_{ij}/a_{ij} and row normalize C_1 to make the rows sum to 1, then the resulting matrix $N^{(1)}$ will have identical rows if and only if A and B are compatible. Analogously if we define C_2 to have elements a_{ij}/b_{ij} and column normalize C_2 to have columns summing to 1, the resulting matrix $N^{(2)}$ will have identical columns if and only if A and B are compatible. A related Kullback-Leibler measure of compatibility is then provided by

$$D = \sum_{i=1}^{I}\sum_{j=1}^{J} \bar{n}_{\cdot j}^{(1)} \log(\bar{n}_{\cdot j}^{(1)}/n_{ij}^{(1)})$$

$$+ \sum_{i=1}^{I}\sum_{j=1}^{J} \bar{n}_{i\cdot}^{(2)} \log(\bar{n}_{i\cdot}^{(2)}/n_{ij}^{(2)}) . \qquad (13.3.18)$$

where $\bar{n}_{\cdot j}^{(1)} = \frac{1}{I}\sum_{i=1}^{I} n_{ij}^{(1)}$ and $\bar{n}_{i\cdot}^{(2)} = \frac{1}{J}\sum_{j=1}^{J} n_{ij}^{(2)}$. Of course other measures of inhomogeneity of the rows of N_1 and the columns of N_2 can be used if desired.

13.4 An Example

Consider conditional distributional matrices A and B as follows:

$$A = \begin{pmatrix} \frac{1}{4} & \frac{1}{3} \\ \frac{1}{4} & \frac{1}{3} \\ \frac{1}{2} & \frac{1}{3} \end{pmatrix}, \quad B = \begin{pmatrix} \frac{1}{2} & \frac{1}{2} \\ \frac{1}{3} & \frac{2}{3} \\ \frac{1}{4} & \frac{3}{4} \end{pmatrix}. \tag{13.4.1}$$

It is not difficult to verify that these are incompatible. We will use this pair A, B to illustrate the spectrum of compatibility measures described in the present paper and to illustrate the various most-nearly compatible matrices P obtained using the different compatibility measures.

The long run distributions for the Markov chains BA' and $A'B$ are respectively given by

$$\nu = \left(\frac{13}{37}, \frac{24}{37} \right) \tag{13.4.2}$$

and

$$\mu = \left(\frac{45}{148}, \frac{45}{148}, \frac{58}{148} \right). \tag{13.4.3}$$

Consequently

$$(\mu_i b_{ij}) = \begin{pmatrix} \frac{45}{296} & \frac{45}{296} \\ \frac{30}{296} & \frac{60}{296} \\ \frac{29}{296} & \frac{87}{296} \end{pmatrix} \tag{13.4.4}$$

and

$$(\nu_j a_{ij}) = \begin{pmatrix} \frac{26}{296} & \frac{64}{296} \\ \frac{26}{296} & \frac{64}{296} \\ \frac{52}{296} & \frac{64}{296} \end{pmatrix}. \tag{13.4.5}$$

Since $(\mu_i b_{ij}) \not\equiv (\nu_j a_{ij})$ we confirm the incompatibility of A and B. The most nearly compatible distribution using the Markov chain approach is then

$$\left(\frac{1}{2} (\mu_i b_{ij} + \nu_j a_{ij}) \right) = \begin{pmatrix} \frac{71}{592} & \frac{109}{592} \\ \frac{56}{592} & \frac{124}{592} \\ \frac{81}{592} & \frac{151}{592} \end{pmatrix}. \tag{13.4.6}$$

An alternative, chosen to have the same marginals as (13.4.4) and (13.4.5) but to be equally distant from (13.4.4) and (13.4.5) in terms of Kullback-Leibler distance is

$$P^* = (1/296) \begin{pmatrix} 35.40 & 54.60 \\ 28.33 & 61.67 \\ 40.27 & 75.73 \end{pmatrix}. \tag{13.4.7}$$

The incompatibility measures (13.3.5)–(13.3) when applied to our example (13.4.1) yield values of

$$M_1 = 0.004136 \tag{13.4.8}$$

$$M_2 = 0.062162 \tag{13.4.9}$$

$$M_3 = 0.143809 \tag{13.4.10}$$

$$M_4 = 0.077703 \tag{13.4.11}$$

If we use incompatibility measures (13.3.9)–(13.3.12), which do not involve the Markov chains $A'B$ and BA', the corresponding values for example (13.4.1) are:

$$Q = 0.010 \tag{13.4.12}$$

$$Q' = 0.241 \tag{13.4.13}$$

$$Q'' = 0.061 \tag{13.4.14}$$

$$Q''' = 1.157 \tag{13.4.15}$$

The corresponding most nearly compatible matrices P are given by

$$P_Q = \begin{pmatrix} 0.119 & 0.194 \\ 0.096 & 0.224 \\ 0.122 & 0.246 \end{pmatrix} \tag{13.4.16}$$

$$P_{Q'} = \begin{pmatrix} 0.207 & 0.207 \\ 0.103 & 0.207 \\ 0.069 & 0.207 \end{pmatrix} \tag{13.4.17}$$

$$P_{Q''} = \begin{pmatrix} 0.117 & 0.148 \\ 0.097 & 0.212 \\ 0.142 & 0.283 \end{pmatrix} \tag{13.4.18}$$

and

$$P_{Q'''} = \begin{pmatrix} 0.001 & 0.001 \\ 0.130 & 0.261 \\ 0.152 & 0.455 \end{pmatrix}, \tag{13.4.19}$$

where in the last two it has been assumed that $p_{ij} > 0.001$, otherwise, Q'' and Q''' are not well defined (division by zero). The Kullback-Leibler measure of compatibility (13.3.13) yields a value of 0.0791, attained when P is given by

$$P_K = \begin{pmatrix} 0.11806 & 0.16282 \\ 0.09507 & 0.20403 \\ 0.14497 & 0.27505 \end{pmatrix}. \tag{13.4.20}$$

Next we turn to the 3 variant forms of ϵ-compatibility, (13.3.14)–(13.3.16). Using Option 1 we find

$$\epsilon_1^* = 0.03472 \tag{13.4.21}$$

and

$$P_1 = \begin{pmatrix} \frac{17}{144} & \frac{3}{16} \\ \frac{1}{12} & \frac{2}{9} \\ \frac{19}{144} & \frac{37}{144} \end{pmatrix} = \begin{pmatrix} 0.11806 & 0.18750 \\ 0.08333 & 0.22222 \\ 0.13194 & 0.25694 \end{pmatrix}. \tag{13.4.22}$$

Option 2 yields

$$\epsilon_2^* = 0.069444 \tag{13.4.23}$$

and

$$P_2 = \begin{pmatrix} \frac{17}{144} & \frac{3}{16} \\ \frac{5}{54} & \frac{23}{108} \\ \frac{19}{144} & \frac{37}{144} \end{pmatrix} = \begin{pmatrix} 0.11806 & 0.18750 \\ 0.09259 & 0.21296 \\ 0.13194 & 0.25694 \end{pmatrix}, \tag{13.4.24}$$

while Option 3 results in

$$\epsilon^* = 0.076087 \tag{13.4.25}$$

and

$$P_3 = \begin{pmatrix} \frac{13}{92} & \frac{13}{92} \\ \frac{5}{46} & \frac{5}{23} \\ \frac{9}{92} & \frac{27}{92} \end{pmatrix} = \begin{pmatrix} 0.14130 & 0.14130 \\ 0.10870 & 0.21739 \\ 0.09783 & 0.29348 \end{pmatrix}. \tag{13.4.26}$$

For completeness we will measure the compatibility of A and B using the measures which involved a positivity assumption. First we need the uniform marginal representations of A and B [as in (13.4.1)]. They are

$$UMR(A) = \begin{pmatrix} 0.1477 & 0.1857 \\ 0.1477 & 0.1857 \\ 0.2047 & 0.1287 \end{pmatrix} \tag{13.4.27}$$

and

$$UMR(B) = \begin{pmatrix} 0.2152 & 0.1181 \\ 0.1589 & 0.1745 \\ 0.1259 & 0.2074 \end{pmatrix}. \tag{13.4.28}$$

Incompatibility is once more confirmed, since $UMR(A) \neq UMR(B)$. A nearly compatible distribution P say P_{UMR} is obtained by averaging (13.4.27) and (13.4.28) and then adjusting this average to have marginals ν and μ [given in (13.4.2) and (13.4.3)]. In this way we get

$$P_{UMR} = \begin{pmatrix} 0.1194 & 0.1847 \\ 0.0959 & 0.2082 \\ 0.1361 & 0.2558 \end{pmatrix}. \tag{13.4.29}$$

Using a simple Euclidean distance measure between matrices, i.e.,

$$d(C,D) = \sqrt{\sum_{i=1}^{I} \sum_{j=1}^{J} (c_{ij} - d_{ij})^2},$$

we obtain as our measure of incompatibility

$$d(UMR(A), UMR(B)) = 0.147547 \tag{13.4.30}$$

The cross-product ratio matrices corresponding to A and B are [recall (13.3.17)]

$$CPR(A) = \begin{pmatrix} 0.5 & 1 \\ 0.5 & 1 \\ 1 & 1 \end{pmatrix} \tag{13.4.31}$$

and

$$CPR(B) = \begin{pmatrix} 3 & 1 \\ 1.5 & 1 \\ 1 & 1 \end{pmatrix}. \tag{13.4.32}$$

Our nearly compatible distribution P_{CPR} is obtained by averaging (13.4.31) and (13.4.32) and adjusting this average to have marginals ν and μ [given by (13.4.2) and (13.4.3)]. Thus we obtain

$$P_{CPR} = \begin{pmatrix} 0.1344 & 0.1696 \\ 0.0948 & 0.2093 \\ 0.1221 & 0.2697 \end{pmatrix}. \tag{13.4.33}$$

The corresponding measure of incompatibility, again using Euclidean distance between matrices, is

$$d(CPR(A), CPR(B)) = 2.69 \tag{13.4.34}$$

Turning to the final incompatibility measure (13.3.18) we find

$$N_1 = \begin{pmatrix} 4/7 & 3/7 \\ 2/5 & 3/5 \\ 2/11 & 9/11 \end{pmatrix} \tag{13.4.35}$$

and

$$N_2 = \begin{pmatrix} 2/13 & 12/29 \\ 3/13 & 9/29 \\ 8/13 & 8/29 \end{pmatrix} \tag{13.4.36}$$

and consequently

$$D = 0.1837 + 0.1426 = 0.3263 \ .$$

Note that this last approach does not provide us with a most nearly compatible P, it just provides a measure of compatibility.

References

1. Arnold, B. C., Castillo, E. and Sarabia, J. M. (1999). *Conditional Specification of Statistical Models*. Springer, New York.

2. Arnold, B. C. and Gokhale, D. V. (1998a). Distributions most nearly compatible with given families of conditional distributions. *Test* **7**, 377–390.

3. Arnold, B. C. and Gokhale, D. V. (1998b). Remarks on incompatible conditional distributions. *Technical Report #260*, Department of Statistics, University of California, Riverside.

4. Arnold, B. C. and Press, S. J. (1989). Compatible conditional distributions. *Journal of the American Statistical Association* **84**, 152–156.

5. Darroch, J. and Ratcliff, D. (1972). Generalized iterative scaling for log-linear models. *Annals of Mathematical Statistics* **43**, 1470–1480.

6. Gokhale, D. V. and Kullback, S. (1978). *The Information in Contingency Tables*. Marcel Dekker, New York.

7. Mosteller, F. (1968). Association and estimation in contingency tables. *Journal of the American Statistical Association* **63**, 1–28.

14

A Characterization of a Distribution Arising from Absorption Sampling

Adrienne W. Kemp
University of St. Andrews, St. Andrews, Scotland

ABSTRACT The paper finds that the distribution arising from direct absorption sampling is characterized as the distribution of $U|U+V=m$ where m is constant and U and V have respectively a q-binomial and a Heine distribution with the same argument parameter. An application is given concerning the total number of cases of employee misbehaviour in two workplaces, given that there is underreporting in the first workplace.

Keywords and phrases Absorption distribution, q-binomial distribution, Heine distribution, basic hypergeometric series, q-series, visibility bias, weighted distribution, weight function

14.1 Introduction

Absorption sampling and the absorption distribution have a long history that dates back to the post 1939-1945 problem studied by Blomqvist (1952), Borenius (1953), and Zacks and Goldfarb (1966). It concerns the number of individuals Y out of n who fail to cross a minefield containing m mines, i.e. the number of particles absorbed when n particles pass through a medium containing m absorption points. Dunkl (1981) studied the equivalent problem of the number of typos Y that are found in n scans of a manuscript containing m typos; see also Johnson and Kotz (1969) and Johnson, Kotz, and Kemp (1992). Dunkl was the first person to recognize the connection between the distribution and the theory of q-hypergeometric functions, i.e. q-series.

Kemp (1998) recast the problem in terms of a closed population of m endangered animals for which a captive breeding programme is planned. She supposed that a fixed amount of effort is available each day for searching for the animals, but that when an animal is found then the

search is abandoned for the rest of that day. She assumed that the
animals have constant and independent probabilities of evading capture
such that the probability that a particular animal evades capture on a
particular day is q. If a fixed number k of individuals is needed and
the variable of interest, X, is the number of search days in excess of k
that are required, then we have inverse absorption sampling. If a fixed
number n of search days is available and the variable of interest, Y, is the
number of individuals found, then we have direct absorption sampling,
as in the Blomqvist and Dunkl scenarios.

Other applications of absorption sampling include the flight of planes
past a number of potentially fatal missile launchers, scans of a computer
listing of commercial transactions for fraudulant transactions, and the
search by parents for the latest electronic toy when only a limited number
of such toys is produced. Rawlings (1997) has generalized the particle
absorption model in such a way that the integer restriction on m is
removed.

Kemp (1998) realized that the relationship between direct and inverse
absorption sampling is analogous to that between direct and inverse binomial sampling. She found that it is straightforward to obtain the
probability mass function (pmf) for inverse absorption sampling and
therefrom the pmf for direct absorption sampling. Initially the probability that the searcher locates one of the required items is $1 - q^m$, but
with each subsequent successful search the probability of finding an item
is reduced, $1 - q^{m-1}$, $1 - q^{m-2}$, etc. The probability generating function
(pgf) for X, the number of searches in excess of k that are required in
order to find a fixed number k of items, is therefore a convolution of k
geometric pgfs, each with support $0, 1, \ldots$,

$$G_X(z) = \prod_{i=1}^{k} \frac{(1 - q^{m-i+1})}{(1 - q^{m-i+1}z)}. \tag{14.1.1}$$

Expansion via the q-binomial theorem gives

$$\Pr[X = x] = q^{x(m-k+1)} \frac{(q^m; q^{-1})_k (q^k; q)_x}{(q; q)_x}, \qquad x = 0, 1, \ldots, \tag{14.1.2}$$

where $(a; q)_0 = 1$, $(a; q)_x = (1 - a)(1 - aq) \cdots (1 - aq^{x-1})$, $0 < q < 1$.

As shown by Kemp (1998), the pmf for Y, the number of items found
given a fixed number n of searches, is then

$$\Pr[Y = y] = \frac{\Pr[X + k = n + 1 | k = y + 1]}{\Pr[\text{last search is successful}]}$$

$$= q^{(n-y)(m-y)} \frac{(1 - q^m) \cdots (1 - q^{m-y})(1 - q^{y+1}) \cdots (1 - q^n)}{(1 - q) \cdots (1 - q^{n-y})(1 - q^{m-y})}$$

$$= q^{(n-y)(m-y)} \frac{(q^{m-y+1};q)_y(q^{n-y+1};q)_y}{(q;q)_y},$$

$$y = 0, 1, \ldots, \min(m, n). \tag{14.1.3}$$

Kemp found that the (direct) absorption distribution is log-concave and unimodal and that it has an increasing failure rate. From (14.1.3) the pgf is

$$G_Y(s) = q^{mn} {}_2\phi_1(q^{-n}, q^{-m}; 0; q, qz), \tag{14.1.4}$$

where ${}_A\phi_B(\cdot)$ is Gasper and Rahman's (1990) (G/R) definition of a basic hypergeometric series (q-series):

$$\begin{aligned}
&{}_A\phi_B(a_1, \ldots, a_A; b_1, \ldots, b_B; q, z) \\
&= \sum_{j=0}^{\infty} \frac{(a_1;q)_j \ldots (a_A;q)_j z^j}{(b_1;q)_j \ldots (b_B;q)_j (q;q)_j} \left[(-1)^j q^{\binom{j}{2}} \right]^{B-A+1}
\end{aligned}$$

The symmetry in m and n is an unexpected feature; we assume from here onwards that $m \leq n$.

The purpose of this paper is to show that when $m \leq n$, m constant, and U and V are independent, then $Y \sim (U | U + V = m)$ has the absorption distribution with pgf (14.1.4) if and only if U has a q-binomial distribution with parameters (λ, q) and V has a Heine distribution with parameters (n, λ, q). The q-binomial distribution was introduced into the literature by Kemp and Newton (1990) and Kemp and Kemp (1991). It has the pgf

$$G_B(z) = \frac{{}_1\phi_0(q^{-n}; -; q, -\theta z)}{{}_1\phi_0(q^{-n}; -; q, -\theta)}. \tag{14.1.5}$$

and pmf

$$\Pr[U = u] = \frac{1}{(-\theta q^{-n}; q)_n} \cdot \frac{(q^{-n};q)_u(-\theta)^u}{(q;q)_u}, \qquad u = 0, 1, \ldots, n. \tag{14.1.6}$$

The Heine distribution was first studied by Benkherouf and Bather (1988); see also Kemp (1992a,b). Its pgf is

$$G_H(z) = \frac{{}_0\phi_0(-; -; q, -\psi z)}{{}_0\phi_0(-; -; q, -\psi)} \tag{14.1.7}$$

and its pmf is

$$\Pr[V = v] = \frac{1}{(-\psi; q)_\infty} \cdot \frac{q^{v(v-1)/2}\psi^v}{(q;q)_v}, \qquad v = 0, 1, \ldots. \tag{14.1.8}$$

An application of the characterization theorem is given in section 3. It concerns the total number of cases of employee misbehaviour (e.g. thefts, misuses of the Internet) in two workplaces belonging to the same company, where the number of cases in each workplace has a Heine (λ, q) distribution but there is underreporting of cases in the first workplace.

14.2   The Characterization Theorem

THEOREM 14.2.1

The distribution with pgf

$$G_Y(z) = q^{mn}{}_2\phi_1(q^{-n}, q^{-m}; 0; q, qz)$$

(i.e. the absorption distribution with pgf (14.1.4)) is the distribution of $Y \sim (U|U + V = m)$, where m is a constant, $m \leq n$, and U and V are independent, iff U has a q-binomial distribution with pgf (14.1.6) and V has a Heine distribution with pgf (14.1.7), where $\theta = \psi = \lambda$.

PROOF If U and V are independent and have distributions (14.1.6) and (8) respectively, with $\theta = \psi = \lambda$, then

$$P[U = u|U + V = m]$$

$$= K^{-1} \frac{(q^{-n}; q)_u(-\lambda)^u}{(q; q)_u} \cdot \frac{q^{(m-u)(m-u-1)/2}\lambda^{m-u}}{(q; q)_{m-u}}$$

$$= K^{-1} \frac{(q^{-n}; q)_u(-1)^u}{(q; q)_u} \cdot \frac{\lambda^m q^{m(m-1)/2 + u(u-1)/2 + (1-m)u}}{1}$$

$$\cdot \frac{(q^{-m}; q)_u(-1)^u q^{mu - u(u-1)/2}}{(q; q)_m}$$

$$= K^{-1} \frac{\lambda^m q^{m(m-1)/2}(q^{-m}; q)_u(q^{-n}; q)_u q^u}{(q; q)_m(q; q)_u}, \qquad u = 0, 1, \ldots, m,$$

where K^{-1} is the normalizing coefficient. The pgf for $Y \sim (U|U + V = m)$ is therefore

$$G_Y(z) = \frac{{}_2\phi_1(q^{-n}, q^{-m}; 0; q, qz)}{{}_2\phi_1(q^{-n}, q^{-m}; 0; q, q)}$$

$$= q^{mn}{}_2\phi_1(q^{-n}, q^{-m}; 0; q, qz),$$

i.e. (14.1.4), by the q-Vandermonde theorem, G/R (1.5.3).   ∎

Patil and Seshadri's (1964) theorem, as amended by Menon (1966), provides a proof of the converse; see also Kagan, Linnik, and Rao (1973). Let U and V be independent discrete random variables and set $c(u, u + v) = \Pr[U = u|U + V = u + v]$. If

$$\gamma(u, v) = \frac{c(u + v, u + v)c(0, v)}{c(u, u + v)c(v, v)} = \frac{h(u + v)}{h(u)h(v)} \tag{14.2.1}$$

where $h(\cdot)$ is a nonnegative function, then

$$f(u) = f(0)h(u)e^{au}, \quad \text{and} \quad g(v) = g(0)k(v)e^{av}$$

where

$$0 \le f(u) = \Pr[U = u], \ 0 \le g(v) = \Pr[V = v], \ k(v) = h(v)c(0,v)/c(v,v),$$

and a is a constant. The proof of the theorem depends on the Cauchy functional equation.

Suppose that $Y \sim (U|U + V = m)$ has the pmf (14.1.3) for the absorption distribution and that $m = \min(m, n)$. Then

$$
\begin{aligned}
c(u, u+v) &= \Pr[U = u|U + V = u + v = m]\\
&= \frac{q^{(n-u)v}(q^{v+1};q)_u(q^{n-u+1};q)_u}{(q;q)_u},
\end{aligned}
$$

$$
\begin{aligned}
c(u+v, u+v) &= \Pr[U = u+v|U + V = u + v = m]\\
&= (q^{n-u-v+1};q)_{u+v},
\end{aligned}
$$

$$c(0, v) = \Pr[U = 0|U + V = v = m] = q^{nv},$$

$$c(v, v) = \Pr[U = v|U + V = v = m] = (q^{n-v+1};q)_v$$

and hence

$$
\begin{aligned}
\gamma(u, v) &= \frac{(q^{n-u-v+1};q)_{u+v} \times q^{nv}}{q^{(n-u)v}(q^{v+1};q)_u(q^{n-u+1};q)_u\{(q;q)_u\}^{-1} \times (q^{n-v+1};q)_v}\\
&= \frac{(q^{-n};q)_{u+v}(-1)^{u+v}}{(q;q)_{u+v}} \cdot \frac{(q;q)_u}{(q^{-n};q)_u(-1)^u} \cdot \frac{(q;q)_v}{(q^{-n};q)_v(-1)^v}.
\end{aligned}
$$

From (14.2.1),

$$h(u) = \frac{(q^{-n};q)_u(-1)^u}{(q;q)_u};$$

also

$$\frac{c(0,v)}{c(v,v)} = \frac{q^{nv}}{(q^{n-v+1};q)_v} = \frac{(-1)^v q^{v(v-1)/2}}{(q^{-n};q)_v}.$$

Hence

$$\Pr[U = u] = \Pr[U = 0]\frac{(q^{-n};q)_u(-1)^u e^{au}}{(q;q)_u}$$

$$\Pr[V = v] = \Pr[V = 0]\frac{q^{v(v-1)/2}e^{av}}{(q;q)_v}.$$

The pgf's for U and V are therefore (14.1.6) and (8), respectively, with $\theta = \psi = e^a = \lambda$, i.e. U has a q-binomial distribution with parameters (n, λ, q) and support $0, 1, \ldots, n$ whilst V has a Heine distribution with parameters (λ, q) and support $0, 1, \ldots$.

14.3 An Application

Consider the number of events per employee (e.g. accidents, absenteeisms, thefts, misuses of the Internet), U^* and V, in two different workplaces that operate independently although they belong to the same company. Let the distributions of the number of events per person in the two workplaces have Heine distributions with the same parameters λ, q. Suppose now that events are underreported in the first workplace with weight function

$$w(u) = (1 - q^n)(1 - q^{n-1}) \cdots (1 - q^{n-u+1}) q^{n(n-u)}, \qquad u \geq 0,$$

i.e. $w(0) = q^{n^2}$, $w(1) = (1 - q^n) q^{n(n-1)}$, ..., $w(n) = (1 - q^n)(1 - q^{n-1}) \ldots (1-q)$, $w(u) = 0$ for $u > n$. The outcome is that low counts are sometimes ignored and high counts greater than n are ignored altogether. This is a visibility bias process; it gives rise to the distribution with pmf

$$\Pr[U = u] = w(u) \Pr[U^* = u] / \sum_{u \geq 0} w(u) \Pr[U^* = u], \qquad u = 0, 1, \ldots, n.$$

When the original distribution in the first workplace is Heine, then the outcome distribution has the pmf

$$\Pr[U = u] = C (q^{n-u+1}; q)_u q^{n(n-u)} \frac{\lambda^u q^{u(u-1)/2}}{(q;q)_u (-\lambda; q)_\infty}, \qquad u = 0, 1, \ldots, n,$$

where

$$C^{-1} = \sum_{u=0}^{n} (q^{n-u+1}; q)_u q^{n(n-u)} \frac{\lambda^u q^{u(u-1)/2}}{(q;q)_u (-\lambda; q)_\infty} = \sum_{u=0}^{n} \frac{(q^{-n}; q)_u (-\lambda)^u q^{n^2}}{(q;q)_u (-\lambda; q)_\infty}$$

$$= \frac{q^{n^2} (-\lambda q^{-n}; q)_n}{(-\lambda; q)_\infty}$$

by Heine's theorem, G/R (1.3.2), giving

$$\Pr[U = u] = \frac{\lambda^u q^{u(u-1)/2} (q^{n-u+1}; q)_u q^{-nu}}{(q;q)_u (-\lambda q^{-n}; q)_n} = \frac{(-\lambda)^x (q^{-n}; q)_u}{(q;q)_u \, {}_1\phi_0(q^{-n}; -; q, -\lambda)},$$

where $u = 0, 1, \ldots, n$. The number of reported events in the first workplace, U, therefore has a q-binomial distribution.

If we assume that underreporting occurs only in the first workplace, then the number of reported events in the second workplace, V, has the Heine distribution with pmf

$$\Pr[V = v] = \frac{q^{v(v-1)/2}\lambda^v}{(-\lambda; q)_\infty (q; q)_v}, \qquad v = 0, 1, \ldots . \qquad (14.3.1)$$

Conditional on a given total number of reported events, m, in the two workplaces, where $m \leq n$, it follows that the number of reported events in the first workplace, $U | U + V = m$, has the absorption distribution with pgf (14.1.4). Note that this is true even when $m > n$ where n is the parameter of the weight function.

References

1. Benkherouf, L. and Bather, J. A. (1988). Oil exploration: sequential decisions in the face of uncertainty. *Journal of Applied Probability* **25**, 529–543.

2. Blomqvist, N. (1952). On an exhaustion process. *Skandinavisk Aktuarietidskrift* **35**, 201–210.

3. Borenius, G. (1953). On the statistical distribution of mine explosions. *Skandinavisk Aktuarietidskrift* **36**, 151–157.

4. Dunkl, C. F. (1981). The absorption distribution and the q-binomial theorem. *Communications in Statistics — Theory and Methods* **A10**, 1915–1920.

5. Gasper, G. and Rahman, M. (1990). *Basic Hypergeometric Series.* Cambridge University Press, Cambridge, England.

6. Johnson, N. L. and Kotz, S. (1969). *Discrete Distributions.* Houghton Mifflin, New York.

7. Johnson, N. L., Kotz, S., and Kemp, A. W. (1992). *Univariate Discrete Distributions,* Second edition. John Wiley & Sons, New York.

8. Kagan, A. M., Linnik, Yu. V., and Rao, C. R. (1973). *Characterization Problems in Mathematical Statistics.* John Wiley & Sons, New York.

9. Kemp, A. W. (1992a). Heine-Euler extensions of the Poisson distribution. *Communications in Statistics — Theory and Methods* **21**, 571–588.

10. Kemp, A. W. (1992b). Steady State Markov chain models for the Heine and Euler distributions. *Journa of Applied Probability* **29**, 869–876.

11. Kemp, A. W. (1998). Absorption sampling and the absorption distribution. *Journa of Applied Probability*, **35**, 1–6.

12. Kemp, A. W. and Kemp, C. D. (1991). Weldon's dice data revisited, *American Statistician*, **45**, 216–222.

13. Kemp, A. W. and Newton, J. (1990). Certain state-dependent processes for dichotomised parasite populations. *Journal of Applied Probability* **27**, 251–258.

14. Menon, M. V. (1966). Characterization theorems for some univariate probability distributions. *Journal of the Royal Statistical Society, Series B* **28**, 143–145.

15. Patil, G. P. and Seshadri, V. (1964). Characterization theorems for some univariate probability distributions. *Journal of the Royal Statistical Society, Series B* **26**, 286–292.

16. Rawlings, D. (1997). Absorption processes: models for q-identities. *Advances in Applied Mathematics* **18**, 133–148.

17. Zacks, S. and Goldfarb, D. (1966). Survival probabilities in crossing a field containing absorption points. *Naval Research Logistics Quarterly* **13**, 35–48.

15

Refinements of Inequalities for Symmetric Functions

Ingram Olkin
Stanford University, Stanford, CA

ABSTRACT The arithmetic-geometric mean inequality has motivated many variations and generalizations. One extension, due to Ky Fan, is an inequality of ratios of arithmetic means versus geometric means. Using a majorization argument we show that this inequality is a special case of a general monotonicity result.

Keywords and phrases Majorization, Schur-convex functions, monotonicity of means moments

The arithmetic-geometric mean inequality has perhaps motivated more variations and extensions than any other inequality. One of these, due to Ky Fan and reported in Beckenbach and Bellman (1961, p. 5), has generated a life of its own.

Let $x_1, \ldots, x_n$ be a sequence of positive real numbers in the interval $(0, \frac{1}{2}]$, $\tilde{x}_i = 1 - x_i$, $i = 1, \ldots, n$, and let the arithmetic, geometric and harmonic means be denoted by

$$A_n = \frac{\sum x_i}{n}, \quad G_n = \Pi x_i^{1/n}, \quad H_n = \frac{n}{\sum x_i^{-1}},$$

$$A'_n = \frac{\sum \tilde{x}_i}{n}, \quad G'_n = \Pi \tilde{x}_i^{1/n}, \quad H'_n = \frac{n}{\sum \tilde{x}_i^{-1}}.$$

The Ky Fan inequality asserts that

$$\frac{G_n}{G'_n} \le \frac{A_n}{A'_n}, \tag{15.1.1}$$

with equality if and only if $x_1 = \cdots = x_n$. There are many proofs of (15.1.1) since it was first introduced. Beckenbach and Bellman (1961) indicate that (15.1.1) is a good candidate for a proof by forward or backward induction. Wang and Wang (1984) show further that

$$\frac{H_n}{H'_n} \leq \frac{G_n}{G'_n}, \tag{15.1.2}$$

with equality if and only if $x_1 = \cdots = x_n$. For another extension and further references see Sándor and Trif (1999).

Inequalities (15.1.1) and (15.1.2) are suggestive of a hierarchy of inequalities. Define the generalized means

$$M(p) = \left(\frac{\sum x_i^p}{n}\right)^{1/p}, \quad M'(p) = \left(\frac{\sum \tilde{x}_i^p}{n}\right)^{1/p}, \tag{15.1.3}$$

then (15.1.1) and (15.1.2) state that

$$\frac{M(-1)}{M'(-1)} \leq \frac{M(0)}{M'(0)} \leq \frac{M(1)}{M'(1)}, \tag{15.1.4}$$

with equality if and only if $x_1 = \cdots = x_n$.

Inequalities (15.1.4) are suggestive of the stronger result that $\frac{M(p)}{M'(p)}$ is increasing in p, and we show below that this is indeed the case. In order to motivate the proof we first discuss the proof of the simpler inequality (15.1.1), as provided in Marshall and Olkin (1979, p. 97–98). The key ingredients are the majorization ordering and the characterization of the order-preserving functions, called Schur-convex functions.

DEFINITION 15.1.1
For $x, y \in \mathcal{R}^n$, $x_{[1]} \geq \cdots \geq x_{[n]}$, $y_{[1]} \geq \cdots \geq y_{[n]}$, $x \prec y$ (read y majorizes x) if $\sum_1^k x_{[i]} \leq \sum_1^k y_{[i]}$, $k = 1, \ldots, n-1$, $\sum_1^n x_{[i]} = \sum_1^n y_{[i]}$.

There are a number of characterizations of Schur-convex functions. One due to Schur in 1923 and enhanced by Ostrowski in 1952 [see Marshall and Olkin (1979, p. 57)] is the following.

PROPOSITION 15.1.2
Let $I \subset \mathcal{R}$ be an open interval and let $\varphi : I^n \to \mathcal{R}$ be continuously differentiable. Necessary and sufficient conditions for φ to be Schur-convex on I^n are (i) φ is symmetric on I^n, and (ii) $\partial\varphi(z)/\partial z_i$ is decreasing in $i = 1, \ldots, n$ for all $z_1 \geq \cdots \geq z_n$, $z \in I^n$.

To prove (15.1.1) we first show that on the set $\{z : 0 < z_i \leq \frac{1}{2}, i = 1, \ldots, n\}$,

$$\varphi(x) = \frac{\Pi(1 - x_i)^{1/n}}{\sum(1 - x_i)} \quad \frac{\sum x_i}{\Pi x_i^{1/n}} \equiv \frac{An}{A'n} \quad \frac{G'n}{Gn} \tag{15.1.5}$$

is symmetric and Schur-convex.

The symmetry is immediate. Schur-convexity is obtained by showing that

$$\frac{\partial \log \varphi(x)}{\partial x_1} - \frac{\partial \log \varphi(x)}{\partial x_2} = \frac{1}{n}\left\{\frac{1}{x_2(1 - x_2)} - \frac{1}{x_1(1 - x_1)}\right\} \geq 0$$

because $[z(1 - z)]^{-1}$ is decreasing in $z \in (0, \frac{1}{2}]$.

Because Schur-convex functions preserve the majorization ordering, and because

$$(x_1, \ldots, x_n) \succ (\bar{x}, \ldots, \bar{x}), \quad \bar{x} = \frac{\sum x_i}{n},$$

it follows from (15.1.5) that

$$\varphi(x_i, \ldots, x_n) \geq \varphi(\bar{x}, \ldots, \bar{x}) = 1,$$

which is inequality (15.1.1).

The following proposition is a result of Marshall, Olkin, and Proschan (1967) on the monotonicity of the ratio of means [see also Marshall and Olkin (1979, p. 130)].

PROPOSITION 15.1.3
If $x_1 \geq \ldots \geq x_n > 0$, $y_i > 0$ and y_i/x_i is decreasing in $i = 1, \ldots, n$, and $\sum x_i = \sum y_i$, then

$$x \equiv (x_1, \ldots, x_n) \prec (y_1, \ldots, y_n) \equiv y.$$

Furthermore, if $w_i \geq, 0, \sum w_i = 1$, then

$$g_w(r) = \left(\frac{\sum w_i x_i^r}{\sum w_i y_i^r}\right)^{1-r}, \quad r \neq 0,$$

$$= \frac{\Pi x_i^{w_i}}{\Pi y_i^{w_i}}, \quad r = 0 \tag{15.1.6}$$

is increasing in r.

In the above context if $x_1 \geq \cdots \geq x_n > 0$ and $y_i = \tilde{x}_i$, then y_i/x_i is decreasing in i, so that $g_w(r)$ is increasing in r. The cases $r = -1, 0, 1$, $w_i = \cdots = w_n = 1/n$ yields (15.1.4). The result of Wang and Wang (1984) with weights also follows from (15.1.6).

We review the essence of the proof of Proposition 15.1.3. A first step is a characterization of majorization, that if $x_i > 0$ and y_i/x_i is decreasing in $i = 1, \ldots, n$, and $\sum x_i = \sum y_i$, that x is majorized by y:

$$x = (x_i, \ldots, x_n) \prec (y_i, \ldots, y_n) = y.$$

If y_i/x_i is decreasing, then

$$\left(\frac{1}{\sum x_j^r}\right) (x_1^r, \ldots, x_n^r) \prec \frac{1}{\left(\sum y_j^r\right)} (y_1^r, \ldots, y_n^r). \tag{15.1.7}$$

Because $\varphi(x) = \sum x_i^t$ is Schur-convex for $t \geq 1$, it follows from (15.1.7) that

$$\frac{\sum x_i^{rt}}{\left(\sum s_j^r\right)^t} \leq \frac{\sum y_i^{rt}}{\left(\sum y_j^r\right)^t}, \quad t \geq 1. \tag{15.1.8}$$

Let $t = \frac{s}{r}$, fix r and s such that $|s| \geq |r|$, $rs > 0$, then from (15.1.8) with $w_1 = \ldots = w_n$, that $g(r) \geq g(s)$ if $s \geq r > 0$ and $g(s) \geq g(r)$ if $s \leq r < 0$. Because g is continuous at 0, we obtain (15.1.6). The insertion of weights requires an additional argument [see Marshall and Olkin (1979, p. 131)].

The majorization (15.1.1) shows that other functions could be used to generate new inequalities. This follows from the fact that for convex g,

$$\frac{1}{\sum g(x_i)} (g(x_1), \ldots, g(x_n)) \prec \frac{1}{\sum g(y_i)} (g(y_1), \ldots, g(y_n)).$$

A more general continuous version of (15.1.5) can be obtained as follows. If F and G are probability distribution functions, $\bar{F} = 1 - F$ and $\bar{G} = 1 - G$ are survival functions, such that $F(0) = G(0) = 0$ and $\bar{F}^{-1}(p)/\bar{G}^{-1}(p)$ is increasing in p, then

$$\left[\frac{\int x^r dG(x)}{\int x^r dF(x)}\right]^{1/r}$$

is increasing in r.

Acknowledgements. The author gratefully acknowledges support of the Alexander von Humboldt Foundation.

References

1. Beckenbach, E. F. and Bellman, R. (1961). *Inequalities.* Springer-Verlag, Berlin.

2. Marshall, A. W. and Olkin, I. (1979). *Inequalities: Theory of Majorization and Its Applications*. Academic Press, New York.

3. Marshall, A. W., Olkin, I., and Proschan, F. (1967). Monotonicity of ratios of means and other applications of majorization. In *Inequalities* (Ed., O. Shisha), pp. 177–190. Academic Press, New York.

4. Sándor, J. and Trif, T. (1999). A new refinement of the Ky Fan inequality. *Mathematical Inequalities and Applications* **2**, 529–533.

5. Wang, W.-L. and Wang, P.-F. (1984). A class of inequalities for symmetric functions. *Acta Mathematica Sinica* **27**, 485–497. In Chinese.

16

General Occupancy Distributions

Ch. A. Charalambides
University of Athens, Athens, Greece

ABSTRACT Consider a supply of balls randomly distributed in n distinguishable urns and assume that the number R of different kinds of balls allocated in the n urns is a Poisson random variable. Further, suppose that the number X of balls of any specific kind distributed in any specific urn is a random variable with probability function $q_x = P(X = x)$, $x = 0, 1, 2, ...$. The probability function and factorial moments of the number K of occupied urns (by at least one ball each), given that $R = r$ different kinds of balls, among which $R_y = r_y$ are of multiplicity y for $y = 1, 2, ..., m$, with $r_1 + r_2 + \cdots + r_m = r$, are distributed in the n urns, are deduced. They are expressed in terms of finite differences of the u-fold convolution of q_x, $x = 0, 1, 2, ...$. Illustrating these results, the cases with q_x, $x = 0, 1, 2, ...$, zero-one Bernoulli and geometric distributions, or more generally binomial and negative binomial distributions, are presented.

Keywords and phrases Finite differences, random occupancy model, committee problems

16.1   Introduction

In discrete probability theory many random phenomena can be described in terms of urn models. This is an advantageous approach since urn models can be easily visualized and are very flexible. In occupancy theory, a collection of m balls of a general specification $(1^{r_1} 2^{r_2} \cdots m^{r_m})$, where $r_j \geq 0$ is the number of balls of multiplicity j, $j = 1, 2, ..., m$, with $r_1 + 2r_2 + \cdots + mr_m = m$, are randomly distributed in n urns of a general specification $(1^{u_1} 2^{u_2} \cdots n^{u_n})$, where $u_i \geq 0$ is the number of urns of multiplicity i, $i = 1, 2, ..., n$, with $u_1 + 2u_2 + \cdots + nu_n = n$. Further, the urns may be of limited or unlimited capacity and the balls in the urns may be ordered or unordered [cf. MacMahon (1960) and Riordan (1958)].

Consider the particular case of n distinguishable urns ($u_1 = n$, $u_i = 0$, $i = 2, 3, ..., n$) and let Z_j be the number of balls distributed in the jth urn, $j = 1, 2, ..., n$. The random variables Z_j, $j = 1, 2, ..., n$ are called random occupancy numbers. The reduction of the joint distribution of the random occupancy numbers $Z_1, Z_2, ..., Z_n$ to a joint conditional distribution of independent random variables $X_1, X_2, ..., X_n$ given that $S_n = X_1 + X_2 + \cdots + X_n = m$ is a powerful technique in the derivation and study of the distribution of the number K_i of urns occupied by i balls each. Barton and David (1959), considering a supply of balls randomly distributed in n distinguishable urns and assuming that the number X of balls distributed in any specific urn ($X_j = X$, $j = 1, 2, ..., n$) is a random variable obeying a Poisson, binomial or negative binomial law, derived the probability function and the factorial moments of the number K_0 of empty urns, in these three cases, given that $S_n = m$; these distributions are the classical, the restricted and the pseudo-contagious occupancy distributions, respectively. It is worth noticing that the assumption that X obeys a Poisson law is equivalent to the assumption that the m balls distributed in the urns are distinguishable. Charalambides (1986, 1997) derived, in the general case $P(X = x) = q_x$, $x = 0, 1, 2, ...$, the probability function and factorial moments of the number $K = n - K_0$ of occupied urns and more generally of the number K_i of urns occupied by i balls each, given that $S_n = m$. Holst (1980), considering a supply of r different kinds of balls randomly distributed in n distinguishable urns and assuming that the number $X_{i,j}$ of balls of the ith kind distributed in the jth urn is a random variable obeying a zero-one Bernoulli law independent of i, derived a representation of the characteristic function of the number K_0 of empty urns, given that $Y_i = y_i$ balls of the ith kind are distributed in the n urns, $i = 1, 2, ..., r$. This representation is used to study the asymptotic distribution of K_0 when $n \to \infty$. Further, the exact probability function and the factorial moments of K_0 given that $Y_i = y_i$, $i = 1, 2, ..., r$, were deduced. The assumption that $X_{i,j}$ obeys a zero-one Bernoulli law is equivalent to the assumption that the capacity of each urn is limited to one ball from each kind.

In the present paper a general occupancy distribution is derived. More precisely, a supply of balls randomly distributed in n distinguishable urns is considered and the number R of different kinds of balls distributed in the n urns is assumed to be a Poisson random variable. Further, it is assumed that the number X of balls of any specific kind distributed in any specific urn is a random variable with a general probability function $P(X = x) = q_x$, $x = 0, 1, 2, ...$. Then, the probability function and the factorial moments of the number K of occupied urns, given that $R = r$ different kinds of balls, among which $R_y = r_y$ are of multiplicity y for $y = 1, 2, ..., m$, with $r_1 + r_2 + \cdots r_m = r$, are distributed in the n urns, are deduced (Section 16.2). Further, some special occupancy distributions are presented (Section 16.3).

16.2   A General Random Occupancy Model

Consider a supply of balls randomly distributed in n distinguishable urns. Assume that the number R of different kinds of balls distributed in the n urns is a Poisson random variable with probability function

$$P(R = r) = e^{-\lambda}\frac{\lambda^r}{r!}, r = 0, 1, 2, ...(0 < \lambda < \infty). \qquad (16.2.1)$$

Further, suppose that the number $X_{i,j}$ of balls of the ith kind distributed in the jth urn is a random variable with known probability function

$$P(X_{i,j} = x) = q_x, \, x = 0, 1, 2, ..., \, j = 1, 2, ..., n. \qquad (16.2.2)$$

Assuming that the occupancy of each urn is independent of the others, the probability function of the number $Y_i = X_{i,1} + X_{i,2} + \cdots + X_{i,n}$ of balls of the ith kind distributed in the n urns, $P(Y_i = y) = q_y(n)$, $y = 0, 1, 2, ..., i = 1, 2, ...,$ is given by the sum

$$q_{y_i}(n) = \sum q_{x_{i,1}} q_{x_{i,2}} \cdots q_{x_{i,n}}, \, i = 1, 2, ..., \qquad (16.2.3)$$

where the summation is extended over all integers $x_{i,j} \geq 0, j = 1, 2, ..., n$ such that $x_{i,1} + x_{i,2} + \cdots + x_{i,n} = y_i$. Also, the total number of balls distributed in the n urns $S_R = Y_1 + Y_2 + \cdots + Y_R$ has a compound Poisson distribution with probability generating function

$$P_{S_R}(t) = e^{-\lambda[1 - P_Y(t)]}, \, P_Y(t) = \sum_{y=0}^{\infty} q_y(n)t^y.$$

Consider, in addition, the number R_y of those variables among $Y_1, Y_2, ...,$ Y_R that are equal to y, for $y = 0, 1, 2, ...$ Clearly, R_y is the number of different kinds of balls of multiplicity y distributed in the n urns and $\sum_{y=0}^{\infty} R_y = R$. Further, $R_y, y = 0, 1, 2, ...$ are stochastically independent random variables with Poisson probability function

$$P(R_y = r) = e^{-\lambda q_y(n)}\frac{[\lambda q_y(n)]^r}{r!}, \, r = 0, 1, 2, ..., \, y = 0, 1, 2, ... \qquad (16.2.4)$$

and $\sum_{y=1}^{\infty} yR_y = S_R$ [cf Feller (1968, pp. 291–292)]. Note that the assumption that R is the number of different kinds of balls distributed in the n urns implies $R_0 = 0$. Also, if the total number of balls distributed in the n urns is assumed to be $S_R = m$, then $R_{m+1} = 0$, $R_{m+2} = 0, \ldots$. The inverse is not necessarily true. In the sequel, this stochastic model is referred as *the general random occupancy model.*

Under this stochastic model, let K be the number of occupied urns (by at least one ball each) and

$$p_k \equiv p_k(r_1, r_2, \ldots, r_m; r, n)$$
$$= P(K = k | R = r, \, R_1 = r_1, \; R_2 = r_2, \ldots, R_m = r_m),$$
$$k = 1, 2, \ldots, n \qquad (16.2.5)$$

with $r_1 + r_2 + \cdots + r_m = r$. Also, with $(u)_j = u(u - 1) \cdots (u - j + 1)$ the (descending) factorial of u of degree j, let

$$\mu_{(j)} \equiv \mu_{(j)}(r_1, r_2, \ldots, r_m; r, n)$$
$$= E[(K)_j | R = r, R_1 = r_1, R_2 = r_2, \ldots, R_m = r_m],$$
$$j = 1, 2, \ldots. \qquad (16.2.6)$$

The probability function (16.2.5) and its factorial moments (16.2.6) may be expressed in terms of finite differences of the convolutions (16.2.1). The derivation of these expressions is facilitated by the following lemma

LEMMA 16.2.1
Consider the general random occupancy model and assume that $R = r$ different kinds of balls, among which $R_y = r_y$ are of multiplicity y for $y = 1, 2, \ldots, m$, with $r_1 + r_2 + \cdots + r_m = r$, are randomly distributed in n distinguishable urns. Let A_j be the event that the jth urn remains empty, $j = 1, 2, \ldots, n$. Then the events $A_1, A_2, \ldots, A_n$ are exchangeable with

$$P(A_{j_1} A_{j_2} \cdots A_{j_k}) = \frac{q_0^{rk} [q_1(n - k)]^{r_1} [q_2(n - k)]^{r_2} \cdots [q_m(n - k)]^{r_m}}{[q_1(n)]^{r_1} [q_2(n)]^{r_2} \cdots [q_m(n)]^{r_m}}$$
$$(16.2.7)$$

for every k-combination $\{j_1, j_2, \ldots, j_k\}$ of the n indices $\{1, 2, \ldots, n\}$ and $k = 1, 2, \ldots, n$.

PROOF Note first that the event $\{R_1 = r_1, R_2 = r_2, \ldots, R_m = r_m\}$ given the event $R = r$, with $r_1 + r_2 + \cdots + r_m = r$, in addition to $R_0 = 0$, implies $R_{m+1} = 0$, $R_{m+2} = 0, \ldots$. Thus, on using (16.2.1) and (16.2.4) and the independence of R_y, $y = 0, 1, 2, \ldots$, it follows that

$$P(R_1 = r_1, R_2 = r_2, ..., R_m = r_m | R = r)$$

$$= P(R_0 = 0) \prod_{y=1}^{m} P(R_y = r_y) \prod_{y=m+1}^{\infty} P(R_y = 0)/P(R = r)$$

$$= \frac{r!}{r_1! r_2! \cdots r_m!} [q_1(n)]^{r_1} [q_2(n)]^{r_2} \cdots [q_m(n)]^{r_m}.$$

Further, consider the number $Z_i = X_{i,j_{k+1}} + X_{i,j_{k+2}} + \cdots + X_{i,j_n}$ of balls of the ith kind distributed in the $n - k$ urns $\{j_{k+1}, j_{k+2}, ..., j_n\}$, $i = 1, 2, ...,$ where $\{j_{k+1}, j_{k+2}, ..., j_n\} \subseteq \{1, 2, ..., n\} - \{j_1, j_2, ..., j_k\}$. Then $P(Z_i = z) = q_z(n - k)$, $z = 0, 1, 2, ...,$ $i = 1, 2, ...$ Assume that $X_{i,j_1} = X_{i,j_2} = \cdots = X_{i,j_k} = 0$, $i = 1, 2, ...$ and let U_z be the number of those variables among $Z_1, Z_2, ..., Z_R$ that are equal to z, for $z = 0, 1, 2, ...$ The random variables U_z, $z = 0, 1, 2, ...$ are stochastically independent with Poisson probability function

$$P(U_z = u) = e^{-\lambda q_z(n-k)} \frac{[\lambda q_z(n-k)]^u}{u!} \quad , u = 0, 1, 2, ..., z = 0, 1, 2, ...$$

and $\sum_{z=0}^{\infty} U_z = R$, $\sum_{z=1}^{\infty} z U_z = S_R$. Then the probability $P(A_{j_1} A_{j_2} \cdots A_{j_k})$ is equal to the conditional probability of the event

$$\{X_{i,j_s} = 0, \ s = 1, 2, ..., k, \ i = 1, 2, ..., R, \ U_1 = u_1,$$

$$U_2 = u_2, ..., U_m = u_m\},$$

with $u_1 + u_2 + \cdots + u_m = r$, given the occurrence of the event

$$\{R_1 = r_1, \ R_2 = r_2, ..., R_m = r_m, \ R = r\},$$

with $r_1 + r_2 + \cdots + r_m = r$, which, in turn, is equal to

$$\frac{P(X_{i,j_s} = 0, \ s = 1, 2, ..., k, \ i = 1, 2, ..., r, \ U_1 = r_1, ..., U_m = r_m | R = r)}{P(R_1 = r_1, R_2 = r_2, ..., R_m = r_m | R = r)}$$

$$= \frac{q_0^{rk} [q_1(n - k)]^{r_1} [q_2(n - k)]^{r_2} \cdots [q_m(n - k)]^{r_m}}{[q_1(n)]^{r_1} [q_2(n)]^{r_2} \cdots [q_m(n)]^{r_m}}.$$

Hence (16.2.7) is established. ∎

THEOREM 16.2.2

(a) Under the general random occupancy model, the conditional probability $p_k \equiv p_k(r_1, r_2, ..., r_m; r, n)$ that k urns are occupied, given that $R = r$

different kinds of balls, among which $R_y = r_y$ are of multiplicity y for $y = 1, 2, ..., m$, with $r_1 + r_2 + \cdots + r_m = r$, are randomly distributed in n distinguishable urns, $k = 1, 2, ..., n$, is given by

$$p_k = \binom{n}{k} \frac{[\Delta_u^k q_0^{r(n-u)} [q_1(u)]^{r_1} [q_2(u)]^{r_2} \cdots [q_m(u)]^{r_m}]_{u=0}}{[q_1(n)]^{r_1} [q_2(n)]^{r_2} \cdots [q_m(n)]^{r_m}}. \qquad (16.2.8)$$

(b) The jth factorial moment $\mu_{(j)} \equiv \mu_{(j)}(r_1, r_2, ..., r_m; r, n)$, $j = 1, 2, ...,$ of the probability function p_k, $k = 1, 2, ..., n$, is given by

$$\mu_{(j)} = (n)_j \frac{[\Delta_u^j g_0^{r(n-u)} [q_1(u)]^{r_1} [q_2(u)]^{r_2} \cdots [q_m(u)]^{r_m}]_{u=n-j}}{[q_1(n)]^{r_1} [q_2(n)]^{r_2} \cdots [q_m(n)]^{r_m}}. \qquad (16.2.9)$$

PROOF (a) Consider the general random occupancy model and assume that $R = r$ different kinds of balls, among which $R_y = r_y$ are of multiplicity y for $y = 1, 2, ..., m$, with $r_1 + r_2 + \cdots + r_m = r$, are randomly distributed in n distinguishable urns. Let A_j be the event that the jth urn remains empty, $j = 1, 2, ..., n$. The conditional probability $p_k \equiv p_k(r_1, r_2, ..., r_m; r, n)$ that k urns are occupied is equal to the probability that exactly $n - k$ among the n events $A_1, A_2, ..., A_n$ occur. Thus, on using the inclusion and exclusion principle and since, by Lemma 2.1, the events $A_1, A_2, ..., A_n$ are exchangeable, it follows that

$$p_k = \binom{n}{n-k} \sum_{s=n-k}^{n} (-1)^{s-n+k} \binom{k}{n-s} P(A_{j_1} A_{j_2} \cdots A_{j_s})$$

$$= \binom{n}{k} \sum_{j=0}^{k} (-1)^{k-j} \binom{k}{j} \frac{q_0^{r(n-j)} [q_1(j)]^{r_1} [q_2(j)]^{r_2} \cdots [q_m(j)]^{r_m}}{[q_1(n)]^{r_1} [q_2(n)]^{r_2} \cdots [q_m(n)]^{r_m}}.$$

The last expression implies (16.2.8).

(b) The ith binomial moment

$$\frac{\nu_{(i)}}{i!} = E\left[\binom{K_0}{i} \middle| R = r, R_1 = r_1, R_2 = r_2, ..., R_m = r_m \right]$$

of the number K_0 of empty urns is equal to the conditional probability that any i urns remain empty given that $R = r$ different kinds of balls, among which $R_y = r_y$ are of multiplicity y for $y = 1, 2, ..., m$, with $r_1 + r_2 + \cdots + r_m = r$, are randomly distributed in the n urns. Thus, by (16.2.7),

$$\frac{\nu_{(i)}}{i!} = \binom{n}{i} \frac{q_0^{r_i} [q_1(n-i)]^{r_1} [q_2(n-i)]^{r_2} \cdots [q_m(n-i)]^{r_m}}{[q_1(n)]^{r_1} [q_2(n)]^{r_2} \cdots [q_m(n)]^{r_m}}.$$

Since

$$\mu_{(j)} = E[(K)_j | R = r, R_1 = r_1, R_2 = r_2, ..., R_m = r_m]$$
$$= E[(n - K_0)_j | R = r, R_1 = r_1, R_2 = r_2, ..., R_m = r_m]$$
$$= (-1)^j E([K_0 - n]_j | R = r, R_1 = r_1, R_2 = r_2, ..., R_m = r_m)$$
$$= (-1)^j \nu_{[j]}(n),$$

on using the expression [Charalambides (1986)]

$$\nu_{[j]}(n) = \sum_{i=0}^{j} (-1)^{j-i} \frac{j!}{i!} \binom{n-i}{j-i} \nu_{(i)}$$

it follows that

$$\mu_{(j)} = (n)_j \sum_{i=0}^{j} (-1)^i \binom{j}{i} \frac{q_0^{ri} [q_1(n-i)]^{r_1} [q_2(n-i)]^{r_2} \cdots [q_m(n-i)]^{r_m}}{[q_1(n)]^{r_1} [q_2(n)]^{r_2} \cdots [q_m(n)]^{r_m}}.$$

The last expression implies (16.2.9).

An interesting corollary of Theorem 16.2.2 is deduced when it is assumed that $R = r$ and $R_s = r$, $R_y = 0$, $y \neq s$. ∎

COROLLARY 16.2.3

(a) Under the general random occupancy model, the conditional probability $p_k(s; r, n)$ that k urns are occupied, given that s balls from each of r kinds are distributed in the n distinguishable urns, $k = 1, 2, ..., n$, is given by

$$p_k(s; r, n) = \binom{n}{k} \frac{[\Delta_u^k q_0^{r(n-u)} [q_s(u)]^r]_{u=0}}{[q_s(n)]^r}. \tag{16.2.10}$$

(b) The jth factorial moment $\mu_{(j)}(s; r, n)$, $j = 1, 2, ...,$ of the probability function $p_k(s; r, n)$, $k = 1, 2, ..., n$, is given by

$$\mu_{(j)}(s; r, n) = (n)_j \frac{[\Delta_u^j g_0^{r(n-u)} [q_s(u)]^r]_{u=n-j}}{[q_s(n)]^r}.$$

16.3   Special Occupancy Distributions

16.3.1   Geometric Probabilities

Assume that the number X of balls of any specific kind distributed in any specific urn obeys a *geometric distribution* with

$$q_x = P(X = x) = pq^x, \ x = 0, 1, 2, \ldots \ (q = 1 - p, \ 0 < p < 1).$$

Its u-fold convolution obeys a negative binomial distribution with

$$q_y(u) = P(Y = y) = \binom{u + y - 1}{y} p^u q^y, \ y = 0, 1, 2, \ldots.$$

Then, according to Lemma 16.2.1, the conditional probability that k specified urns remain empty given that $R = r$ different kinds of balls, among which $R_y = r_y$ are of multiplicity y for $y = 1, 2, \ldots, m$, with $r_1 + r_2 + \cdots + r_m = r$, are randomly distributed in n distinguishable urns, $k = 1, 2, \ldots, n$, is given by

$$P(A_{j_1} A_{j_2} \cdots A_{j_k}) = \frac{\prod\limits_{y=1}^{m} \left(\dbinom{n - k + y - 1}{y} \right)^{r_y}}{\prod\limits_{y=1}^{m} \left(\dbinom{n + y - 1}{y} \right)^{r_y}}.$$

Therefore, the general random occupancy model, with the assumption of geometric probabilities, is equivalent to the occupancy model, where a fixed number m of balls of the general specification $(1^{r_1} 2^{r_2} \cdots m^{r_m})$, with $r_1 + 2r_2 + \cdots + mr_m = m$, $r_1 + r_2 + \cdots + r_m = r$, are randomly distributed in n distinguishable urns [cf Riordan (1958, Chapter 5)].

Thus, by Theorem 16.2.2, the occupancy probability function

$$p_k = P(K = k | R = r, R_1 = r_1, R_2 = r_2, \ldots, R_m = r_m), \ k = 1, 2, \ldots, n$$

is given by

$$p_k = \binom{n}{k} \frac{\left[\Delta_u^k \prod\limits_{y=1}^{m} \left(\dbinom{u + y - 1}{y} \right)^{r_y} \right]_{u=0}}{\prod\limits_{y=1}^{m} \left(\dbinom{n + y - 1}{y} \right)^{r_y}}$$

$$= \binom{n}{k} \sum_{j=0}^{k} (-1)^{k-j} \binom{k}{j} \frac{\prod\limits_{y=1}^{m} \left(\dbinom{j + y - 1}{y} \right)^{r_y}}{\prod\limits_{y=1}^{m} \left(\dbinom{n + y - 1}{y} \right)^{r_y}}. \qquad (16.3.1)$$

The closely related occupancy probability function of the number K_0 of empty urns, under the occupancy model where a fixed number m of balls of the general specification $(1^{r_1}2^{r_2}\cdots m^{r_m})$, with $r_1 + 2r_2 + \cdots + mr_m = m$, $r_1 + r_2 + \cdots + r_m = r$, are randomly distributed in n distinguishable urns, was derived by Riordan (1958, p. 96).

The jth factorial moment

$$\mu_{(j)} = E[(K)_j \mid R = r, R_1 = r_1, R_2 = r_2, ..., R_m = r_m], \ j = 1, 2, ...$$

is given by

$$\mu_{(j)} = (n)_j \frac{\left[\Delta_u^j \prod_{y=1}^{m} \binom{u+y-1}{y}^{r_y}\right]_{u=n-j}}{\prod_{y=1}^{m} \binom{n+y-1}{y}^{r_y}}$$

$$= (n)_j \sum_{i=0}^{j} (-1)^{j-i} \binom{j}{i} \frac{\prod_{y=1}^{m} \binom{n-j+i+y-1}{y}^{r_y}}{\prod_{y=1}^{m} \binom{n+y-1}{y}^{r_y}}. \quad (16.3.2)$$

In particular, by Corollary 16.2.3, the conditional probability $p_k(s; r, n)$, $k = 1, 2, ..., n$, that k urns are occupied, given that s balls from each of r kinds are distributed in n distinguishable urns, is given by

$$p_k(s; r, n) = \binom{n}{k} \frac{\left[\Delta_u^k \binom{u+s-1}{s}^{r}\right]_{u=0}}{\binom{n+s-1}{s}^r}$$

$$= \binom{n}{k} \sum_{j=0}^{k} (-1)^{k-j} \binom{k}{j} \frac{\binom{j+s-1}{s}^{r}}{\binom{n+s-1}{s}^r}. \quad (16.3.3)$$

The jth factorial moment $\mu_{(j)}(s; r, n)$, $j = 1, 2, ..., n$, of this distribution is

$$\mu_{(j)}(s; r, n) = (n)_j \frac{\left[\Delta_u^j \binom{u+s-1}{s}^{r}\right]_{u=n-j}}{\binom{n+s-1}{s}^r},$$

$$= (n)_j \sum_{i=0}^{j} (-1)^{j-i} \binom{j}{i} \frac{\left(\dfrac{n-j+i+s-1}{s} \right)^r}{\left(\dfrac{n+s-1}{s} \right)^r}.$$

(16.3.4)

The following two remarks concerning extensions of and/or imposition of restrictions on this random occupancy model are worth noting. The replacement of the assumption that the number X obeys a geometric distribution by the assumption that it obeys the more general negative binomial distribution with

$$q_x = P(X = x) = \binom{s+x-1}{x} p^s q^x, \quad x = 0, 1, 2, ...,$$

divides each urn into s distinguishable compartments. In this case

$$q_y(u) = P(Y = y) = \binom{su + y - 1}{y} p^{su} q^y, \quad y = 0, 1, 2, ...$$

and

$$p_k = \binom{n}{k} \frac{\left[\Delta_u^k \prod_{y=1}^{m} \left(\dfrac{su + y - 1}{y} \right)^{r_y} \right]_{u=0}}{\prod_{y=1}^{m} \left(\dfrac{sn + y - 1}{y} \right)^{r_y}}$$

$$= \binom{n}{k} \sum_{j=0}^{k} (-1)^{k-j} \binom{k}{j} \frac{\prod\limits_{y=1}^{m} \left(\dfrac{su + y - 1}{y} \right)^{r_y}}{\prod\limits_{y=1}^{m} \left(\dfrac{sn + y - 1}{y} \right)^{r_y}} \qquad (16.3.5)$$

with

$$\mu_{(j)} = (n)_j \frac{\left[\Delta_u^j \prod_{y=1}^{m} \left(\dfrac{su + y - 1}{y} \right)^{r_y} \right]_{u=n-j}}{\prod_{y=1}^{m} \left(\dfrac{sn + y - 1}{y} \right)^{r_y}}$$

$$= (n)_j \sum_{i=0}^{j} (-1)^{j-i} \binom{j}{i} \frac{\prod\limits_{y=1}^{m} \left(\dfrac{s(n - j + i) + y - 1}{y} \right)^{r_y}}{\prod\limits_{y=1}^{m} \left(\dfrac{sn + y - 1}{y} \right)^{r_y}}.$$

(16.3.6)

Further, a restriction on the capacity of each urn or each compartment can be expressed by choosing a suitable truncated version of the distribution of X. Thus, the assumption of geometric probabilities truncated to the right at the point s, that is, with

$$q_x = P(X = x) = (1 - q^{s+1})^{-1} p q^x, \ x = 0, 1, 2, ..., s,$$

whence

$$q_y(u) = P(Y = y) = L(y, u, s)(1 - q^{s+1})^{-u} p^u q^y, \ y = 0, 1, 2, ..., su$$

where

$$L(y, u, s) = \sum_{j=0}^{u} (-1)^j \binom{u}{j} \binom{u + y - j(s+1) - 1}{u - 1},$$

imposes the restriction that the capacity of each urn is limited to s balls from each kind. The corresponding occupancy distribution can be similarly deduced.

Example 1

Formation of committees from grouped candidates. Assume that there are m vacant positions on r committees. More specifically $r_y \geq 0$ committees have y vacant positions, $y = 1, 2, ..., m$, so that $r_1 + 2r_2 + \cdots + m r_m = m$ and $r_1 + r_2 + \cdots + r_m = r$. Further, assume that for these m positions there are m priority-ordered nominations by each of n different groups. In a random selection of the m members of the r committees, find the probability that k of the n groups are represented on at least one committee [cf. Johnson and Kotz (1977, Chapter 3)].

The m vacant positions on the r committees may be considered as m balls of the specification $(1^{r_1} 2^{r_2} \cdots m^{r_m})$, with $r_1 + 2r_2 + \cdots + m r_m = m$, $r_1 + r_2 + \cdots + r_m = r$, and the n different groups of candidates as n distinguishable urns. Thus, the placement of a ball of the ith kind to the jth urn corresponds to the selection of the jth group to occupy a position on the ith committee. Since it is assumed that the nominations are priority-ordered, the selection of the jth group uniquely determines the candidate who occupies a position on the ith committee. Further, there is no restriction on the number of balls of same kind that each urn may accommodate. Therefore the probability that k of the n groups are represented on at least one committee is given by (16.3.1). In the particular case of the formation of r committees each of size s, the probability that k of the n groups are represented on at least one committee is given by (16.3.3). ⬚

16.3.2   Bernoulli Probabilities

Assume that the number X of balls of any specific kind distributed in any specific urn obeys a *zero-one Bernoulli distribution* with

$$q_x = P(X = x) = p^x q^{1-x}, \ x = 0,1 \ (q = 1 - p, \ 0 < p < 1).$$

Its u-fold convolution obeys a binomial distribution with

$$q_y(u) = P(Y = y) = \binom{u}{y} p^y q^{u-y}, \ y = 0, 1, ..., u.$$

Then, according to Lemma 16.2.1, the conditional probability that k specified urns remain empty given that $R = r$ different kinds of balls, among which $R_y = r_y$ are of multiplicity y for $y = 1, 2, ..., m$, with $r_1 + r_2 + \cdots + r_m = r$, are randomly distributed in n distinguishable urns, $k = 1, 2, ..., n$, is given by

$$P(A_{j_1} A_{j_2} \cdots A_{j_k}) = \frac{\prod\limits_{y=1}^{m} \binom{n-k}{y}^{r_y}}{\prod\limits_{y=1}^{m} \binom{n}{y}^{r_y}}.$$

Therefore, the general random occupancy model, with the assumption of Bernoulli probabilities, is equivalent to the occupancy model where a fixed number m of balls of the general specification $(1^{r_1} 2^{r_2} \cdots m^{r_m})$, with $r_1 + 2r_2 + \cdots + mr_m = m$, $r_1 + r_2 + \cdots + r_m = r$, are randomly distributed in n distinguishable urns, each of capacity limited to one ball from each kind.

Thus, by Theorem 16.2.2, the occupancy probability function is given by

$$p_k = \binom{n}{k} \frac{\left[\Delta_u^k \prod\limits_{y=1}^{m} \binom{u}{y}^{r_y} \right]_{u=0}}{\prod\limits_{y=1}^{m} \binom{n}{y}^{r_y}}$$

$$= \binom{n}{k} \sum_{j=0}^{k} (-1)^{k-j} \binom{k}{j} \frac{\prod\limits_{y=1}^{m} \binom{j}{y}^{r_y}}{\prod\limits_{y=1}^{m} \binom{n}{y}^{r_y}}. \tag{16.3.7}$$

Its jth factorial moment is given by

$$\mu_{(j)} = (n)_j \frac{\left[\Delta_u^j \prod_{y=1}^{m} \binom{u}{y}^{r_y}\right]_{u=n-j}}{\prod_{y=1}^{m} \binom{n}{y}^{r_y}}$$

$$= (n)_j \sum_{i=0}^{j} (-1)^{j-i} \binom{j}{i} \frac{\prod_{y=1}^{m} \binom{n-j+i}{y}^{r_y}}{\prod_{y=1}^{m} \binom{n}{y}^{r_y}}. \qquad (16.3.8)$$

In particular, by Corollary 16.2.3, the conditional probability $p_k(s; r, n)$ that k urns are occupied, given that s balls from each of r kinds are distributed in n distinguishable urns, each of capacity limited to one ball from each kind, is given by

$$p_k(s; r, n) = \binom{n}{k} \frac{\left[\Delta_u^k \binom{u}{s}^{r}\right]_{u=0}}{\binom{n}{s}^{r}}$$

$$= \binom{n}{k} \sum_{j=0}^{k} (-1)^{k-j} \binom{k}{j} \frac{\binom{j}{s}^{r}}{\binom{n}{s}^{r}} \qquad (16.3.9)$$

and

$$\mu_{(j)}(s; r, n) = (n)_j \frac{\left[\Delta_u^j \binom{u}{s}^{r}\right]_{u=n-j}}{\binom{n}{s}^{r}}$$

$$= (n)_j \sum_{i=0}^{j} (-1)^{j-i} \binom{j}{i} \frac{\binom{n-j+i}{s}^{r}}{\binom{n}{s}^{r}}. \qquad (16.3.10)$$

As in the case of the random occupancy model with geometric probabilities, extensions of and/or imposition of restrictions on this random occupancy model can be achieved by assuming that X obeys a binomial or a suitable truncated binomial distribution. Notice that the assumption that X obeys the binomial distribution

$$q_x = \binom{s}{x} p^x q^{s-x}, \quad x = 0, 1, 2, ..., s \ (q = 1 - p, \ 0 < p < 1)$$

corresponds to the assumption that each urn is divided into s distinguishable compartments each with capacity limited to one ball. In this case

$$q_y(u) = \binom{su}{y} p^y q^{su-y}, \quad y = 0, 1, 2, ..., su$$

and

$$p_k = \binom{n}{k} \frac{\left[\Delta_u^k \prod_{y=1}^{m} \binom{su}{y}^{r_y} \right]_{u=0}}{\prod_{y=1}^{m} \binom{sn}{y}^{r_y}}$$

$$= \binom{n}{k} \sum_{j=0}^{k} (-1)^{k-j} \binom{k}{j} \frac{\prod_{y=1}^{m} \binom{sj}{y}^{r_y}}{\prod_{y=1}^{m} \binom{sn}{y}^{r_y}}, \tag{16.3.11}$$

with

$$\mu_{(j)} = (n)_j \frac{\left[\Delta_u^j \prod_{y=1}^{m} \binom{su}{y}^{r_y} \right]_{u=n-j}}{\prod_{y=1}^{m} \binom{sn}{y}^{r_y}}$$

$$= (n)_j \sum_{i=0}^{j} (-1)^{j-i} \binom{j}{i} \frac{\prod_{y=1}^{m} \binom{s(n-j+i)}{y}^{r_y}}{\prod_{y=1}^{m} \binom{sn}{y}^{r_y}}. \tag{16.3.12}$$

Example 2
Formation of committees from individual candidates. As in Example 1, assume that there are m vacant positions on r committees, among which $r_y \geq 0$ committees have y vacant positions, $y = 1, 2, ..., m$, so that $r_1 + 2r_2 + \cdots + mr_m = m$ and $r_1 + r_2 + \cdots + r_m = r$. Further, assume that for these m positions there are n individual nominations. In a random

selection of the m members of the r committees, find the probability that k of the n candidates will participate to at least one committee [cf Johnson and Kotz (1977, Chapter 3)].

As in Example 1, the m vacant positions on the r committees may be considered as m balls of the specification $(1^{r_1} 2^{r_2} \cdots m^{r_m})$, with $r_1 + 2r_2 + \cdots + mr_m = m$, $r_1 + r_2 + \cdots + r_m = r$, and the n individual candidates as n distinguishable urns. Thus, the placement of a ball of the i-th kind to the j-th urn corresponds to the selection of the j-th candidate for a position on the i-th committee. We assume that no candidate can be selected to more than one positions of the same committee, while it can participate in more than one committee. This assumption implies that each cell may accommodate at most one object from each kind. Consequently, the probability that k of the n candidates will participate to at least one committee is given by (16.3.7). In the particular case of the formation of r committees each of size s, the probability that k of the n candidates are represented on at least one committee is given by (16.3.9). It is worth noticing that this committee problem can be rephrased as a multiple capture-recapture problem [cf Holst (1980)]. The probability function of the number of different individuals observed in the census, which was also derived by Berg (1974), is given by (16.3.7). ⬚

Example 3

Formation of committees from grouped candidates with unordered nominations. As in Example 1, assume that there are m vacant positions on r committees, among which $r_y \geq 0$ committees have y vacant positions, $y = 1, 2, ..., m$, so that $r_1 + 2r_2 + \cdots + mr_m = m$ and $r_1 + r_2 + \cdots + r_m = r$. Further, suppose that for the m positions there are s unordered nominations by each of n different groups. In a random selection of the m members of the r committees, find the probability that k of the n groups are represented on at least one committee.

In this case the n different groups each with s nominations may be considered as n distinguishable urns each with s distinguishable compartments of capacity limited to one ball. Then the probability that k of the n groups are represented on at least one committee is given by (16.3.11). ⬚

Acknowledgement This research was partially supported by the University of Athens Research Special Account under grant 70/4/3406.

References

1. Barton, D. E. and David, F. N. (1959). Contagious occupancy. *Journal of the Royal Statistical Society, Series B* **21**, 120–130.

2. Berg, S. (1974). Factorial series distributions with applications to capture-recapture problems. *Scandinavian Journal of Statistics* **1**, 145–152.

3. Charalambides, Ch. A. (1986). Derivation of probabilities and moments of certain generalized discrete distributions via urn models. *Communications in Statistics—Theory and Methods* **15**, 677–696.

4. Charalambides, Ch. A. (1997). A unified derivation of occupancy and sequential occupancy distributions. In *Advances in Combinatorial Methods and Applications to Probability and Statistics* (Ed. N. Balakrishnan), pp. 259–273. Birkhäuser, Boston.

5. Feller, W. (1968). *An Introduction to Probability Theory and its Applications*, Volume 1, Third Edition. John Wiley & Sons, New York.

6. Holst, L. (1980). On matrix occupancy, committee, and capture-recapture problems. *Scandinavian Journal of Statistics* **7**, 139–146.

7. Johnson, N. L. and Kotz, S. (1977). *Urn Models and their Application*. John Wiley & Sons, New York.

8. MacMahon, P. A. (1960). *Combinatory Analysis*, Volumes 1 and 2. Chelsea, New York.

9. Riordan, J. (1958). *An Introduction to Combinatorial Analysis*. John Wiley & Sons, New York.

17

A Skew t Distribution

M. C. Jones
The Open University, Milton Keynes, UK

ABSTRACT I first consider the formulation and properties of a new skew family of distributions on the real line which includes the (symmetric) Student t distributions as special cases. The family can be obtained through a simple generalization of a representation of the t distribution in Cacoullos (1965). I go on to explore briefly two ways in which bivariate distributions with skew t marginals can be developed.

Keywords and phrases Beta distribution, Dirichlet distribution, multivariate t and beta distributions, skewness, Student t distribution

17.1 Introduction

Before I embark on the paper proper, I must record my pleasure at being able to attend and enjoy the conference and to contribute to this volume in honour of Professor Cacoullos. Quite a lot of my own work has built on seminal papers of Professor Cacoullos from the 1960s. In particular, given my own emphasis over the years on kernel smoothing methodology [e.g. Wand and Jones (1995)], the reader might have expected me to take as my starting point Cacoullos (1966), the first paper on multivariate kernel density estimation. However, the current paper actually has links with a quite different source, namely Cacoullos (1965). (I am pleased that Professor Politis spoke at the conference on multivariate kernel density estimation!).

A fuller title for this paper might be "A New Univariate Skew t Distribution and Some Bivariate Extensions". Its primary purpose is to provide a family of distributions with support the whole real line that includes members with nonzero skewness but also includes the symmetric Student t distributions as special cases. The reasons for this exercise are twofold: to provide distributions with both skewness and heavy tails for use in robustness studies (both frequentist and Bayesian); and to provide further models for data. The second purpose of the paper is to outline briefly two extensions of the skew t distribution to the bivariate case.

At the time of the conference, a fuller account of the new skew t family could be found in my unpublished manuscript "A Skew Extension of the t Distribution". Since then, it has been decided — in collaboration with a colleague — to add a number of things to that paper, particularly to do with inference and a number of data applications of the distribution, and the new, in preparation, version will be referred to as Jones and Faddy (2000).

A plan of the current paper is as follows. The new (univariate) skew t distribution will be derived in Section 17.2. Some of its properties will be presented in Section 17.3. A first bivariate extension of the skew t distribution will be outlined in Section 17.4 and a second, quite different, bivariate extension will be outlined in Section 17.5. The figures in the paper were prepared using the UNIRAS graphics system in FORTRAN.

17.2   Derviation of Skew t Density

Let us proceed by analogy with the beta distribution. Consider as a starting point the *symmetric* beta distribution on $[-1, 1]$. (Readers more familiar with the beta distribution residing on $[0, 1]$ simply need to apply a linear transformation to change support.) The symmetric beta distribution is one in which the two parameters of the Beta(a, b) distribution are equal (to a, say). The distribution is symmetric about zero, with density proportional to $(1 - x^2)^{a-1}$. (In fact, it forms a family of kernels for univariate kernel estimation!)

Now, factorise $1 - x^2$ into $(1 + x)(1 - x)$. Power each factor up. But why should each factor have the same power? Why not give each factor a different power? Doing so gives a density proportional to $(1 + x)^{a-1}(1 - x)^{b-1}$ which is precisely the usual beta distribution on $[-1, 1]$. We have succeeded in going from a symmetric starting point to a more general skew family encompassing the symmetric distribution by a little mathematical sleight of hand! But we have not yet succeeded in coming up with something new.

However, we can apply much the same trick to the t distribution. In our teaching, we tend to shy away from giving the t density, and this prompted one Open University student to state that it must be so *horrific* that we don't write it down! However, as the beta distribution is essentially based on $1 - x^2$, so the horrific t distribution is essentially based on $1+x^2$. In fact, I will write the t density on $2a$ degrees of freedom (for convenience) as being proportional to $(1 + x^2/(2a))^{-(a+1/2)}, x \in \Re$. Now factorise this. No, I shan't venture into complex numbers. Instead, write

$$\left(1 + \frac{x^2}{2a}\right)^{-1} = 1 - \frac{x^2}{2a + x^2}$$

which factorises into

$$\left(1 + \frac{x}{\sqrt{2a + x^2}}\right)\left(1 - \frac{x}{\sqrt{2a + x^2}}\right).$$

Each of these can be given different powers $a + 1/2$ and $b + 1/2$ so that, with a simple rescaling taking $2a$ to $a + b$ for clarity, we have a density $f(x; a, b)$ proportional to

$$\left(1 + \frac{x}{\sqrt{a + b + x^2}}\right)^{a+1/2}\left(1 - \frac{x}{\sqrt{a + b + x^2}}\right)^{b+1/2}. \qquad (17.2.1)$$

This is the proposal for a *skew t distribution*. Its normalising factor is tractable and turns out to be

$$\{B(a, b)\sqrt{a + b}\ 2^{a+b-1}\}^{-1}$$

where $B(\cdot, \cdot)$ is the beta function. When $a = b$, f reduces to the t distribution on $2a$ degrees of freedom. Note also that $f(x; b, a) = f(-x; a, b)$.

Again, this all seems to be just so much mathematical legerdemain. Well, there is an alternative way of deriving distribution (17.2.1). It is by transformation of a beta random variable. To this end, if B has the Beta(a, b) distribution on $[-1, 1]$, then

$$T = \frac{\sqrt{a + b}\ B}{\sqrt{1 - B^2}} \qquad (17.2.2)$$

has the skew t distribution (17.2.1).

And this is the link with Cacoullos (1965) because formula (17.2.2) can also be written in terms of a pair (X, Y) of independent χ^2 random variables with $2a$ and $2b$ degrees of freedom, respectively:

$$T = \frac{\sqrt{a + b}\ (X - Y)}{2\sqrt{XY}}. \qquad (17.2.3)$$

Cacoullos (1965) is the paper setting out the special case of this relationship between a symmetric t random variable and a pair of independent chi-squares with the same degrees of freedom, say $2a$, or equivalently by writing $F = X/Y$ which has the F distribution on $2a, 2a$ degrees of freedom and of which the right-hand side of (17.2.3) is a simple function. Cacoullos (1965) remains very important as *the* paper setting out the relationship in the symmetric case, although it may well have been known to others beforehand. In a corrigendum to the paper, Professor Pratt, the journal editor, remarks "The editor is beginning to think he and the six or so experts he consulted are alone in not having been aware of this relation". I hope that I and the few people who have heard me talk on this subject are not alone in not being aware of the generalised relationship (17.2.3) and hence the skew t distribution!

Cacoullos (1999) generalises his 1965 paper in a different way.

17.3   Properties of Skew t Distribution

Figure 17.1 shows a variety of skew t densities each with $b = 2$. These densities have been standardised to have zero mean and unit variance. The symmetric density belongs to the Student t distribution on $2b = 4$ degrees of freedom. The other six densities, displaying increasing amounts of skewness, correspond to a doubling of a (which was initially 2) each time.

The moments of the skew t distribution are available in closed form and a general formula for $E(T^r)$, $r = 1, 2, ...$, is given in Jones and Faddy (2000). The rth moment exists provided both a and b exceed $r/2$. The mean and variance (used in standardisation above) are

$$E(T) = \frac{(a+b)^{1/2}(a-b)}{2} \frac{\Gamma(a-1/2)\Gamma(b-1/2)}{\Gamma(a)\Gamma(b)},$$

where $\Gamma(\cdot)$ is the gamma function, and

$$V(T) = \frac{(a+b)}{4}\left[\frac{\{(a-b)^2 + (a-1) + (b-1)\}}{(a-1)(b-1)} - \left\{\frac{(a-b)\Gamma(a-1/2)\Gamma(b-1/2)}{\Gamma(a)\Gamma(b)}\right\}^2\right].$$

An expression for the skewness can also be written down.

Straightforward differentiation shows that the skew t density is always unimodal with mode at

$$\frac{(a-b)\sqrt{a+b}}{\sqrt{2a+1}\sqrt{2b+1}}.$$

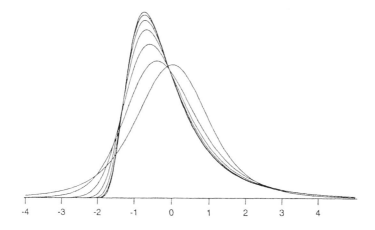

FIGURE 17.1
Densities (17.2.1), standardised to have zero mean and unit
variance, for $a = 2^i b$, $i = 0, ..., 6$, having increasing amounts of
skewness with i, in the case $b = 2$

Many (but not all) aspects of the properties of the skew t distribution depend on a and b through $\nu = a + b$ and $\lambda = a - b$. It is tempting to reparametrise the skew t distribution in terms of ν and λ which have appealing interpretations as 'essential' degrees of freedom and a parameter controlling skewness, respectively, although I shall not do so here.

It is suggested by Figure 17.1 that the densities converge to a limiting case as $a \to \infty$ with b held fixed. In fact, this obtains provided the mean and variance, which behave as a and a^2, respectively, as a increases, are standardised away. Jones and Faddy (2000) show that this limiting distribution is that of $\sqrt{2/Z}$ where Z follows the chi-squared distribution on $2b$ degrees of freedom. It is reasonable to think of this distribution as displaying the greatest skewness that a skew t distribution can possess: see Jones and Faddy (2000) for more on this skewness. If b is allowed to grow to ∞ with a, apparently in any fashion, the limiting distribution is the normal (as is clearly the case when $a = b \to \infty$).

Further properties of the skew t distribution are immediate because of its link to the beta distribution. For example, random variates can readily be produced, the skew t distribution function can be written in terms of the beta distribution function, and there are further transformation links with other distributions. See Jones and Faddy (2000, Section 17.5).

17.4   A First Bivariate Skew t Distribution

This section concerns the first of two suggestions for bivariate distributions with skew t marginals which I shall briefly pursue. This first bivariate extension actually works in the general multivariate case (although it is not clear whether it is of much practical importance in three or more dimensions) and is part of a wider approach to building multivariate distributions with given marginals that I explore elsewhere, Jones (2000).

The general, very simple, idea is that of replacing a marginal distribution of a given multivariate distribution with a desired marginal by multiplication i.e. really, of replacing the marginal distribution when representing the multivariate distribution by the product of marginal and conditional distributions. Specialising to the bivariate case, if $f(x, y)$ is a bivariate density with X-marginal density f_X, say, then

$$f_1(x, y) = g(x)f(x, y)/f_X(x) \qquad (17.4.1)$$

will have X-marginal g instead of f_X (provided the support of g is a subset of the support of f_X).

Now take f to be the density of the spherically symmetric bivariate t distribution on ν degrees of freedom, say; $f \propto (1+\nu^{-1}(x^2+y^2))^{-(\nu/2+1)}$. Replace its X-marginal f_X, which, like its Y-marginal, is the ordinary t distribution on ν degrees of freedom, by multiplying by g taken to be the density of the skew t distribution with parameters a and c, say, and dividing by f_X. The result is the distribution with density

$$\frac{\Gamma((\nu + 2)/2)}{\Gamma((\nu+1)/2)B(a,c)(a+c)^{1/2}2^{a+c-1}\sqrt{\nu\pi}}$$

$$\times \frac{(1 + \nu^{-1}x^2)^{(\nu+1)/2}\left(1 + \frac{x}{(a+c+x^2)^{1/2}}\right)^{a+1/2}\left(1 - \frac{x}{(a+c+x^2)^{1/2}}\right)^{c+1/2}}{(1+\nu^{-1}(x^2 + y^2))^{\nu/2+1}}.$$

$$(17.4.2)$$

Here, a, c and ν are all positive.

Distribution (17.4.2) has a number of interesting properties. First, it has a skew t X-marginal distribution by construction. Second, the conditional distribution of Y given $X = x$ is the symmetric t distribution on $\nu + 1$ degrees of freedom, rescaled by a factor of $\{(\nu + 1)^{-1}(x^2 + \nu)\}^{1/2}$; this property is shared with the original symmetric bivariate t distribution. Third, the Y-marginal of distribution (17.4.2) is also symmetric. Note that it is not itself a t distribution. However, in many

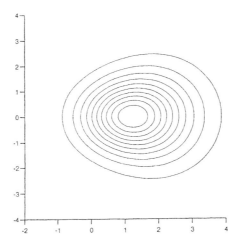

FIGURE 17.2
The bivariate density (17.4.2) with $a = 13$, $\nu = 10$ **and** $c = 7$

cases, this marginal is well approximated by a t distribution with degrees of freedom given by matching variances with the Y-marginal distribution [Jones (2000)]. Fourth, distribution (17.4.2) retains the zero correlation of the underlying symmetric bivariate t distribution. In fact, it also retains the same local dependence function [Holland and Wang (1987) and Jones (1996)] as the underlying bivariate t distribution. [All these properties are special instances of properties consequent on applying (17.4.1) to a general spherically symmetric f, Jones (2000).]

An example of density (17.4.2) is shown in Figure 17.2. It has $a = 13$, $\nu = 10$ and $c = 7$. This means it has a $t(13, 7)$ X-marginal, scaled symmetric t on 11 degrees of freedom $Y|X$ conditionals, and a symmetric Y-marginal well approximated by a t distribution on about 6 degrees of freedom.

17.5 A Second Bivariate Skew t Distribution

An alternative suggestion for a bivariate skew t distribution, this time with both marginals skew t, is the subject of this final section. The paper containing fuller details on this topic is not yet in a very advanced state of preparation and so does not merit a proper citation.

The starting point here is the bivariate Dirichlet distribution which has three nonnegative parameters a, b, and c and sample space the triangle

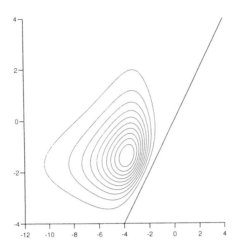

FIGURE 17.3
The bivariate density (17.5.2) with $a = 1$, $b = 2$ and $c = 3$

$u > 0$, $v > 0$, $u + v < 1$. This has beta marginals, specifically $U \sim$ Beta$(a, b + c)$ and $V \sim$ Beta$(b, a + c)$, here dwelling on the more usual space $[0, 1]$. Note that at most one of these two beta marginals can be symmetric.

Now make the "usual" transformations to the marginals of this distribution to make them into skew t distributions! In fact, by setting B in (17.2.2) equal to $2U - 1$ and $-(2V - 1)$, in turn (recall that B has support $[-1, 1]$ and U, V have support $[0, 1]$), things are set up in a particularly convenient way:

$$X = \frac{\sqrt{d}(2U - 1)}{2\sqrt{U(1 - U)}}, \quad Y = \frac{\sqrt{d}(1 - 2V)}{2\sqrt{V(1 - V)}}. \tag{17.5.1}$$

Here, $d = a + b + c$. X and Y have joint density

$$\frac{d\Gamma(d + 1)}{2^{d-1}\Gamma(a)\Gamma(b)\Gamma(c)} \left(1 + \frac{x}{\sqrt{d + x^2}}\right)^{a-1} \left(1 - \frac{y}{\sqrt{d + y^2}}\right)^{b-1}$$

$$\times \frac{1}{(d + x^2)^{3/2}(d + y^2)^{3/2}} \left(\frac{y}{\sqrt{d + y^2}} - \frac{x}{\sqrt{d + x^2}}\right)^{c-1} \tag{17.5.2}$$

This has support — and this is not immediately obvious — the interesting sample space $y > x$.

In direct correspondence with the beta marginals of the Dirichlet distribution, the skew t marginals of distribution (17.5.2) are $t(a, b+c)$ and $t(b, a + c)$, respectively, and again at most one of these two marginals can be symmetric. An example of distribution (17.5.2) is given in Figure 17.3. This has $a = 1$, $b = 2$ and $c = 3$; its marginals are therefore $t(1, 5)$ and $t(2, 4)$, respectively. (Actually, the distribution's location has been shifted for convenience.)

Fuller properties of this skew t-Dirichlet distribution are being investigated for publication elsewhere.

References

1. Cacoullos, T. (1965). A relation between t and F-distributions. *Journal of the American Statistical Association* **60**, 528-531; Corrigendum, **60**, 1249.

2. Cacoullos, T. (1966). Estimation of a multivariate density. *Annals of the Institute of Statistical Mathematics* **18**, 179-189.

3. Cacoullos, T. (1999). On the Pitman-Morgan-Wilks tests of homoscedasticity for the bivariate normal distribution. To appear.

4. Holland, P. W. and Wang, Y. J. (1987). Dependence function for continuous bivariate densities. *Communications in Statistics — Theory and Methods* **16**, 863-876.

5. Jones, M. C. (1996). The local dependence function. *Biometrika* **83**, 899-904.

6. Jones, M. C. (2000). Marginal replacement in multivariate densities, with application to skewing spherically symmetric distributions. To appear.

7. Jones, M. C. and Faddy, M. J. (2000). A skew extension of the t distribution, with applications. In preparation.

8. Wand, M. P. and Jones, M. C. (1995). *Kernel Smoothing.* Chapman and Hall, London.

18

On the Posterior Moments for Truncation Parameter Distributions and Identifiability by Posterior Mean for Exponential Distribution with Location Parameter

Yimin Ma and N. Balakrishnan
McMaster University, Hamilton, ON

ABSTRACT The exact analytical expressions of posterior moments for the two general truncation parameter likelihood functions with arbitrary prior distributions are obtained by using the sufficient statistics for these truncation parameter distributions. In particular, the explicit forms for posterior mean and variance are given. Some examples are also discussed to illustrate the obtained results. Next, an explicit expression for mixing distribution in terms of posterior mean for exponential distribution with location parameter is obtained and then the identifiability by posterior mean is established. As an example, the gamma mixing distribution family is used to illustrate the results obtained.

Keywords and phrases Posterior moments, prior distributions, sufficient statistics, truncation parameters, exponential distribution, identifiability, location parameter, mixing distribution, posterior mean

18.1 Introduction

Let $\ell(y|\theta)$ be the likelihood function of an independent and identically distributed sample $y = (x_1, \ldots, x_n)$ from distribution $p(x|\theta)$, then the Bayes estimator and posterior risk, under squared error loss, are posterior mean $E(\theta|y)$ and posterior variance $\text{var}(\theta|y)$ respectively. For the normal likelihood function with known variance and an arbitrary prior distribution, the explicit expressions for the posterior mean and variance are derived by Pericchi and Smith (1992). For an arbitrary location parameter likelihood function and the normal prior distribution, the exact form of posterior mean is given by Polson (1991). Pericchi, Sansó, and Smith (1993) also discussed the posterior cumulant relations in Bayesian inference assuming the exponential family forms either on the likelihood or on the prior. They all mentioned that analytical Bayesian computations without the assumption of normality, either on likelihood function or on prior distribution, are very difficult and needed investigation.

In this paper, we consider the likelihood functions of random samples from the following two different types of truncation parameter distributions:

Type I truncation parameter density

$$p_1(x_1|\theta_1) = h_1(x_1)/k_1(\theta_1), \quad a < \theta_1 \le x_1 < b; \qquad (18.1.1)$$

Type II truncation parameter density

$$p_2(x_2|\theta_2) = h_2(x_2)/k_2(\theta_2), \quad a < x_2 \le \theta_2 < b, \qquad (18.1.2)$$

where $h_1(x_1)$ and $h_2(x_2)$ are positive, continuous and integrable over (θ_1, b) and (a, θ_2), respectively, for θ_i, $i = 1, 2$, in the interval (a, b), $-\infty \le a < b \le \infty$.

In the first part of this paper, we find the exact analytical expressions of posterior moments for the two different truncation parameter likelihood functions with arbitrary prior distributions. In Section 18.2, the explicit forms for posterior moments are derived by using the sufficient statistics for these truncation parameter distributions. In particular, the explicit expressions of posterior mean and variance are given, respectively, for these two truncation parameter distribution models. In Section 18.3, two examples representing the two different truncation parameter distribution models are discussed to illustrate the obtained results.

Next, suppose that (X, θ) is a random vector in which the conditional distribution of X given θ is $f(x|\theta)$ and θ has a mixing (prior) distribution $G(\theta)$, then the mixture (marginal) distribution of X is given by

$$f_G(x) = \int f(x|\theta)dG(\theta).$$

The original concept of identifiability means that there is a one-to-one correspondence between the mixing distribution G and the mixture distribution f_G; see Teicher (1961, 1963) for a general introduction. In recent years, however, there are some other research works about the one-to-one correspondence between the mixing distribution G and the posterior mean $E(\theta|x)$, instead of the mixture distribution f_G; most of these papers have dealt with different distributions from the exponential distribution family. For example, see Korwar (1975), Cacoullos and Papageorgiou (1984), Papageorgiou (1985), Cacoullos (1987), Kyriakoussis and Papageorgiou (1991), Arnold, Castillo, and Sarabia (1993), Papageorgiou and Wesolowski (1997), and some others. Exception is the paper by Gupta and Wesolowski (1997), who have established not only the identifiability but also an explicit relationship between the mixing distribution and the posterior mean for the uniform distribution.

In the second part of this paper, we consider the exponential distribution with location parameter θ given by

$$f(x|\theta) = \exp\{-(x-\theta)\} \qquad 0 < \theta < x < \infty. \tag{18.1.3}$$

This exponential distribution with location (or threshold) parameter θ arises in many areas of applications including reliability and life-testing, survival analysis, and engineering problems; for example, see Balakrishnan and Basu (1995). In the literature, the location parameter is often interpreted to be the minimum guaranteed life time.

We assume that the location parameter θ has a continuous mixing distribution $dG(\theta) = g(\theta)d\theta$ on $(0,\infty)$, so that the mixture distribution of X is given by

$$f_G(x) = f(x) = \int_0^x e^{-(x-\theta)} g(\theta) d\theta. \tag{18.1.4}$$

We then consider the identifiability of mixtures by posterior mean for the exponential distribution in (18.1.3) with continuous mixing distributions. In Section 18.4, an explicit relation between the mixing distribution G and the general posterior mean $E(w(\theta)|x)$, where $w(\theta)$ is a differentiable function, is derived and then the identifiability of the continuous mixture by general posterior mean $E(w(\theta)|x)$ is established. In particular, an explicit expression for the mixing distribution in terms of the posterior mean $E(\theta|x)$ is presented. In Section 18.5, the gamma mixing distribution family is used to demonstrate the identifiability results for the exponential distribution (18.1.3) established in this paper.

18.2 Posterior Moments

Let $y_i = (x_{i1}, \ldots, x_{in})$ denote the independent and identically distributed samples of size n from truncation parameter distributions $p_i(x_i|\theta_i)$, $i =$

1, 2, given by (18.1.1) and (18.1.2), respectively, then $T_1 = x_{1(1)}$ and $T_2 = x_{2(n)}$ are sufficient statistics for θ_i, $i = 1, 2$, respectively, where $x_{1(1)}$ is the smallest order statistic from $y_1 = (x_{11}, \ldots, x_{1n})$ and $x_{2(n)}$ the largest order statistic from $y_2 = (x_{21}, \ldots, x_{2n})$. The conditional density functions of T_1 and T_2 can be easily derived as follows [Arnold, Balakrishnan, and Nagaraja (1992, p. 12)]:

$$
\begin{aligned}
f_{T_1}(t_1|\theta_1) &= n\{1 - P_1(t_1|\theta_1)\}^{n-1} p_1(t_1|\theta_1) \\
&= n\{k_1(t_1)\}^{n-1} h_1(t_1)/\{k_1(\theta_1)\}^n \\
&= H_1(t_1)/K_1(\theta_1), \quad a < \theta_1 \leq t_1 < b, \qquad (18.2.1)
\end{aligned}
$$

$$
\begin{aligned}
f_{T_2}(t_2|\theta_2) &= n\{P_2(t_2|\theta_1)\}^{n-1} p_2(t_2|\theta_2) \\
&= n\{k_2(t_2)\}^{n-1} h_2(t_2)/\{k_2(\theta_2)\}^n \\
&= H_2(t_2)/K_2(\theta_2), \quad a < t_2 \leq \theta_2 < b, \qquad (18.2.2)
\end{aligned}
$$

where $H_i(t_i) = n\{k_i(t_i)\}^{n-1} h_i(t_i)$ and $K_i(\theta_i) = \{k_i(\theta_i)\}^n$, $i = 1, 2$, and the second equalities in (18.2.1) and (18.2.2) are obtained by using relations

$$
k_1(\theta_1) = \int_{\theta_1}^{b} h_1(x_1) dx_1
$$

and

$$
k_2(\theta_2) = \int_{a}^{\theta_2} h_2(x_2) dx_2,
$$

respectively. Note that the conditional density functions of T_1 and T_2 are still type I and type II truncation parameter densities respectively.

It is assumed in this paper the truncation parameters θ_i, $i = 1, 2$, have arbitrary prior distributions $\pi_i(\theta_i)$ on (a, b) with $\pi_i(a) = 0$ and $\pi_i(b) = 1$, respectively, then the marginal distributions of T_i, $i = 1, 2$, are given by

$$
f_{\pi_1}(t_1) = f_1(t_1) = \int_{a}^{t_1} \{H_1(t_1)/K_1(\theta_1)\} d\pi_1(\theta_1), \qquad (18.2.3)
$$

$$
f_{\pi_2}(t_2) = f_2(t_2) = \int_{t_2}^{b} \{H_2(t_2)/K_2(\theta_2)\} d\pi_2(\theta_2). \qquad (18.2.4)
$$

From these density functions (18.2.1)–(18.2.4), we are able to demonstrate the explicit expressions of posterior moments in terms of the marginal distributions of T_i, $i = 1, 2$, respectively, for the two different truncation parameter likelihood functions with arbitrary prior distributions.

We first demonstrate the exact relations between the posterior means of general functions $g_i(\theta_i)$ and the marginal distributions $f_i(t_i)$, $i = 1, 2$, respectively, in Theorems 18.2.1 and 18.2.2, then we can easily give the explicit expressions for posterior moments and posterior means and variances in Corollaries 18.2.3–18.2.6.

THEOREM 18.2.1
For the type I truncation parameter distribution model, if $g_1(\cdot)$ is differentiable and

$$\int_a^{t_1} |g_1'(s_1)| \{ H_1(t_1)/H_1(s_1) \} dF_1(s_1) \; < \; \infty,$$

then we have

$$E(g_1(\theta_1) \mid y_1)$$
$$= E(g_1(\theta_1) \mid t_1) = g_1(t_1) - \frac{\int_a^{t_1} g_1'(s_1)\{H_1(t_1)/H_1(s_1)\} dF_1(s_1)}{f_1(t_1)}$$
$$= g_1(t_1) - \frac{\int_a^{t_1} g_1'(s_1)[\{k_1(t_1)\}^{n-1} h_1(t_1)/\{k_1(s_1)\}^{n-1} h_1(s_1)] dF_1(s_1)}{f_1(t_1)},$$

where $f_1(\cdot)$ is given by (18.2.3) and $F_1(\cdot)$ the corresponding cumulative distribution function.

PROOF By Fubini's theorem

$$\int_a^{t_1} g_1'(s_1) \frac{H_1(t_1)}{H_1(s_1)} dF_1(s_1)$$
$$= \int_a^{t_1} g_1'(s_1) \frac{H_1(t_1)}{H_1(s_1)} f_1(s_1) ds_1$$
$$= \int_a^{t_1} g_1'(s_1) \frac{H_1(t_1)}{H_1(s_1)} \left\{ \int_a^{s_1} \frac{H_1(s_1)}{K_1(\theta_1)} d\pi_1(\theta_1) \right\} ds_1$$
$$= \int_a^{t_1} \int_{\theta_1}^{t_1} g_1'(s_1) \frac{H_1(t_1)}{K_1(\theta_1)} ds_1 d\pi_1(\theta_1)$$
$$= g_1(t_1) f_1(t_1) - \int_a^{t_1} g_1(\theta_1) f_1(t_1 \mid \theta_1) d\pi_1(\theta_1),$$

then we obtain

$$E(g_1(\theta_1) \mid t_1) = \frac{\int_a^{t_1} g_1(\theta_1) f_1(t_1 \mid \theta_1) d\pi_1(\theta_1)}{f_1(t_1)}$$

$$= g_1(t_1) - \frac{\int_a^{t_1} g_1'(s_1)\{H_1(t_1)/H_1(s_1)\} dF_1(s_1)}{f_1(t_1)}.$$

$\blacksquare$

THEOREM 18.2.2

For the type II truncation parameter distribution model, if $g_2(\cdot)$ is differentiable and

$$\int_{t_2}^b |g_2'(s_2)| \mid \{H_2(t_2)/H_2(s_2)\} dF_2(s_2) < \infty,$$

then we have

$$E(g_2(\theta_2) \mid y_2)$$

$$= E(g_2(\theta_2) \mid t_2) = g_2(t_2) + \frac{\int_{t_2}^b g_2'(s_2)\{H_2(t_2)/H_2(s_2)\} dF_2(s_2)}{f_2(t_2)}$$

$$= g_2(t_2) + \frac{\int_{t_2}^b g_2'(s_2)[\{k_2(t_2)\}^{n-1}h_2(t_2)/\{k_2(s_2)\}^{n-1}h_2(s_2)] dF_2(s_2)}{f_2(t_2)},$$

where $f_2(\cdot)$ is given by (18.2.4) and $F_2(\cdot)$ the corresponding cumulative distribution function.

PROOF It is similar as the proof of Theorem 18.2.1 and omitted. $\blacksquare$

COROLLARY 18.2.3

For the type I truncation parameter distribution model, we have

$$E(\theta_1^r \mid y_1) = E(\theta^r \mid t_1) = t_1^r - \frac{\int_a^{t_1} rs_1^{r-1}\{H_1(t_1)/H_1(s_1)\} dF_1(s_1)}{f_1(t_1)}$$

$$= t_1^r - \frac{\int_a^{t_1} rs_1^{r-1}[\{k_1(t_1)\}^{n-1}h_1(t_1)/\{k_1(s_1)\}^{n-1}h_1(s_1)] dF_1(s_1)}{f_1(t_1)},$$

where $f_1(\cdot)$ is given by (18.2.3) and $F_1(\cdot)$ the corresponding cumulative distribution function.

COROLLARY 18.2.4

For the type II truncation parameter distribution model, we have

$$E(\theta_2^r|y_2) = E(\theta_2^r|t_2) = t_2^r + \frac{\int_{t_2}^b rs_2^{r-1}\{H_2(t_2)/H_2(s_2)\}dF_2(s_2)}{f_2(t_2)}$$

$$= t_2^r + \frac{\int_{t_2}^b rs_2^{r-1}[\{k_2(t_2)\}^{n-1}h_2(t_2)/\{k_2(s_2)\}^{n-1}h_2(s_2)]dF_2(s_2)}{f_2(t_2)},$$

where $f_2(\cdot)$ is given by (18.2.4) and $F_2(\cdot)$ the corresponding cumulative distribution function.

COROLLARY 18.2.5

For the type I truncation parameter distribution model, the posterior mean and variance are given by, respectively,

$$E(\theta_1|y_1) = E(\theta_1|t_1) = t_1 - \frac{u_1(t_1)}{f_1(t_1)},$$

$$var(\theta_1|y_1) = var(\theta_1|t_1) = \frac{2t_1u_1(t_1) - v_1(t_1)}{f_1(t_1)} - \frac{u_1^2(t_1)}{f_1^2(t_1)},$$

where

$$u_1(t_1) = \int_a^{t_1} \{H_1(t_1)/H_1(s_1)\}dF_1(s_1),$$

$$v_1(t_1) = \int_a^{t_1} 2s_1\{H_1(t_1)/H_1(s_1)\}dF_1(s_1).$$

COROLLARY 18.2.6

For the type II truncation parameter distribution model, the posterior mean and variance are given by, respectively,

$$E(\theta_2|y_2) = E(\theta_2|t_2) = t_2 + \frac{u_2(t_2)}{f_2(t_2)},$$

$$var(\theta_2|y_2) = var(\theta_2|t_2) = \frac{v_2(t_2) - 2t_2u_2(t_2)}{f_2(t_2)} - \frac{u_2^2(t_2)}{f_2^2(t_2)},$$

where

$$u_2(t_2) = \int_{t_2}^b \{H_2(t_2)/H_2(s_2)\}dF_2(s_2),$$

$$v_2(t_2) = \int_{t_2}^b 2s_2\{H_2(t_2)/H_2(s_2)\}dF_2(s_2).$$

18.3   Examples

Example 1

Let $x_{11},\dots,x_{1n}$ be independent and identically distributed according to the truncated exponential distribution as follows

$$p_1(x_1|\theta_1) = \lambda\exp\{-\lambda(x_1-\theta_1)\}, \quad -\infty < \theta_1 \le x_1 < \infty, \ \lambda > 0,$$

it is type I truncation parameter density with $h_1(x_1) = \lambda e^{-\lambda x_1}$, $k_1(\theta_1) = e^{-\lambda\theta_1}$ and $T_1 = \min(x_{11},\dots,x_{1n})$ is the sufficient statistic for θ_1. Then we have by Corollary 18.2.3,

$$E(\theta_1^r|t_1) = t_1^r - \frac{\int_{-\infty}^{t_1} rs_1^{r-1}\{e^{-n\lambda(t_1-s_1)}\}dF_1(s_1)}{f_1(t_1)}$$

and when $r = 1$, the posterior mean is given by

$$E(\theta_1|t_1) = t_1 - \frac{\int_{-\infty}^{t_1} e^{-n\lambda(t_1-s_1)}dF_1(s_1)}{f_1(t_1)}; \qquad (18.3.1)$$

for the special case $n = 1$, $\lambda = 1$, this result of (18.3.1) is exactly the same as that given by Fox (1978).   □

Example 2

Let $x_{21},\dots,x_{2n}$ be independent and identically distributed according to the power function distribution as follows

$$p_2(x_2|\theta_2) = \alpha x_2^{\alpha-1}/\theta_2^\alpha, \quad 0 < x_2 \le \theta_2 < \infty, \ \alpha > 0,$$

it is type II truncation parameter density with $h_2(x_2) = \alpha x_2^{\alpha-1}$, $k_2(\theta_2) = \theta_2^\alpha$ and $T_2 = \max(x_{21},\dots,x_{2n})$ is the sufficient statistic for θ_2. Then we have by Corollary 18.2.4,

$$E(\theta_2^r|t_2) = t_2^r + \frac{\int_{t_2}^{\infty} rs_2^{r-1}(t_2^{n\alpha-1}/s_2^{n\alpha-1})dF_2(s_2)}{f_2(t_2)}$$

and when $r = 1$, the posterior mean is given by

$$E(\theta_2|t_2) = t_2 + \frac{\int_{t_2}^{\infty}(t_2^{n\alpha-1}/s_2^{n\alpha-1})dF_2(s_2)}{f_2(t_2)}; \qquad (18.3.2)$$

for the special case $n = 1$, $\alpha = 1$, this result of (18.3.2) is exactly the same as that given by Fox (1978) for uniform distribution. ▯

18.4 Identifiability by Posterior Mean

Before we present the main results about the identifiability by posterior mean for the exponential distribution with location parameter, we will introduce some results existing already in the literature which will be very useful for the developments here.

LEMMA 18.4.1
For the exponential distribution in (18.1.3), we have

$$G(t) = F(t) + f(t), \qquad (18.4.1)$$

where $f(t)$ is given by (18.1.4) and $F(t)$ is the corresponding cumulative distribution function.

PROOF This relation was obtained by Blum and Susarla (1981) and by Prasad and Singh (1990). ∎

LEMMA 18.4.2
For the exponential distribution in (18.1.3), if $w(\theta)$ is a differentiable function, then

$$M(x) = E(w(\theta)|x) = w(x) - \frac{\int_0^x w'(s)e^{-x}e^s f(s)ds}{f(x)} . \qquad (18.4.2)$$

In particular, when $w(\theta) = \theta$,

$$m(x) = E(\theta|x) = x - \frac{\int_0^x e^{-x}e^s f(s)ds}{f(x)} , \qquad (18.4.3)$$

where $f(x)$ is given by (18.1.4).

PROOF The expression in (18.4.2) is presented in Theorem 18.2.1, and the special case (18.4.3) was given earlier by Fox (1978). ∎

Next, we will present an explicit expression for the mixing distribution in terms of the general posterior mean $M(x)$ for the exponential distribution in the following theorem, and then the identifiability by the general posterior mean is established.

THEOREM 18.4.3

For the exponential distribution in (18.1.3), if the general posterior mean $M(x)$ is differentiable, then the mixing density function is given by

$$g(t) = C\frac{M'(t)}{w(t) - M(t)}\exp\left\{-t + \int\frac{M'(t)dt}{w(t) - M(t)}\right\}, \qquad (18.4.4)$$

where C is the norming constant. Therefore, the mixing density is uniquely determined by the general posterior mean.

PROOF From Lemma 18.4.2, we have

$$e^x f(x)\{w(x) - M(x)\} = \int_0^x w'(s)e^s f(s)ds.$$

By differentiating both sides, we obtain

$$f'(x)\{w(x) - M(x)\} = f(x)\{M'(x) - w(x) + M(x)\}$$

and

$$\int\frac{f'(x)}{f(x)}dx = \int\left(\frac{M'(x)}{w(x) - M(x)} - 1\right)dx$$

which yields

$$f(x) = C\exp\left\{-x + \int\frac{M'(x)dx}{w(x) - M(x)}\right\}$$

where C is the norming constant. Then by Lemma 18.4.1, the mixing density function is given by

$$g(t) = f(t) + f'(t)$$

$$= C\frac{M'(t)}{w(t) - M(t)}\exp\left\{t + \int\frac{M'(t)dt}{w(t) - M(t)}\right\}.$$

∎

When $w(\theta) = \theta$, $M(x) = E(\theta|x) = m(x)$, we have the following corollary.

COROLLARY 18.4.4

For the exponential distribution in (18.1.3), if the posterior mean $m(x)$ is differentiable, then the mixing density is given by

$$g(t) = C\frac{m'(t)}{t - m(t)}\exp\left\{t + \int\frac{m'(t)dt}{t - m(t)}\right\}, \qquad (18.4.5)$$

where C is the norming constant.

Note that Gupta and Wesolowski (1997) obtained a similar expression for the uniform distribution by considering the definition of posterior mean directly. Our expressions are obtained by using the relationship between the mixing distribution and the mixture distribution and the relationship between the posterior mean and the mixture distribution. Certainly, the procedure used here is easier to understand and also easy to go through.

18.5 An Illustrative Example

Finally, we consider here discuss an example to illustrate the results obtained in the last section. Suppose that the posterior mean $m(x)$ is given by

$$m(x) = \left(1 - \frac{1}{1+\alpha}\right) x, \qquad x > 0, \ \alpha > 0.$$

Then by (18.4.5), we have

$$
\begin{aligned}
g(t) &= C \frac{\left(1 - \frac{1}{1+\alpha}\right)}{\frac{1}{1+\alpha} t} \exp\left\{-t + \int \frac{\alpha\, dt}{t}\right\} \\
&= C\, e^{-t}\, t^{\alpha-1} \\
&= \frac{t^{\alpha-1}\, e^{-t}}{\Gamma(\alpha)}, \qquad t > 0, \ \alpha > 0.
\end{aligned}
$$

This is the gamma distribution which simply means that there is a one-to-one relationship between the gamma mixing distribution family and the posterior mean $m(x) = \left(1 - \frac{1}{1+\alpha}\right) x$.

References

1. Arnold, B. C., Balakrishnan, N., and Nagaraja, H. N. (1992). *A First Course in Order Statistics*. John Wiley & Sons, New York.

2. Arnold, B. C., Castillo, E., and Sarabia, J. M. (1993). Conditionally specified models: Structure and inference. In *Multivariate Analysis: Future Directions* **2** (Eds., C. M. Cuadras and C. R. Rao), 441–450. Elsevier, Amsterdam.

3. Balakrishnan, N. and Basu, A. P. (Eds.) (1995). *The Exponential Distribution: Theory, Methods and Applications.* Gordon and Breach, Langhorne, PA.

4. Blum, J. R. and Susarla, V. (1981). Maximal derivation theory of some estimators of prior distribution functions. *Annals of the Institute of Statistical Mathematics* **33**, 425–436.

5. Cacoullos, T. (1987). Characterizing priors by posterior expectations in multiparameter exponential families. *Annals of the Institute of Statistical Mathematics* **39**, 399–405.

6. Cacoullos, T. and Papageorgiou, H. (1984). Characterizations of mixtures of continuous distributions by their posterior means. *Scandinavian Actuarial Journal* **8**, 23–30.

7. Fox, R. (1978). Solutions to empirical Bayes squared error loss estimation problems. *Annals of Statistics* **6**, 846–853.

8. Gupta, A. K. and Wesolowski, J. (1997). Uniform mixtures via posterior means. *Annals of the Institute of Statistical Mathematics* **49**, 171–180.

9. Korwar, R. M. (1975). On characterizing some discrete distributions by linear regression. *Communications in Statistics* **4**, 1133–1147.

10. Kyriakoussis, A. and Papageorgiou, H. (1991). Characterizations of logarithmic series distributions. *Statistica Neerlandica* **45**, 1–8.

11. Papageorgiou, H. (1985). On characterizing some discrete distributions by a conditional distribution and a regression function. *Biometrical Journal* **27**, 473–479.

12. Papageorgiou, H. and Wesolowski, J. (1997). Posterior mean identifies the prior distribution in NB and related models. *Statistics & Probability Letters* **36**, 127–134.

13. Pericchi, L. R., Sansó, B., and Smith, A. F. M. (1993). Posterior cumulant relationships in Bayesian inference involving the exponential family. *Journal of the American Statistical Association* **88**, 1419–1426.

14. Pericchi, L. R. and Smith, A. F. M. (1992). Exact and approximate posterior moments for a normal location parameter. *Journal of the Royal Statistical Society, Series B* **54**, 793–804.

15. Polson, N. G. (1991). A representation of the posterior mean for a location model. *Biometrika* **78**, 426–430.

16. Prasad, B. and Singh, R. S. (1990). Estimation of prior distribution and empirical Bayes estimation in a non-exponential family. *Journal of Statistical Planning and Inference* **24**, 81–86.

17. Teicher, H. (1961). Identifiability of mixtures. *Annals of Mathematical Statistics* **32**, 244–248.

18. Teicher, H. (1963). Identifiability of finite mixtures. *Annals of Mathematical Statistics* **34**, 1265–1269.

19

Distributions of Random Volumes without Using Integral Geometry Techniques

A. M. Mathai
McGill University, Montreal, Quebec, Canada

ABSTRACT The usual techniques available in the literature in deriving integer moments and distributions of random volumes of random geometrical configurations are the integral geometry techniques. The author has earlier introduced a method based on functions of matrix argument and Jacobians of matrix transformations and derived arbitrary moments, not just integer moments, and the exact distributions of the volume contents of random p-parallelotopes and p-simplices in Euclidean n-space, $p \leq n + 1$, for the following situations: (1) all the points are either inside a hypersphere of unit radius or have general distributions associated with them such as generalized type-1 beta, type-2 beta or Gaussian, (2) some points are uniformly distributed over the surface of a hypersphere of radius 1 and the remaining have general distributions as in (1). In all these derivations no result from integral geometry is used. In the present paper the case when the points have some general matrix-variate distributions as well as the case when some points are restricted to the surface of a hypersphere and the remaining points have general classes of distributions as in (1) will be considered. The exact arbitrary moments and the exact distributions of the volume content of a random parallelotope will be derived without using any result from integral geometry. The densities of random distances, areas and volumes in several particular cases will be written in terms of Bessel, Whittaker, and Gauss' hypergeometric functions.

Keywords and phrases Random parallelotope, random simplex, random content, moments, exact distribution, Jacobians, matrix transformations, uniform random points, type-1, type-2 and Gaussian random points, matrix-variate distributions

19.1   Introduction

Let O be the origin of a rectangular coordinate system in the Euclidean n-space R^n. Let $P_1, \ldots, P_r$ be r points in this system. Consider the vectors $O\vec{P_1}, \ldots, O\vec{P_r}$ for $r \leq n$. Let these vectors be linearly independent. Then these ordered vectors can generate an r-parallelotope or an r-simplex. If we consider $r \leq n + 1$ then the origin can be shifted to one of the points and then the situation reduces to the one above. Let $\nabla_{r,n}$ and $\Delta_{r,n}$ denote the r-contents or the volume contents of these r-parallelotope and r-simplex respectively. Then $\nabla_{r,n} = r! \, \Delta_{r,n}$. Integer moments of $\nabla_{r,n}$, thereby those of $\Delta_{r,n}$, are available in Miles (1971) and the many references therein. Many results from integral geometry are used in those papers in deriving these results. In the present paper we will consider a method based on Jacobians of matrix transformations and functions of matrix argument. For the various types of matrix transformations and the associated Jacobians see the recent book: Mathai (1997). By using the technique of Jacobians of matrix transformations Mathai (1998) derived several results when all the points are inside a hypersphere of radius 1 or have some general matrix-variate distributions or independently and identically distributed with some general distributions associated with each point. In a subsequent work the author has looked into the problem when some of the points are uniformly distributed onver the surface of a hypersphere thereby some of the vectors $O\vec{P_i}$, $i = 1, \ldots, q$, $q \leq r$ have unit length, that is, $\|O\vec{P_i}\| = 1$, $i = 1, \ldots, q$, and the remaining points have general distributions associated with them. In the present paper these results will be extended to cover the situation when the first q vectors $O\vec{P_1}, \ldots, O\vec{P_q}$, $q \leq r$ have lengths unities, these q points are distributed over the surface of the n-sphere but all points together have some general distributions and the situations when the points are independently distributed and having some general distributions.

Let X be a $r \times n$, $r \leq n$, matrix of real variables where let X_j, the j-th row of X represent the point P_j, $j = 1, \ldots, r$. When $Y = XX'$, where X' denotes the transpose of X, has a mtrix-variate distribution one can evaluate the moments and distributions of the volume content $\nabla_{r,n}$ of the r-parallelotope. In terms of a determinant

$$\nabla_{r,n} = |XX'|^{1/2} = |Y|^{1/2}, \tag{19.1.1}$$

where $Y = Y' > 0$ almost surely. In this case the rows of X need not be independently distributed as long as Y has a tractable matrix-variate distribution. For such an approach see the last chapters of Mathai

(1997). Several classes of matrix-variate distributions are listed in the book. Then when Y has a general matrix-variate distribution one can evaluate arbitrary moments and the exact density of $\nabla_{r,n}$. In the present paper we consider general matrix-variate distributions as well as the situation where the X_js are independently distributed, subject to the restriction that some of the points are on the surface of an n-sphere of radius 1. In order to tackle the situation when some of the random points are restricted to the surface of an n-sphere various results on Jacobians available in the literature are to be extended to cover such a situation.

Some of the results that we need, and which can be proved easily, will be listed here as lemmas without proofs. All the variables appearing are real unless stated otherwise. The parameters could be complex. Let

$$\Omega_j = \{-1 \le t_{jk} \le 1, k = 1, \ldots, j, 0 \le t_{j1}^2 + \ldots + t_{jj}^2 \le 1\}$$
$$w_j = \{0 \le z_k \le 1, k = 1, \ldots, j, 0 \le z_1 + \ldots + z_j \le 1\}$$

and

$$w_j^* = \{-1 \le t_{jk} \le 1, k = 1, \ldots, j, t_{j1}^2 + \ldots + t_{jj}^2 = 1\}. \quad (19.1.2)$$

w_j^* describes the surface of a j-sphere of radius unity. That is,

$$t_{jj} = \pm\sqrt{1 - t_{j1}^2 - \ldots - t_{jj-1}^2}.$$

The integral over w_j^* is the inegral over the surface of a j-sphere of radius unity and the associated differential element is

$$2[1 - t_{j1}^2 - \ldots - t_{jj-1}^2]^{-1/2} dt_{j1} \ldots dt_{jj-1}. \quad (19.1.3)$$

Type–1 Dirichlet integral is given in Lemma 19.1.1 and Lemmas 19.1.2, 19.1.3, 19.1.4 and 19.1.5 can be established with the help of the type–1 Dirichlet integral.

LEMMA 19.1.1

$$\int_{w_j} z_1^{\alpha_1 - 1} \ldots z_j^{\alpha_j - 1} (1 - z_1 - \ldots - z_j)^{\alpha_{j+1} - 1} dz_1 \ldots dz_j$$

$$= \frac{\Gamma(\alpha_1) \ldots \Gamma(\alpha_{j+1})}{\Gamma(\alpha_1 + \ldots + \alpha_{j+1})} \quad for \ \Re(\alpha_k) > 0, \ k = 1, \ldots, j+1$$

where $\Re(\cdot)$ denotes the real part of $(\cdot)$.

LEMMA 19.1.2

$$\int_{\Omega_j} (t_{jj}^2)^{\gamma_j} \, \mathrm{d}t_{j1} \ldots \mathrm{d}t_{jj}$$

$$= \int_{\omega_j} z_1^{\frac{1}{2}-1} \ldots z_{j-1}^{\frac{1}{2}-1} z_j^{\gamma_j+\frac{1}{2}-1} (1 - z_1 - \ldots - z_j)^{1-1} \mathrm{d}z_1 \ldots \mathrm{d}z_j$$

$$= \frac{[\Gamma(\frac{1}{2})]^{j-1} \Gamma(\gamma_j + \frac{1}{2})}{\Gamma(\frac{j}{2} + 1 + \gamma_j)} \quad \text{for } \Re(\gamma_j) > -\frac{1}{2}.$$

LEMMA 19.1.3
The volume of an n-sphere of unit radius is $\pi^{n/2}/\Gamma\left(\frac{n}{2}+1\right)$. This follows from Lemma 19.1.2 by taking $\gamma_j = 0$ and replacing j by n.

LEMMA 19.1.4
$(t_{j1}^2)^{\delta_{j1}} \ldots (t_{jj}^2)^{\delta_{jj}}$ integrated over the surface of a j-sphere of unit radius is given by

$$\int_{\omega_j^*} (t_{j1}^2)^{\delta_{j1}} \ldots (t_{jj}^2)^{\delta_{jj}} \, \mathrm{d}t_{j1} \ldots \mathrm{d}t_{jj-1}$$

$$= 2 \int_{\omega_j^*} (t_{j1}^2)^{\delta_{j1}} \ldots (t_{jj-1}^2)^{\delta_{jj}-1} [1 - t_{j1}^2 - \ldots - t_{jj-1}^2]^{\delta_{jj} - \frac{1}{2}} \mathrm{d}t_{j1} \ldots \mathrm{d}t_{jj-1}$$

$$= \frac{2\left\{\prod_{k=1}^j \Gamma\left(\frac{1}{2} + \delta_{jk}\right)\right\}}{\Gamma\left(\frac{j}{2} + \delta_{j1} + \ldots + \delta_{jj}\right)} \quad \text{for } \Re(\delta_{jk}) > -\frac{1}{2}, \ k = 1, \ldots, j.$$

The result follows from first multiplying by the differential element over the surface of a j-sphere and then from using Lemmas 19.1.2 and 19.1.1.

LEMMA 19.1.5
The surface area of an n-sphere of unit radius, $2\pi^{n/2}/\Gamma(n/2)$, is available from Lemma 19.1.4 by putting $\delta_{jk} = 0$, $k = 1, \ldots, j$ and then replacing j by n.

LEMMA 19.1.6
Type–2 Dirichlet integral is given by

$$\int_{z_1>0,\ldots,z_n>0} z_1^{\alpha_1-1} \ldots z_n^{\alpha_n-1} (1 + z_1 + \ldots + z_n)^{-(\alpha_1+\ldots+\alpha_{n+1})} \mathrm{d}z_1 \ldots \mathrm{d}z_n$$

$$= \frac{\Gamma(\alpha_1) \ldots \Gamma(\alpha_{n+1})}{\Gamma(\alpha_1 + \ldots + \alpha_{n+1})} \quad \text{for } \Re(\alpha_k) > 0, \ k = 1, \ldots, n+1.$$

LEMMA 19.1.7

$$\int_{-\infty < t_{jk} < \infty, \ k=1,\ldots,j} (t_{j1}^2)^{\gamma_{j1}} \ldots (t_{jj}^2)^{\gamma_{jj}} (1 + t_{j1}^2 + \ldots + t_{jj}^2)^{-\delta_j}$$

$$dt_{j1} \ldots dt_{jj}$$

$$= \frac{\left\{ \prod_{k=1}^{j} \Gamma \left(\frac{1}{2} + \gamma_{jk} \right) \right\} \Gamma \left(\delta_j - \gamma_{j1} - \ldots - \gamma_{jj} - \frac{j}{2} \right)}{\Gamma(\delta_j)}$$

for $\Re(\delta_j) > \Re(\sum_{k=1}^{j} \gamma_{jk}) + \frac{j}{2}, \ \Re(\delta_j) > 0.$

If $Y = XX'$ where some of the rows of X have Euclidean lengths unity, the corresponding points are on the surface of an n-sphere, and further the corresponding diagonal elements in Y are unities. Note that Y is $r \times r$ and X is $r \times n$, $r \le n$. This requires modifications of the Jacobians of matrix transformations when some of the diagonal elements of the matrix are unities. These results are established by the author and these will be stated as lemmas here.

LEMMA 19.1.8
Let $Y = (y_{ij})$ *be an* $r \times r$ *symmetric positive definite matrix of functionally independent real variables with* $y_{11} = 1 = \ldots = y_{qq}, \ q \le r$. *Let* T *be a lower triangular matrix with positive diagonal elements and of functionally independent real variables. Then*

$$Y = TT' \Rightarrow$$

$$dY = 2^{r-q} \left\{ \prod_{j=2}^{q} t_{jj}^{r-j} \right\} \left\{ \prod_{j=q+1}^{r} t_{jj}^{r-j+1} \right\} dT$$

where dY *and* dT *denote the wedge products of the differentials in* Y *and* T *respectively, with* $t_{jj} = \sqrt{1 - t_{j1}^2 - \ldots - t_{jj-1}^2}, \ j = 2, \ldots, q$ *and if the last* q *diagonal elements of* Y *are unities then*

$$dY = 2^{r-q} \left\{ \prod_{j=1}^{r-q} t_{jj}^{r-j+1} \right\} \left\{ \prod_{j=r-q+1}^{r-1} t_{jj}^{r-j} \right\} dT$$

with

$$t_{jj} = \sqrt{1 - t_{j1}^2 - \ldots - t_{jj-1}^2}, \ j = r - q + 1, \ldots, r - 1.$$

For the proof of the next result see Mathai (1997).

LEMMA 19.1.9
Let $X = (x_{ij})$ be an $r \times n$, $n \geq r$, matrix of functionally independent real variables, T be an $r \times r$ lower triangular matrix with positive diagonal elements, U_1 be an $r \times n$ semiorthonormal matrix, $U_1 U_1' = I_r$, U_2 be an $(n - r) \times n$ semiorthonormal matrix, $U_2 U_2' = I_{n-r}$, $U_1 U_2' = 0$ and let $U = \begin{pmatrix} U_1 \\ U_2 \end{pmatrix}$ be an $n \times n$ full orthonormal matrix with u_k denoting the k-th column of U. Then

$$X = TU_1 \Rightarrow dX = \left\{ \prod_{j=1}^{r} t_{jj}^{n-j} \right\} dT \, dU_1$$

where

$$dU_1 = \wedge_{j=1}^{r} \wedge_{k=j+1}^{n} (du_j') u_k$$

and (du_j') denoting the vector of differentials in u_j'.

The following lemma can be established by integrating $e^{-\text{tr}(XX')}$ directly over X as well as by using Lemma 19.1.9 and then equating the two quantities, where $\text{tr}(\cdot)$ denotes the trace of the matrix $(\cdot)$.

LEMMA 19.1.10

$$\int dU_1 = \frac{2^r \pi^{nr/2}}{\Gamma_r \left(\frac{n}{2} \right)}$$

where, for example,

$$\Gamma_r(\alpha) = \pi^{\frac{r(r-1)}{4}} \Gamma(\alpha) \Gamma \left(\alpha - \frac{1}{2} \right) \ldots \Gamma \left(\alpha - \frac{r-1}{2} \right),$$

$$\Re(\alpha) > \frac{r-1}{2}. \qquad (19.1.4)$$

In Lemma 19.1.9 let the first q rows of X be such that $\sum_{k=1}^{n} x_{ik}^2 = 1$, $i = 1, \ldots, q$ so that $t_{11} = 1$, $t_{jj} = \sqrt{1 - t_{j1}^2 - \ldots - t_{jj-1}^2}$, $j = 2, \ldots, q$. Then we have the following result:

LEMMA 19.1.11
In Lemma 9 let the first q rows of X be such that $\sum_{k=1}^{n} x_{ik}^2 = 1$, $i = 1, \ldots, q$ so that $t_{11} = 1$, $t_{jj} = \sqrt{1 - t_{j1}^2 - \ldots - t_{jj-1}^2}$, $j = 2, \ldots, q$ then

$$X = TU_1 \Rightarrow dX = c \left\{ \prod_{j=1}^{r} t_{jj}^{n-j} \right\} dT dU_1$$

where $T_i T_i' = 1$, $i = 1, \ldots, q$, $T_i = (t_{i1}, t_{i2}, \ldots, t_{ii}, 0, \ldots, 0)$ *and c is a constant.*

By integrating $e^{-\text{tr}(XX')}$ over X, subject to the condition $\sum_{k=1}^{n} x_{ik}^2 = 1$, $i = 1, \ldots, q$, directly over X as well as by using Lemma 19.1.11 we can establish the following result:

LEMMA 19.1.12
The constant c in Lemma 19.1.11 is unity.

19.2 Evaluation of Arbitrary Moments of the Random Volumes

With the help of the above lemmas we are in a position to evaluate arbitrary moments and the exact density of $\nabla_{r,n}$ for various types of situations. As mentioned earlier when the $r \times r$ real symmetric positive definite matrix Y has a matrix-variate distribution the problem becomes simpler and the necessary mathematics is available in Mathai (1997). For the sake of illustration we will consider two results on the distribution of the volume content, $|Y| = |XX'| = \nabla_{r,n}^2$, of the r-parallelotope in n-space for the case when all the points are freely varying and for the case when some of the points are restricted to be on the surface of an n-sphere. Then we will focus at the situation where the rows of X, where $Y = XX'$, are independently distributed and having some general distributions, and at the same time the first q rows are restricted to be of lengths unities.

19.2.1 Matrix-Variate Distributions for X

Let the joint density of X, where $Y = XX'$, be given by

$$f(Y) = f(XX') = c\,|XX'|^{\alpha} e^{-\text{tr}(XX')},$$

$$Y = Y' > 0, \ \Re(\alpha) > -\frac{n}{2} + \frac{r-1}{2} \qquad (19.2.1)$$

where c is the normalizing constant. Consider the transformation from Lemma 19.1.9

$$X = TU_1 \Rightarrow dX = \left(\prod_{j=1}^{r} t_{jj}^{n-j}\right) dT\, dU_1,$$

and from Lemma 19.1.10

$$\int dU_1 = \frac{2^r \pi^{nr/2}}{\Gamma_r\left(\frac{n}{2}\right)}.$$

From the fact that the total probability is 1, that is, $\int_X f(XX')dX = 1$, we have

$$1 = \int_X f(XX')dX$$

$$= c\frac{2^r \pi^{nr/2}}{\Gamma_r\left(\frac{n}{2}\right)} \int_T |TT'|^{\alpha + \frac{n-j}{2}} e^{-\mathrm{tr}(TT')} dT$$

$$= c\frac{2^r \pi^{nr/2}}{\Gamma_r\left(\frac{n}{2}\right)} \left\{ \prod_{i<j} \int_{-\infty}^{\infty} e^{-t_{ij}^2} dt_{ij} \right\} \left\{ \prod_{j=1}^{r} \int_0^{\infty} (t_{jj}^2)^{\alpha + \frac{n-j}{2}} e^{-t_{jj}^2} dt_{jj} \right\}.$$

But

$$\prod_{j=1}^{r} \int_0^{\infty} (t_{jj}^2)^{\alpha + \frac{n-j}{2}} e^{-t_{jj}^2} dt_{jj} = \frac{1}{2^r} \prod_{j=1}^{r} \Gamma\left(\alpha + \frac{n-j+1}{2}\right),$$

for $\Re(\alpha) > -\frac{n}{2} + \frac{r-1}{2}$ and

$$\prod_{i<j} \int_{-\infty}^{\infty} e^{-t_{ij}^2} dt_{ij} = \pi^{r(r-1)/4}.$$

Substituting these back we have

$$c = \frac{\Gamma_r\left(\frac{n}{2}\right)}{\pi^{nr/2}\Gamma_r\left(\alpha + \frac{n}{2}\right)}.$$

The general hth moment of the volume content of the r-parallelotope in n-space, that is $E|XX'|^{h/2}$, is available by replacing α by $\alpha + \frac{h}{2}$ and then taking the ratio of the normalizing constants. This will be stated as a therem

THEOREM 19.2.1

When the $r \times n$ matrix X representing the r random points $P_1, \ldots, P_r$ in n-space have a matrix-variate density as given in (19.2.1) then the hth moment of the volume content of the r-parallelotope created by the linearly independent vectors $\vec{OP_1}, \ldots, \vec{OP_r}$, where O denotes the origin, is given by

$$E(\nabla_{r,n})^h = E|XX'|^{\frac{h}{2}} = \frac{\Gamma_r\left(\alpha + \frac{n+h}{2}\right)}{\Gamma_r\left(\alpha + \frac{n}{2}\right)}$$

$$\text{for } \Re(h) > -n - 2\Re(\alpha). \qquad (19.2.2)$$

Now, let us examine the situation when the first q, $q \leq r$, points are on the surface of a hypersphere of radius 1. In this case the first q vectors have lengths unities. That is,

$$\|\vec{OP_j}\| = 1, \quad j = 1, \ldots, q. \tag{19.2.3}$$

Suppose that the $r \times n$, $n \geq r$, matrix, representing the r random points, still has the matrix-variate density given in (19.2.1) subject to the conditions in (19.2.3). Let the normalizing constant in this case be denoted by c_1. Then from Lemmas 19.1.9, 19.1.10 and 19.1.4 we have the following: Let $T_j = (t_{j1}, \ldots, t_{jj}, 0, \ldots, 0)$ be the jth row of T. Then

$$1 = \int_X f(XX') dX$$

$$= c_1 \left(\int dU_1 \right) e^{-q} \left\{ \prod_{j=2}^{q} \int_{T_j T_j' = 1} (t_{jj}^2)^{\alpha + \frac{n-j}{2}} dT_j \right\}$$

$$\times \left\{ \prod_{j=q+1}^{r} \int_0^\infty (t_{jj}^2)^{\alpha + \frac{n-j}{2}} e^{-t_{jj}^2} dt_{jj} \right\} \left\{ \prod_{i<j=q+1}^{r} \int_{-\infty}^\infty e^{-t_{ij}^2} dt_{ij} \right\}$$

$$= c_1 \frac{2^r \pi^{nr/2}}{\Gamma_r \left(\frac{n}{2}\right)} e^{-q} \pi^{\frac{q(q-1)}{4}} \left\{ \frac{\prod_{j=1}^{q} \Gamma \left(\alpha + \frac{n-j+1}{2} \right)}{\Gamma^q \left(\alpha + \frac{n}{2} \right)} \right\}$$

$$\times \frac{1}{2^{r-q}} \left\{ \prod_{j=q+1}^{r} \Gamma \left(\alpha + \frac{n-j+1}{2} \right) \right\} \pi^{(r-q)(r+q-1)/4}$$

$$= c_1 2^q e^{-q} \frac{\pi^{nr/2}}{\Gamma_r \left(\frac{n}{2}\right)} \frac{\Gamma_r \left(\alpha + \frac{n}{2} \right)}{\Gamma^q \left(\alpha + \frac{n}{2} \right)} \text{ for } \Re(\alpha) > -\frac{n}{2} + \frac{r-1}{2}, \quad n \geq r.$$

Now, the h-th moment of the volume content of the r-parallelotope in n-space is available by replacing α by $\alpha + \frac{h}{2}$ and then taking the ratios of the normalizing constants c_1. This will be stated as a theorem.

THEOREM 19.2.2
Let the r random points $P_1, \ldots, P_r$ in n-space represented by the $r \times n$ matrix X, $n \geq r$, have a joint density as given in (19.2.1) subject to the condition that the first q points are on the surface of a hypersphere of radius 1 or subject to the condition in (19.2.3) and let the points be linearly independent. Then the hth moment of the volume content of the r-parallelotope determined by the vectors $\vec{OP_1}, \ldots, \vec{OP_r}$, where O indicates the origin, is given by

$$E(\nabla_{r,n})^h = E|XX'|^{\frac{h}{2}} = \frac{\Gamma_r\left(\alpha + \frac{n+h}{2}\right)}{\Gamma_r\left(\alpha + \frac{n}{2}\right)} \frac{\Gamma^q\left(\alpha + \frac{n}{2}\right)}{\Gamma^q\left(\alpha + \frac{n+h}{2}\right)},$$

$$\Re(h) > -n - 2\Re(\alpha). \qquad (19.2.4)$$

Similar procedure can be applied to derive the h-th arbitrary moments of the volume contents of the r-parallelotopes when X, the matrix representing the r random points, has other types of matrix-variate distributions in the case of unrestricted points or in the case where some of the points are restricted to be on a hypersphere. Observe that the volume content of the r-simplex determined by the r points $P_1, \ldots, P_r$, denoted by $\Delta_{r,n}$, is connected to the volume content of the r-parallelotope by the relation

$$\Delta_{r,n} = \frac{1}{r!} \nabla_{r,n}.$$

Thus the moments of the volume contents of the r-simplices are available from those of the r-parallelotopes and hence a separate discussion is not necessary.

Before concluding this section let us examine the distribution of $\nabla_{r,n}^2$ coming from Theorems 19.2.1 and 19.2.2. From (19.2.2), by replacing h by $2h$,

$$E(\nabla_{r,n}^2)^h = \frac{\Gamma_r\left(\alpha + \frac{n}{2} + h\right)}{\Gamma_r\left(\alpha + \frac{n}{2}\right)} = \prod_{j=1}^{r} \frac{\Gamma\left(\alpha + \frac{n-j+1}{2} + h\right)}{\Gamma\left(\alpha + \frac{n-j+1}{2}\right)}$$

$$= E(y_1^h)E(y_2^h)\ldots E(y_r^h)$$

where $y_1, \ldots, y_r$ are independent real gamma random variables with the parameters $\left(\alpha + \frac{n-j+1}{2}, 1\right)$, $j = 1, \ldots, r$. Hence, structurally, $\nabla_{r,n}^2$ is a product of r independent real gamma random variables. The exact density of such a structure can be written in terms of a G-function of the type $G_{0,r}^{r,0}(\cdot)$, for the details see Mathai (1993). Let $x = \nabla_{r,n}^2$ and let the density be denoted by $f_r(x)$. Then writing $h = s - 1$,

$$f_r(x) = \frac{x^{-1}}{\prod_{j=1}^{r} \Gamma\left(\alpha + \frac{n-j+1}{2}\right)} G_{0,r}^{r,0}\left(x \Big|_{\alpha + \frac{n-j+1}{2},\ j=1,2,\ldots,r}\right). \quad (19.2.5)$$

Particular cases can be written in terms of known elementary functions. For example, for $r = 1$,

$$f_1(x) = \frac{1}{\Gamma\left(\alpha + \frac{n}{2}\right)} x^{\alpha + \frac{n}{2} - 1} e^{-x}, \quad 0 < x < \infty. \qquad (19.2.6)$$

For $r = 2$ one can combine the gammas with the help of the duplication formula for gamma functions, namely

$$\Gamma(2z) = \pi^{-\frac{1}{2}} 2^{2z-1} \Gamma(z) \Gamma\left(z + \frac{1}{2}\right). \tag{19.2.7}$$

Then for $r = 2$

$$E(x^h) = \frac{\Gamma(2\alpha + n - 1 + 2h)}{2^{2h} \Gamma(2\alpha + n - 1)}.$$

Thus $u = 2x^{\frac{1}{2}}$ has a gamma density, denoted by $f_2(u)$ where

$$f_2(u) = \frac{1}{\Gamma(2\alpha + n - 1)} u^{2\alpha + n - 2} e^{-u}, \qquad u > 0. \tag{19.2.8}$$

For $r = 4$ also one can combine the gammas two by two by using the formula (19.2.7). Then

$$E(4x^{\frac{1}{2}})^h = \frac{\Gamma(2\alpha + n - 1 + h)\Gamma(2\alpha + n - 3 + h)}{\Gamma(2\alpha + n - 1)\Gamma(2\alpha + n - 3)}.$$

Let $v = 4x^{\frac{1}{2}}$. Then the density of v, denoted by $f_4(v)$, is given by the following:

$$f_4(v) = \frac{v^{-1}}{\Gamma(2\alpha + n - 1)\Gamma(2\alpha + n - 3)} G_{0,2}^{2,0}\left(v\big|_{2\alpha+n-1, 2\alpha+n-3}\right).$$

This G-function can be simplified in terms of a Bessel function as follows:

$$f_4(v) = \frac{2\, v^{2\alpha+n-3}}{\Gamma(2\alpha + n - 1)\Gamma(2\alpha + n - 3)} K_2\left(2v^{\frac{1}{2}}\right), \qquad v > 0 \tag{19.2.9}$$

where $K_2(\cdot)$ is a Bessel function, see Mathai (1993, p. 130).

In the case when the first q points are restricted to be on the surface of a hypersphere of unit radius then the hth moment is coming from (19.2.4). That is,

$$E(\nabla_{r,n}^2)^h = d \frac{\prod_{j=2}^r \Gamma\left(\alpha + \frac{n-j+1}{2} + h\right)}{\Gamma^{q-1}\left(\alpha + \frac{n}{2} + h\right)}$$

$$= d \left\{ \prod_{j=2}^q \frac{\Gamma\left(\alpha + \frac{n-j+1}{2} + h\right)}{\Gamma\left(\alpha + \frac{n}{2} + h\right)} \right\}$$

$$\times \left\{ \prod_{j=q+1}^r \Gamma\left(\alpha + \frac{n-j+1}{2} + h\right) \right\}$$

$$= E(u_2^h) \dots E(u_q^h) E(v_1^h) \dots E(v_{r-q}^h)$$

where d is the normalizing constant such that the h-th moment is 1 when $h = 0$, $u_2, \ldots, u_q$ are independent real type-1 beta random variables with the parameters $\left(\alpha + \frac{n-j+1}{2}, \frac{j-1}{2}\right)$, $j = 2, \ldots, q$ and $v_1, \ldots, v_{r-q}$ are independent real gamma random variables with the parameters $\left(\alpha + \frac{n-j+1}{2}, 1\right)$, $j = q+1, \ldots, r$. Then structurally $\nabla_{r,n}^2$ in this case is a product of the form $u_2 \ldots u_q v_1 \ldots v_{r-q}$. The exact density can be written as a G-function of the type $G_{q-1,r-1}^{r-1,0}(\cdot)$. Let $y = \nabla_{r,n}^2$ in (19.2.4) and let the density of y be denoted by $f_{r,q}(y)$. Then for the general r and q, $q \leq r$, the exact density of y is the following:

$$f_{r,q}(y) = \frac{\Gamma^{q-1}\left(\alpha + \frac{n}{2}\right)}{\left\{\prod_{j=2}^{r} \Gamma\left(\alpha + \frac{n-j+1}{2}\right)\right\}} y^{-1} \, G_{q-1,r-1}^{r-1,0}\left(y \left|\begin{array}{l} \alpha+\frac{n}{2}, \ldots, \alpha+\frac{n}{2} \\ \alpha+\frac{n-j+1}{2}, \; j=2,\ldots,r \end{array}\right.\right),$$

$$y > 0. \qquad (19.2.10)$$

Particular cases can be written up in terms of elementary functions. For example, for $q = 1, r = 2$, y has a gamma density with the parameters $\left(\alpha + \frac{n-1}{2}, 1\right)$. That is,

$$f_{2,1}(y) = \frac{1}{\Gamma\left(\alpha + \frac{n-1}{2}\right)} x^{\alpha+\frac{n-3}{2}} e^{-y}, \quad y > 0. \qquad (19.2.11)$$

For $q = 2, r = 2$, y has a type-1 beta density with the parameters $\left(\alpha + \frac{n-1}{2}, \frac{1}{2}\right)$. That is,

$$f_{2,2}(y) = \frac{\Gamma\left(\alpha + \frac{n}{2}\right)}{\Gamma\left(\alpha + \frac{n-1}{2}\right) \Gamma\left(\frac{1}{2}\right)} y^{\alpha+\frac{n-3}{2}} (1-y)^{-\frac{1}{2}}, \quad 0 < y < 1. $$

$$(19.2.12)$$

For $q = 2, r = 3$ the density of y is equivalent to that of the density of a product of a real type-1 beta random variable and a real gamma random variable where the variables are independent. This can be worked out by using the transformation of variables technique. This is also directly available from the G-function representation in (19.2.10). A G-function of the type $G_{1,2}^{2,0}(\cdot)$ can be written in terms of a Whittaker function $W_{.,.}(\cdot)$ by using the formula on page 130 of Mathai (1993), that is,

$$G_{1,2}^{2,0}\left(z \left|\begin{array}{l} \alpha \\ \beta, \gamma \end{array}\right.\right) = z^{\frac{\beta+\gamma-1}{2}} e^{-z/2} W_{\frac{1+\beta+\gamma-2\alpha}{2}, \frac{\beta-\gamma}{2}}(z).$$

In our case the parameters are $\alpha \to \alpha + \frac{n}{2}$, $\beta \to \alpha + \frac{n-1}{2}$, $\gamma \to \alpha + \frac{n-2}{2}$. Substituting these the density is the following:

$$f_{3,2}(y) = \frac{\Gamma\left(\alpha + \frac{n}{2}\right)}{\Gamma\left(\alpha + \frac{n-1}{2}\right)\Gamma\left(\alpha + \frac{n-2}{2}\right)} y^{\alpha + \frac{n}{2} - \frac{7}{4}} e^{-\frac{y}{2}} W_{-\frac{1}{4}, \frac{1}{4}}(y), \quad y > 0.$$

$$(19.2.13)$$

For $r = 3, q = 3$, that is, when all the three points are restricted to be on the surface of a hypersphere then the density is equivalent to that of a product of two independent real type-1 beta random variables and from (19.2.10) the density reduces to the following:

$$f_{3,3}(y) = \frac{\Gamma^2\left(\alpha + \frac{n}{2}\right)}{\Gamma\left(\alpha + \frac{n-1}{2}\right)\Gamma\left(\alpha + \frac{n-1}{2}\right)} y^{-1} G_{2,2}^{2,0}\left(y \Big|_{\alpha + \frac{n-1}{2}, \alpha + \frac{n-2}{2}}^{\alpha + \frac{n}{2}, \alpha + \frac{n}{2}}\right).$$

This G-function can be written in terms of a Gauss' hypergeometric function by using the formula on page 130 of Mathai (1993), namely

$$G_{2,2}^{2,0}\left(z \Big|_{\beta - \frac{1}{2}, \beta - 1}^{\beta, \beta}\right) = \frac{z^{\beta - 1}(1 - z)^{\frac{1}{2}}}{\Gamma\left(\frac{3}{2}\right)}$$

$$\times \ _2F_1\left(\frac{1}{2}, \frac{1}{2}; \frac{3}{2}; 1 - z\right), \quad |z| < 1.$$

Therefore the density in this case reduces to the following:

$$f_{3,3}(y) = \frac{\Gamma^2\left(\alpha + \frac{n}{2}\right)}{\Gamma\left(\alpha + \frac{n-1}{2}\right)\Gamma\left(\alpha + \frac{n-2}{2}\right)\Gamma\left(\frac{3}{2}\right)} y^{\alpha + \frac{n}{2} - 2}(1 - y)^{\frac{1}{2}}$$

$$\times \ _2F_1\left(\frac{1}{2}, \frac{1}{2}; \frac{3}{2}; 1 - y\right), \quad 0 < y < 1. \qquad (19.2.14)$$

When $r = 3$ and $q = 1$ the G-function in (19.2.10) reduces to the form $G_{0,2}^{2,0}(\cdot)$. This form is already mentioned earlier, and this G-function can be written in terms of a Bessel function by using the formula

$$G_{0,2}^{2,0}\left(z \Big|_{\alpha, \beta}\right) = 2z^{(\alpha + \beta)/2} K_{\alpha - \beta}\left(2z^{\frac{1}{2}}\right).$$

19.2.2 Type–1 Beta Distribution for X

Here we consider another matrix-variate distribution of the beta type. Let the $r \times n, r \leq n$ matrix X have a matrix-variate density of the type

$$f_X(X) = k|XX'|^\alpha |I - XX'|^{\beta - \frac{r+1}{2}} \qquad (19.2.15)$$

for $O < XX' < I$ where k is the normalizing constant, I is the identity matrix and all the eigenvalues of XX' are between 0 and 1. Let $X = TU_1$ as in Section 19.2.1. Then the joint density of T and U_1, denoted by $f(T, U_1)$, is given by

$$f(T, U_1) dT dU_1 = k |TT'|^\alpha |I - TT'|^{\beta - \frac{r+1}{2}} \left\{ \prod_{j=1}^r (t_{jj}^2)^{\frac{n-j}{2}} \right\} dT dU_1.$$

Integrating out U_1 the density of T, denoted by $f_T(T)$, is therefore

$$f_T(T) dT = k \frac{2^r \pi^{\frac{n}{2}}}{\Gamma_r \left(\frac{n}{2}\right)} |TT'|^\alpha |I - TT'|^{\beta - \frac{r+1}{2}} \left\{ \prod_{j=1}^r (t_{jj}^2)^{\frac{n-j}{2}} \right\} dT.$$

From the general result corresponding to Lemma 19.1.8 we have

$$Y = TT' \Rightarrow dY = 2^r \left\{ \prod_{j=1}^r (t_{jj}^2)^{\frac{r-j+1}{2}} \right\} dT.$$

Then the density of Y, denoted by $f_Y(Y)$, is available as

$$f_Y(Y) = k \frac{\pi^{\frac{nr}{2}}}{\Gamma_r \left(\frac{n}{2}\right)} |Y|^{\alpha + \frac{n}{2} - \frac{r+1}{2}} |I - Y|^{\beta - \frac{r+1}{2}}, \quad O < Y < I.$$

Integrating Y with the help of a real matrix-variate type-1 beta integral and then substituting for k we have the density of Y given by

$$f_Y(Y) = \frac{\Gamma_r \left(\alpha + \beta + \frac{n}{2}\right)}{\Gamma_r(\beta) \Gamma_r \left(\alpha + \frac{n}{2}\right)} |Y|^{\alpha + \frac{n}{2} - \frac{r+1}{2}} |I - Y|^{\beta - \frac{r+1}{2}},$$

$$\text{for } O < Y < I, \ \Re(\alpha) > -\frac{n}{2} + \frac{r-1}{2}, \ \Re(\beta) > \frac{r-1}{2}.$$

The h-th moment of the r-content of the r-parallelotope, $\nabla_{r,n}$, is available by integrating out over the density $f_X(X)$.

THEOREM 19.2.3
Let the $r \times n$, $r \le n$ matrix representing the r linearly independent random points in n-space have a matrix-variate density as in (19.2.15). Then the hth moment of $\nabla_{r,n}$ is given by

$$E(\nabla_{r,n})^h = E|XX'|^{h/2} = \frac{\Gamma_r \left(\alpha + \frac{n+h}{2}\right)}{\Gamma_r \left(\alpha + \frac{n}{2}\right)} \frac{\Gamma_r \left(\alpha + \beta + \frac{n}{2}\right)}{\Gamma_r \left(\alpha + \beta + \frac{n+h}{2}\right)},$$

$$\Re(h) > -2\Re(\alpha) - n + p - 1.$$

$$(19.2.16)$$

If $\beta = \frac{r+1}{2}$ then the density in (19.2.15) is a matrix-variate Pareto type. In this case also if all the points are freely varying then the hth moment of the r-content of the r-parallelotope is available from (19.2.16) by putting $\beta = \frac{r+1}{2}$. But if some of the points are restricted to be on the surface of the n-sphere then the range of integration for the individual variables seems to be difficult to determine. From Lemma 19.1.8 note that the Jacobian cannot be rewritten in terms of the determinant of Y and hence integration with the help of matrix-variate integrals seems to fail here.

Now we will consider a few cases when the rows of X are statistically independently distributed and having spherically symmetric distributions. These cases in the general forms are discussed by the author previously and hence the results will be listed here with the minimum number of steps for the sake of completeness. Some special cases will be enumerated in terms of hypergeometric functions which were not done earlier.

19.2.3 The Case when the Rows of X are Independently Distributed

Let X be the $r \times n$, $n \geq r$ matrix of real variables, T the lower triangular matrix with positive diagonal elements and U_1 the semiorthonormal matrix as defined before. Let $X_j = (x_{j1}, \ldots, x_{jn})$ be the jth row of X. Then the representation

$$X = TU_1,$$

gives

$$X_j = t_{j1} u_{(1)} + t_{j2} u_{(2)} + \ldots + t_{jj} u_{(j)}$$

where $u_{(i)}$ is the i-th row of U_1, and therefore

$$X_j X_j' = t_{j1}^2 + \ldots + t_{jj}^2, \ j = 1, \ldots, r,$$

since $u_{(i)} u_{(i)}' = 1$, $u_{(i)} u_{(k)}' = 0$, $i \neq k$. If X_j, $j = 1, \ldots, r$ are statistically independently distributed with the n-variate density of X_j a function of $X_j X_j'$ then the density is a function of $t_{j1}^2 + \ldots + t_{jj}^2$ and $\{t_{j1}, \ldots, t_{jj}\}$, $j = 1, \ldots, r$ are independently distributed, and

$$E(\nabla_{r,n}^h) = E|XX'|^{h/2} = E|TT'|^{h/2}$$

$$= k_r \frac{2^r \pi^{nr/2}}{\Gamma_r\left(\frac{n}{2}\right)} \int_T \left\{ \prod_{j=1}^r (t_{jj}^2)^{\frac{h+n-j}{2}} \right\} g_j(t_{j1}^2 + \ldots + t_{jj}^2) \mathrm{d}T$$

$$(19.2.17)$$

where k_r is the normalizing constant in the density of X and $g_j(X_j X_j')$ denotes the density of X_j. If the first q points are distributed independently and uniformly over the surface of an n-sphere of unit radius then $g_j(\cdot)$ is a constant which gets absorbed in k_r for $j = 1, \ldots, q$. Then for $j = 1, \ldots, q$ the integrals over the t_{ij}'s are such that $t_{11} = 1$, $t_{jj} > 0$, $t_{jj} = \sqrt{1 - (t_{j1}^2 + \ldots + t_{jj-1}^2)}$. Then from Lemma 19.1.4 the integrals over the j-sphere for $j = 2, \ldots, q$ gives

$$\prod_{j=2}^{q} \left[\frac{\Gamma^{j-1}\left(\frac{1}{2}\right) \Gamma\left(\frac{h+n-j+1}{2}\right)}{\Gamma\left(\frac{h+n}{2}\right)} \right] = \pi^{q(q-1)/4} \frac{\left\{\prod_{j=1}^{q} \Gamma\left(\frac{h+n-j+1}{2}\right)\right\}}{\Gamma^{q}\left(\frac{h+n}{2}\right)}.$$

(19.2.18)

The remaining integrals and k_r depend upon $g_j(\cdot)$, $j = q+1, \ldots, r$. We will consider a few cases here.

Let the n-sphere be of radius unity. Out of the r independently distributed random points let the first q be uniform over the surface and the remaining $r - q$, $q \le r$ be uniform inside the sphere. From the volume and surface contents we have

$$k_r = \frac{\Gamma^{q}\left(\frac{n}{2}\right) \Gamma^{r-q}\left(\frac{n}{2} + 1\right)}{2^{q} \pi^{nr/2}}.$$

(19.2.19)

From Lemma 19.1.2 the integral inside the j-sphere, $j = q+1, \ldots, r$ and for $t_{jj} > 0$, gives

$$\prod_{j=q+1}^{r} \left\{ \frac{\Gamma^{j-1}\left(\frac{1}{2}\right) \Gamma\left(\frac{n-j+h+1}{2}\right)}{2\Gamma\left(\frac{n+h}{2} + 1\right)} \right\}$$

$$= \frac{\pi^{(r-q)(r+q-1)/4} \prod_{j=q+1}^{r} \Gamma\left(\frac{n+h-j+1}{2}\right)}{2^{r-q} \Gamma^{r-q}\left(\frac{n+h}{2} + 1\right)}$$

(19.2.20)

Combining the results from (19.2.17) to (19.2.20) we have the following:

THEOREM 19.2.4
When the r points $P_1, \ldots, P_r$ in R^n are independently and uniformly distributed in an n-sphere of unit radius such that $P_1, \ldots, P_q$ are uniform over the surface and $P_{q+1}, \ldots, P_r$ are uniform inside the sphere, $q \le r$, then the h-th moment of the r-content $\nabla_{r,n}$ of the r-parallelotope generated by the vectors $\vec{OP_1}, \ldots, \vec{OP_r}$, where O indicates the center of the sphere, is given by

$$E(\nabla_{r,n}^h) = E|XX'|^{h/2}$$

$$= \left(\frac{n}{n+h}\right)^{r-q} \frac{\Gamma_r\left(\frac{n+h}{2}\right)}{\Gamma_r\left(\frac{n}{2}\right)} \frac{\Gamma^r\left(\frac{n}{2}\right)}{\Gamma^r\left(\frac{n+h}{2}\right)} \quad \text{for } \Re(h) > -n+r-1.$$

19.2.4 Type-1 Beta Distributed Independent Rows of X

Let $X_j, j = 1, \ldots, q$ be uniform on the surface of an n-sphere of radius unity and let $X_j, j = q+1, \ldots, p$ have the densities

$$f_j(X_j) = d_j(1 - X_j X_j')^{\beta_j}, \quad X_j X_j' < 1, \qquad \Re(\beta_j) > -1. \quad (19.2.21)$$

Then from Lemma 19.1.1

$$d_j = \frac{\Gamma\left(\beta_j + \frac{n}{2} + 1\right)}{\pi^{n/2}\Gamma(\beta_j + 1)}, \qquad \Re(\beta_j) > -1. \quad (19.2.22)$$

From Lemma 19.1.2 the integrals over t_{ij}'s, $i \geq j$, $t_{jj} > 0$, $i = q + 1, \ldots, r$, in the h-th moment of $|XX'|^{1/2}$ gives the product

$$\frac{1}{2^{r-q}} \prod_{j=q+1}^{r} \left\{ \frac{\Gamma^{j-1}\left(\frac{1}{2}\right) \Gamma\left(\frac{n-j+h+1}{2}\right) \Gamma(\beta_j + 1)}{\Gamma\left(\frac{n+h}{2} + 1 + \beta_j\right)} \right\}$$

$$= \frac{\pi^{(r-q)(r+q-1)/4}}{2^{r-q}} \prod_{j=q+1}^{r} \left\{ \frac{\Gamma\left(\frac{n-j+h+1}{2}\right) \Gamma(\beta_j + 1)}{\Gamma\left(\frac{n+h}{2} + 1 + \beta_j\right)} \right\}.$$

$$(19.2.23)$$

The normalizing constant in this case, again denoted by k_r, is

$$k_r = \frac{\Gamma^q\left(\frac{n}{2}\right)}{2^q \pi^{qn/2}} \left\{ \prod_{j=q+1}^{r} \frac{\Gamma\left(\beta_j + \frac{n}{2} + 1\right)}{\pi^{n/2}\Gamma(\beta_j + 1)} \right\}. \quad (19.2.24)$$

From (19.2.17) to (19.2.24) we have the following result:

THEOREM 19.2.5
When the r independent random points $P_1, \ldots, P_r$ are such that the first q of them are uniform over the surface of an n-sphere of radius unity

*and the remaining $r - q$ points are type-1 beta distributed as in (19.2.21)
then the hth moment of the r content of the r-parallelotope created by
the vectors $\vec{OP}_1, \ldots, \vec{OP}_r$, where O indicates the center of the sphere, is
given by*

$$E(\nabla_{r,n}^h) = E|XX'|^{h/2}$$

$$= \frac{\Gamma_r\left(\frac{n+h}{2}\right)}{\Gamma_r\left(\frac{n}{2}\right)} \frac{\Gamma^q\left(\frac{n}{2}\right)}{\Gamma^q\left(\frac{n+h}{2}\right)} \left\{ \prod_{j=q+1}^{r} \frac{\Gamma\left(\beta_j + \frac{n}{2} + 1\right)}{\Gamma\left(\beta_j + \frac{n+h}{2} + 1\right)} \right\}$$

for $\Re(h) > -n + r - 1$, $-n - 2 - 2\Re(\beta_j)$, $\Re(\beta_j) > -1$, $j = q+1, \ldots, r$.

19.2.5   Type-2 Beta Distributed Independent Rows of X

Assume that the first q points out of the r independently distributed
points are uniform over the surface of an n-sphere and the remaining
$r - q$ have the densities

$$g_j(X_j) = \eta_j(1 + X_j X_j')^{-(\alpha_j + \frac{n}{2})}, \qquad \Re(\alpha_j) > 0, \; j = q+1, \ldots, r$$

$$X_j X_j' = x_{j1}^2 + \ldots + x_{jn}^2.$$

From Lemma 19.1.7

$$\eta_j = \frac{\Gamma\left(\alpha_j + \frac{n}{2}\right)}{\pi^{n/2}\Gamma\left(\alpha_j\right)}. \tag{19.2.25}$$

The integral over the t_{ij}'s, in the h-th moment of $|XX'|^{1/2}$, for $j = q+1, \ldots, r$ is also available from Lemma 19.1.7 by replacing γ_j by $\frac{n-j+h}{2}$
and δ_j by $\alpha_j + \frac{n}{2}$ for $j = q+1, \ldots, r$ and then taking the product,
multiplied by the normalizing constant η_j and observing that $t_{jj} > 0$ for
all j. That is,

$$\prod_{j=q+1}^{r} \eta_j \int_{T_j} (t_{jj}^2)^{\frac{n-j+h}{2}} (1 + t_{j1}^2 + \ldots + t_{jj}^2)^{-(\alpha_j + \frac{n}{2})} dT_j$$

$$= \frac{\pi^{(r-q)(r+q-1)/4}}{\pi^{(r-q)n/2} 2^{r-q}} \left\{ \prod_{j=q+1}^{r} \frac{\Gamma\left(\alpha_j - \frac{h}{2}\right)\Gamma\left(\frac{n-j+h+1}{2}\right)}{\Gamma(\alpha_j + \frac{n}{2})} \right\}.$$

$$\tag{19.2.26}$$

Then from (19.2.17)–(19.2.19) and (19.2.24)–(19.2.26) we have the following result:

THEOREM 19.2.6

Out of the r independently distributed random points $P_1, \ldots, P_r$ in R^n if the first q are uniform over the surface of an n-sphere of unit radius and the remaining $r-q$ have the densities $g_j(X_j)$, $j = q+1, \ldots, r$ of (19.2.25) then the hth moment of the r-content $\nabla_{r,n}$ of the r-parallelotope created by the r vectors $\vec{OP_1}, \ldots, \vec{OP_r}$, where O indicates the center of the sphere, is given by

$$E(\nabla_{r,n}^h) = E|XX'|^{h/2}$$

$$= \frac{\Gamma_r\left(\frac{n+h}{2}\right)}{\Gamma_r\left(\frac{n}{2}\right)} \frac{\Gamma^q\left(\frac{n}{2}\right)}{\Gamma^q\left(\frac{n+h}{2}\right)} \left\{ \prod_{j=q+1}^{r} \frac{\Gamma\left(\alpha_j - \frac{h}{2}\right)}{\Gamma(\alpha_j)} \right\}$$

for $-n + r - 1 < \Re(h) < \Re(2\alpha_j)$, $\Re(\alpha_j) > 0$, $j = q+1, \ldots, r$.

19.2.6 Independently Gaussian Distributed Points

Out of the r independently distributed random points in R^n if the first q are uniform over an n-sphere of unit radius and the remaining $r - q$ have the densities

$$h_j(X_j) = \frac{\lambda^{n/2}}{\pi^{n/2}} e^{-(\lambda X_j X_j')}, \quad X_j = (x_{j1}, \ldots, x_{jn}), \quad -\infty < x_{jk} < \infty,$$

$$j = q+1, \ldots, r, \ r \leq n \tag{19.2.27}$$

then the integral over the t_{ij}s, with $t_{jj} > 0$, in the hth moment of $|XX'|^{1/2}$ for $i = q+1, \ldots, r$, $j = 1, \ldots, n$ gives

$$\prod_{j=q+1}^{r} \frac{\lambda^{n/2}}{\pi^{n/2}} \int \cdots \int (t_{jj}^2)^{\frac{n-j+h}{2}} e^{-\lambda(t_{j1}^2 + \ldots + t_{jj}^2)} dt_{j1} \ldots dt_{jj}$$

$$= \left(\frac{\lambda}{\pi}\right)^{(r-q)n/2} \lambda^{-(r-q)(n+h)/2} \pi^{(r-q)(r+q-1)/4}$$

$$\times \frac{1}{2^{r-q}} \left\{ \prod_{j=q+1}^{r} \Gamma\left(\frac{n-j+h+1}{2}\right) \right\}. \tag{19.2.28}$$

The integral over the surface of the n-sphere for the first q points yields

$$\frac{\Gamma^q(n/2)}{2^q \pi^{qn/2}} \frac{\pi^{q(q-1)/4}}{\Gamma^q\left(\frac{n+h}{2}\right)} \left\{ \prod_{j=1}^{q} \Gamma\left(\frac{n+h-j+1}{2}\right) \right\} \tag{19.2.29}$$

and

$$\int dU_1 = \frac{2^r \pi^{nr/2}}{\Gamma_r\left(\frac{n}{2}\right)}.$$ (19.2.30)

From (19.2.28)–(19.2.30) we have the following:

THEOREM 19.2.7
*Consider r independently distributed random points $P_1, \ldots, P_r$ in R^n
where the first q of them are uniform over the surface of an n-sphere with
center O and radius unity and the remaining $r - q$ have the Gaussian
densities given in (19.2.27). Then the hth moment of the r-content $\nabla_{r,n}$
of the r-parallelotope created by the vectors $\vec{OP}_1, \ldots, \vec{OP}_r$ is given by*

$$E(\nabla_{r,n}^h) = E|XX'|^{h/2}$$

$$= \frac{1}{\lambda^{(r-q)h/2}} \frac{\Gamma^q\left(\frac{n}{2}\right)}{\Gamma^q\left(\frac{n+h}{2}\right)} \frac{\Gamma_r\left(\frac{n+h}{2}\right)}{\Gamma_r\left(\frac{n}{2}\right)}, \quad \Re(h) > -n + r - 1.$$

Note that two gammas can be cancelled from $E(\nabla_{r,n}^h)$ above. Also observe that in the density of (19.2.27) we could have replaced λ by λ_j for
each j still the procedure goes through.

19.2.7   Distributions of the r-Contents

The moment expression in Theorem 19.2.4 can be written in the following
form for $E(\nabla_{r,n}^2)^h = E(\nabla_{r,n})^{2h}$.

$$E(\nabla_{r,n}^2)^h = a \left\{ \prod_{j=2}^{r} \frac{\Gamma(\gamma_j + h)}{\Gamma(\gamma_j + \delta_j + h)} \right\}, \quad \gamma_j = \frac{n}{2} - \frac{j-1}{2}, \ j = 2, \ldots, r$$

(19.2.31)

where δ_j, $j = 1, \ldots, r$ is the sequence

$$\left\{ \frac{1}{2}, \frac{2}{2}, \ldots, \frac{q-1}{2}, \frac{q}{2} + 1, \ldots, \frac{r-1}{2} + 1 \right\}$$

and a is a normalizing constant which is available by putting $h = 0$ in
the gamma product and taking the reciprocal of the resulting quantity.
Hence the density of $\nabla_{r,n}^2$ can be written as a G-function of the form
$G_{r-1,r-1}^{r-1,0}$ from where series representation is available, see the details

of the procedure from Mathai (1993). The density can also be looked upon as the density of a product of independent real type–1 beta random variables.

In Theorem 19.2.5, that is the type-1 beta distributed case, if the parameters β_j's are replaced by zeros then the hth moment of $\nabla^2_{r,n}$ agrees with that in (19.2.31). Hence the hth moment of $\nabla^2_{r,n}$ in this case can also be written as (19.2.31) with γ_j, $j = 1, \ldots, r$ remaining the same whereas δ_j, $j = 2, \ldots, q$ remain the same but for $j = q+1, \ldots, r$ add $\beta_{q+1}, \ldots, \beta_r$ to the parameters in (19.2.31). Hence the density of $\nabla^2_{r,n}$ in this case also is a G-function of the form $G^{r-1,0}_{r-1,r-1}$.

In Theorem 19.2.6 the hth moment of $\nabla^2_{r,n}$ can be written in the following form:

$$E(\nabla^2_{r,n})^h = b \frac{\left\{ \prod_{j=2}^{r-1} \Gamma(\gamma_j + h) \right\} \left\{ \prod_{j=q+1}^{r} \Gamma(1 - \delta_j - h) \right\}}{\left\{ \prod_{j=1}^{q-1} \Gamma(\rho_j + h) \right\}}$$

(19.2.32)

where b is the normalizing constant and $\gamma_j = \frac{n}{2} - \frac{j-1}{2}, j = 2, \ldots, r$, $\delta_j = 1 - \alpha_j$, $j = q+1, \ldots, r$, $\rho_j = \frac{n}{2}$, $j = 1, \ldots, q-1$. The density can be evaluated as a G-function of the form $G^{r-1,r-q}_{r-1,r-1}$. When evaluating the series form, part of the function is coming from the residues corresponding to the gammas $\Gamma(\gamma_j + h)$ and the remaining from the analytic continuation or from the gammas $\Gamma(1 - \delta_j - h)$, see the details from Mathai (1993). Since the particular cases of this situation do not seem to be available in the literature we will list a few cases here. Let $z = \nabla^2_{r,n}$ in Theorem 19.2.6 and let the density of z be denoted by $g_{r,q}(z)$.

For $r = 2$, $q = 2$

$$E(z^h) = \frac{\Gamma\left(\frac{n}{2}\right)}{\Gamma\left(\frac{n-1}{2}\right)} \frac{\Gamma\left(\frac{n-1}{2} + h\right)}{\Gamma\left(\frac{n}{2} + h\right)}.$$

Here z has a type-1 beta density with the parameters $\left(\frac{n-1}{2}, \frac{1}{2}\right)$. In this case there is no type-2 distributed point. The density is given by

$$g_{2,2}(z) = \frac{\Gamma\left(\frac{n}{2}\right)}{\Gamma\left(\frac{n-1}{2}\right)\Gamma\left(\frac{1}{2}\right)} z^{\frac{n-3}{2}}(1 - z)^{-\frac{1}{2}}, \ 0 < z < 1. \quad (19.2.33)$$

For $r = 2$, $q = 1$

$$E(z^h) = \frac{1}{\Gamma\left(\frac{n-1}{2}\right)\Gamma(\alpha_2)} \Gamma\left(\frac{n-1}{2} + h\right) \Gamma(\alpha_2 - h).$$

The density is equivalent to that of the ratio z_1/z_2 where z_1 and z_2 are independently distributed real gamma variables with the parameters $\frac{n-1}{2}$ and α_2 respectively or the moment can also be looked upon as the h-th moment of a type–2 beta random variable. Hence the density is the following:

$$g_{2,1}(z) = \frac{\Gamma\left(\alpha_2 + \frac{n-1}{2}\right)}{\Gamma\left(\frac{n-1}{2}\right)\Gamma(\alpha_2)} z^{\frac{n-3}{2}}(1+z)^{-\left(\alpha_2 + \frac{n-1}{2}\right)}, \quad 0 < z < \infty.$$

$$(19.2.34)$$

For $r = 3$, $q = 3$ the density is equivalent to the density of a product of two independent real type-1 beta random variables with the parameters $\left(\frac{n-1}{2}, \frac{1}{2}\right), \left(\frac{n-2}{2}, 1\right)$ and the density coincides with the $f_{3,3}(y)$ given in (2.14). For $r = 3$, $q = 2$ the hth moment is of the form

$$E(z^h) = \frac{\Gamma\left(\frac{n}{2}\right)}{\Gamma\left(\frac{n-1}{2}\right)\Gamma\left(\frac{n-2}{2}\right)\Gamma(\alpha_3)} \frac{\Gamma\left(\frac{n-1}{2} + h\right)}{\Gamma\left(\frac{n}{2} + h\right)}\Gamma\left(\frac{n-2}{2} + h\right)\Gamma(\alpha_3 - h).$$

This is equivalent to the hth moment of a product of independent real random variables where one is a type-1 beta and the other is a type-2 beta. The hth moment can be simplified to the form

$$E(z^h) = \frac{\left(\frac{n}{2} - 1\right)}{\Gamma\left(\frac{n-1}{2}\right)\Gamma(\alpha_3)} \frac{\Gamma\left(\frac{n-1}{2} + h\right)\Gamma(\alpha_3 - h)}{\left(\frac{n}{2} - 1 + h\right)}.$$

There is one pole at $h = -\frac{n}{2} + 1$ and poles at $h = -\frac{n-1}{2} - \nu$, $\nu = 0, 1, \ldots$ for $0 < z < 1$ and poles at $h = \alpha_3 + \nu$, $\nu = 0, 1, \ldots$ for $z > 1$. Evaluating as the sum of the residues at these poles and then simplifying we have the following density where $_2F_1(\cdot)$ denotes a Gauss' hypergeometric function.

$$g_{3,2}(z) = \gamma \begin{cases} \sqrt{\pi}\,\Gamma\left(\alpha_3 + \frac{n}{2} - 1\right) z^{\frac{n-4}{2}} \\ -2\Gamma\left(\alpha_3 + \frac{n-1}{2}\right) z^{\frac{n-3}{2}}\gamma_1(z), \ 0 < z < 1 \\ \frac{\Gamma\left(\alpha_3 + \frac{n-1}{2}\right)\Gamma\left(\alpha_3 + \frac{n-2}{2}\right)}{\Gamma\left(\alpha_3 + \frac{n}{2}\right)} z^{-\alpha_3 - 1}\gamma_2(z), \ z > 1 \end{cases} \quad (19.2.35)$$

for $\Re(\alpha_3) > -\frac{n}{2} + 1$ where

$$\gamma = \frac{\left(\frac{n}{2} - 1\right)}{\Gamma\left(\frac{n-1}{2}\right)\Gamma(\alpha_3)},$$

$$\gamma_1(z) = \,_2F_1\left(\alpha_3 + \frac{n-1}{2}, \frac{1}{2}; \frac{3}{2}; -z\right),$$

and
$$\gamma_2(z) = {}_2F_1\left(\alpha_3 + \frac{n-1}{2}, \alpha_3 + \frac{n-2}{2}; \alpha_3 + \frac{n}{2}; -\frac{1}{z}\right).$$

For $r = 3$, $q = 1$ the hth moment has the form

$$E(z^h) = \left[\Gamma\left(\frac{n-1}{2}\right)\Gamma\left(\frac{n-2}{2}\right)\Gamma(\alpha_2)\Gamma(\alpha_3)\right]^{-1}$$
$$\times \Gamma\left(\frac{n-1}{2}+h\right)\Gamma\left(\frac{n-2}{2}+h\right)\Gamma(\alpha_2 - h)\Gamma(\alpha_3 - h).$$

This can also be looked upon as the hth moment of a product of two independent real type-2 random variables. Evaluating by the residue calculus the density when the poles are simple, that is, for $\alpha_2 - \alpha_3 \neq \pm\nu$, $\nu = 0, 1, \ldots$, is the following:

$g_{3,1}(z)$
$$= \delta \begin{cases} \sqrt{\pi}\,\Gamma\left(\alpha_2 + \frac{n-2}{2}\right)\Gamma\left(\alpha_3 + \frac{n-2}{2}\right)z^{\frac{n-4}{2}}\delta_1(z) \\ -2\sqrt{\pi}\,\Gamma\left(\alpha_2 + \frac{n-1}{2}\right)\Gamma\left(\alpha_3 + \frac{n-1}{2}\right)z^{\frac{n-3}{2}}\delta_2(z),\ 0 < z < 1 \\ \Gamma\left(\alpha_2 + \frac{n-1}{2}\right)\Gamma\left(\alpha_2 + \frac{n-2}{2}\right)\Gamma(\alpha_3 - \alpha_2)z^{-\alpha_2-1}\delta_3(z) \\ +\Gamma\left(\alpha_3 + \frac{n-1}{2}\right)\Gamma\left(\alpha_3 + \frac{n-2}{2}\right)\Gamma(\alpha_2 - \alpha_3)z^{-\alpha_3-1}\delta_4(z),\ z > 1 \end{cases}$$
$$(19.2.36)$$

where

$$\delta^{-1} = \Gamma\left(\frac{n-1}{2}\right)\Gamma\left(\frac{n-2}{2}\right)\Gamma(\alpha_2)\Gamma(\alpha_3),$$

$$\delta_1(z) = {}_2F_1\left(\alpha_2 + \frac{n-2}{2}, \alpha_3 + \frac{n-2}{2}; \frac{1}{2}; z\right)$$

$$\delta_2(z) = {}_2F_1\left(\alpha_2 + \frac{n-1}{2}, \alpha_3 + \frac{n-1}{2}; \frac{3}{2}; z\right),$$

$$\delta_3(z) = {}_2F_1\left(\alpha_2 + \frac{n-1}{2}, \alpha_2 + \frac{n-2}{2}; \alpha_2 - \alpha_3 + 1; \frac{1}{z}\right),$$

and

$$\delta_4(z) = {}_2F_1\left(\alpha_3 + \frac{n-1}{2}, \alpha_3 + \frac{n-2}{2}; \alpha_3 - \alpha_2 + 1; \frac{1}{z}\right).$$

For $r > 3$ the poles of the integrand will be of higher order and hence the density will involve psi and zeta functions. Hence more special cases will not be listed here.

In Theorem 19.2.7 the h-th moment of $\lambda^{(r-q)}\nabla^2_{r,n}$ is of the form

$$E(\lambda^{r-q}\nabla^2_{r,n})^h = c\frac{\prod_{j=2}^r \Gamma(\gamma_j + h)}{\prod_{j=1}^{q-1} \Gamma(\delta_j + h)} \qquad (19.2.37)$$

where c is the normalizing constant, $\gamma_j = \frac{n}{2} - \frac{j-1}{2}$, $j = 2,\ldots,r$ and $\delta_j = \frac{n}{2}$, $j = 1,\ldots,q-1$. The density corresponding to (19.2.37) is a G-function of the form $G^{r-1,0}_{q-1,r-1}$. Series representation of such a G-function is also available from Mathai (1993). Several special cases of this structure are already done in Section 19.2.1.

In all the above cases some of the special cases can be evaluated in terms of elementary special functions as done in Section 19.2.1. For general parameters the series forms of the densities will be of logarithmic type involving psi and generalized zeta functions due to the presence of poles of higher orders in the integrand.

References

1. Mathai, A. M. (1998). Random p-content of a p-parallelopote in Euclidean n-space. *Advances in Applied Probability* **32(2)**, (to appear).

2. Mathai, A. M. (1997). *Jacobians of Matrix Transformations and Functions of Matrix Argument.* World Scientific Publishing, New York.

3. Mathai, A. M. (1993). *A Handbook of Generalized Special Functions for Statistical and Physical Sciences.* Oxford University Press, Oxford.

4. Miles, R. E. (1971). Isotropic random simplices. *Advances in Applied Probability* **3**, 353–382.

PART IV

TIME SERIES, LINEAR, AND NON-LINEAR MODELS

20

Cointegration of Economic Time Series

T. W. Anderson
Stanford University, Stanford, CA

ABSTRACT Regression models and simultaneous equations models are reviewed. The Limited Information Maximum Likelihood estimator of a structural equation is a vector annihilated by a reduced rank regression estimator of the reduced form. This estimator is based on the canonical correlations and vectors of the dependent and independnt variables. Cointegated models are introduced as nonstationary autoregessive processes having some stationary linear functions; the error-correction form of the model has a coefficient matrix of lower rank. The reduced rank regression estimator of this matrix is in part a superefficient estimator. Its large-sample behavior and that of the likelihood ratio criterion for testing rank are described.

Keywords and phrases Regression models, autoregressive processes, simultaneous equations, reduced rank regression estimator, nonstationary processes, error-correction form

20.1 Introduction

The purpose of this paper is to introduce the econometric techniques suitable for analyzing cointegrated economic time series. The relevant series may be nonstationary with some linear combinations being stationary. The observed series may display trends, inflation, and/or increasing volatility, but some relations between them may hold constant. Among the objectives is the estimation of these relations. The methodology is an extension of the techniques used in regression analysis and for simultaneous equations models.

For background we review the basic ideas of regression analysis and the single-equation methods for simultaneous equation models. The latter are examples of "reduced rank regression," which forms the basis for cointegration analysis. The models we treat are the simplest that show the essential characteristics of cointegration; they are autoregressions of order 1. The means are supposed to have been eliminated; no intercept is included.

A method of detrending some nonstationary series is to difference the series [Box and Jenkins (1971)]. A linear trend is eliminated by a single differencing, for instance. The kind of nonstationary series considered in this paper can be reduced to stationarity by one differencing. We look for linear combinations of the observed variables which are stationary without differencing.

A more detailed treatment is given in Anderson (1999). For a fuller study see Johansen (1995) and Dhrymes (1998). The idea of cointegration was introduced by Granger (1981).

20.2   Regression Models

Possibly the most frequent statistical analysis in econometrics is regression analysis. The data for which this methodology is appropriate consists of a set of p economic or dependent or endogenous variables observed over time constituting a sequence of vectors $\mathbf{Y}_t = (Y_{1t}, \ldots, Y_{pt})'$ and a set of q explanatory or independent or predetermined variables constituting a sequence of vectors $\mathbf{X}_t = (X_{1t}, \ldots, X_{qt})'$, $t = 1, 2, \ldots, T$. The explanatory variables may be noneconomic or exogenous variables or the economic variables observed earlier in time.

A linear regression model is

$$\mathbf{Y}_t = \mathbf{B}\mathbf{X}_t + \mathbf{Z}_t, \tag{20.2.1}$$

where $\mathbf{Z}_t = (Z_{1t}, \ldots, Z_{pt})'$ is a vector of unobserved random variables or disturbances that are uncorrelated with $\mathbf{X}_t, \mathbf{X}_{t-1}, \ldots$. In formal terms the assumptions are $\mathcal{E}\mathbf{Z}_t = \mathbf{0}$,

$$\mathcal{E}\mathbf{Z}_t\mathbf{Z}_t' = \mathbf{\Sigma}_{ZZ}, \qquad \mathcal{E}\mathbf{Z}_t\mathbf{X}_t' = \mathbf{0}. \tag{20.2.2}$$

The statistical analysis concerns $\mathbf{B}$, the matrix of regression.

We observe $\mathbf{Y}_t, \mathbf{X}_t$, $t = 1, \ldots, T$. We shall assume $\mathcal{E}\mathbf{X}_t = \mathbf{0}$. In practice observations would be replaced by the observations minus the respective sample means. For the sake of simplifying the exposition we assume $\mathcal{E}\mathbf{X}_t = \mathbf{0}$.

Define the sample covariance matrices

$$\mathbf{S}_{YY} = \frac{1}{T}\sum_{t=1}^{T}\mathbf{Y}_t\mathbf{Y}_t', \qquad p \times p, \tag{20.2.3}$$

$$\mathbf{S}_{YX} = \frac{1}{T}\sum_{t=1}^{T}\mathbf{Y}_t\mathbf{X}_t', \qquad p \times q, \tag{20.2.4}$$

$$\mathbf{S}_{XX} = \frac{1}{T}\sum_{t=1}^{T}\mathbf{X}_t\mathbf{X}_t', \qquad q \times q. \tag{20.2.5}$$

The Least Squares Estimator of the matrix of regression coefficients is

$$\hat{\mathbf{B}} = \mathbf{S}_{YX}\mathbf{S}_{XX}^{-1}. \tag{20.2.6}$$

When the $\mathbf{X}_t$'s are nonstochastic, an element of $\hat{\mathbf{B}}$ is a minimum variance unbiased estimator of the corresponding element of $\mathbf{B}$.

The residuals

$$\hat{\mathbf{Z}}_t = \mathbf{Y}_t - \hat{\mathbf{B}}\mathbf{X}_t \tag{20.2.7}$$

are approximations to the unobserved disturbances $\mathbf{Z}_t$. The matrix

$$\mathbf{S}_{\hat{Z}\hat{Z}} = \frac{1}{T}\sum_{t=1}^{T}(\mathbf{Y}_t - \hat{\mathbf{B}}\mathbf{X}_t)(\mathbf{Y}_t - \hat{\mathbf{B}}\mathbf{X}_t)' \tag{20.2.8}$$

is an estimator of the covariance matrix $\boldsymbol{\Sigma}_{ZZ}$.

Statistical inference is carried out in terms of $\hat{\mathbf{B}}$ and $\mathbf{S}_{\hat{Z}\hat{Z}}$. For example, if the $\mathbf{X}_t$'s are nonstochastic, the variance of an element $\hat{\beta}_{ij}$ of $\hat{\mathbf{B}}$ is $\frac{1}{T}\sigma_{zz}^{ii}(\mathbf{S}_{XX}^{-1})_{jj}$, where $(\mathbf{S}_{XX}^{-1})_{jj}$ is the j,j-th element of the inverse of $\mathbf{S}_{XX}$. This variance is estimated by $\frac{1}{T}s_{\hat{Z}\hat{Z}}^{ii}(\mathbf{S}_{XX}^{-1})_{jj}$.

Whether $\mathbf{X}_1, \ldots, \mathbf{X}_T$ are stochastic or nonstochastic, the estimators $\hat{\mathbf{B}}$ and $\mathbf{S}_{\hat{Z}\hat{Z}}$ can be used. If $\mathbf{S}_{XX} \to \boldsymbol{\Sigma}_{XX}$, a positive definite matrix, the asymptotic theory holds.

20.3   Simultaneous Equation Models

For many econometric purposes the regression model is too simple. An example is the elementary market situation in which the observed price and quantity of a good is a result of the behaviors of consumers and producers; that is, the observed price and quantity are the equilibrium solution of a demand schedule and a supply function.

The general (linear) simultaneous equations model is

$$\mathbf{A}\mathbf{Y}_t = \boldsymbol{\Xi}\mathbf{X}_t + \mathbf{U}_t, \qquad |\mathbf{A}| \neq 0, \tag{20.3.1}$$

where as before $\mathbf{Y}_t$ is a vector of endogenous variables, $\mathbf{X}_t$ a vector of predetermined variables, and $\mathbf{U}_t$ is a vector of unobservable disturbances

[Koopmans (1950)]. A component equation may describe the behavior of some set of economic agents. For example, it might be the demand function of consumers. The solution of (3.1) for $\mathbf{Y}_t$ yields a regression model (20.2.1), known as the reduced form. The relations between the quantities in (20.2.1) and in (20.3.1) is

$$\Xi = \mathbf{AB}, \qquad \mathbf{U}_t = \mathbf{AZ}_t. \tag{20.3.2}$$

To make inferences about the parameters of the simultaneous equation model, $\mathbf{A}, \Xi$, and $\boldsymbol{\Sigma}_{UU} = \mathcal{E}\mathbf{U}_t\mathbf{U}'_t$ we could first estimate the parameters of the regression model and than solve $\hat{\Xi} = \hat{\mathbf{A}}\hat{\mathbf{B}}$ for $\hat{\Xi}$ and $\hat{\mathbf{A}}$. However, to carry out this program requires some a prior knowledge of Ξ and $\mathbf{A}$; that is, each structural equation must be identified.

As an over-simplified example, suppose that in an agricultural market the demand equation for hogs, the first component equation of (20.3.1), does not depend on the other endogenous variables and does not depend on the predetermined variables; that is, the first row of Ξ consists of 0s and the first row of $\mathbf{A}$ has $p-2$ 0s. The structural equation is

$$\alpha_{11}(\text{hog price}) + \alpha_{12}(\text{hog quantity}) = u_{1t}. \tag{20.3.3}$$

The first row of $\boldsymbol{\Omega} = \mathbf{AB}$ can be written as

$$\boldsymbol{\xi}_{(1)} = (0,0,\ldots,0) = \boldsymbol{\alpha}_{(1)}\mathbf{B}$$

$$= (\alpha_{11},\alpha_{12},0\ldots0)\begin{bmatrix} \beta_{11} & \cdots & \beta_{1q} \\ \vdots & & \\ \beta_{p1} & \cdots & \beta_{pq} \end{bmatrix}$$

$$= (\alpha_{11},\alpha_{12})\begin{bmatrix} \beta_{11}\cdots\beta_{1q} \\ \beta_{21}\cdots\beta_{2q} \end{bmatrix}. \tag{20.3.4}$$

To solve (20.3.4) for $(\alpha_{11},\alpha_{12})$ we need

$$\text{rank}\begin{bmatrix} \beta_{11},\ldots,\beta_{1q} \\ \beta_{21},\ldots,\beta_{2q} \end{bmatrix} = 1. \tag{20.3.5}$$

To use the sample equivalent of (20.3.4) for estimation of $(\alpha_{11},\alpha_{12})$ we want the estimator of

$$\begin{bmatrix} \beta_{11},\ldots,\beta_{1q} \\ \beta_{21},\ldots,\beta_{2q} \end{bmatrix} \tag{20.3.6}$$

to be of rank 1, say

$$\begin{bmatrix} \bar{\beta}_{11},\ldots,\bar{\beta}_{1q} \\ \bar{\beta}_{21},\ldots,\bar{\beta}_{2q} \end{bmatrix}. \tag{20.3.7}$$

Then the estimator of $(\alpha_{11}, \alpha_{12})$ is the solution to

$$(\bar{\alpha}_{11}, \bar{\alpha}_{12}) \begin{bmatrix} \bar{\beta}_{11}, \dots, \bar{\beta}_{1q} \\ \bar{\beta}_{21}, \dots, \bar{\beta}_{2q} \end{bmatrix} = 0. \qquad (20.3.8)$$

The solution $(\bar{\alpha}_{11}, \bar{\alpha}_{12})$ is the Limited Information Maximum Likelihood (LIML) Estimator [Anderson-Rubin (1949)]. It is the maximum likelihood estimator using only the information in (20.3.4) when the $\mathbf{U}_t$s are normally distributed. The estimator of the matrix (20.3.6) is a special case of the reduced rank regression estimator, which will be explained below.

20.4 Canonical Analysis and the Reduced Rank Regression Estimator

It is enlightening to put the reduced rank regression estimator in terms of a canonical analysis based on the regression model (20.2.1). We can write an analysis of variance table for the random vector $\mathbf{Y}_t$, the sample vector $\mathbf{Y}_t$, and a linear combination of the sample vector $\phi'\mathbf{Y}_t$.

Analysis of Variance

	R.V.	Sample	$\phi'\mathbf{Y}_t$
Effect	$\mathbf{B}\Sigma_{XX}\mathbf{B}'$	$\hat{\mathbf{B}}\mathbf{S}_{XX}\hat{\mathbf{B}}'$	$\phi'\mathbf{B}\Sigma_{XX}\mathbf{B}'\phi$
Error	Σ_{ZZ}	$\mathbf{S}_{\hat{Z}\hat{Z}}$	$\phi'\Sigma_{ZZ}\phi$
Total	Σ_{YY}	$\mathbf{S}_{YY}$	$\phi'\Sigma_{YY}\phi$

The sum of an effect variance and the corresponding error variance is the total variance. The effect variance can be interpreted as the signal variance and the error variance as the noise variance. Then

$$\frac{\phi'\mathbf{B}\Sigma_{XX}\mathbf{B}'\phi}{\phi'\Sigma_{\hat{Z}\hat{Z}}\phi} = \frac{\text{signal}}{\text{noise}} \text{ of } \phi'\mathbf{Y}_t. \qquad (20.4.1)$$

We may ask for those linear combinations with the largest signal-to-noise ratios; that is,

$$\max_{\phi} \frac{\phi'\mathbf{B}\Sigma_{XX}\mathbf{B}'\phi}{\phi'\Sigma_{ZZ}\phi}, \qquad \phi'\Sigma_{\hat{Z}\hat{Z}}\phi = 1. \qquad (20.4.2)$$

The maximizing ϕ satisfies

$$\mathbf{B}\Sigma_{XX}\mathbf{B}'\phi = \theta\Sigma_{ZZ}\phi, \qquad (20.4.3)$$

where θ satisfies

$$|\mathbf{B}\Sigma_{XX}\mathbf{B}' - \theta\Sigma_{ZZ}| = 0. \qquad (20.4.4)$$

Let the solutions to (20.4.4) be

$$\theta_1 \geq \theta_2 \geq \cdots \geq \theta_p \geq 0 \tag{20.4.5}$$

and the corresponding solutions to (20.4.3) and $\phi'\Sigma_{ZZ}\phi = 1$ be

$$\phi_1, \phi_2, \ldots, \phi_p, \qquad \phi_{ii} > 0. \tag{20.4.6}$$

Then

$$\phi_i \mathbf{B} \Sigma_{XX} \mathbf{B}' \phi_j = \phi_i' \Sigma_{ZZ} \phi_j \theta_j = \theta_i, \qquad i = j,$$
$$= 0, \qquad i \neq j. \tag{20.4.7}$$

The rank of $\mathbf{B}$ is k if and only if $\theta_k > 0$ and

$$\theta_{k+1} = \cdots = \theta_p = 0. \tag{20.4.8}$$

In the sample a similar algebra can be carried out. The sample signal-to-noise ratio is maximized by $\mathbf{f}'\mathbf{Y}_t$ for $\mathbf{f}$ satisfying

$$\hat{\mathbf{B}} \mathbf{S}_{XX} \hat{\mathbf{B}}' \mathbf{f} = t \mathbf{S}_{\hat{Z}\hat{Z}} \mathbf{f}, \qquad \mathbf{f}' \mathbf{S}_{\hat{Z}\hat{Z}} \mathbf{f} = 1, \tag{20.4.9}$$

for t satisfying

$$|\hat{\mathbf{B}} \mathbf{S}_{XX} \hat{\mathbf{B}}' - t \mathbf{S}_{\hat{Z}\hat{Z}}| = 0. \tag{20.4.10}$$

Let the solutions to (20.4.11) be

$$t_1 > t_2 > \cdots > t_p \qquad (p + q \leq N), \tag{20.4.11}$$

$$\mathbf{f}_1, \mathbf{f}_2, \ldots, \mathbf{f}_p, \qquad f_{ii} > 0. \tag{20.4.12}$$

The Reduced Rank Regression estimator of $\mathbf{B}$ for prescribed rank k is

$$\hat{\mathbf{B}}_k = \mathbf{S}_{\hat{Z}\hat{Z}} \mathbf{F}_1 \mathbf{F}_1' \hat{\mathbf{B}}, \tag{20.4.13}$$

where

$$\mathbf{F}_1 = (\mathbf{f}_1, \ldots, \mathbf{f}_k) \tag{20.4.14}$$

Anderson (1951). The estimator $\hat{\mathbf{B}}_k$ is a maximum likelihood estimator if the disturbances are normally distributed.

The equation (20.4.10) can be transformed to

$$|\mathbf{S}_{XY} \mathbf{S}_{YY}^{-1} \mathbf{S}_{YX} - r^2 \mathbf{S}_{XX}| = 0, \tag{20.4.15}$$

where $r^2 = t/(1+t)$, and (20.4.9) can be transformed to

$$\mathbf{S}_{XY} \mathbf{S}_{YY}^{-1} \mathbf{S}_{YX} \hat{\gamma} = r^2 \mathbf{S}_{XX} \hat{\gamma}. \tag{20.4.16}$$

Then the reduced rank regression estimator of $\mathbf{B}$ can be written as

$$\hat{\mathbf{B}}_k = \mathbf{S}_{YX}\hat{\mathbf{\Gamma}}_1\hat{\mathbf{\Gamma}}_1', \tag{20.4.17}$$

where $\hat{\mathbf{\Gamma}}_1 = (\hat{\gamma}_1, \ldots, \hat{\gamma}_k)$ and $\hat{\gamma}_i$ is normalized as $\hat{\gamma}_i\mathbf{S}_{XX}\hat{\gamma}_i = 1$ for $i = 1, \ldots, k$ $(r_1^2 > \cdots > r_p^2)$.

If the rank of $\mathbf{B}$ is k, then

$$\phi_j'\mathbf{B} = \mathbf{0}, \qquad j = k+1, \ldots, p. \tag{20.4.18}$$

In particular, if $k = p - 1$, $\phi_p'\mathbf{B} = \mathbf{0}$, and the maximum likelihood estimator of ϕ_p is $\mathbf{f}_p$ (or a multiple of it). The LIML estimator of a structural equation is $\mathbf{f}_p$ when the relevant part of the reduced form is (20.2.1).

The likelihood ratio criterion for testing the hypothesis that rank $\mathbf{B}$ is k is

$$2 \log \lambda = N \sum_{j=k+1}^{p} \log(1 + t_j) \sim N \sum_{j=k+1}^{p} t_j \tag{20.4.19}$$

[Anderson (1951)]. If the null hypothesis is true, $2 \log \lambda \xrightarrow{d} \chi^2_{(p-k)(q-k)}$.

20.5 Autoregressive Processes

Many sets of economic variables constitute autoregressive processes. In (20.2.1) replace $\mathbf{X}_t$ by $\mathbf{Y}_{t-1}$ to obtain

$$\mathbf{Y}_t = \mathbf{B}\mathbf{Y}_{t-1} + \mathbf{Z}_t. \tag{20.5.1}$$

We have suppressed a constant term (intercept) in (20.5.1) as well as exogenous variables. The model (20.5.1) defines a stationary process if the roots λ_i, $i = 1, \ldots, p$, of

$$|\mathbf{B} - \lambda\mathbf{I}| = 0 \tag{20.5.2}$$

satisfy $|\lambda_i| < 1$. Then

$$\mathbf{Y}_t = \sum_{s=0}^{\infty} \mathbf{B}^s \mathbf{Z}_{t-s}, \tag{20.5.3}$$

and the covariance matrix of $\mathbf{Y}_t$ is

$$\mathbf{\Sigma}_{YY} = \sum_{s=0}^{\infty} \mathbf{B}^s \mathbf{\Sigma}_{ZZ} \mathbf{B}'^s. \tag{20.5.4}$$

Given a sample $\mathbf{Y}_0, \mathbf{Y}_1, \ldots, \mathbf{Y}_T$, the least squares estimator of $\mathbf{B}$ is (20.2.6) with $\mathbf{S}_{YX}$ and $\mathbf{S}_{XX}$ replaced by $T^{-1} \sum_{t=1}^{T} \mathbf{Y}_t \mathbf{Y}'_{t-1}$ and $T^{-1} \sum_{t=1}^{T} \mathbf{Y}_{t-1} \mathbf{Y}'_{t-1}$, respectively. The residuals and the estimator of $\mathbf{\Sigma}_{ZZ}$ are given by (20.2.7) and (20.2.8) with $\mathbf{X}_t$ replaced by $\mathbf{Y}_{t-1}$. The canonical analysis and reduced rank regression estimator $\hat{\mathbf{B}}_k$ are as described in Section 20.4 with the obvious changes.

20.6   Nonstationary Models

A particular nonstationary model occurs when all the roots of $|\mathbf{B} - \lambda \mathbf{I}| = 0$ are equal to 1. This is the case when $\mathbf{B} = \mathbf{I}$. Then

$$\mathbf{Y}_t = \mathbf{Y}_{t-1} + \mathbf{Z}_t, \qquad t = 1, 2, \ldots. \qquad (20.6.1)$$

The initial vector $\mathbf{Y}_0$ can be random or nonstochastic. For convenience we assume that $\mathbf{Y}_0 = \mathbf{0}$. Then

$$\mathbf{Y}_t = \sum_{s=0}^{t-1} \mathbf{Z}_{t-s}, \qquad t = 1, 2, \ldots, \qquad (20.6.2)$$

$$\mathcal{E} \mathbf{Y}_t \mathbf{Y}'_t = \mathcal{E} \sum_{r,s=0}^{t-1} \mathbf{Z}_{t-r} \mathbf{Z}'_{t-s} = t \mathbf{\Sigma}_{ZZ}. \qquad (20.6.3)$$

The model (20.6.1) defines a random walk. In another terminology we say the $\{\mathbf{Y}_t\}$ is integrated of order 1, denoted $\{\mathbf{Y}_t\} \in I(1)$. The sequence $\{\mathbf{Z}_t\}$ is integrated of order 0, denoted $\{\mathbf{Z}_t\} \in I(0)$. Note that the first difference of $\mathbf{Y}_t$

$$\Delta \mathbf{Y}_t = \mathbf{Y}_t - \mathbf{Y}_{t-1} = \mathbf{Z}_t; \qquad (20.6.4)$$

hence $\{\Delta \mathbf{Y}_t\} \in I(0)$.

Differencing has been used in time series analysis for another model – a model with a linear trend

$$\mathbf{Y}_t = \boldsymbol{\alpha} + \boldsymbol{\beta} t + \mathbf{Z}_t; \qquad (20.6.5)$$

that is, $\mathcal{E} \mathbf{Y}_t = \boldsymbol{\alpha} + \boldsymbol{\beta} t$. Here

$$\mathcal{E} \Delta \mathbf{Y}_t = \boldsymbol{\alpha} + \boldsymbol{\beta} t - [\boldsymbol{\alpha} + \boldsymbol{\beta}(t-1)]$$

$$= \boldsymbol{\beta}. \qquad (20.6.6)$$

Since $\Delta\mathbf{Y}_t$ has no trend, the differencing operation can be considered as a detrending device. This is an important feature of the Box-Jenkins time series analysis packages. The "error-correction form" of (20.5.1) is

$$\Delta\mathbf{Y}_t = \mathbf{\Pi}\mathbf{Y}_{t-1} + \mathbf{Z}_t, \tag{20.6.7}$$

where

$$\mathbf{\Pi} = \mathbf{B} - \mathbf{I} \tag{20.6.8}$$

In the pure $I(1)$ model $\mathbf{\Pi} = \mathbf{0}$.

20.7 Cointegrated Models

Granger (1981) suggested that although many economic time series appear as $I(1)$, linear combinations of these variables may be $I(0)$; that is, some relations between variables are stationary. He called models for such variables "cointegrated."

Suppose $\{\mathbf{Y}_t\} \in I(1)$. Consider a linear combination

$$\boldsymbol{\omega}'\mathbf{Y}_t = \boldsymbol{\omega}'\mathbf{B}\mathbf{Y}_{t-1} + \boldsymbol{\omega}'\mathbf{Z}_t. \tag{20.7.1}$$

If $\boldsymbol{\omega}'\mathbf{B} = \boldsymbol{\omega}'$, then

$$\boldsymbol{\omega}'\mathbf{Y}_t = \boldsymbol{\omega}'\mathbf{Y}_{t-1} + \boldsymbol{\omega}'\mathbf{Z}_t \in I(1). \tag{20.7.2}$$

If $\boldsymbol{\omega}'\mathbf{B} = \lambda\boldsymbol{\omega}, \ |\lambda| < 1$, then

$$\boldsymbol{\omega}'\mathbf{Y}_t = \lambda\boldsymbol{\omega}'\mathbf{Y}_{t-1} + \boldsymbol{\omega}'\mathbf{Z}_t \in I(0). \tag{20.7.3}$$

Let

$$\boldsymbol{\omega}_i'\mathbf{B} = \lambda_i\boldsymbol{\omega}_i'; \tag{20.7.4}$$

that is

$$\boldsymbol{\omega}_i'(\mathbf{B} - \lambda_i\mathbf{I}) = \mathbf{0}, \tag{20.7.5}$$

where λ_i is a root of $|\mathbf{B} - \lambda\mathbf{I}| = 0$. Suppose $|\lambda_i| < 1, \ i = 1, \ldots, k, \ \lambda_i = 1, \ i = k+1, \ldots, p$. We can write $|\mathbf{B} - \lambda\mathbf{I}| = 0$ as

$$0 = |\mathbf{B} - \mathbf{I} - (\lambda - 1)\mathbf{I}| = |\mathbf{\Pi} - (\lambda - 1)\mathbf{I}|. \tag{20.7.6}$$

Thus 0 is a characteristic root of $\mathbf{\Pi} = \mathbf{B} - \mathbf{I}$ of multiplicity $n = p - k$ and there are k roots of $\mathbf{\Pi}$ different from 0. The rank of $\mathbf{\Pi}$ is k.

The model (20.5.1) in the error-correction form (20.6.7) is of the form (20.2.1) with $\mathbf{Y}_t$ replaced by $\Delta\mathbf{Y}_t$, $\mathbf{B}$ replaced by $\mathbf{\Pi}$, and $\mathbf{X}_t$ replaced by $\mathbf{Y}_{t-1}$. The least squares estimators of $\mathbf{\Pi}$ and $\mathbf{\Sigma}_{ZZ}$ are

$$\hat{\mathbf{\Pi}} = \frac{1}{T}\sum_{t=1}^{T}\Delta\mathbf{Y}_t\mathbf{Y}_{t-1}'\left(\frac{1}{T}\sum_{t=1}^{T}\mathbf{Y}_{t-1}\mathbf{Y}_{t-1}'\right)^{-1}, \tag{20.7.7}$$

$$\mathbf{S}_{\hat{Z}\hat{Z}} = \frac{1}{T}\sum_{t-1}^{T}\hat{\mathbf{Z}}_t\hat{\mathbf{Z}}_t' = \frac{1}{T}\sum_{t=1}^{T}\Delta\mathbf{Y}_t\Delta\mathbf{Y}_t' - \hat{\mathbf{\Pi}}\frac{1}{T}\sum_{t=1}^{T}\mathbf{Y}_{t-1}\mathbf{Y}_{t-1}'\hat{\mathbf{\Pi}}', \quad (20.7.8)$$

respectively, where $\hat{\mathbf{Z}}_t = \mathbf{Y}_t - \hat{\mathbf{B}}\mathbf{Y}_{t-1} = \Delta\mathbf{Y}_t - \hat{\mathbf{\Pi}}\mathbf{Y}_{t-1}$, $t = 1, \dots, T$. The reduced rank regression estimator of $\mathbf{\Pi}$ of rank k is

$$\hat{\mathbf{\Pi}}_k = \mathbf{S}_{\hat{Z}\hat{Z}}\mathbf{F}_1\mathbf{F}_1'\hat{\mathbf{\Pi}} = \frac{1}{T}\sum_{t=1}^{T}\Delta\mathbf{Y}_t\mathbf{Y}_{t-1}'\hat{\mathbf{\Gamma}}_1\hat{\mathbf{\Gamma}}_1', \quad (20.7.9)$$

where $\mathbf{F}_1$ is defined by (20.4.14) and $\mathbf{f}_1, \dots, \mathbf{f}_k$ are the solutions to (20.4.9) with $\mathbf{X}_t$ replaced by $\mathbf{Y}_{t-1}$ and $\mathbf{B}$ by $\mathbf{\Pi}$ and $\hat{\mathbf{\Gamma}}_1 = (\hat{\boldsymbol{\gamma}}_1, \dots, \hat{\boldsymbol{\gamma}}_k)$ and $\hat{\boldsymbol{\gamma}}_1, \dots, \hat{\boldsymbol{\gamma}}_k$ are the solutions to (20.4.16) with $\mathbf{Y}_t$ replaced by $\Delta\mathbf{Y}_t$ and $\mathbf{X}_t$ by $\mathbf{Y}_{t-1}$ [Johansen (1988)]. Note that because we number the roots in descending order, the n smaller roots of (20.4.10) estimate the n characteristic roots of $\mathbf{\Pi}$ of 0, which correspond to the n characteristic roots of $\mathbf{B}$ of 1.

The likelihood ratio criterion to test the null hypothesis that the rank of $\mathbf{\Pi}$ is k is (20.4.9).

20.8   Asymptotic Distribution of Estimators and Test Criterion

The behavior of the statistics in the cointegrated model is different from that in the stationary model because of the random walk aspect. To study the behavior we want to distinguish between the stationary and nonstationary dimensions. Let

$$\mathbf{\Omega}_1 = (\boldsymbol{\omega}_1, \dots, \boldsymbol{\omega}_k), \qquad \mathbf{\Omega}_2 = (\boldsymbol{\omega}_{k+1}, \dots, \boldsymbol{\omega}_p), \quad (20.8.1)$$

$\mathbf{\Omega} = (\mathbf{\Omega}_1, \mathbf{\Omega}_2)$. The vectors $\boldsymbol{\omega}_{k+1}, \dots, \boldsymbol{\omega}_p$ correspond to the characteristic roots 0 of $\mathbf{\Pi}$ (and 1 of $\mathbf{B}$). Then

$$\mathbf{\Omega}_1'\mathbf{\Pi} = (\mathbf{\Lambda}_1 - \mathbf{I})\mathbf{\Omega}_1', \qquad \mathbf{\Omega}_2'\mathbf{\Pi} = \mathbf{0}. \quad (20.8.2)$$

Define

$$\mathbf{X}_{1t} = \mathbf{\Omega}_1'\mathbf{Y}_t, \qquad \mathbf{W}_{1t} = \mathbf{\Omega}_1'\mathbf{Z}_t, \quad (20.8.3)$$

$$\mathbf{X}_{2t} = \mathbf{\Omega}_2'\mathbf{Y}_t, \qquad \mathbf{W}_{2t} = \mathbf{\Omega}_2'\mathbf{Z}_t, \quad (20.8.4)$$

$\mathbf{X}_t = (\mathbf{X}_{1t}', \mathbf{X}_{2t}')'$, $\mathbf{W}_t = (\mathbf{W}_{1t}', \mathbf{W}_{2t}')'$. Then

$$\mathbf{X}_t = \mathbf{\Psi}\mathbf{X}_{t-1} + \mathbf{W}_t, \quad (20.8.5)$$

$$\Delta\mathbf{X}_t = \mathbf{\Upsilon}\mathbf{X}_{t-1} + \mathbf{W}_t, \quad (20.8.6)$$

where $\boldsymbol{\Psi} = \boldsymbol{\Omega}'\mathbf{B}\boldsymbol{\Omega}'^{-1}$ and $\boldsymbol{\Upsilon} = \boldsymbol{\Omega}'\boldsymbol{\Pi}\boldsymbol{\Omega}'^{-1} = \boldsymbol{\Psi} - \mathbf{I}$.
Define

$$\mathbf{S}_{\Delta\Delta} = \frac{1}{T}\sum_{t=1}^{T}\Delta\mathbf{X}_t\Delta\mathbf{X}_t', \tag{20.8.7}$$

$$\mathbf{S}_{\Delta-} = \frac{1}{T}\sum_{t=1}^{T}\Delta\mathbf{X}_t\mathbf{X}_{t-1}, \tag{20.8.8}$$

$$\mathbf{S}_{--} = \frac{1}{T}\sum_{t=1}^{T}\mathbf{X}_{t-1}\mathbf{X}_{t-1}'. \tag{20.8.9}$$

The least squares estimator of $\boldsymbol{\Upsilon}$ is $\hat{\boldsymbol{\Upsilon}} = \mathbf{S}_{\Delta-}\mathbf{S}_{--}^{-1}$ and of $\boldsymbol{\Psi}$ is $\hat{\boldsymbol{\Psi}} = \hat{\boldsymbol{\Upsilon}} + \mathbf{I}$. The reduced rank regression estimator of $\boldsymbol{\Upsilon}$ is $\hat{\boldsymbol{\Upsilon}}_k = \mathbf{S}_{\Delta-}\mathbf{G}_1\mathbf{G}_1'$, where $\mathbf{G}_1 = (\mathbf{g}_1, \ldots, \mathbf{g}_k)$ and $\mathbf{g}_j$ is a solution to

$$\mathbf{S}_{-\Delta}\mathbf{S}_{\Delta\Delta}^{-1}\mathbf{S}_{\Delta-}\mathbf{g} = r^2\mathbf{S}_{--}\mathbf{g}, \qquad \mathbf{g}'\mathbf{S}_{--}\mathbf{g} = 1. \tag{20.8.10}$$

and r^2 is a solution to

$$|\mathbf{S}_{-\Delta}\mathbf{S}_{\Delta\Delta}^{-1}\mathbf{S}_{\Delta-} - r^2\mathbf{S}_{--}| = 0. \tag{20.8.11}$$

To describe the asymptotic distributions of the estimators we introduce the vector Brownian motion process. Let $\mathbf{V}_1, \mathbf{V}_2, \ldots$ be independently identically distributed with $\mathcal{E}\mathbf{V}_t = \mathbf{0}$, $\mathcal{E}\mathbf{V}_t\mathbf{V}_t' = \boldsymbol{\Sigma}$. For $0 \le u \le 1$ consider

$$\frac{1}{\sqrt{T}}\sum_{t=1}^{[Tu]}\mathbf{V}_t = \sqrt{u}\frac{1}{\sqrt{Tu}}\sum_{t=1}^{[Tu]}\mathbf{V}_t, \tag{20.8.12}$$

which converges in distribution to $N(\mathbf{0}, u\boldsymbol{\Sigma})$ (Central Limit Theorem). ($[Tu]$ denotes the integer part of Tu.) Then for $0 \le w \le u \le 1$ $(1/\sqrt{T})\sum_{t=1}^{[Tu]}\mathbf{V}_t - (1/\sqrt{T})\sum_{t=1}^{[Tw]}\mathbf{V}_t$ converges in distribution to $N(\mathbf{0}, (u-w)\boldsymbol{\Sigma})$ and is asymptotically independent of (20.8.12). The sequences satisfy a "tightness condition." We write

$$\frac{1}{\sqrt{T}}\sum_{t=1}^{[Tu]}\mathbf{V}_t \overset{w}{\to} \mathbf{V}(u), \qquad 0 \le u \le 1. \tag{20.8.13}$$

(The sequence $\{T^{-\frac{1}{2}}\mathbf{V}_t\}$ converges weakly to $\{\mathbf{V}(u)\}$.) If $\boldsymbol{\Sigma} = \mathbf{I}$, we term $\mathbf{V}(u)$ the standard Brownian motion. For more detail see Johansen (1995), Appendix B.7, for example.

Consider the least squares estimator of $\boldsymbol{\Psi}$. Let $\boldsymbol{\Psi} = (\boldsymbol{\Psi}_1, \boldsymbol{\Psi}_2)$, where $\boldsymbol{\Psi}_1$ has k columns, and similarly partition $\hat{\boldsymbol{\Psi}}, \boldsymbol{\Upsilon}$, and $\hat{\boldsymbol{\Upsilon}}$. Then

$$T(\hat{\boldsymbol{\Psi}}_2 - \boldsymbol{\Psi}_2)$$

$$= T(\hat{\boldsymbol{\Upsilon}}_2 - \boldsymbol{\Upsilon}_2) \xrightarrow{d} \int_0^1 d\mathbf{W}(u)\mathbf{W}_2'(u) \left[\int_0^1 \mathbf{W}_2(u)\mathbf{W}_2'(u)\,du\right]^{-1},$$

$$(20.8.14)$$

where $\mathbf{W}(u) = [\mathbf{W}_1'(u), \mathbf{W}_2'(u)]'$ is the Brownian motion process for $\boldsymbol{\Sigma} = \boldsymbol{\Sigma}_{WW}$. Also $\sqrt{T}(\hat{\boldsymbol{\Psi}}_1 - \boldsymbol{\Psi}_1) = \sqrt{T}(\hat{\boldsymbol{\Upsilon}}_1 - \boldsymbol{\Upsilon}_1)$ has a limiting normal distribution with the covariance of the limiting distribution of $\sqrt{T}(\hat{\psi}_{ij} - \psi_{ij})$ and $\sqrt{T}(\hat{\psi}_{kl} - \psi_{kl})$ being $\sigma_{ik}^w[(\boldsymbol{\Sigma}_{--}^{22})^{-1}]_{jl}$. Note that the elements of $\hat{\boldsymbol{\Psi}} - \boldsymbol{\Psi} = \hat{\boldsymbol{\Upsilon}} - \boldsymbol{\Upsilon}$ that involves the stationary part of $\mathbf{X}(t)$ are normalized by $\sqrt{T}$ and have the usual kind of asymptotic normal distribution, but the part corresponding to the nonstationary part is normalized by T; $\hat{\boldsymbol{\Psi}}_2$ is a superefficient estimator. The effect of the covariance of $\mathbf{X}_{2t}$ growing is to make the estimator more accurate.

The least squares estimators of $\mathbf{B}$ and $\boldsymbol{\Pi}$ are $\hat{\mathbf{B}} = (\boldsymbol{\Omega}')^{-1}\hat{\boldsymbol{\Psi}}\boldsymbol{\Omega}'$ and $\hat{\boldsymbol{\Pi}} = (\boldsymbol{\Omega}')^{-1}\boldsymbol{\Upsilon}\boldsymbol{\Omega}'$, respectively. The asymptotic behavior of $\hat{\mathbf{B}}$ and $\hat{\boldsymbol{\Pi}}$ can be derived from that of $\hat{\boldsymbol{\Psi}}$ and $\hat{\boldsymbol{\Upsilon}}$.

If the rank of $\boldsymbol{\Upsilon}$ is k, then $r_i^2 \xrightarrow{p} 0$, $i = k+1, \ldots, p$. In fact $d_{k+1} = Tr_{k+1}^2, \ldots, d_p = Tr_p^2$ have a limiting distribution, which is the distribution of the $p - k$ roots of

$$\left|\left[\int_0^1 \mathbf{B}_2(u)\,d\mathbf{B}_2'(u)\int_0^1 d\mathbf{B}_2(v)\mathbf{B}_2'(v) - d\int_0^1 \mathbf{B}_2(u)\mathbf{B}_2'(u)\,du\right]\right| = 0$$

$$(20.8.15)$$

[Johansen (1988)]. Then the limiting distribution of the test criterion $-2\log\lambda$ is the distribution of

$$\text{tr}\int_0^1 d\mathbf{B}_2(v)\mathbf{B}_2'(v)\left[\int_0^1 \mathbf{B}_2(u)\mathbf{B}_2'(u)\,du\right]^{-1}\int_0^1 \mathbf{B}_2(u)\,d\mathbf{B}_2'(u). \quad (20.8.16)$$

This distribution may be compared to the limiting $\chi_{(p-k)^2}^2$ for the stationary process.

The asymptotic behavior of the larger roots of (20.8.11) depends mainly on the stationary part of $\mathbf{X}_t$, which satisfies

$$\Delta\mathbf{X}_{1t} = \boldsymbol{\Upsilon}_{11}\mathbf{X}_{1,t-1} + \mathbf{W}_{1t}. \quad (20.8.17)$$

The k larger roots of (20.8.11) converge in probability to the roots of

$$|\boldsymbol{\Upsilon}_{11}\boldsymbol{\Sigma}_{--}^{11}\boldsymbol{\Upsilon}_{11}' - \rho^2(\boldsymbol{\Upsilon}_{11}\boldsymbol{\Sigma}_{--}^{11}\boldsymbol{\Upsilon}_{11}' + \boldsymbol{\Sigma}_{WW}^{11.2})| = 0, \quad (20.8.18)$$

where $\boldsymbol{\Sigma}_{--}^{11} = \sum_{s=0}^{\infty} \boldsymbol{\Psi}_{11}^s \boldsymbol{\Sigma}_{WW}^{11} \boldsymbol{\Psi}_{11}'$ and

$$\boldsymbol{\Sigma}_{WW}^{11.2} = \boldsymbol{\Sigma}_{WW}^{11} - \boldsymbol{\Sigma}_{WW}^{12}(\boldsymbol{\Sigma}_{WW}^{22})^{-1}\boldsymbol{\Sigma}_{WW}^{21}. \quad (20.8.19)$$

Let the roots of (20.8.18) be $\rho_1^2, \ldots, \rho_k^2$. Then limiting distribution of $\sqrt{T}(s_1^2 - \rho_1^2), \ldots, \sqrt{T}(s_k^2 - \rho_k^2)$ has been given by Anderson (1999); it is too complicated to give here.

Let the solutions to

$$\boldsymbol{\Upsilon}_{11} \boldsymbol{\Sigma}_{--}^{11} \boldsymbol{\Upsilon}_{11} \boldsymbol{\gamma}_1 = \rho^2 (\boldsymbol{\Upsilon}_{11} \boldsymbol{\Sigma}_{--}^{11} \boldsymbol{\Upsilon}_{11} + \boldsymbol{\Sigma}_{WW}^{11.2}) \boldsymbol{\gamma}_1, \qquad \boldsymbol{\gamma}_1' \boldsymbol{\Sigma}_{--}^{11} \boldsymbol{\gamma}_1 = 1 \tag{20.8.20}$$

be $\boldsymbol{\gamma}_{11}, \ldots, \boldsymbol{\gamma}_{1k}$, and define $\boldsymbol{\Gamma}_{11} = (\boldsymbol{\gamma}_{11}, \ldots, \boldsymbol{\gamma}_{1k})$. Then the solution to (20.8.10) for $r^2 = r_i^2$ is an estimator of $(\boldsymbol{\gamma}_{1i}', \mathbf{0})'$, $i = 1, \ldots, k$.

To describe the statistical behavior of the characteristic vectors corresponding to the larger roots it will be convenient to normalize them differently. Note that the reduced rank regression estimator $\hat{\boldsymbol{\Upsilon}}_k = \mathbf{S}_{\Delta-} \mathbf{G}_1 \mathbf{G}_1'$ can be written equivalently as

$$\hat{\boldsymbol{\Upsilon}}_k = \mathbf{S}_{\Delta-} \mathbf{G}_1 \mathbf{G}_{11}' (\mathbf{G}_1 \mathbf{G}_{11}^{-1})', \tag{20.8.21}$$

where $\mathbf{G}_1 = (\mathbf{G}_{11}', \mathbf{G}_{21}')'$. Then $\mathbf{G}_1 \mathbf{G}_{11}^{-1}$ estimates $(\mathbf{I}, \mathbf{0})'$, and

$$T \mathbf{G}_{21} \mathbf{G}_{11}^{-1} \xrightarrow{d} \left[\int_0^1 \mathbf{W}_2(u) \mathbf{W}_2'(u)\, du \right]^{-1} \times \int_0^1 \mathbf{W}_2(u)\, d\mathbf{W}_{1.2}'(u), \tag{20.8.22}$$

where

$$\mathbf{W}(u) = \begin{bmatrix} \mathbf{W}_1(u) \\ \mathbf{W}_2(u) \end{bmatrix}, \quad \begin{matrix} k \\ p - k \end{matrix} \tag{20.8.23}$$

$$\mathbf{W}_{1.2}(u) = \mathbf{W}_1(u) - \boldsymbol{\Sigma}_{WW}^{12} (\boldsymbol{\Sigma}_{WW}^{22})^{-1} \mathbf{W}_2(u). \tag{20.8.24}$$

Note that this estimator is super-efficient; the error is of the order of $1/T$.

References

1. Anderson, T. W. (1951). Estimating linear restrictions on regression coefficients for multivariate normal distributions. *Annals of Mathematical Statistics* **22**, 327–351. [Correction, *Annals of Statistics* **8**, (1980), p. 1400.]

2. Anderson, T. W. (1999). Canonical analysis, reduced rank regression, and cointegration of time series.

3. Anderson, T. W. and Rubin, H. (1949). Estimation of the parameters of a single equation in a complete system of stochastic equations. *Annals of Mathematical Statistics* **20**, 46–63.

4. Box, G. E. P. and Jenkins, G. M. (1976). *Time Series: Forecasting and Control.* Holden-Day, San Francisco.

5. Dhrymes, P. (1988). *Time Series, Unit Roots, and Cointegration.* Academic Press, San Diego.

6. Granger, C. W. J. (1981). Some properties of time series data and their use in econometric model specification. *Journal of Econometrics* **16**, 121–130.

7. Johansen, S. (1988). Statistical analysis of cointegration vectors. *Journal of Economic Dynamics and Control* **12**, 231–254.

8. Johansen, S. (1995). *Likelihood-based Inference in Cointegrated Vector Autoregressive Models.* Oxford University Press, Oxford.

9. Koopmans, T. C. (1950). *Statistical Inference in Dynamic Economic Models.* John Wiley & Sons, New York.

21

On Some Power Properties of Goodness-of-Fit Tests in Time Series Analysis

Efstathios Paparoditis
University of Cyprus, Nicosia, Cyprus

ABSTRACT The power properties of a goodness-of-fit test for time series models are investigated. The test statistic is an L_2 functional of the smoothed difference between the sample spectral density (periodogram) rescaled by the hypothesized spectral density and the expected value of this ratio under the null hypothesis. Different classes of local alternatives are considered and comparisons of the test proposed to some known goodness-of-fit tests in time series analysis are made.

Keywords and phrases Goodness-of-fit, local alternatives, periodogram, spectral density, time series models

21.1 Testing Spectral Density Fits

Testing the fit of a time series model is an important diagnostic step in time series analysis. In this paper some power properties of general goodness-of-fit tests are investigated, i.e., of tests which can be applied when no a priori information is available about the kind of departure of the true correlation structure from that postulated under the null.

The class of stochastic processes $\{X_t, t \in \mathbf{Z}\}$ considered satisfies the following assumption.

(A1) $X_t = \sum_{j=-\infty}^{\infty} \psi_j \varepsilon_{t-j}$ where $\psi_0 = 1$, $\sum_{j=-\infty}^{\infty} |j|^{1/2+\beta} |\psi_j| < \infty$ for some $\beta > 0$ and $\{\varepsilon_t\}$ is a sequence of independent identically distributed random variables with mean zero, positive variance σ_ε^2 and $E\varepsilon_1^8 < \infty$.

A way to describe the autocorrelation structure of such a process is by means of its spectral density function f defined by

$$f(\lambda) = \sigma_\varepsilon^2 (2\pi)^{-1} | \sum_{j=-\infty}^{\infty} \psi_j e^{-i\lambda j}|^2 \qquad \text{for } \lambda \in [-\pi, \pi].$$

Let $\widetilde{\mathcal{F}}$ be the set of spectral densities of the class of stochastic processes described in (A1) and let $\mathcal{F} \subset \widetilde{\mathcal{F}}$ such that

$$\mathcal{F} = \Big\{ f : \ f \text{ is first order Lipschitz continuous and } \inf_{\lambda \in [0, \pi]} f(\lambda) > 0 \Big\}.$$

Assume that observations $X_1, X_2, \ldots, X_T$ of such a process are available and that a model f_0 from this class has been selected to describe the dependence structure of $\{X_t\}$. We are then interested in testing the hypothesis

$$H_0 : f = f_0 \quad \text{against} \quad H_1 : f \neq f_0.$$

In order to make the presentation self-contained we initially briefly discuss the idea underlying the test proposed by Paparoditis (1997) and summarize some basic results. Consider the space $(L_2[-\pi, \pi], < \cdot, \cdot >)$, with the definition $< f, g >= (2\pi)^{-1} \int_{-\pi}^{\pi} f(x) g(x) dx$ of the inner product. Denote by $I(\lambda)$ the sample spectral density (periodogram) given by

$$I(\lambda) = \frac{1}{2\pi T} \Big| \sum_{t=1}^{T} X_t e^{-i\lambda t} \Big|^2. \tag{21.1.1}$$

Usually the periodogram is calculated at the Fourier frequencies $\lambda_j = 2\pi j / T$, $j = -N, -N+1, \ldots, N-1, N$ where $N = [(T-1)/2]$ and $[\ \cdot\ ]$ denotes integer part.

Consider the rescaled periodogram statistic $Q_T(\cdot)$ given by

$$Q_T(\lambda) = \frac{I(\lambda)}{f_0(\lambda)} - 1.$$

It is well known (cf. Priestley (1981), p. 418) that if $f \in \mathcal{F}$ then $E[Q_T(\lambda)] = q(\lambda) + O(\log(T) T^{-1})$ uniformly in λ where

$$q(\cdot) := \frac{f(\cdot)}{f_0(\cdot)} - 1 \begin{cases} = 0 & \text{under } H_0 \\ \\ \neq 0 & \text{under } H_1. \end{cases} \tag{21.1.2}$$

The idea used to test the goodness-of-fit of the spectral density function f_0 is to estimate the function $q(\cdot)$ nonparametrically using the observable random variables $Q_T(\lambda_j)$, $j = -N, -N+1, \ldots, N$ and to compare the result obtained with the zero function, i.e., the (asymptotically) expected form of $q(\cdot)$ under H_0. To be more specific, denote by $\hat{q}_h(\cdot)$ the nonparametric (kernel) estimator of $q(\cdot)$ given by

$$\hat{q}_h(\lambda) = \frac{1}{T} \sum_{j=-N}^{N} K_h(\lambda - \lambda_j) Q_T(\lambda_j) \qquad (21.1.3)$$

where $K_h(\cdot) = h^{-1}K(\cdot/h)$. Here

(A2) $K(\cdot)$ is a bounded, symmetric and nonnegative kernel with compact support $[-\pi, \pi]$ satisfying $(2\pi)^{-1} \int K(x)dx = 1$.

Furthermore, we assume that

(A3) The bandwidth h satisfies $hT^\delta \to 1$ for some $0 < \delta < 1$.

The statistic used to test the null hypothesis that $f = f_0$ is then based on the L_2 norm of $\hat{q}_h(\cdot)$, i.e.,

$$S_{T,h} = Th^{1/2} \int_{-\pi}^{\pi} \left\{ \hat{q}_h(\lambda) \right\}^2 d\lambda. \qquad (21.1.4)$$

Note that $\hat{q}_h(\lambda)$ is a mean square consistent estimator of $q(\lambda)$ since under the assumption made and by standard arguments (cf. for instance, Priestley (1981)) we have that

$$\mathrm{E}[\hat{q}_h(\lambda)] = q(\lambda) + o(1)$$

and

$$\mathrm{Var}[\hat{q}_h(\lambda)] = O(T^{-1}h^{-1}).$$

Thus we expect under H_0 a small value of $\hat{q}_h(\lambda)$, i.e., a large value of $S_{T,h}$ will argue against the null hypothesis. More precisely, the following theorem has been established in Paparoditis (1997).

THEOREM 21.1.1
Assume (A1)-(A3) and let $\varepsilon_t \sim N(0, \sigma^2)$.

(i)   Under H_0 and as $T \to \infty$,

$$\mathcal{L}\left(\frac{S_{T,h} - \mu_h(K)}{\sigma(K)} \right) \to N(0, 1),$$

where

$$\mu_h(K) = 2\pi h^{-1/2} \|K\|^2, \qquad (21.1.5)$$

$\sigma(K) = \sqrt{\sigma^2(K)}$ and

$$\sigma^2(K) = \pi^{-1} \int_{-2\pi}^{2\pi} \left\{ \int_{-\pi}^{\pi} K(u)K(u+x)du \right\}^2 dx. \qquad (21.1.6)$$

(ii)   Under H_1 and if f is the true spectral density, then as $T \to \infty$,

$$T^{-1}h^{-1/2}S_{T,h} \to 2\pi \left\| \frac{f}{f_0} - 1 \right\|^2 \quad \text{in probability,}$$

where $\| \cdot \|$ denotes the L_2 norm in $L_2[-\pi, \pi]$.

Note that the asymptotic null distribution of $S_{T,h}$ does not depend on any unknown parameters or functionals of the process. Modified appropriately, the above test can be also applied to test the hypothesis that f belongs to a finite dimensional parametric class of spectral densities, i.e., that

$$H_0 : f \in \mathcal{F}_\Theta \quad \text{against} \quad H_1 : f \notin \mathcal{F}_\Theta$$

where

$$\mathcal{F}_\Theta = \{ f(\cdot, \boldsymbol{\theta}) \text{ and } \boldsymbol{\theta} \in \Theta \subset \mathbf{R}^\mathbf{p} \}.$$

Briefly, let $\hat{\boldsymbol{\theta}}$ be an estimator of $\boldsymbol{\theta}$. The test statistic used to test the composite hypothesis is given by

$$S_{T,h}(\hat{\boldsymbol{\theta}}) = Th^{1/2} \int_{-\pi}^{\pi} \left\{ \hat{q}_h(\lambda, \hat{\boldsymbol{\theta}}) \right\}^2 d\lambda \qquad (21.1.7)$$

where

$$\hat{q}_h(\lambda, \hat{\boldsymbol{\theta}}) = \frac{1}{T} \sum_{j=-N}^{N} K_h(\lambda - \lambda_j) \left(\frac{I(\lambda_j)}{f(\lambda_j, \hat{\boldsymbol{\theta}})} - 1 \right). \qquad (21.1.8)$$

It has been shown that if $\hat{\boldsymbol{\theta}}$ is a $\sqrt{T}$-consistent estimator of $\boldsymbol{\theta}$, like the maximum likelihood or the Whittle estimator, then under the null hypothesis $S_{T,h}(\hat{\boldsymbol{\theta}}) = S_{T,h}(\boldsymbol{\theta}_0) + o_P(1)$ where $\boldsymbol{\theta}_0 \in \Theta$ is the true parameter vector; cf. Paparoditis (1997). That is, the (asymptotic) null distribution of the test statistic $S_{T,h}(\hat{\boldsymbol{\theta}})$ is not affected if the unknown parameter vector $\boldsymbol{\theta}$ is replaced by a $\sqrt{T}$-consistent estimator, i.e., this distribution is identical to that of testing a simple hypothesis given in Theorem 21.1.1 (i). However, the limiting behavior of $S_{T,h}(\hat{\boldsymbol{\theta}})$ under the alternative depends on the particular method applied to obtain the parameter estimator $\hat{\boldsymbol{\theta}}$, i.e., in this case we have

$$T^{-1}h^{-1/2}S_{T,h}(\hat{\boldsymbol{\theta}}) \;\to\; 2\pi \left\| \frac{f(\cdot)}{f(\cdot,\boldsymbol{\theta}^\star)} - 1 \right\|^2 \quad \text{in probability}$$

as $T \to \infty$, where $\boldsymbol{\theta}^\star \in \boldsymbol{\Theta}$ is the minimizer of the distance function used to obtain the estimator $\hat{\boldsymbol{\theta}}$. To give an example, if $\hat{\boldsymbol{\theta}}$ is the Whittle estimator then

$$\boldsymbol{\theta}^\star \;=\; \arg\min_{\boldsymbol{\theta} \in \boldsymbol{\Theta}} \frac{1}{4\pi} \int_{-\pi}^{\pi} \left\{ f(\lambda,\boldsymbol{\theta}) + \frac{f(\lambda)}{f(\lambda,\boldsymbol{\theta})} \right\} d\lambda;$$

cf. Dzhaparidze (1986). In the following we confine ourselves to the case of testing a full specified hypothesis in order to simplify calculations and to discuss the power properties of the $S_{T,h}$ test. For more details on testing a composite hypothesis see Paparoditis (1997).

21.2   Local Power Considerations

As Theorem 21.1.1 shows the test based on $S_{T,h}$ is a consistent test, i.e., a test which has power against any alternative having spectral density $f \in \mathcal{F}$ such that $f \neq f_0$. Furthermore, the power of the test is a function of the L_2 distance between the true spectral density f and the hypothesized density f_0 and approaches unity as $T \to \infty$. Apart from these general properties, however, a more informative analysis of the power behavior of $S_{T,h}$ can be made by means of local power considerations. The idea is to consider a sequence of stochastic process denoted by $\{X_{T,t}, t \in \mathbf{Z}\}_{T \in \mathbf{N}}$ with a corresponding sequence of spectral densities $\{f_T(\cdot)\}_{T \in \mathbf{N}}$, $f_T \in \mathcal{F}$ that converge to the null density f_0 as the sample size T increases. The interest here is focused on the maximal rate at which f_T is allowed to converge to f_0 such that the power of the test is bounded away from the significance level and from unity.

To be more specific, let $\{X_{0,t}, t \in \mathbf{Z}\}$ be the process under the null hypothesis with spectral density f_0. Consider the sequence of stochastic processes $\{X_{T,t}, t \in \mathbf{Z}\}_{T \in \mathbf{N}}$ defined by

$$X_{T,t} \;=\; \sum_{j=-\infty}^{\infty} \psi_{T,j} X_{0,t-j} \tag{21.2.1}$$

where the sequence $\{\psi_{T,j}, j \in \mathbf{Z}\}_{T \in \mathbf{N}}$ satisfies

$$\sum_{j=-\infty}^{\infty} |j|^{1/2+\beta} |\psi_{T,j}| < C$$

and the constant C does not depend on T. Since the process $\{X_{T,t}\}$ is obtained by applying a linear filter to the process $\{X_{0,t}\}$ its spectral density is given by

$$f_T(\lambda) = f_0(\lambda) |\Psi_T(\lambda)|^2 \qquad (21.2.2)$$

where

$$|\Psi_T(\lambda)|^2 = \left| \sum_{j=-\infty}^{\infty} \psi_{T,j} \exp\{-i\lambda j\} \right|^2. \qquad (21.2.3)$$

In particular, if $\psi_{T,j} = \delta_{0j}$ with δ_{0j} Kronecker's delta, we have $X_{T,t} = X_{0,t}$ and, therefore, the case of the null hypothesis. Assume that the sequence $\{\psi_{T,j}, j \in \mathbf{Z}\}$ is such that

$$\left| \Psi_T(\lambda) \right|^2 = 1 + c_T \omega\left(\frac{\lambda - \lambda^\star}{b_T}\right) + o(c_T b_T) \qquad (21.2.4)$$

where c_T and b_T are real valued and positive sequences approaching zero as $T \to \infty$, $\lambda^\star$ is a fixed frequency and the $o(c_T b_T)$ term is uniformly in λ. Furthermore, $\omega(\cdot)$ is assumed to be a twice continuously differentiable and bounded function. Under these assumptions $\sup_\lambda ||\Psi_T(\lambda)|^2 - 1| \to 0$ as $T \to \infty$, the spectral density $f_T(\cdot)$ of the process $\{X_{T,t}\}$ is given by

$$f_T(\lambda) = f_0(\lambda)\left[1 + c_T \omega\left(\frac{\lambda - \lambda^\star}{b_T}\right)\right] + o(c_T b_T) \qquad (21.2.5)$$

and

$$\sup_{\lambda \in [-\pi,\pi]} \left| \frac{f_T(\lambda)}{f_0(\lambda)} - 1 \right| \to 0 \qquad \text{as } T \to \infty.$$

Using this approach different classes of local alternatives can be considered. For instance if we choose $b_T \equiv 1$ we are in the case of the so-called Pitman alternatives where the local alternatives approach $f_0(\cdot)$ at the rate c_T. Depending on the particular form of $\omega(\cdot)$ these departures can be thought as being of more global nature in the sense that they not become centered around any particular frequency as the sample increases. On the contrary, if the sequence b_T is also allowed to approach zero as $T \to \infty$ then the departure of f_T from the hypothesized spectral density f_0 becomes more and more centered around the frequency $\lambda^\star$ as T increases. This is the class of sharp peak local alternatives introduced by Rosenblatt (1975) in the context of density testing based on a sample of i.i.d. data; see also Ghosh and Huang (1991). The following theorem characterizes the behavior of the $S_{T,h}$ test statistic for the class of local alternatives given in (21.2.5).

THEOREM 21.2.1

Assume that the true spectral density f_T satisfies (21.2.5) and that the assumptions of Theorem 21.1.1 are fulfilled.

(i)   *If $b_T \equiv 1$ and $c_T = T^{-1/2+\delta/4}$, then as $T \to \infty$,*

$$\mathcal{L}\left(\frac{S_{T,h} - \mu_h(K)}{\sigma(K)}\right) \;\Rightarrow\; N\left(\frac{\nu}{\sigma(K)},\, 1\right),$$

where $\nu = 2\pi\|\omega\|^2$ and $\mu_h(K)$ and $\sigma^2(K)$ are given in (21.1.5) and (21.1.6) respectively.

(ii)   *If $c_T = T^{-\eta}$ and $b_T = T^{-\gamma}$ for some $\eta > 0$ and $\gamma > 0$ such that*

$$\delta/2 \;=\; 1 - 2\eta - \gamma \qquad \text{and} \qquad \gamma < \delta,$$

then the conclusion of Part (i) of the theorem is true.

PROOF   Note first that by definition we have that $X_{T,j} = \sum\limits_{j=-\infty}^{\infty} a_{T,j}\varepsilon_{t-j}$

where $\{a_{T,j}, j \in \mathbf{Z}\}$ is an absolute summable sequence. By Theorem 10.3.1 of Brockwell and Davis (1991) we then have for $j = 1, 2, \ldots, N$ that $I_T(\lambda_j) = f_T(\lambda_j)U_j + R_T(\lambda_j)$ where $I_T(\cdot)$ denotes the periodogram of a realization of length T of the process $\{X_{t,T}\}$, U_j is a sequence of independent standard exponential distributed random variables and the remainder satisfies $\max_{\lambda_j \in [0,\pi]} E|R_T(\lambda_j)|^2 = O(T^{-1})$. Substituting this expression for $I_T(\lambda_j)$ in $S_{T,h}$ we get after some algebra that

$$S_{T,h} = T^{-1}h^{1/2}\int_{-\pi}^{\pi}\left\{\sum_{j=-N}^{N} K_h(\lambda - \lambda_j)\frac{f_T(\lambda_j)}{f_0(\lambda_j)}(U_j - 1)\right\}^2 d\lambda$$

$$+ T^{-1}h^{1/2}\int_{-\pi}^{\pi}\left\{\sum_{j=-N}^{N} K_h(\lambda - \lambda_j)\left(\frac{f_T(\lambda_j)}{f_0(\lambda_j)} - 1\right)\right\}^2 d\lambda \;+\; o_P(1)$$

$$= L_{1,T} \;+\; L_{2,T} \;+\; o_P(1)$$

with an obvious notation fot $L_{1,T}$ and $L_{2,T}$. Since $\sup_\lambda |f_T(\lambda)/f_0(\lambda) - 1| \to 0$ we get that

$$L_{1,T} \;=\; T^{-1}h^{1/2}\int_{-\pi}^{\pi}\left\{\sum_{j=-N}^{N} K_h(\lambda - \lambda_j)(U_j - 1)\right\}^2 d\lambda \;+\; o_P(1)$$

and by Lemma 8.3 of Paparoditis (1997) we have that

$$\mathcal{L}\left(T^{-1}h^{1/2}\int_{-\pi}^{\pi}\Big\{\sum_{j=-N}^{N}K_h(\lambda-\lambda_j)(U_j-1)\Big\}^2 d\lambda - \mu_h\right) \Rightarrow N(0,\sigma_K^2).$$

Furthermore, using (21.2.5) we get for the term $L_{2,T}$ that

$$L_{2,T} = c_T^2 T h^{1/2}\int_{-\pi}^{\pi}\Big\{\int_{-\pi}^{\pi}K_h(\lambda-x)\omega((x-\lambda^*)/b_T)dx\Big\}^2 d\lambda$$

$$+o(b_T c_T^2 T h^{1/2})$$

$$= b_T c_T^2 T h^{1/2}\int_{-\pi}^{\pi}\Big\{\int_{-\pi}^{\pi}K(u)\omega(y-uhb_T^{-1})du\Big\}^2 dy$$

$$+o(b_T c_T^2 T h^{1/2}).$$

Now, if $b_T \equiv 1$ then $L_{2,T} \to \int_{\pi}^{\pi}\omega^2(\lambda)d\lambda$ in probability provided $c_T = T^{-1/2+\delta/4}$. For $b_T = T^{-\gamma}$ the conditions $\delta/2 = 1-2\eta-\gamma$ and $hb_T^{-1} \to 0$ should be satisfied in order for the same convergence to be true. ∎

21.3   Comparisons

Recall that the $S_{T,h}$ test is based on the property that if the postulated model is correct then the (asymptotic) expected value of the ratio between the sample spectral density (periodogram) and the postulated spectral density equals one. Thus the proposed test statistic measures how close is the rescaled periodogram $I(\lambda)/f_0(\lambda)$ to the (sample) spectral density of a white noise process. An alternative way to evaluate the basic statistic $I(\lambda)/f_0(\lambda)$ for testing purposes is by using the cumulative ratio

$$G_T(x) = \int_0^x \frac{I(\lambda)}{f_0(\lambda)}d\lambda. \tag{21.3.1}$$

The null hypothesis that $f = f_0$ can then be tested using the normed difference $G_T(\tau\pi) - \tau G_T(\pi)$ where the variable τ varies in the interval $[0,1]$. In particular, a statistic useful for testing this null hypothesis is given by

$$C_T = (2\pi^2)^{-1}T\int_0^1\Big\{G_T(\tau\pi) - \tau G_T(\pi)\Big\}^2 d\tau, \tag{21.3.2}$$

see Dzhaparidze (1986). Note that if $\{X_t\}$ satisfies assumption (A1) then the process $\zeta_T(\tau) = (\pi\sqrt{2})^{-1}\sqrt{T}(G_T(\tau\pi) - \tau G_T(\pi))$ converges weakly to the Brownian Bridge $U(\tau)$, $0 \leq \tau \leq 1$, Dzhaparidze (1986), Proposition 5.1. Thus

$$\mathcal{L}(C_T) \Rightarrow \mathcal{L}\left(\int_0^1 U^2(\tau)d\tau\right). \tag{21.3.3}$$

Note that the asymptotic null distribution of the C_T test is identical to that of the Cramér-von Mises test based on the integrated squared difference between the standardized sample spectral distribution function and the standardized spectral distribution function of the model under H_0; cf. Anderson (1993).

For the class of local alternatives discussed in this paper the asymptotic behavior of the C_T test is established in the following theorem; cf. Dzhaparidze (1986).

THEOREM 21.3.1
Assume that (A1) is fulfilled and that the true spectral density f_T satisfies (21.2.5).

(i)   If $b_T \equiv 1$ and $c_T = T^{-1/2}$, then as $T \to \infty$,

$$\mathcal{L}(C_T) \Rightarrow \mathcal{L}\left(\int_0^1 (U(\tau) + \Omega(\tau))^2 d\tau\right),$$

where for $\tau \in [0,1]$

$$\Omega(\tau) = \frac{1}{\sqrt{2\pi}}\left[\int_o^{\tau\pi} \omega(\lambda)d\lambda - \tau \int_0^\pi \omega(\lambda)d\lambda\right].$$

(ii)   If $b_T = T^{-\gamma}$ for some $\gamma > 0$ and $c_T = T^{-\eta}$, $\eta > 0$ such that $\eta + \gamma = 1/2$ then the concluson of Part (i) of the theorem is true.

Part (i) of the above theorem has been established by Dzhaparidze (1986), Proposition 5.2, while part (ii) is proved along the same lines.

Theorem 21.3.1 shows that for both classes of local alternatives considered the C_T test is $\sqrt{T}$-consistent, i.e., it has a non trivial power for local deviations approaching the null at a rate as fast as $T^{-1/2}$. Now, compare this with the asymptotic power of the $S_{T,h}$ test stated in Theorem 21.2.1.

Consider first the class of sharp peak alternatives. For this class of local alternatives and for a suitable choice of δ, the $S_{T,h}$ test can be made more powerful than the C_T test. To see this consider the points (δ, γ, η) in the set

$$\Delta = \{(\delta, \eta, \gamma) : \ \delta > 0, \eta > 0, \gamma > 0, \ \delta/2 = 1 - \gamma - 2\eta,$$

$$\gamma < \delta \ \text{ and } \ \gamma > \delta/2\}.$$

For $(\delta, \eta, \gamma) \in \Delta$ we then have

$$\eta + \gamma = \frac{1}{2} + \frac{1}{2}(\gamma - \frac{\delta}{2}) > \frac{1}{2},$$

i.e., the $S_{T,h}$ test is more powerful than the C_T test since it detects local alternatives converging to the null at a rate faster that $T^{-1/2}$.

Consider next the class of Pitman alternatives. Since for $\delta > 0$ we have $T^{-1/2+\delta/4} > T^{-1/2}$ the $S_{T,h}$ test detects alternatives approaching the spectral density f_0 at a rate smaller than the 'parametric' rate $T^{-1/2}$. Note that for $\delta = 1/5$ which is the asymptotically optimal rate for choosing δ in the sense of minimizing the mean square error of estimating $q(\cdot)$, we have $T^{-1/2+\delta/4} = T^{-9/20}$ which is slightly smaller than $T^{-1/2}$. Thus it seems that a price have to be paid for using a nonparametric kernel estimator in $S_{T,h}$, i.e., for smoothing locally the ratio $I(\lambda)/f_0(\lambda)$ to obtain $\hat{q}_h(\lambda)$ instead of integrating this ratio as this is the case of the statistic $G_T(\tau\pi)$. However, this is not entirely true. As we will see, despite its $\sqrt{T}$-consistency the C_T test can have even for the class of Pitman alternatives very low power in detecting deviations from the null if they are due to nonvanishing autocorrelation at high lags. In this case the $S_{T,h}$ test appears to be more powerful. To understand this statement consider first the following proposition which gives a comparable representation of both statistics under the null hypothesis.

PROPOSITION 21.3.2
Assume that the process $\{X_t\}$ satisfies assumption (A1) and that the null hypothesis is true. Then

$$C_T = \frac{T}{\pi^2} \sum_{s=1}^{T-1} \frac{1}{s^2} \frac{\hat{\gamma}_\varepsilon^2(s)}{\sigma_\varepsilon^4} + o_P(1). \tag{21.3.4}$$

If (A2) is also fulfilled, then

$$S_{T,h} = 4\pi T h^{1/2} \sum_{s=1}^{T-1} \left(\frac{1}{2\pi} \int_{-\pi}^{\pi} K(u) \cos(uhs) du \right)^2 \frac{\hat{\gamma}_\varepsilon^2(s)}{\sigma_\varepsilon^4} + o_P(1), \tag{21.3.5}$$

where $\hat{\gamma}_\varepsilon(s) = T^{-1} \sum_{t=1}^{T-s} \varepsilon_t \varepsilon_{t+s}$ is the sample autocovariance at lag s based on the i.i.d. series $\varepsilon_1, \varepsilon_2, \dots, \varepsilon_T$.

PROOF Suppose that $f = f_0$. By Lemma 1 of Dzhaparidze (1986), p 301, we have that $I(\lambda)/f_0(\lambda) = 2\pi I_\varepsilon(\lambda)/\sigma_\varepsilon^2 + R_T(\lambda)/f_0(\lambda)$ where $\sqrt{T} \max_{0 \le \lambda \le \pi} |R_T(\lambda)| = o_P(1)$ and $I_\varepsilon(\lambda) = (2\pi T)^{-1} \sum_{t=1}^{T} \varepsilon_t \exp\{i\lambda t\}$ is the periodogram of the series $\varepsilon_1, \varepsilon_2, \dots, \varepsilon_T$. Since $f_0 \in \mathcal{F}$ it follows for every $\tau \in [0,1]$ that $G_T(\tau\pi) = 2\pi\sigma_\varepsilon^{-2}\int_0^{\tau\pi} I_\varepsilon(\lambda)d\lambda + o_P(T^{-1/2})$. Substituting $I_\varepsilon(\lambda) = (2\pi)^{-1}\sum_{s=-T+1}^{T-1}\hat\gamma_\varepsilon(s)\cos(\lambda h)$ we get

$$G_T(\tau\pi) = \frac{1}{\sigma_\varepsilon^2}\left(\hat\gamma_\varepsilon(0)\tau\pi + 2\sum_{s=1}^{T-1}\frac{\sin(\tau\pi s)}{s}\hat\gamma_\varepsilon(h)\right) + o_P(T^{-1/2}).$$

Therefore,

$$C_T = \frac{T}{2\pi^2}\int_0^1\left(2\sum_{s=1}^{T-1}\frac{\sin(\tau\pi s)}{s\,\sigma_\varepsilon^2}\,\hat\gamma_\varepsilon(s)\right)^2 d\tau + o_P(1)$$

$$= \frac{T}{\pi^2}\sum_{s=1}^{T-1}\frac{1}{s^2}\frac{\hat\gamma_\varepsilon^2(s)}{\sigma_\varepsilon^4} + o_P(1).$$

To obtain the asymptotic expression for the $S_{T,h}$ statistic note first that by substituting the relation $I(\lambda_j)/f_0(\lambda_j) = 2\pi I_\varepsilon(\lambda_j)/\sigma_\varepsilon^2 + R_T(\lambda_j)/f_0(\lambda_j)$ we get

$$S_{T,h} = \frac{\sqrt{h}}{T}\int_{-\pi}^{\pi}\left\{\sum_{j=-N}^{N}K_h(\lambda-\lambda_j)\left(\sum_{s=-T+1}^{T-1}\frac{\hat\gamma_\varepsilon(s)}{\sigma_\varepsilon^2}-1\right)\right\}^2 d\lambda$$

$$+\frac{\sqrt{h}}{T}\int_{-\pi}^{\pi}\left\{\sum_{j=-N}^{N}K_h(\lambda-\lambda_j)\frac{R_T(\lambda_j)}{f_0(\lambda_j)}\right\}^2 d\lambda$$

$$+\frac{2\sqrt{h}}{T}\int_{-\pi}^{\pi}\left\{\sum_{j=-N}^{N}K_h(\lambda-\lambda_j)\left(\frac{2\pi I_\varepsilon(\lambda_j)}{\sigma_\varepsilon^2}-1\right)\right\}$$

$$\times\left\{\sum_{i=-N}^{N}K_h(\lambda-\lambda_i)\frac{R_T(\lambda_i)}{f_0(\lambda_i)}\right\}d\lambda$$

$$= M_{1,T} + M_{2,T} + M_{3,T}$$

with an obvious notation for $M_{i,T}$, $i = 1,2,3$. Now, since $\max_{0\le\lambda\le\pi}|R_T(\lambda)| = o_P(T^{-1/2})$ we get $M_{2,T} = o_P(h^{1/2})$ while using Lemma 8.4 of Paparoditis (1997) we conclude that $M_{3,T} = o_P(1)$. Furthermore,

$$M_{1,T} = \frac{\sqrt{h}}{T}\int_{-\pi}^{\pi}\left\{2\sum_{j=-N}^{N}K_h(\lambda-\lambda_j)\sum_{s=1}^{T-1}\frac{\hat\gamma_\varepsilon(s)}{\sigma_\varepsilon^2}\cos(\lambda_j h)\right\}^2 d\lambda + o_P(1)$$

where the second term on the right hand side of the above expression is due to the fact that $\hat\gamma_\varepsilon(0) - \sigma_\varepsilon^2 = O(T^{-1/2})$ and that for an i.i.d. sequence

$\text{Cov}(\hat{\gamma}_\varepsilon(s_1), \hat{\gamma}_\varepsilon(s_2)) = \delta_{s_1, s_2} O(T^{-1})$; cf. Brockwell and Davis (1991). In particular, and using these relations we have

$$\frac{\sqrt{h}}{T} \int_{-\pi}^{\pi} \left\{ \sum_{j=-N}^{N} K_h(\lambda - \lambda_j) \left(\frac{\hat{\gamma}_\varepsilon(0)}{\sigma_\varepsilon^2} - 1 \right) \right\}^2 d\lambda = O_P(h^{1/2})$$

and

$$\frac{\sqrt{h}}{T} \int_{-\pi}^{\pi} \left\{ 2 \sum_{j=-N}^{N} K_h(\lambda - \lambda_j) \sum_{s=1}^{T-1} \frac{\hat{\gamma}_\varepsilon(s)}{\sigma_\varepsilon^2} \cos(\lambda_j h) \right\}$$

$$\times \left\{ \sum_{j=-N}^{N} K_h(\lambda - \lambda_j) \left(\frac{\hat{\gamma}_\varepsilon(0)}{\sigma_\varepsilon^2} - 1 \right) \right\} d\lambda$$

$$= 2T^{-1} h^{1/2} O_P(T^{1/2}) O_P(T^{1/2})$$

$$= O_P(h^{1/2}).$$

Approximating the Riemann sum by the corresponding integral and using the symmetry of $K(\cdot)$ and a trigonometric identity for $\cos(s\lambda - suh)$ we finally get that

$$M_{1,T} = 4Th^{1/2} \int_{-\pi}^{\pi} \left\{ \sum_{s=1}^{T-1} \left(\frac{1}{2\pi} \int_{-\pi}^{\pi} K_h(\lambda - x) \cos(xs) dx \right) \frac{\hat{\gamma}_\varepsilon(s)}{\sigma_\varepsilon^2} \right\}^2 d\lambda$$

$$+ o_P(1)$$

$$= 4Th^{1/2} \int_{-\pi}^{\pi} \left\{ \sum_{s=1}^{T-1} \left(\frac{1}{2\pi} \int_{-\pi}^{\pi} K(u) \cos(s\lambda - suh) du \right) \frac{\hat{\gamma}_\varepsilon(s)}{\sigma_\varepsilon^2} \right\}^2 d\lambda$$

$$+ o_P(1)$$

$$= 4Th^{1/2} \int_{-\pi}^{\pi} \left\{ \sum_{s=1}^{T-1} \left(\frac{1}{2\pi} \int_{-\pi}^{\pi} K(u) \cos(suh) du \right) \cos(s\lambda) \frac{\hat{\gamma}_\varepsilon(s)}{\sigma_\varepsilon^2} \right\}^2 d\lambda$$

$$+ o_P(1)$$

$$= 4\pi Th^{1/2} \sum_{s=1}^{T-1} \left(\frac{1}{2\pi} \int_{-\pi}^{\pi} K(u) \cos(suh) du \right)^2 \frac{\hat{\gamma}_\varepsilon^2(s)}{\sigma_\varepsilon^4} + o_P(1)$$

where the last equality follows by the orthogonality properties of $\cos(\cdot)$. ∎

The important aspect of Proposition 21.3.2 is that both statistics considered can be expressed (asymptotically) as a weighted sum of the squared error autocorrelations $\widehat{\gamma}_\varepsilon(s)/\sigma_\varepsilon^2$, $s = 1, 2, \ldots, T-1$. (Note that $\widehat{\gamma}_\varepsilon(s)/\sigma_\varepsilon^2$ is not exactly the sample autocorrelation based on $\varepsilon_1, \varepsilon_2, \ldots, \varepsilon_T$ since in the denominator the true variance of the error process $\sigma_\varepsilon^2 \equiv \gamma_\varepsilon(0)$ is used instead of its estimator $\widehat{\gamma}_\varepsilon(0)$). Now, the weighting sequence of the C_T test is just $1/s^2$ while that of $S_{T,h}$ is $< K(\cdot), \cos(sh\cdot) >^2$, i.e., it depends on the particular kernel K and the smoothing bandwidth h used.

By the severe downweight of the squared autocorrelations $\widehat{\gamma}_\varepsilon(s)/\sigma_\varepsilon^2$ at high lags appearing in the series representation of the C_T statistic we might expect that this test will have low power if the departure from the null is due to significant autocorrelation at high lags.

To proceed with the Pitman alternatives, and if the true spectral density equals

$$f_T(\lambda) = f_0(\lambda)\left(1 + T^{-1/2}\omega(\lambda)\right) + o(T^{-1/2})$$

we have by Theorem 21.3.1 that $\mathcal{L}(C_T) \to \mathcal{L}(\int_0^1 (U(\tau) + \Omega(\tau))^2 d\tau)$. It follows then that in this case

$$\mathcal{L}(C_T) \to \mathcal{L}\left(\sum_{j=1}^\infty \frac{(Z_j + \mu_j)^2}{(j\pi)^2}\right), \qquad (21.3.6)$$

where the Z_j are independent $N(0,1)$ distributed and $\mu_j = \sqrt{2}\int_0^1 \Omega(\tau)$ $\sin(j\pi\tau)d\tau$, Shorack and Wellner (1986). Thus local deviations to the null contribute to the test statistic by means of the components μ_j, $j = 1, 2, \ldots$. Note that the components μ_j are related as follows to the function $\omega(\cdot)$ appearing in the local alternatives considered. Using integration by parts and the definition of $\Omega(\tau)$ we get

$$\mu_j = \sqrt{2}\int_0^1 \Omega(\tau)\sin(j\pi\tau)d\tau$$

$$= \frac{\sqrt{2}}{j\pi}\int_0^1 \cos(j\pi\tau)d\Omega(\tau)$$

$$= \frac{1}{j\sqrt{\pi}}\int_0^1 \cos(j\pi\tau)\omega(\pi\tau)d\tau.$$

Recall that

$$\kappa(j) = 2\pi\int_0^1 \cos(j\pi\tau)\omega(\pi\tau)d\tau = 2\int_0^\pi \cos(j\lambda)\omega(\lambda)d\lambda$$

is the autocavariance at lag j of $w(\cdot)$ if the real-valued function $w(\cdot)$ is a spectral density, i.e., if $w(\cdot)$ is nonnegative and symmetric on $[-\pi, \pi]$ and $\int_{[-\pi,\pi]} w(\lambda)d\lambda < \infty$; cf. Brockwell and Davis (1991).

Now, if the deviations from the null are due to significant values of μ_j for $j > 2$ we expect a lower power of the C_T test due to the downweighting of these components in the above series representation of this test. In fact equation (21.3.6) enable us to relate the power behavior of the C_T test to that of the Cramér-von Mises test in testing hypothesis on the distribution function in the context of independent samples and to reproduce several of the results obtained there, see Eubank and LaRiccia (1992). For instance, and using arguments similar to those presented there, we get

$$\inf_{\|w\|^2=1} \lim_{T\to\infty} P\Big(C_T \geq c_{T,\alpha}\Big|f_T(\cdot) = f_0(\cdot)(1 + T^{-1/2}w(\cdot))\Big) = \alpha$$

where $c_{T,\alpha}$ denotes the α-quantile of the distribution of C_T and

$$P(A|f_T(\cdot) = f_0(\cdot)(1 + T^{-1/2}w(\cdot))$$

the probability of the event A under the assumption that the true spectral density is given by $f_T(\cdot) = f_0(\cdot)(1 + T^{-1/2}w(\cdot))$. Losely speaking, this shows that if the deviation from the null is placed at a sufficiently high value of μ_j, the power of the C_T test can degenerate to its level. In contrast to the $S_{T,h}$ test the power of the C_T test is, therefore, not a monotone increasing function of $\|w\|$.

We conclude the comparison between both test by means of a small simulation experiment. Series of length $T = 128$ from the process $X_t = \varepsilon_t + \theta\varepsilon_{t-q}$ have been generated where $\varepsilon_t \sim N(0, \sigma_\varepsilon^2)$ with $\sigma_\varepsilon^2 = 1/(1 + \theta^2)$ and $\theta = 0.4$. The null hypothesis is $H_0 : f = 1/(2\pi)$, i.e., the hypothesis that the observed series is a white noise sequence. Several values of q have been used in order to study the power behavior of both tests for different alternatives. Note that $\text{Var}(X_t) = 1$ and that the autocorrelation function of the process under the alternative is given by $E(X_tX_{t+s}) = \delta_{s,q}\theta/(1+\theta^2)$ for $s \in \mathbf{N}$, i.e., the value of q corresponds to the lag of the autocorrelation which causes the departure from the null.

The proportion of rejections in 1000 samples are reported in Table 21.1 for two different values of the level α. The critical values of the C_T test were taken from Shorack and Wellner (1986) while those of the $S_{T,h}$ test have been calculated using the bootstrap procedure proposed by Paparoditis (1997) based on 1000 bootstrap samples. The later test has been applied using the Bartlett-Priestley kernel and two different values of the smoothing bandwidth h. The $S_{T,h}$ test has been also applied using a cross-validation criterion to select the smoothing parameter h.

TABLE 21.1
Proportion of rejections in 1000 samples of the hypothesis of white noise

		$\alpha = 0.05$ $q=0$	$q=1$	$q=3$	$q=6$	$\alpha = 0.10$ $q=0$	$q=1$	$q=3$	$q=6$
C_T		0.046	0.984	0.205	0.101	0.088	0.993	0.397	0.187
S_{T,h^*}		0.051	0.932	0.612	0.362	0.110	0.983	0.821	0.592
$S_{T,h}$	$h = 0.08$	0.045	0.622	0.511	0.323	0.095	0.830	0.734	0.540
	$h = 0.12$	0.047	0.815	0.639	0.241	0.091	0.940	0.839	0.435

In particular, and following Beltrão and Bloomfield (1987) we select h as the minimizer of the function

$$CV(h) = \frac{1}{N} \sum_{j=1}^{N} \left\{ \log \hat{r}_{-j}(\lambda_j) + \frac{J(\lambda_j)}{\hat{r}_{-j}(\lambda_j)} \right\}$$

where $J(\lambda_j) = I(\lambda_j)/f_0(\lambda_j)$,

$$\hat{r}_{-j}(\lambda_j) = T^{-1} \sum_{s \in \mathbf{N_j}} K_h(\lambda_j - \lambda_s) I(\lambda_s)/f_0(\lambda_s)$$

and $\mathbf{N_j} = \{\mathbf{s} : -\mathbf{N} \le \mathbf{s} \le \mathbf{N}$ and $\mathbf{j} - \mathbf{s} \ne \pm\mathbf{j} \bmod \mathbf{N}\}$. That is, $\hat{r}_{-j}(\cdot)$ is the kernel estimator of $f(\cdot)/f_0(\cdot)$ when we delete the jth point. The $S_{T,h}$ test applied with a data driven bandwidth selection rule is denoted by S_{T,h^*}.

The results of Table 21.1 confirm our asymptotic analysis. As it is seen the tests maintain the level and the C_T test outperforms the $S_{T,h}$ test for $q = 1$. Recall that for this value of q the departure from the null is caused by a nonvanishing first order autocorrelation. However, the power of the C_T test drops off drastically for the values $q = 3$ and $q = 6$ since for these values of q the departure from the null is due to a significant autocorrelation at higher lags, e.g., lag 3 and 6 respectively. For these values of q the $S_{T,h}$ test clearly outperforms the C_T test and its power drops more gradually as q increases.

Acknowledgements The author is grateful to the referee for his helpful comments.

References

1. Anderson, T. W. (1993). Goodness of fit tests for spectral distributions. *Annals of Statistics* **21**, 830–847.

2. Beltrão, K. I. and Bloomfield, P. (1987). Determining the bandwidth of a kernel spectrum estimate. *Journal of Time Series Analysis* **8**, 21–38.

3. Brockwell, P. and Davis, R. (1991). *Time Series: Theory and Methods*, Second Edition. Springer-Verlag, New York.

4. Dzhaparidze, K. (1986). *Parameter Estimation and Hypothesis Testing in Spectral Analysis of Stationary Time Series*. Springer-Verlag, New York.

5. Eubank, R. L. and LaRiccia, V. N. (1992). Asymptotic comparison of Cramér-von Mises and nonparametric function estimation techniques for testing goodness-of-fit. *Annals of Statistics* **20**, 2071–2086.

6. Ghosh, B. K. and Huang, W.-M. (1991). The power and optimal kernel of the Bickel-Rosenblatt test for goodness of fit. *Annals of Statistics* **19**, 999–1009.

7. Paparoditis, E. (1997). Spectral density based goodness-of-fit tests in time series analysis. *Scandinavian Journal of Statistics*, forthcoming.

8. Priestley, M. B. (1981). *Spectral Analysis and Time Series*. Academic Press, New York.

9. Rosenblatt, M. (1975). A quadratic measure of deviation of two-dimensional density estimates and a test of independence. *Annals of Statistics* **3**, 1–14.

10. Shorack, G. R. and Wellner, J. A. (1986). *Empirical Processes with Applications to Statistics*, John Wiley & Sons, New York.

22

Linear Constraints in a Linear Model

Somesh Das Gupta
Indian Statistical Institute, Calcutta, India

ABSTRACT It is shown by geometric arguments that the reduction of a linear model $\mu = E(Y) = A\theta$ by the linear constraints $L'\theta = 0$ can be equivalently obtained by considering the linear constraints $L'_1\theta = 0$, where the columns of L_1 span the space $\mathcal{M} \cap \mathcal{L}$, $\mathcal{M}$ and $\mathcal{L}$ being the column spaces of A' and L, respectively. If $t = \dim(\mathcal{M} \cap \mathcal{L})$ is zero, then the model is unchanged by the linear constraints. It is not necessary to assume that the rows of $L'\theta$ are linearly estimable, to start with. It may be noted that even when all the rows of $L'\theta$ are not linearly estimable, there may exist some linear combinations of the rows of $L'\theta$ which are linearly estimable.

Keywords and phrases Linear models, linear constraints, effectiveness of linear constraints

---

22.1 Introduction

Consider the linear model

$$EY = \mu = A\theta, \qquad (22.1.1)$$

where $A : n \times m$ is a known matrix of rank r, and $\theta \in \mathbb{R}^m$, along with the linear constraints given by

$$L'\theta = 0, \qquad (22.1.2)$$

where $L : m \times s$ is a known matrix of rank s.

The constrained linear model has been considered by Rao (1973), and Wang and Chao (1994), in particular. The algebraic treatment presented by the above authors fails to clarify the role or impact of the constraints adequately. Rao (1973) has posed the reduced model, combining (22.1.1) and (22.1.2), as

$$\mu = AM\tau, \ \tau \in \mathbb{R}^q, \tag{22.1.3}$$

where $M : m \times q$ is a matrix of rank $q = m - s$ such that $L'M = 0$. Although the problem of identifiability of θ can be resolved by considering A-equivalence (i.e. θ_1 and θ_2 are equivalent, if $A\theta_1 = A\theta_2$), it is not clear how the impact of (22.1.2) is reflected in (22.1.3).

Usually the rows of $L'\theta$ are assumed to be linearly estimable. An equivalent assumption is $\mathcal{N} \subset \mathcal{D}$, or the inverse image of $A(\mathcal{D})$ on $\mathbb{R}^m$ is $\mathcal{D}$, where

$$\mathcal{D} = \{\theta \in \mathbb{R}^m : L'\theta = 0\} = \{\theta \in \mathbb{R}^m : \theta = M\tau, \ \tau \in \mathbb{R}^q\}, \tag{22.1.4}$$

$$\mathcal{N} = \{\theta \in \mathbb{R}^m : \ A\theta = 0\} \tag{22.1.5}$$

As a consequence of this assumption, it is possible to express the reduced model given by

$$\mu \in \mathcal{V}_0 = \{\mu \in \mathbb{R}^n : \mu = A\theta, L'\theta = 0\} \tag{22.1.6}$$

as $\mu \in \mathcal{V} = \mathcal{C}(A)$, the column space of A, along with $\mu \perp \mathcal{V}_1$, where $\mathcal{V}_1$ is a subspace of $\mathcal{V}$ satisfying $\mathcal{V}_1 = \mathcal{C}(B), E(B'Y) = L'\theta$. The above assumption ensures the existence of such a matrix B.

In case all or some of the rows of $L'\theta$ are not linearly estimable, it is not apparently possible to get such a simple reduction. It is seen that (22.1.3) leads to a similar specification; however, the relationship between $\mathcal{V}_1$ and L cannot be clearly revealed through (22.1.3). We shall present a geometric description of the role or impact of the linear constraints, and relate $\mathcal{V}_1$ to L geometrically.

22.2 Geometric Interpretation of the Role of the Linear Constraints

Let

$$\mathcal{M} = \mathcal{C}(A'), \ \mathcal{L} = \mathcal{C}(L), \ \mathcal{M}_1 = \mathcal{M} \cap \mathcal{L}, \tag{22.2.1}$$

$$t = dim(\mathcal{M} \cap \mathcal{L}), \tag{22.2.2}$$

and $\mathcal{M}_0$ be the orthogonal complement of $\mathcal{M}_1$ in $\mathcal{M}$. It is clear that $0 \le t \le \min(s, r)$. We assume $r > 0, s > 0$ excluding the trivial cases. Our main result is given as follows.

THEOREM 22.2.1
(a) If $t = 0$,

$$\begin{aligned} \{\mu \in \mathbb{R}^n : \mu = A\theta, \ \theta \perp \mathcal{L}\} &= \{\mu \in \mathbb{R}^n : \mu = A\theta, \ \theta \in \mathcal{M}\} \\ &= \{\mu \in \mathbb{R}^n : \mu = A\theta, \ \theta \in \mathbb{R}^m\} \end{aligned} \tag{22.2.3}$$

(b) If $0 < t < s$, then $dim(\mathcal{V}_0) = r - t > r - s$, and

$$\begin{aligned} \{\mu \in \mathbb{R}^n : \mu = A\theta, \ \theta \perp \mathcal{L}\} &= \{\mu \in \mathbb{R}^n : \mu = A\theta, \ \theta \in \mathcal{M}_0\} \\ &= \{\mu \in \mathbb{R}^n : \mu = A\theta, \ \theta \perp \mathcal{M}_1\} \end{aligned} \tag{22.2.4}$$

(c) If $t = s$, then $dim(\mathcal{V}_0) = r - s, \mathcal{M}_1 = \mathcal{L}$, and

$$\{\mu \in \mathbb{R}^n : \mu = A\theta, \ \theta \perp \mathcal{L}\} = \{\mu \in \mathbb{R}^n : \mu = A\theta, \ \theta \in \mathcal{M}_0\} \tag{22.2.5}$$

Note 1
If $t = 0$, the model $\mu = A\theta, \theta \in \mathbb{R}^m$, is unchanged by the linear constraints $L'\theta = 0$. We call this case "ineffective". In this case, no nontrivial linear combination of $L'\theta$ is linearly estimable. If $0 < t < s$, some of the rows of $L'\theta$ are not linearly estimable. But there exist some linear combinations of the rows of $L'\theta$ which are linearly estimable. We call this case "partially effective", since the reduction in the model by the constraints $\theta \perp \mathcal{L}$ is also accomplished by the constraint $\theta \perp \mathcal{M}_1, \mathcal{M}_1$ being a proper vector subspace of $\mathcal{L}$; note that $dim(\mathcal{V}_0) = r - t > r - s$, s being the rank of L. If $t = s$, the resulting reduction in the model cannot be accomplished by the constraint $\theta \perp \mathcal{L}_1$ for any proper vector subspace $\mathcal{L}_1$ of $\mathcal{L}$; note that $dim(\mathcal{V}_0) = r - s$. We call this case "fully effective".

Following Rao (1973) it may be seen that

$$t = \text{rank } (A) + \text{rank } (L) - \text{rank } (A':L), \qquad (22.2.6)$$

and

$$rank(AM) = r - t, \qquad (22.2.7)$$

where M is defined in (22.1.3). Note that $t > 0$ if $m < r + s$
The above theorem is proved using the following lemmas.

LEMMA 22.2.2

$$\mathcal{M} \cap \mathcal{L} = \{\theta \in \mathbb{R}^m : \theta \in \mathcal{M}, \ \theta \perp P_{\mathcal{M}}(\mathcal{D})\}, \qquad (22.2.8)$$

where $P_{\mathcal{M}}$ is the operator (or the symmetric idempotent matrix) corresponding to the orthogonal projection on $\mathcal{M}$, and $\mathcal{D}$ is given in (22.1.4).

PROOF First note that any vector $d \in \mathbb{R}^m$ can be expressed as $d = \alpha + \gamma$, $\alpha \in \mathcal{M}, \gamma \in \mathcal{N}$. Moreover, $P_{\mathcal{M}}(d) = \alpha$. For a non-zero vector α_0

$$\alpha_0 \in \mathcal{M} \cap \mathcal{L} \Leftrightarrow \alpha_0 \in \mathcal{M}, \ \alpha_0 \perp \mathcal{D} \Leftrightarrow \alpha_0 \in \mathcal{M}, \ \alpha_0 \perp P_{\mathcal{M}}(d) \ \text{ for all } d \in \mathcal{D},$$

$$(22.2.9)$$

since $d - P_{\mathcal{M}}(d) \in \mathcal{N}$ and $\alpha_0 \in \mathcal{M} \Leftrightarrow \alpha_0 \perp \mathcal{N}$. ∎

Lemma 22.2.2 simply states that $P_{\mathcal{M}}(\mathcal{D}) = \mathcal{M}_0$, the orthogonal complement of $\mathcal{M}_1$ in $\mathcal{M}$. The following lemma is an easy consequence of the fact that $d - P_{\mathcal{M}}(d) \in \mathcal{N}$ for any $d \in \mathbb{R}^m$.

LEMMA 22.2.3
For any vector subspace $\mathcal{D}$ in $\mathbb{R}^m$

$$\{\mu \in \mathbb{R}^n : \mu = A\theta, \theta \in \mathcal{D}\} = \{\mu \in \mathbb{R}^n : \mu = A\theta, \ \theta \in P_{\mathcal{M}}(\mathcal{D})\}$$

$$(22.2.10)$$

Under the linear constraints $L'\theta = 0$ the effective domain of θ is $\mathcal{D}$. The following lemmas give descriptions of the structure of the set $\mathcal{D}$, which are of interest on their own. The vector space spanned by a set of vectors U_i's is denoted by $\mathcal{S}\{U_i's\}$.

LEMMA 22.2.4
Suppose $\mathcal{M} \cup \mathcal{L} = \mathcal{L}$, i.e $\mathcal{L} \subset \mathcal{M}$. Then

$$\mathcal{D} = \mathcal{M}_0 \oplus \mathcal{N}, \tag{22.2.11}$$

where $\mathcal{M}_0$ is the orthogonal complement of $\mathcal{L}$ (or $\mathcal{M}_1$) in $\mathcal{M}$.

PROOF First note that any vector $\theta \in \mathcal{N}$ or any vector $\theta \in \mathcal{M}_0$ satisfies $L'\theta = 0$. Next note that $dim(\mathcal{D}) = m - s$, and $dim(\mathcal{M}_0 \oplus \mathcal{N}) = (r - s) + (m - r) = (m - s)$. ∎

LEMMA 22.2.5
Suppose $0 < t \le \min(r, s)$. Then

$$\mathcal{D} = \mathcal{S}\{\alpha_{t+1}^* + \gamma_{t+1}^*, \ldots, \alpha_r^* + \gamma_r^*\} \oplus \mathcal{N}_2 \tag{22.2.12}$$

where $\{\alpha_{t+1}^*, \ldots, \alpha_r^*\}$ is an orthonormal basis of $\mathcal{M}_0$, γ_i's are specific vectors in a vector subspace $\mathcal{N}_1$ of $\mathcal{N}$ of dimension $s - t$, and $\mathcal{N}_2$ is the orthogonal complement of $\mathcal{N}_1$ in $\mathcal{N}$.

PROOF Let $\{\alpha_1^*, \ldots, \alpha_t^*, \ldots, \alpha_r^*\}$ be an orthonormal basis of $\mathcal{M}$ so that $\{\alpha_1^*, \ldots, \alpha_t^*\}$ is a basis of $\mathcal{M}_1 = \mathcal{M} \cap \mathcal{L}$, and $\{\alpha_1^*, \ldots, \alpha_t^*, \alpha_{t+1}^* + \gamma_{t+1}^*, \ldots, \alpha_s^* + \gamma_s^*\}$ is a basis of $\mathcal{L}$, where α_i's are in $\mathcal{M}$ and γ_i's are in $\mathcal{N}$.
Suppose $\sum_{t+1}^{s} d_i \gamma_i = 0$ for some d_i's. Then $\sum_{t+1}^{s} d_i(\alpha_i + \gamma_i) = \sum_{t+1}^{s} d_i \alpha_i$, which belongs to $\mathcal{M}_1$. By the choice of the above basis of $\mathcal{L}$, it follows that $\sum_{t+1}^{s} d_i \alpha_i = 0$. Consequently, $\sum_{t+1}^{s} d_i(\alpha_i + \gamma_i) = 0$, which implies that d_i's are all 0. Hence γ_i's are linearly independent.
Write

$$U_1 = (\alpha_1^*, \ldots, \alpha_t^*), \ U_2 = (\alpha_{t+1}^*, \ldots, \alpha_r^*), \ V = (\gamma_{t+1}, \ldots, \gamma_s).$$

Then
$$(\alpha_{t+1}, \ldots, \alpha_s) = U_1 C_1 + U_2 C_2$$

for some matrices C_1 and C_2. It can be checked that the columns of

$$U_2 - V(V'V)^{-1} C_2'$$

are orthogonal to each of the vectors in the above basis of $\mathcal{L}$. Write

$$-V(V'V)^{-1} C_2' = (\gamma_{t+1}^*, \ldots, \gamma_r^*)$$

It is easy to see that the vectors $\{\alpha_{t+1}^* + \gamma_{t+1}^*, \ldots, \alpha_r^* + \gamma_r^*\}$ are linearly independent.

Let $\mathcal{N}_2$ be the orthogonal complement of $\mathcal{S}\{\gamma_{t+1}, \ldots, \gamma_s\} \equiv \mathcal{N}_1$ in $\mathcal{N}$. Then $\dim(\mathcal{N}_2) = (m - r) - (s - t)$, and $\mathcal{N}_2$ is orthogonal to $\mathcal{L}$, as well as to $\mathcal{S}\{\alpha_{t+1}^* + \gamma_{t+1}^*, \ldots, \alpha_r^* + \gamma_r^*\}$. This yields the desired result, since

$$\dim (\mathcal{D}) = m - s = \{(m - r) - (s - t)\} + (r - t).$$

                                                                    ■

PROOF OF THE THEOREM The theorem essentially follows from Lemma 22.2.2 and Lemma 22.2.3 or from Lemmas 22.2.3-22.2.5, as shown below.

(a) It is clear from Lemma 22.2.3 that the reduced model can be expressed as

$$\{\mu \in \mathbb{R}^n : \mu = A\theta, \ \theta \in P_{\mathcal{M}}(\mathcal{D})\},$$

where $P_{\mathcal{M}}(\mathcal{D}) = \mathcal{M}_0$, the orthogonal complement of $\mathcal{M} \cap \mathcal{L}$ in $\mathcal{M}$. By Lemma 22.2.2, $\dim P_{\mathcal{M}}(\mathcal{D}) = r$. Thus $\mathcal{M}_0 = \mathcal{M}$. Lemma 22.2.3 now yields the result.

(b) $0 < t < s$. Let $\{l_1, \ldots, l_s\}$ be a basis of $\mathcal{L}$ such that $\{l_1, \ldots, l_t\}$ is a basis of $\mathcal{M}_1 = \mathcal{M} \cap \mathcal{L}$. Write

$$L_1 = (l_1 \ldots l_t), \ L_2 = (l_{t+1} \ldots l_s).$$

Then

$$L'\theta = 0 \ \Leftrightarrow \ L_1'\theta = 0, \ L_2'\theta = 0.$$

Let $\mathcal{D}_1$ and $\mathcal{D}_2$ be the sets of $\theta \in \mathbb{R}^m$ for which $L_1'\theta = 0$ and $L_2'\theta = 0$, respectively. Note that $\mathcal{C}(L_1) \subset \mathcal{M}$ and $\mathcal{C}(L_2) \cap \mathcal{M} = \{0\}$. By Lemmas 22.2.4 and 22.2.5, we find that

$$P_{\mathcal{M}}(\mathcal{D}_1) = \mathcal{M}_0, \ P_{\mathcal{M}}(\mathcal{D}_2) = \mathcal{M}.$$

It is easy to check that

$$P_{\mathcal{M}}(\mathcal{D}_1 \cap \mathcal{D}_2) = P_{\mathcal{M}}(\mathcal{D}_1) \cap P_{\mathcal{M}}(\mathcal{D}_2) = \mathcal{M}_0.$$

To see the above, note that it follows from the structure of $\mathcal{D}_1$ and $\mathcal{D}_2$, described in (22.2.11) and (22.2.12), that given $\alpha \in P_{\mathcal{M}}(\mathcal{D}_1) \cap P_{\mathcal{M}}(\mathcal{D}_2)$ there exists a $\theta \in \mathcal{D}_1 \cap \mathcal{D}_2$ for which $P_{\mathcal{M}}(\theta) = \alpha$.

This yields the result.

(c) If $t = s, \mathcal{M}_1 = \mathcal{L}$. Lemma 22.2.3 yields the result. It can also be seen from Lemma 22.2.4 that $P_{\mathcal{M}}(\mathcal{D}) = \mathcal{M}_0$.

Note 2

Rao (1973) and Wang and Chao (1994) have considered the problem of minimizing $(Y - A\theta)'(Y - A\theta)$ using Lagrange multipliers for the constraints $L'\theta = 0$. It is possible to find constraints $L_1'\theta = 0$ and $L_2'\theta = 0$ such that $\mathcal{C}(L_1) \in \mathcal{M}$, $\mathcal{C}(L_2) \cap \mathcal{M} = \{0\}$, and

$$L'\theta = 0 \Leftrightarrow L_1'\theta = 0, \; L_2'\theta = 0.$$

Using the structure of L_2, as given in the proof of Lemma 4, it can be seen that the value of $A\hat{\theta}$, $\hat{\theta}$ being a solution of the normal equations with the constraints $L_1'\theta = 0$, is the same as $A\hat{\hat{\theta}}$ where $\hat{\hat{\theta}}$ is a solution of normal equations with constraints $L_1'\theta = 0$ and $L_2'\theta = 0$, or equivalently $L'\theta = 0$. This is the algebraic aspect of our result.

Note 3

Roy (1958) considered only the linear hypothesis $H_0 : L'\theta = 0$ such that the rows of $L'\theta$ are all linearly estimable. He called such a hypothesis "testable". However, it is not necessary to make this distinction. If the constraints $L'\theta = 0$ are ineffective then the hypothesis H_0 is vacuous. On the other hand, if $L'\theta = 0$ is partially effective, then H_0 is equivalent to $L_1'\theta = 0$ where $\mathcal{C}(L_1) = \mathcal{M} \cap \mathcal{L}$.

Note 4

The results presented in this note merely gives geometric description of the role of the linear constraints, and apparently provide no help in the calculations needed for relevant statistical inference. Given a linear parametric function $l'\theta$, we generally check whether $l'\theta$ is linearly estimable from the simple or patterned structure of the design matrix

A. Otherwise, one has to resort to the sweep-out method on $(A' \colon l)$ to find out whether $l \in \mathcal{C}(A')$. We can also use the structure of A or the sweep-out method to find the value of t, and in case $0 < t \leq s$, we can also get a basis of $\mathcal{M} \cap \mathcal{L}$. This, in turn, will avoid solving the normal equations with Lagrange multipliers.

Note 5

Eaton *et al.* (1969) have considered the following model

$$\{\mu \in \mathbb{R}^n : T\mu = S\beta, \; \beta \in \mathbb{R}^p\}, \tag{22.2.13}$$

where $T : m \times n$ and $S : m \times p$ are known matrices; they have introduced this model as a generalization of the classical linear model. With reference to the constrained linear model considered in this note, it can be seen that there exist linear transformations

$$T : \mu \in \mathbb{R}^n \to \theta \in \mathcal{M}, \ S : \beta \in \mathbb{R}^{r-t} \to \theta \in \mathcal{M}_0$$

such that (22.2.13) holds. As a matter of fact, (22.2.13) is equivalent to $\mu \in \mathcal{V}_1 \oplus \mathcal{V}_2$, where $\mathcal{V}_1$ is a known vector subspace of $\mathbb{R}^n$ and $\mathcal{V}_2$ is a known proper vector subspace of the orthogonal complement of $\mathcal{V}_1$ in $\mathbb{R}^n$. It can be seen that any linear model can be expressed in the above form, and (22.2.13) cannot be considered as a generalization of the classical linear model.

Note 6

Suppose $E(Y) = A\theta$ and $Cov(Y) = \Sigma$, which is singular of rank q, and known except for a scale multiple. Without any loss of generality we may assume that

$$\Sigma = \begin{bmatrix} \sigma^2 I_q & 0 \\ 0 & 0 \end{bmatrix}$$

Let

$$Y = \begin{bmatrix} Y_1 \\ Y_2 \end{bmatrix} \begin{matrix} q \\ n-q \end{matrix}, \ A = \begin{bmatrix} A_1 \\ A_2 \end{bmatrix} \begin{matrix} q \\ n-q \end{matrix}.$$

then the model reduces to $E(Y_1) = A_1\theta$, $Cov(Y_1) = \sigma^2 I_q$, $Y_2 = A_2\theta$. A solution of $Y_2 = A_2\theta$ can be expressed as $\theta_0 + \tau$, where θ_0 is a particular solution and $A_2\tau = 0$. Write $Y_1^* = Y_1 - A_1\theta_0$. Then the model reduces further to

$$E(Y_1^*) = A_1\tau, \ Cov(Y_1^*) = \sigma^2 I_q, \ A_2\tau = 0.$$

This is a linear model with linear constraints.

Lastly, we present three simple examples to illustrate our result.

Example 1

$\mu_1 = \theta_1 + \theta_2$, $\mu_2 = 2(\theta_1 + \theta_2)$; θ_1, θ_2 in $\mathbb{R}$. Let the constraint be $\theta_1 - \theta_2 = 0$. The model is unaffected by the constraint. ⬛

Example 2

$\mu_1 = \theta_1 + \theta_2$, $\mu_2 = \theta_3 + \theta_4$; θ_i's are in $\mathbb{R}$. Let the constraints be $\theta_1 = \theta_2 = \theta_3 = 0$. Note that $\theta_1, \theta_2, \theta_3$ are not linearly estimable, but $\theta_1 + \theta_2$ is. We get the some reduction of the model by the constraint $\theta_1 + \theta_2 = 0$. ⬛

Example 3

Consider the block-treatment experiment with the additive model

$$\mu_{ij} = \gamma + \tau_i + \beta_j; \ i = 1, \ldots, t; \ j = 1, \ldots, b.$$

Then it is well known that the constraints $\sum_i \tau_i = 0$, $\sum_j \beta_j = 0$ do not change the model. One can check that no non-trivial linear combination of $\sum_i \tau_i$ and $\sum_j \beta_j$ is linearly estimable. ∎

References

1. Eaton, M. L., Eckles, J. E. and Morris, C. N. (1969). Estimation for a generalization of the usual linear statistical model. The Rand Corporation Memo. RM-6078-PR, Santa Monica, California.

2. Rao, C. R. (1973). *Linear Statistical Inference and Its Applications*. John Wiley & Sons, New York.

3. Roy, S. N. (1958). *Some Aspects of Multivariate Analysis*. Asia Publishing House, Calcutta.

4. Wang, S. G. and Chao, S. C. (1994). *Advanced Linear Models*. Marcel Dekker, New York.

23

M-methods in Generalized Nonlinear Models

Antonio I. Sanhueza and Pranab K. Sen
Universidad de La Frontera, Temuco, Chile
University of North Carolina, Chapel Hill, NC

ABSTRACT In generalized nonlinear (regression) models (GNLM), maximum likelihood (ML) or quasi-likelihood (QL) methods based statistical inference procedures are generally not robust against outliers or error contamination. In fact, nonrobustness is more likely to be compounded by the choice of link functions as needed for GNLM's. For this reason, robust procedures based on appropriate M-statistics are considered here. Our proposed M-estimators and M-tests, formulated along the lines of generalized least squares procedures, are relatively more robust and their (asymptotic) properties are studied.

Keywords and phrases Asymptotic normality, efficiency, estimating equations, generalized additive models, M-estimators, M-tests, robustness, uniform asymptotic linearity

23.1   Introduction

Let us consider the (univariate) nonlinear regression model

$$Y_i = f(\mathbf{x}_i, \boldsymbol{\beta}) + e_i, \; i = 1, ..., n \tag{23.1.1}$$

where Y_i are the observable random variables (r.v.), $\mathbf{x}_i = (x_{1i}, x_{2i}, ..., x_{mi})^t$ are known regression constants, $\boldsymbol{\beta} = (\beta_1, \beta_2, ..., \beta_p)^t$ is a vector of unknown parameters, $f(,)$ is a (nonlinear) function (of $\boldsymbol{\beta}$) of specified form; and the errors e_i are assumed to be independent r.v.s with mean 0, but not necessarily identically distributed. We assume that the distribution of Y_i is in the exponential family with density:

$$g_{Y_i}(y, \theta_i, \phi) = c(y, \theta_i) \, exp\{(y \, \theta_i - b(\theta_i))/a(\phi)\}, \; i = 1, ..., n, \tag{23.1.2}$$

where $\phi \ (> 0)$ is a nuisance scale parameter, the θ_i are the parameters of interest, and the functional forms of $a(\cdot), b(\cdot)$ and $c(\cdot)$ are assumed to be known. In this context, the mean and variance of Y_i are easily shown [see McCullagh and Nelder (1989, pp. 28–29)] to be:

$$\mu_i = E Y_i = b'(\theta_i) = \mu_i(\theta_i) \qquad (23.1.3)$$

$$Var \, Y_i = a(\phi) \, (\partial/\partial\theta_i)\mu_i(\theta_i) = a(\phi) \, v_i(\mu_i(\theta_i)). \qquad (23.1.4)$$

In that way, we let:
$$b'(\theta_i) = f(\mathbf{x}_i, \, \boldsymbol{\beta})$$

and say that Y_i follows a generalized nonlinear model (GNLM). Hence, the variance of Y_i is given by:

$$Var \, Y_i = a(\phi) \, v_i(f(\mathbf{x}_i, \, \boldsymbol{\beta})),$$

which remains the same for all $i = 1, ..., n$.

If we let f be a monotone differentiable function, such that:

$$f(\mathbf{x}_i, \boldsymbol{\beta}) = f(\mathbf{x}_i^t \boldsymbol{\beta}),$$

the r.v. Y_i follows the generalized linear model (GLM) introduced by Nelder and Wedderburn (1972). Also, when

$$f(\mathbf{x}_i, \boldsymbol{\beta}, \mathbf{Z}_i) = f(\mathbf{x}_i^t \boldsymbol{\beta}) + \sum_{j=1}^{q} h_j(Z_{ij})$$

where the h_j are smooth (possibly nonlinear) functions dependent only upon the coordinate Z_{ij} of the auxiliary explanatory vector $\mathbf{Z}_i$, we note that Y_i follows the generalized additive model (GAM) [Hastie and Tibshirani (1990)].

In the context of the exponential family, the estimating equation for the maximum likelihood estimator (MLE) $\hat{\boldsymbol{\beta}}_n$ in model (23.1.1) is the form:

$$\sum_{i=1}^{n} \frac{1}{v_i(f(\mathbf{x}_i, \, \hat{\boldsymbol{\beta}}_n))} \, (Y_i - f(\mathbf{x}_i, \, \hat{\boldsymbol{\beta}}_n)) \, \mathbf{f}_{\boldsymbol{\beta}}(\mathbf{x}_i, \boldsymbol{\beta}) = \mathbf{0}, \qquad (23.1.5)$$

where $\mathbf{f}_{\boldsymbol{\beta}}(\mathbf{x}_i, \boldsymbol{\beta}) = (\partial/\partial\boldsymbol{\beta})f(\mathbf{x}_i, \, \boldsymbol{\beta})$, which are exactly of the form of the weighted least square equations for model (23.1.1). Thus, we can use the iteratively reweighted least square (IRWLS) procedure to implement maximum likelihood estimation of $\boldsymbol{\beta}$.

We may define the expected value and variance of Y_i as in (23.1.3) and (23.1.4), respectively, where $v_i(\cdot)$ in (23.1.4) have known form and ϕ is a possibly unknown positive scalar constant, but we make no specific assumption on the distribution of Y_i. In this context, we can make use of the quasilikelihood (QL) approach [Wedderburn (1974) and McCullagh (1983)], and see that the QL equations for estimating β in (23.1.1) are the same as given in (23.1.5). This allows us to incorporate those cases where the functional form of the variance of the response variable is known, though it may depend on the parameters of interest in the model.

Incorporating the ML and QL equations, we formulate suitable M-estimators of β, and study their (asymptotic) properties (including consistency and asymptotic normality). M-tests are also considered for testing suitable hypotheses on β. Two computational algorithms for M-estimators based on Newton-Rapson and Fisher's scoring methods are presented. In Section 2, preliminary definitions and regularity conditions are presented. Section 3 deals with the asymptotic distribution theory of M-estimators of β, and in this respect, a uniform asymptotic linearity result on M-statistics (in the regression parameter) is presented in detail. Section 4 is devoted to related M-tests, and two iterative computational methods are considered.

23.2   Definitions and Assumptions

Considering that the density function of Y_i in (23.1.1) given by (23.1.2) leads to weighted least square, we define an M-estimator of β as the minimization:

$$\hat{\beta}_n = Arg \cdot min\left\{\sum_{i=1}^{n} \frac{1}{h[(v_i(f(\mathbf{x}_i,\boldsymbol{\beta}))]} h^2(Y_i - f(\mathbf{x}_i,\boldsymbol{\beta})) : \boldsymbol{\beta} \in \boldsymbol{\Theta} \subseteq \Re^p\right\}$$

(23.2.1)

where $h(\cdot)$ is a real valued function, and $\boldsymbol{\Theta}$ is a compact subset of $\Re^p$.

Assuming that the variance function in the equation (23.2.1) is fixed and $\psi(z) = (\partial/\partial z)h^2(z)$, we have that the estimating equation for the minimization in (23.2.1) is given by:

$$\sum_{i=1}^{n} \lambda(\mathbf{x}_i, Y_i, \hat{\boldsymbol{\beta}}_n) = 0 \qquad\qquad (23.2.2)$$

where

$$\lambda(\mathbf{x}_i, Y_i, \boldsymbol{\beta}) = \frac{1}{h[v(f(\mathbf{x}_i, \boldsymbol{\beta}))]} \psi(Y_i - f(\mathbf{x}_i, \boldsymbol{\beta})) \mathbf{f}_{\boldsymbol{\beta}}(\mathbf{x}_i, \boldsymbol{\beta}). \qquad (23.2.3)$$

In particular, if we let $h(z) = z$, we have the ML and QL equations for estimating β as given by (23.1.5). In the conventional setup of robust methods [Huber (1981), Hampel *et al.* (1985), Jurečková and Sen (1996), and others], we will primarily use bounded and monotone functions $h(\cdot)$; the so called Huber-score function corresponds to

$$h(z) = \begin{cases} \frac{1}{\sqrt{2}} z & \text{if } |z| \leq k \\ \{k[|z| - \frac{k}{2}]\}^{\frac{1}{2}} & \text{if } |z| > k \end{cases} \qquad (23.2.4)$$

for suitable chosen k $(0 < k < \infty)$. Whenever the errors e_i in (23.1.1) have a symmetric distribution, we may choose k as a suitable percentile point of this law (say the 90th or 95th percentile), and let h be as in (23.2.4). However, whenever the errors may not have a symmetric law, we choose two values k_1 and k_2 $(k_1 < 0 < k_2)$, and let

$$h(z) = \begin{cases} \frac{1}{\sqrt{2}} z & \text{if } k_1 \leq z \leq k_2 \\ \{k_1[z - \frac{k_1}{2}]\}^{\frac{1}{2}} & \text{if } z \leq k_1 \, (< 0) \\ \{k_2[z - \frac{k_2}{2}]\}^{\frac{1}{2}} & \text{if } z \geq k_2 \, (> 0), \end{cases}$$

where k_1 and k_2 are chosen so that $E\psi(Y_i - f(\mathbf{x}_i, \boldsymbol{\beta})) = 0$. This can always be made if the distribution of e_i is known up to a scalar factor (that may depend on $\mathbf{x}_i^t \boldsymbol{\beta}$).

We make the following sets of regularity assumptions concerning (A) the score function ψ, (B) the function $f(\cdot)$, and (C) the functions $k_1(\cdot)$ and $k_2(\cdot)$, where

$$k_1(x_i, \boldsymbol{\beta}) = 1/h[v(f(\mathbf{x}_i, \boldsymbol{\beta}))], \qquad (23.2.5)$$

and

$$k_2(x_i, \boldsymbol{\beta}) = \frac{h'[v(f(\mathbf{x}_i, \boldsymbol{\beta}))] \, v'(f(\mathbf{x}_i, \boldsymbol{\beta})}{h^2[v(f(\mathbf{x}_i, \boldsymbol{\beta}))]}. \qquad (23.2.6)$$

[A1]:
ψ is nonconstant, absolutely continuous and differentiable with respect to $\boldsymbol{\beta}$.

[A2]: Let $e = Y - f(x, \boldsymbol{\beta})$,
(i) $E\psi^2(e) < \infty$, and $E\psi(e) = 0$.

(ii) $E|\psi'(e)|^{1+\delta} < \infty$ for some $0 < \delta \leq 1$, and $E\psi'(e) = \gamma \, (\neq 0)$

[A3]:
(i) $\lim_{\delta \to 0} E\left\{ \text{Sup}_{\|\Delta\| \leq \delta} \left| \psi(Y - f(\mathbf{x}, \boldsymbol{\beta} + \Delta)) - \psi(Y - f(\mathbf{x}, \boldsymbol{\beta})) \right| \right\} = 0$

(ii) $\lim_{\delta \to 0} E\left\{ \text{Sup}_{\|\Delta\| \leq \delta} \left| \psi'(Y - f(\mathbf{x}, \boldsymbol{\beta} + \Delta)) - \psi'(Y - f(\mathbf{x}, \boldsymbol{\beta})) \right| \right\} = 0$

[B1]:
$f(\mathbf{x}, \boldsymbol{\beta})$ is continuous and twice differentiable with respect to $\boldsymbol{\beta} \in \boldsymbol{\Theta}$, where $\boldsymbol{\Theta}$ is a compact subset of $\Re^p$.

[B2]:
(i) $\lim_{n \to \infty} \frac{1}{n} \boldsymbol{\Gamma}_{1n}(\boldsymbol{\beta}) = \boldsymbol{\Gamma}_1(\boldsymbol{\beta})$, where

$$\boldsymbol{\Gamma}_{1n}(\boldsymbol{\beta}) = \sum_{i=1}^{n} \{ \frac{1}{h[v(f(\mathbf{x}_i, \boldsymbol{\beta}))]} \mathbf{f}_{\boldsymbol{\beta}}(\mathbf{x}_i, \boldsymbol{\beta}) \, \mathbf{f}_{\boldsymbol{\beta}}^t(\mathbf{x}_i, \boldsymbol{\beta}) \},$$

and $\boldsymbol{\Gamma}_1(\boldsymbol{\beta})$ is a positive definite matrix.

(ii) $\lim_{n \to \infty} \frac{1}{n} \boldsymbol{\Gamma}_{2n}(\boldsymbol{\beta}) = \boldsymbol{\Gamma}_2(\boldsymbol{\beta})$, where

$$\boldsymbol{\Gamma}_{2n}(\boldsymbol{\beta}) = \sum_{i=1}^{n} \{ \frac{u(\mathbf{x}_i)}{h^2[v(f(\mathbf{x}_i, \boldsymbol{\beta}))]} \mathbf{f}_{\boldsymbol{\beta}}(\mathbf{x}_i, \boldsymbol{\beta}) \, \mathbf{f}_{\boldsymbol{\beta}}^t(\mathbf{x}_i, \boldsymbol{\beta}) \},$$

and $\boldsymbol{\Gamma}_2(\boldsymbol{\beta})$ is a positive definite matrix.

(iii) $max \left\{ \frac{u(\mathbf{x}_i)}{h^2[v(f(\mathbf{x}_i, \boldsymbol{\beta}))]} \mathbf{f}_{\boldsymbol{\beta}}^t(\mathbf{x}_i, \boldsymbol{\beta}) \, (\boldsymbol{\Gamma}_{2n}(\boldsymbol{\beta}))^{-1} \mathbf{f}_{\boldsymbol{\beta}}(\mathbf{x}_i, \boldsymbol{\beta}) \right\} \longrightarrow 0$, as $n \to \infty$

[B3]:
(i) $\lim_{\delta \to 0} \text{Sup}_{\|\boldsymbol{\Delta}\| \leq \delta} \left| (\partial/\partial\beta_j) f(\mathbf{x}, \boldsymbol{\beta} + \boldsymbol{\Delta}) \, (\partial/\partial\beta_k) f(\mathbf{x}, \boldsymbol{\beta} + \boldsymbol{\Delta}) - \right.$

$$\left. (\partial/\partial\beta_j) f(\mathbf{x}, \beta) \, \partial/\partial\beta_k) f(\mathbf{x}, \beta) \right| = 0; \ k = 1, ..., p.$$

(ii) $\lim_{\delta \to 0} \text{Sup}_{\|\boldsymbol{\Delta}\| \leq \delta} \left| (\partial^2/\partial\beta_j \, \partial\beta_k) f(\mathbf{x}, \boldsymbol{\beta} + \boldsymbol{\Delta}) - (\partial^2/\partial\beta_j \, \partial\beta_k) f(\mathbf{x}, \boldsymbol{\beta}) \right| = 0$ for $j, k = 1, ..., p.$

[C]:
(i) $\lim_{\delta \to 0} \text{Sup}_{\|\boldsymbol{\Delta}\| \leq \delta} \left| k_1(\mathbf{x}, \boldsymbol{\beta} + \boldsymbol{\Delta}) - k_1(\mathbf{x}, \boldsymbol{\beta}) \right| = 0$, uniformly in $\mathbf{x}$.

(ii) $\lim_{\delta \to 0} \text{Sup}_{\|\boldsymbol{\Delta}\| \leq \delta} \left| k_2(\mathbf{x}, \boldsymbol{\beta} + \boldsymbol{\Delta}) - k_2(\mathbf{x}, \boldsymbol{\beta}) \right| = 0$, uniformly in $\mathbf{x}$.

23.3   Asymptotic Results

In the first result we shall prove the uniform asymptotic linearity on M-statistics given in the following theorem.

THEOREM 23.3.1
Under the conditions [**A1**]-[**A3**], [**B1**]-[**B3**], *and* [**C**] *we have that:*

$$Sup_{\|\mathbf{t}\|\leq C}\left\|\frac{1}{\sqrt{n}}\sum_{i=1}^{n}\{\lambda(\mathbf{x}_i, Y_i, \boldsymbol{\beta}+n^{-\frac{1}{2}}\mathbf{t}) - \lambda(\mathbf{x}_i, Y_i, \boldsymbol{\beta})\} + \frac{\gamma}{n}\boldsymbol{\Gamma}_{1n}(\boldsymbol{\beta})\mathbf{t}\right\|$$

$$= o_p(1) \qquad (23.3.1)$$

as $n \to \infty$, *where* $\lambda(\mathbf{x}_i, Y_i, \boldsymbol{\beta})$ *was defined in (23.2.3).*

PROOF We consider the jth element of the vector $\lambda(\mathbf{x}_i, Y_i, \boldsymbol{\beta})$ denoted for

$$\lambda_j(\mathbf{x}_i, Y_i, \boldsymbol{\beta}) = \frac{1}{h[v(f(\mathbf{x}_i, \boldsymbol{\beta}))]}\psi(Y_i - f(\mathbf{x}_i, \boldsymbol{\beta}))\, f_{\beta_j}(\mathbf{x}_i, \boldsymbol{\beta}), j = 1, ..., p,$$

where $f_{\beta_j}(\mathbf{x}_i, \boldsymbol{\beta}) = (\partial/\partial\beta_j)f(\mathbf{x}_i, \boldsymbol{\beta})$.

By using linear Taylor expansion, we have:

$$\lambda_j(\mathbf{x}_i, Y_i, \boldsymbol{\beta}+n^{-\frac{1}{2}}\mathbf{t}) - \lambda_j(\mathbf{x}_i, Y_i, \boldsymbol{\beta})$$

$$= \frac{1}{\sqrt{n}}\sum_{k=1}^{p}t_k\{(\partial/\partial\beta_k)\lambda_j(\mathbf{x}_i, Y_i, \boldsymbol{\beta})\}$$

$$+ \frac{1}{\sqrt{n}}\sum_{k=1}^{p}t_k\{(\partial/\partial\beta_k)\lambda_j(\mathbf{x}_i, Y_i, \boldsymbol{\beta}+\frac{h\mathbf{t}}{\sqrt{n}}) - (\partial/\partial\beta_k)\lambda_j(\mathbf{x}_i, Y_i, \boldsymbol{\beta})\},$$

where

$$(\partial/\partial\beta_k)\lambda_j(\mathbf{x}_i, Y_i, \boldsymbol{\beta})$$
$$= k_1(\mathbf{x}_i, \boldsymbol{\beta})\{\psi(Y_i - f(\mathbf{x}_i, \boldsymbol{\beta}))\,(\partial^2/\partial\beta_k\,\partial\beta_j)f(\mathbf{x}_i, \boldsymbol{\beta})$$
$$- \psi'(Y_i - f(\mathbf{x}_i, \boldsymbol{\beta}))\, f_{\beta_j}(\mathbf{x}_i, \boldsymbol{\beta})f_{\beta_k}(\mathbf{x}_i, \boldsymbol{\beta})\}$$
$$- k_2(\mathbf{x}_i, \boldsymbol{\beta})\,\psi(Y_i - f(\mathbf{x}_i, \boldsymbol{\beta}))\, f_{\beta_j}(\mathbf{x}_i, \boldsymbol{\beta})\, f_{\beta_k}(\mathbf{x}_i, \boldsymbol{\beta})$$

and $k_1(\mathbf{x}_i, \boldsymbol{\beta})$ and $k_2(\mathbf{x}_i, \boldsymbol{\beta})$ are given in (23.2.5) and (23.2.6), respectively.

Then for each $j = 1, ..., p$ we have that:

$$
\text{Sup}_{\|\mathbf{t}\|\le C}\left|\frac{1}{\sqrt{n}}\sum_{i=1}^{n}\{\lambda_j(\mathbf{x}_i,Y_i,\boldsymbol{\beta}+\frac{h\,\mathbf{t}}{\sqrt{n}})-\lambda_j(\mathbf{x}_i,Y_i,\boldsymbol{\beta})\}\right.
$$

$$
\left.+\frac{\gamma}{n}\sum_{i=1}^{n}\sum_{k=1}^{p}\{t_k\,\frac{1}{h[v(f(\mathbf{x}_i,\boldsymbol{\beta}))]}\,f_{\beta_j}(\mathbf{x}_i,\boldsymbol{\beta})f_{\beta_k}(\mathbf{x}_i,\boldsymbol{\beta})\}\right|
$$

$$
\le \text{Sup}_{\|\mathbf{t}\|\le C}\left|\frac{1}{n}\sum_{i=1}^{n}\sum_{k=1}^{p}t_k\{(\partial/\partial\beta_k)\lambda_j(\mathbf{x}_i,Y_i,\boldsymbol{\beta}+\frac{h\,\mathbf{t}}{\sqrt{n}})\right.
$$

$$
\left.-(\partial/\partial\beta_k)\lambda_j(\mathbf{x}_i,Y_i,\boldsymbol{\beta})\}\right|
$$

$$
+\text{Sup}_{\|\mathbf{t}\|\le C}\left|\frac{1}{n}\sum_{i=1}^{n}\sum_{k=1}^{p}\{t_k(\partial/\partial\beta_k)\lambda_j(\mathbf{x}_i,Y_i,\boldsymbol{\beta})\}\right.
$$

$$
\left.+\frac{\gamma}{n}\sum_{i=1}^{n}\sum_{k=1}^{p}t_k\,\frac{1}{h[v(f(\mathbf{x}_i,\boldsymbol{\beta}))]}\,f_{\beta_j}(\mathbf{x}_i,\boldsymbol{\beta})f_{\beta_k}(\mathbf{x}_i,\boldsymbol{\beta})\right|,
$$

where

$$
\text{Sup}_{\|\mathbf{t}\|\le C}\left|\frac{1}{n}\sum_{i=1}^{n}\sum_{k=1}^{p}t_k\{(\partial/\partial\beta_k)\lambda_j(\mathbf{x}_i,Y_i,\boldsymbol{\beta}+\frac{h\,\mathbf{t}}{\sqrt{n}})\right.
$$

$$
\left.-(\partial/\partial\beta_k)\lambda_j(\mathbf{x}_i,Y_i,\boldsymbol{\beta})\}\right|
$$

$$
\le \frac{1}{n}C\sum_{i=1}^{n}\sum_{k=1}^{p}\text{Sup}_{\|\mathbf{t}\|\le C}\left|(\partial/\partial\beta_k)\lambda_j(\mathbf{x}_i,Y_i,\boldsymbol{\beta}+\frac{h\,\mathbf{t}}{\sqrt{n}})\right.
$$

$$
\left.-(\partial/\partial\beta_k)\lambda_j(\mathbf{x}_i,Y_i,\boldsymbol{\beta})\right|.
$$

Now, based on conditons [**A3**] (i)–(ii), [**B3**] (i)–(ii), and [**C3**] (i)–(ii) we may prove that:

$$
E\left\{\text{Sup}_{\|\mathbf{t}\|\le C}\left|(\partial/\partial\beta_k)\lambda_j(\mathbf{x}_i,Y_i,\boldsymbol{\beta}+\frac{h\,\mathbf{t}}{\sqrt{n}})\right.\right.
$$

$$
\left.\left.-(\partial/\partial\beta_k)\lambda_j(\mathbf{x}_i,Y_i,\boldsymbol{\beta})\right|\right\}\longrightarrow 0,\quad \forall i\le n,
$$

and

$$E\left\{\mathrm{Sup}_{\|\mathbf{t}\|\leq C}\left|\frac{1}{n}\sum_{i=1}^{n}\sum_{k=1}^{p}t_k\left\{(\partial/\partial\beta_k)\lambda_j(\mathbf{x}_i,Y_i,\boldsymbol{\beta}+\frac{h\,\mathbf{t}}{\sqrt{n}})\right.\right.\right.$$

$$\left.\left.\left.-(\partial/\partial\beta_k)\lambda_j(\mathbf{x}_i,Y_i,\boldsymbol{\beta})\right\}\right|\right\}\longrightarrow 0.$$

Also,

$$Var\left\{\mathrm{Sup}_{\|\mathbf{t}\|\leq C}\left|\frac{1}{n}\sum_{i=1}^{n}\sum_{k=1}^{p}t_k\{(\partial/\partial\beta_k)\lambda_j(\mathbf{x}_i,Y_i,\boldsymbol{\beta}+\frac{h\,\mathbf{t}}{\sqrt{n}})\right.\right.$$

$$\left.\left.-(\partial/\partial\beta_k)\lambda_j(\mathbf{x}_i,Y_i,\boldsymbol{\beta})\}\right|\right\}$$

$$\leq\frac{1}{n^2}C^2\sum_{i=1}^{n}\left(Var\{\sum_{k=1}^{p}\mathrm{Sup}_{\|\mathbf{t}\|\leq C}\left|(\partial/\partial\beta_k)\lambda_j(\mathbf{x}_i,Y_i,\boldsymbol{\beta}+\frac{h\,\mathbf{t}}{\sqrt{n}})\right.\right.$$

$$\left.\left.-(\partial/\partial\beta_k)\lambda_j(\mathbf{x}_i,Y_i,\boldsymbol{\beta})\right|\}\right)$$

$$\leq C^2 K/n\longrightarrow 0.$$

Therefore:

$$\mathrm{Sup}_{\|\mathbf{t}\|\leq C}\left|\frac{1}{n}\sum_{i=1}^{n}\sum_{k=1}^{p}t_k\{(\partial/\partial\beta_k)\lambda_j(\mathbf{x}_i,Y_i,\boldsymbol{\beta}+\frac{h\,\mathbf{t}}{\sqrt{n}})\right.$$

$$\left.-(\partial/\partial\beta_k)\lambda_j(\mathbf{x}_i,Y_i,\boldsymbol{\beta})\}\right|=o_p(1). \tag{23.3.2}$$

In addition, we may have,

$$\mathrm{Sup}_{\|\mathbf{t}\|\leq C}\left|\frac{1}{n}\sum_{i=1}^{n}\sum_{k=1}^{p}t_k(\partial/\partial\beta_k)\lambda_j(\mathbf{x}_i,Y_i,\boldsymbol{\beta})\right.$$

$$\left.+\frac{\gamma}{n}\sum_{i=1}^{n}\sum_{k=1}^{p}t_k\frac{1}{h[v(f(\mathbf{x}_i,\boldsymbol{\beta}))]}f_{\beta_j}(\mathbf{x}_i,\boldsymbol{\beta})f_{\beta_k}(\mathbf{x}_i,\boldsymbol{\beta})\right|$$

$$=\mathrm{Sup}_{\|\mathbf{t}\|\leq C}\left|\frac{1}{n}\sum_{i=1}^{n}\sum_{k=1}^{p}\{t_k\,\psi(Y_i-f(\mathbf{x}_i,\boldsymbol{\beta}))\,[k_1(\mathbf{x}_i,\boldsymbol{\beta})\right.$$

$$\times(\partial^2/\partial\beta_k\,\partial\beta_j)f(\mathbf{x}_i,\boldsymbol{\beta})$$

$$-k_2(\mathbf{x}_i,\boldsymbol{\beta})\,f(\mathbf{x}_i,\boldsymbol{\beta}))\,f_{\beta_j}(\mathbf{x}_i,\boldsymbol{\beta})\,f_{\beta_k}(\mathbf{x}_i,\boldsymbol{\beta})]\}$$

$$-\frac{1}{n}\sum_{i=1}^{n}\sum_{k=1}^{p}\{t_k\, k_1(\mathbf{x}_i,\boldsymbol{\beta})\,[\psi'(Y_i-f(\mathbf{x}_i,\boldsymbol{\beta}))-\gamma\,]$$

$$\times f_{\beta_j}(\mathbf{x}_i,\boldsymbol{\beta})\,f_{\beta_k}(\mathbf{x}_i,\boldsymbol{\beta})\}\Big|$$

$$\le C\sum_{k=1}^{p}\Big|\frac{1}{n}\sum_{i=1}^{n}\psi(Y_i-f(\mathbf{x}_i,\boldsymbol{\beta}))\,[k_1(\mathbf{x}_i,\boldsymbol{\beta})\,(\partial^2/\partial\beta_k\,\partial\beta_j)f(\mathbf{x}_i,\boldsymbol{\beta})$$

$$-k_2(\mathbf{x}_i,\boldsymbol{\beta})\,f_{\beta_j}(\mathbf{x}_i,\boldsymbol{\beta})\,f_{\beta_k}(\mathbf{x}_i,\boldsymbol{\beta})]\Big|$$

$$+C\sum_{k=1}^{p}\Big|\frac{1}{n}\sum_{i=1}^{n}k_1(\mathbf{x}_i,\boldsymbol{\beta})\,[\psi'(Y_i-f(\mathbf{x}_i,\boldsymbol{\beta}))-\gamma\,]$$

$$\times f_{\beta_j}(\mathbf{x}_i,\boldsymbol{\beta})\,f_{\beta_k}(\mathbf{x}_i,\boldsymbol{\beta})\Big|$$

which by using the Markov WLNN and conditions [**A2**] (i)–(ii) yields:

$$\frac{1}{n}\sum_{i=1}^{n}\{\psi(Y_i-f(\mathbf{x}_i,\boldsymbol{\beta}))\,[k_1(\mathbf{x}_i,\boldsymbol{\beta})\,(\partial^2/\partial\beta_k\,\partial\beta_j)f(\mathbf{x}_i,\boldsymbol{\beta})$$

$$-k_2(\mathbf{x}_i,\boldsymbol{\beta})\,f_{\beta_j}(\mathbf{x}_i,\boldsymbol{\beta})\,f_{\beta_k}(\mathbf{x}_i,\boldsymbol{\beta})]\}=o_p(1),$$

and

$$\frac{1}{n}\sum_{i=1}^{n}\{\,k_1(\mathbf{x}_i,\boldsymbol{\beta})\,[\psi'(Y_i-f(\mathbf{x}_i,\boldsymbol{\beta}))-\gamma\,]\,f_{\beta_j}(\mathbf{x}_i,\boldsymbol{\beta})\,f_{\beta_k}(\mathbf{x}_i,\boldsymbol{\beta})\}=o_p(1).$$

Thus:

$$\mathrm{Sup}_{\|\mathbf{t}\|\le C}\Big|\frac{1}{n}\sum_{i=1}^{n}\sum_{k=1}^{p}t_k(\partial/\partial\beta_k)\lambda_j(\mathbf{x}_i,Y_i,\boldsymbol{\beta})$$

$$+\frac{\gamma}{n}\sum_{i=1}^{n}\sum_{k=1}^{p}t_k\,\frac{1}{h[v(f(\mathbf{x}_i,\boldsymbol{\beta}))]}\,f_{\beta_j}(\mathbf{x}_i,\boldsymbol{\beta})f_{\beta_k}(\mathbf{x}_i,\boldsymbol{\beta})\Big|=o_p(1).$$

$$(23.3.3)$$

Therefore, from (23.3.2) and (23.3.3) we may conclude that:

$$\text{Sup}_{\|\mathbf{t}\|\leq C}\left|\frac{1}{n}\sum_{i=1}^{n}\{\lambda_j(\mathbf{x}_i,Y_i,\boldsymbol{\beta}+\frac{h\mathbf{t}}{\sqrt{n}})-\lambda_j(\mathbf{x}_i,Y_i,\boldsymbol{\beta})\}\right.$$

$$\left.+\frac{\gamma}{n}\sum_{i=1}^{n}\sum_{k=1}^{p}t_k\frac{1}{h[v(f(\mathbf{x}_i,\boldsymbol{\beta}))]}f_{\beta_j}(\mathbf{x}_i,\boldsymbol{\beta})f_{\beta_k}(\mathbf{x}_i,\boldsymbol{\beta})\right|=o_p(1),\ j=1,...,p.$$

∎

We may now consider the following theorem:

THEOREM 23.3.2
Under the conditions [A1]-[A3], [B1]-[B3], *and* [C] *there exists a sequence* $\hat{\boldsymbol{\beta}}_n$ *of solutions of* (23.2.2) *such that:*

$$n^{\frac{1}{2}}\|\hat{\boldsymbol{\beta}}_n-\boldsymbol{\beta}\|=O_p(1) \tag{23.3.4}$$

$$\hat{\boldsymbol{\beta}}_n=\boldsymbol{\beta}+\frac{1}{n\gamma}(\frac{1}{n}\boldsymbol{\Gamma}_{1n}(\boldsymbol{\beta}))^{-1}\sum_{i=1}^{n}\lambda(\mathbf{x}_i,Y_i,\boldsymbol{\beta})+o_p(n^{-\frac{1}{2}}). \tag{23.3.5}$$

PROOF From Theorem 1.3.1 we have that the system of equations:

$$\sum_{i=1}^{n}\lambda_j(\mathbf{x}_i,Y_i,\boldsymbol{\beta}+n^{-\frac{1}{2}}\mathbf{t})=0$$

has a root t_n that lies in $\|t\|\leq C$ with probability exceeding $1-\epsilon$ for $n\geq n_0$. Then $\hat{\boldsymbol{\beta}}_n=\boldsymbol{\beta}+n^{-\frac{1}{2}}t_n$ is a solution of the equations in (23.2.2) satisfying:

$$P(\|n^{\frac{1}{2}}(\hat{\boldsymbol{\beta}}_n-\boldsymbol{\beta})\|\leq C)\geq 1-\epsilon\ \text{for}\ n\geq n_0.$$

Inserting $\mathbf{t}\longrightarrow n^{\frac{1}{2}}(\hat{\boldsymbol{\beta}}_n-\boldsymbol{\beta})$ in (23.3.1), we have that the expression in (23.3.5). ∎

In the following theorem we shall prove the asymptotic normality.

THEOREM 23.3.3
Under the conditions [A1], [A2] *(i)-(ii)*, [B1], [B2] *(i)-(ii)*, *we have that:*

$$\frac{1}{\sqrt{n}}\sum_{i=1}^{n}\lambda(\mathbf{x}_i,Y_i,\boldsymbol{\beta})\longrightarrow N_p(0,\sigma_\psi^2\boldsymbol{\Gamma}_2(\boldsymbol{\beta})),\ \text{as}\ n\to\infty \tag{23.3.6}$$

PROOF We consider an arbitrary linear compound:

$$Z_n^* = \eta^t \frac{1}{\sqrt{n}} \sum_{i=1}^n \lambda(\mathbf{x}_i, Y_i, \boldsymbol{\beta}), \ \eta \in \Re^p,$$

and have that:

$$Z_n^* = \frac{1}{\sqrt{n}} \sum_{i=1}^n \frac{1}{h[v(f(\mathbf{x}_i, \boldsymbol{\beta}))]} \psi(Y_i - f(\mathbf{x}_i, \boldsymbol{\beta})) \eta^t \mathbf{f_\beta}(\mathbf{x}_i, \boldsymbol{\beta}) = \sum_{i=1}^n c_{ni} Z_i,$$

where

$$c_{ni} = \frac{\sigma_\psi}{\sqrt{n}} \frac{u(\mathbf{x}_i)}{h[v(f(\mathbf{x}_i, \boldsymbol{\beta}))]} \eta^t \mathbf{f_\beta}(\mathbf{x}_i, \boldsymbol{\beta}),$$

and

$$Z_i = \psi(Y_i - f(\mathbf{x}_i, \boldsymbol{\beta}))/(\sigma_\psi \sqrt{u(\mathbf{x}_i)}).$$

Then by using the Hájek-Šidak Central Limit Theorem, we may show that Z_n^* converges in law to a normal distribution, as $n \to \infty$. In order to use this Theorem we need to verify the regularity condition about c_{ni}, which is given by

$$\max_{1 \leq i \leq n} c_{ni}^2 / \sum_{i=1}^n c_{ni}^2 \longrightarrow 0,$$

as $n \to \infty$, and it can be reformulated by requiring that as $n \to \infty$,

$$\sup_{\eta \in \Re^p} \left[\max_{1 \leq i \leq n} \eta^t \frac{u(\mathbf{x}_i)}{h^2[v(f(\mathbf{x}_i, \boldsymbol{\beta}))]} \mathbf{f_\beta}(\mathbf{x}_i, \boldsymbol{\beta}) \mathbf{f_\beta^t}(\mathbf{x}_i, \boldsymbol{\beta}) \eta \ / \ \eta^t \, \boldsymbol{\Gamma}_{2n}(\boldsymbol{\beta}) \, \eta \right]$$

$$\longrightarrow 0.$$

Now, in view of the Courant's Theorem, we have that:

$$\sup_{\eta \in \Re^p} \left[\eta^t \frac{u(\mathbf{x}_i)}{h^2[v(f(\mathbf{x}_i, \boldsymbol{\beta}))]} \mathbf{f_\beta}(\mathbf{x}_i, \boldsymbol{\beta}) \mathbf{f_\beta^t}(\mathbf{x}_i, \boldsymbol{\beta}) \eta \ / \ \eta^t \, \boldsymbol{\Gamma}_{2n}(\boldsymbol{\beta}) \, \eta \right]$$

$$= ch_1 \left\{ \frac{u(\mathbf{x}_i)}{h^2[v(f(\mathbf{x}_i, \boldsymbol{\beta}))]} \mathbf{f_\beta}(\mathbf{x}_i, \boldsymbol{\beta}) \mathbf{f_\beta^t}(\mathbf{x}_i, \boldsymbol{\beta}) (\boldsymbol{\Gamma}_{2n}(\boldsymbol{\beta}))^{-1} \right\}$$

$$= \frac{u(\mathbf{x}_i)}{h^2[v(f(\mathbf{x}_i, \boldsymbol{\beta}))]} \mathbf{f_\beta^t}(\mathbf{x}_i, \boldsymbol{\beta}) (\boldsymbol{\Gamma}_{2n}(\boldsymbol{\beta}))^{-1} \mathbf{f_\beta}(\mathbf{x}_i, \boldsymbol{\beta}),$$

so this condition is reduced to the condition [**B2**] (iii) (Noether's condition). Thus, we conclude that:

$$Z_n^* / [\sum_{i=1}^n c_{ni}^2]^{\frac{1}{2}} \longrightarrow N(0, 1), \text{ as } n \to \infty$$

and by using the Cramer-Wold Theorem and condition [**B2**] (ii) we may prove the expression in (23.3.6). ∎

COROLLARY 23.3.4
Under the conditions [**A1**]-[**A3**], [**B1**]-[**B3**],

$$\sqrt{n}\,(\hat{\boldsymbol{\beta}}_n - \boldsymbol{\beta}) \longrightarrow N_p\left(\mathbf{0},\, \gamma^{-2}\,\sigma_\psi^2\,\boldsymbol{\Gamma}_1^{-1}(\boldsymbol{\beta})\,\boldsymbol{\Gamma}_2(\boldsymbol{\beta})\,\boldsymbol{\Gamma}_1^{-1}(\boldsymbol{\beta})\right) \qquad (23.3.7)$$

PROOF From Theorem 1.3.2 we have that:

$$\sqrt{n}\,(\hat{\boldsymbol{\beta}}_n - \boldsymbol{\beta}) = (\frac{1}{n}\,\boldsymbol{\Gamma}_n(\boldsymbol{\beta}))^{-1}\,\frac{1}{\sqrt{n}}\,\sum_{i=1}^{n}\lambda(\mathbf{x}_i, Y_i, \boldsymbol{\beta}) + o_p(1)$$

Then from Theorem 1.3.3 and the Slutsky Theorem we may have the expression in (23.3.7). ∎

COROLLARY 23.3.5
Under the conditions [**A1**]-[**A3**], [**B1**]-[**B3**],

$$\hat{\boldsymbol{\Gamma}}^{-\frac{1}{2}}\,\sqrt{n}\,(\hat{\boldsymbol{\beta}}_n - \boldsymbol{\beta}) \longrightarrow N_p\left(\mathbf{0},\, I_p\right), \qquad (23.3.8)$$

where

$$\hat{\boldsymbol{\Gamma}} = \hat{\gamma}^{-2}\,\hat{\sigma}_\psi^2\,(\frac{1}{n}\,\boldsymbol{\Gamma}_{1n}(\hat{\boldsymbol{\beta}}_n))^{-1}\,(\frac{1}{n}\,\boldsymbol{\Gamma}_{2n}(\hat{\boldsymbol{\beta}}_n))\,(\frac{1}{n}\,\boldsymbol{\Gamma}_{1n}(\hat{\boldsymbol{\beta}}_n))^{-1}, \qquad (23.3.9)$$

and $\hat{\gamma}$ and $\hat{\sigma}_\psi^2$ are consistent estimators of γ and σ_ψ^2, respectively.

PROOF Using expression (23.3.7) and the Slutsky Theorem we have the expression in (23.3.8). ∎

COROLLARY 23.3.6
Under the conditions [**A1**]-[**A3**], [**B1**]-[**B3**],

$$n\,(\hat{\boldsymbol{\beta}}_n - \boldsymbol{\beta})^t\,\hat{\boldsymbol{\Gamma}}^{-1}\,(\hat{\boldsymbol{\beta}}_n - \boldsymbol{\beta}) \longrightarrow \chi_p^2 \qquad (23.3.10)$$

PROOF By using expression (23.3.8) and the Cochran Theorem, we prove the expression in (23.3.10). ∎

We may use (23.3.10) to provide a confidence set for $\boldsymbol{\beta}$. Let $B_n(\alpha) = \{\boldsymbol{\beta} : n\,(\hat{\boldsymbol{\beta}}_n - \boldsymbol{\beta})^t\,\hat{\boldsymbol{\Gamma}}^{-1}\,(\hat{\boldsymbol{\beta}}_n - \boldsymbol{\beta}) \le \chi_{p,\alpha}^2\}$, where $\chi_{p,\alpha}^2$ is the upper $100\%\alpha$ point of the chi-square distribution with p degrees of freedom. Then, $P\{\boldsymbol{\beta} \in B_n(\alpha)\,/\,\boldsymbol{\beta}\} \longrightarrow 1 - \alpha$, as n increases.

23.4 Test of Significance and Computational Algorithm

23.4.1 Subhypothesis Testing

Let us consider the partition of $\boldsymbol{\beta}$ as

$$\begin{pmatrix} \boldsymbol{\beta}_1 \\ \boldsymbol{\beta}_2 \end{pmatrix}$$

where $\boldsymbol{\beta}_1$ is a vector of $r \times 1$ and $\boldsymbol{\beta}_2$ is a vector of $(p-r) \times 1$. We want to test the hypothesis $H_o : \boldsymbol{\beta}_1 = 0$ vs $H_1 : \boldsymbol{\beta}_1 \neq 0$.

From (23.3.9) we can write $\hat{\boldsymbol{\Gamma}} = \hat{\gamma}_n^{-2}\, \hat{\sigma}_\psi^2\, \boldsymbol{\Sigma}$, where

$$\hat{\boldsymbol{\Sigma}} = (\frac{1}{n}\,\boldsymbol{\Gamma}_{1n}(\hat{\boldsymbol{\beta}}_n))^{-1}\,(\frac{1}{n}\,\boldsymbol{\Gamma}_{2n}(\hat{\boldsymbol{\beta}}_n))\,(\frac{1}{n}\,\boldsymbol{\Gamma}_{1n}(\hat{\boldsymbol{\beta}}_n))^{-1},$$

$$\hat{\gamma}_n = \frac{1}{n}\sum_{i=1}^{n}\psi'(Y_i - f(\mathbf{x}_i, \hat{\boldsymbol{\beta}}_n)),$$

and

$$\hat{\sigma}_\psi^2 = \frac{1}{n-p}\sum_{i=1}^{n}\psi^2(Y_i - f(\mathbf{x}_i, \hat{\boldsymbol{\beta}})).$$

We can also partition $\hat{\boldsymbol{\Sigma}}^{-1}$ as:

$$\hat{\boldsymbol{\Sigma}}^{-1} = \begin{pmatrix} \mathbf{V}^{11} & \mathbf{V}^{12} \\ \mathbf{V}^{21} & \mathbf{V}^{22} \end{pmatrix}$$

and use the Cochran theorem to prove that:

$$n\,(\hat{\boldsymbol{\beta}}_{n1} - \boldsymbol{\beta}_1)^t\,(\hat{\gamma}_n^{-2}\,\hat{\sigma}_\psi^2)^{-1}\,\mathbf{V}^{11}\,(\hat{\boldsymbol{\beta}}_{n1} - \boldsymbol{\beta}_1) \longrightarrow \chi_r^2.$$

Then, for testing the null hypothesis we can define the Wald-type M-test as:

$$W = n\,(\hat{\boldsymbol{\beta}}_{n1})^t\,(\hat{\gamma}_n^{-2}\,\hat{\sigma}_\psi^2)^{-1}\,\mathbf{V}^{11}\,\hat{\boldsymbol{\beta}}_{n1}$$

which under H_o follows (asymptotically) a χ_r^2 distribution.

23.4.2 Nonlinear Hypothesis Testing

Let us consider the nonlinear hypothesis

$$H_o : a(\boldsymbol{\beta}) = 0 \quad \text{vs} \quad H_1 : a(\boldsymbol{\beta}) \neq 0$$

where a is a real valued (nonlinear) function (of $\boldsymbol{\beta}$). By using Corollary 1.3.5 and the delta method we have that:

$$\{\mathbf{a}_{\boldsymbol{\beta}}^t(\hat{\boldsymbol{\beta}}_n)\,\hat{\boldsymbol{\Gamma}}\,\mathbf{a}_{\boldsymbol{\beta}}(\hat{\boldsymbol{\beta}}_n)\}^{-\frac{1}{2}}\,n^{\frac{1}{2}}\,(a(\hat{\boldsymbol{\beta}}_n) - a(\boldsymbol{\beta})) \longrightarrow N(0, 1),$$

where $\mathbf{a}_{\boldsymbol{\beta}}(\boldsymbol{\beta}) = (\partial/\partial\boldsymbol{\beta})a(\boldsymbol{\beta})$. Also, by using the Cochran theorem:

$$n\left(a(\hat{\boldsymbol{\beta}}_n) - a(\boldsymbol{\beta})\right)^t \{\mathbf{a}_{\boldsymbol{\beta}}^t(\hat{\boldsymbol{\beta}}_n) \, \hat{\boldsymbol{\Gamma}} \, \mathbf{a}_{\boldsymbol{\beta}}(\hat{\boldsymbol{\beta}}_n)\}^{-1} \left(a(\hat{\boldsymbol{\beta}}_n) - a(\boldsymbol{\beta})\right) \longrightarrow \chi_1^2$$

Then, for testing the null hypothesis we can use the Wald-type M-test:

$$W = n\left(a(\hat{\boldsymbol{\beta}}_n)\right)^t \{\mathbf{a}_{\boldsymbol{\beta}}^t(\hat{\boldsymbol{\beta}}_n) \, \hat{\boldsymbol{\Gamma}} \, \mathbf{a}_{\boldsymbol{\beta}}(\hat{\boldsymbol{\beta}}_n)\}^{-1} \left(a(\hat{\boldsymbol{\beta}}_n)\right)$$

which under H_o follows a χ_1^2.

23.4.3 Computational Algorithm

In order to solve the equations in (23.2.2), we propose two iterative methods based on the Taylor expansion around some initial guess $\hat{\boldsymbol{\beta}}_n^{(0)}$.
We define the following matrices:

$$A(\boldsymbol{\beta}) = \sum_{i=1}^{n} \psi(Y_i - f(\mathbf{x}_i, \boldsymbol{\beta})) \, k_1(\mathbf{x}_i, \boldsymbol{\beta}) \, f_{\boldsymbol{\beta}}(\mathbf{x}_i, \boldsymbol{\beta}),$$

$$W(\boldsymbol{\beta}) = Diag\left(k_1(\mathbf{x}_1, \boldsymbol{\beta}), ..., k_1(\mathbf{x}_n, \boldsymbol{\beta})\right),$$

$$W_1(\boldsymbol{\beta}) = Diag\left(\psi((Y_1 - f(\mathbf{x}_1, \boldsymbol{\beta}))\right.$$
$$\left. \times k_2(\mathbf{x}_1, \boldsymbol{\beta}), ..., \psi((Y_n - f(\mathbf{x}_n, \boldsymbol{\beta}))k_2(\mathbf{x}_n, \boldsymbol{\beta})\right),$$

$$W_2(\boldsymbol{\beta}) = Diag\left(\psi'((Y_1 - f(\mathbf{x}_1, \boldsymbol{\beta}))\right.$$
$$\left. \times k_1(\mathbf{x}_1, \boldsymbol{\beta}), ..., \psi'((Y_n - f(\mathbf{x}_n, \boldsymbol{\beta})) \, k_1(\mathbf{x}_n, \boldsymbol{\beta})\right),$$

$$X(\boldsymbol{\beta}) = \left(f_{\boldsymbol{\beta}}(\mathbf{x}_1, \boldsymbol{\beta}), ..., f_{\boldsymbol{\beta}}(\mathbf{x}_n, \boldsymbol{\beta})\right)^t,$$

$$\Psi(\mathbf{Y} - f(\mathbf{x}, \boldsymbol{\beta})) = \left(\psi(Y_1 - f(\mathbf{x}_1, \boldsymbol{\beta})), ..., \psi(Y_n - f(\mathbf{x}_n, \boldsymbol{\beta}))\right)^t$$

$$U(\boldsymbol{\beta}) = A(\boldsymbol{\beta}) - X^t(\boldsymbol{\beta})\left[W_1(\boldsymbol{\beta}) - W_2(\boldsymbol{\beta})\right]X(\boldsymbol{\beta}) \qquad \text{and}$$

$$V(\boldsymbol{\beta}) = X^t(\boldsymbol{\beta}) \, W(\boldsymbol{\beta}) \, X(\boldsymbol{\beta}).$$

The first method is similar to the Newton-Raphson procedure, which is given by:

$$\begin{cases} \hat{\boldsymbol{\beta}}_n^{(0)} = \hat{\boldsymbol{\beta}}_{ML} \\[2mm] \hat{\boldsymbol{\beta}}_n^{(l+1)} = \hat{\boldsymbol{\beta}}_n^{(l)} - \{U(\hat{\boldsymbol{\beta}}_n^{(l)})\}^{-1} X^t(\hat{\boldsymbol{\beta}}_n^{(l)}) \, W(\hat{\boldsymbol{\beta}}_n^{(l)}) \, \Psi(\mathbf{Y} - f(\mathbf{x}, \hat{\boldsymbol{\beta}}_n^{(l)})). \end{cases}$$

We can replace $U(\boldsymbol{\beta})$ by its expected value,

$$EU(\boldsymbol{\beta}) = -\gamma \, V(\boldsymbol{\beta}),$$

and propose the following algorithm:

$$
\begin{cases}
\hat{\boldsymbol{\beta}}_n^{(0)} = \hat{\boldsymbol{\beta}}_{ML} \\
\hat{\boldsymbol{\beta}}_n^{(l+1)} = \hat{\boldsymbol{\beta}}_n^{(l)} + (\hat{\gamma}_n^l)^{-1}\left\{V(\hat{\boldsymbol{\beta}}_n^{(l)})\right\}^{-1} X^t(\hat{\boldsymbol{\beta}}_n^{(l)})\,W(\hat{\boldsymbol{\beta}}_n^{(l)}) \\
\qquad\qquad\qquad\qquad \times \Psi(\mathbf{Y} - f(\mathbf{x}, \hat{\boldsymbol{\beta}}_n^{(l)})),
\end{cases}
$$

where

$$
\hat{\gamma}_n^{(l)} = \frac{1}{n}\sum_{i=1}^{n} \psi'(Y_i - f(\mathbf{x}_i, \hat{\beta}_n^{(l)})).
$$

This procedure is similar to the Fisher-scoring method.

Acknowledgements Thanks are due to the reviewers for their helpful comments on the manuscript. The first author was partially supported by Grant #97036 from the International Clinical Epidemiology Network (INCLEN).

References

1. Beal, S. L. and Sheiner, L. B. (1988). Heteroscedastic nonlinear regression. *Technometrics* **30**, 327–338.

2. Bickel, P. J. (1973). On some analogues to linear combinations of order statistics in the linear model. *Annals of Statistics* **1**, 597–616.

3. Bickel, P. J. (1975). One-step Huber estimates in the linear model. *Journal of the American Statistical Association* **70**, 428–434.

4. Carroll, R. J. and Ruppert, D. (1982). Robust estimation in heteroscedastic linear models. *Annals of Statistics* **10**, 429–441.

5. Dutter, R. (1975). Robust regression: different approaches to numerical solutions and algorithms. Research report 6, Fachgruppe fuer statistik, ETH, Zurich.

6. Gallant, A. R. (1975). Seemingly unrelated nonlinear regressions. *Journal of Econometrics* **3**, 35–50.

7. Gallant, A. R. (1987). *Nonlinear Statistical Models*. John Wiley & Sons, New York.

8. Genning, C., Chinchilli, V. M., and Carter, W. H. (1989). Response surface analysis with correlated data: A nonlinear model approach. *Journal of the American Statistical Association* **84**, 805–809.

9. Giltinan, D. M. and Ruppert, D. (1989). Fitting heteroscedastic regression models to individual pharmacokinetic data using standard

statistical software. *Journal of Pharmacokinetics and Biopharmaceutics* **17**, 601-614.

10. Hampel, F. R., Rousseeuw, P. J., Ronchetti, E. and Stabel, W. (1986). *Robust Statistics — The Approach Based on Influence Functions.* John Wiley & Sons, New York.

11. Hartley, H. O. (1961). The modified Gauss-Newton method for the fitting of nonlinear regression functions by least squares. *Technometrics* **3**, 269–280.

12. Hastie, T. J. and Tibshirani, R. J. (1990). *Generalized Additive Models.* Chapman and Hall, London.

13. Hill, R. W. and Holland, P. W. (1977). Two robust alternatives to least squares regression. *Journal of the American Statistical Association* **72**, 828–833.

14. Holland, P. W. and Welsch, R. E. (1977). Robust regression using iteratively reweighted least squares. *Communications in Statistics* **A6**, 813–827.

15. Huber, P. J. (1964). Robust estimator of a location parameter. *Annals of Mathematical Statistics* **35**, 73–101.

16. Huber, P. J. (1973). Robust regression: asymptotics, conjetures and Monte-Carlo. *Annals of Statistics* **1**, 799–821.

17. Huber, P. J. (1981). *Robust Statistics.* John Wiley & Sons, New York.

18. Jaeckel, L. A. (1972). Estimating regression coefficients by minimizing the dispersion of the residuals. *Annals of Mathematical Statistics* **43**, 1449–1458.

19. Jurečková, J. and Sen, P. K. (1996). *Robust Statistical Procedures, Asymtotics and Interrelations.* John Wiley & Sons, New York.

20. Klein, R. and Yohai, V. J. (1981). Asymptotic behavior of iterative M-estimators for the linear model. *Communications in Statistics* **A10**, 2373–2388.

21. Maronna, R. A. (1976). Robust M-estimators of multivariate location and scatter. *Annals of Statistics* **4**, 163–169.

22. Marquardt, D. W. (1963). An algorithm for least-squares estimation of nonlinear parameters. *Journal of the Society for Industrial and Applied Mathematics* **11**, 431–441.

23. McCullagh, P. (1983). Quasilikelihood functions. *Annals of Statistics* **11**, 59–60.

24. McCullagh, P. and Nelder, J. A. (1989). *Generalized Linear Models* Second Edition. Chapman and Hall, UK.

25. Nelder, J. A. and Wedderburn, R. W. M. (1972). Generalized linear models. *Journal of the Royal Statistical Society, Series A* **135**, 370–384.

26. Schrader, R. M. and Hettmansperger, T. P. (1980). Robust analysis of variance based upon a likelihood ratio criterion. *Biometrika* **67**, 93–101.

27. Sen, P. K. (1982). On M-tests in linear models. *Biometrika* **69**, 245–248.

28. Sen, P. K. (1998). Generalized linear and additive models: Robustness perspectives. *Brazilian Journal of Probability and Statistics* **12**, 91–112.

29. Sen, P. K. and Singer, J. M. (1993). *Large Sample Methods in Statistics: An Introduction with Applications*. Chapman and Hall, New York.

30. Singer, J. M. and Sen, P. K. (1985). M-methods in multivariate linear models. *Journal of Multivariate Analysis* **17**, 168–184.

31. van Houwelingen, J. C. (1988). Use and abuse of variance models in regression. *Biometrics* **43**, 1073–1081.

32. Wedderburn, R. W. M. (1974). Quasi-likelihood functions, generalized linear models, and the Gauss-Newton method. *Biometrika* **61**, 439–447.

33. Yohai, V. J. and Maronna, R. A. (1979). Asymptotic behaviour of M-estimators for the linear model. *Annals of Statistics* **7**, 258-268.

34. Zeger, S. L., Liang, K. Y., and Albert, P. S. (1988). Models for longitudinal data: A general estimating equation approach. *Biometrics* **44**, 1049–1060.

PART V

Inference and Applications

24

Extensions of a Variation of the Isoperimetric Problem

Herman Chernoff
Harvard University, Cambridge, MA

ABSTRACT A problem in the efficient storage of noisy information in two dimensional space leads to a gaussian variation of the classical Isoperimetric Problem. In this variation the area and perimeter are modified by the bivariate normal distribution. This problem corresponds to the efficient storage of the noisy information with one bit; i.e. dividing the plane into two sets. The extension considered here corresponds to dividing the plane into n sets. For Example 3 bits of storage corresponds to $n = 8$.

Keywords and phrases Bivariate normal distribution, inequality, isoperimetric problem, information retrieval, information storage

24.1 Introduction

Since this symposium was organized in honor of Professor Theophilos Cacoullos, it seemed appropriate to resurrect a dormant problem which gave rise to an inequality which he generalized, and he and his students studied in considerable detail.

This inequality, which I published in Chernoff (1981), and was an example of a more general result published independently by Brascomb and Lieb by Lieb in 1976, states that if X is normally distributed with mean 0 and variance 1, $g(x)$ is absolutely continuous, and the variance of $g(X)$ is finite, then $E[g'(X)]^2 \geq \mathrm{Var}[g(X)]$. The proof I offered was rather trivial after considerable effort was expended in demonstrating that $g(X)$ could be expanded in Hermite polynomials.

At the time I had derived the inequality, I had two visitors at MIT. They were Louis H. Y. Chen and Theo Cacoullos. I complained to them that the derivation did not seem appropriate and that there must be a better approach. Both Chen and Cacoullos accepted my challenge after leaving MIT and independently derived very interesting different generalizations using different approaches. The bibliography lists a couple of the earlier papers by Cacoullos and Chen. Incidentally, I am almost convinced, that had I the patience to do the scholarly labor, some form of the inequality could be found in the nineteenth century literature.

My object here is to describe the original context from which I came across the inequality, a problem in information retrieval, and how it leads to a variation of the classical isoperimetric problem. Since my retirement, I have had the urge to clean up some problems that I came across in the past and never seemed to have the time to address. One of these is an extension of the variation of the isoperimetric problem.

24.2   Information Retrieval Problem

Around 1973 I attended a lecture on information retrieval in which a problem posed by the FBI fingerprint files was mentioned. It was claimed that the FBI has a library of about 40 million sets of fingerprints, that 30 thousand sets of fingerprints arrive daily, and that each set must be compared with those in the file to see if the subject is already represented there, possibly under another name. The task was said to be accomplished with the help of 15 thousand clerks.

At that time attempts were being made to use computers to automate the process of storing and retrieving the relevant information in the fingerprints. I spoke with some of the people at NIST who were supposed to be leaders in the research effort leading to automation and found that their attempts were rather *ad hoc* and that a serious theoretical framework was lacking. In particular, the role of measurement error was not properly factored into any theoretical approach.

When I discovered that the FBI had files on 5 thousand subjects where two sets of fingerprints were kept for each subject, that seemed to be an excellent source of information on measurement error. My attempts to gain access to those files were discouraged. This led me to terminate my research in this area where the major issues seemed to be measurement error and the high dimensionality of the data. My major publication in this area is Chernoff (1980)

24.3 Information Retrieval without Measurement Error

In the case where there is no measurement error, there is an effective method of storing large amounts of high dimensional data called *hash coding*. To explain this valuable method, consider the problem of setting up a library of $n = 50$ million distinct entries, each represented by a variable X consisting of $m = 100$ binary digits. The object is to store this information efficiently so as to be able to determine whether a target Y coincides with an X in the library.

One possible approach is to set up a storage space consisting of $2^m = 2^{100}$ locations, each identified by a distinct 100 digit address. Than we put each X in the location with address X. When Y is presented one need only go to the location with address Y to see if it is occupied or not. The problem with this approach is that a storage device with 2^{100} addresses is much too large to be practical.

The alternative hash coding approach is to use a storage device with $2n = 100$ million locations, each of which can hold a hundred digit number. Select some random uniformly distributed real numbers, $r_1, r_2, \ldots$, on $(0, 1)$. Construct a *hashing* function $f(x, r)$ which assigns a location from 1 to $2n$ for each x and r. The hashing function should be selected to be sensitive to changes in the components of x, and more or less uniformly distributed.

The library is constructed as follows. First X_1 goes into location $f(X_1, r_1)$. Having assigned locations for $X_1, X_2, \ldots, X_k$, put x_{k+1} into location $f(X_{k+1}, r_1)$ if it is empty. If it is not empty, try $f(X_{k+1}, r_2)$, $f(X_{k+1}, r_3), \ldots$ until an empty location is found to hold X_{k+1}.

To determine whether Y is in the library, we need to compare Y with the contents of $f(Y, r_1), f(Y, r_2), \ldots$ until a match or an empty location is found.

When the library is constructed, most of the locations $f(X_i, r_1)$ will be empty at first, but as the library fills up, one would expect to find occupied locations about half the time. Then we would have, on the average, about 2 tries to find an empty location. Similarly in matching the target Y, we will have to search, on the average, about 2 times before finding an empty spot if Y is not in the library. If it should turn out that the cost of storage is very high, we need not use $2n$ locations. Using $1.5n$ locations would be adequate, but our method would require more comparisons in constructing the library and searching for matches to the target. Thus the relative costs of computing and storage space determine what is a good balance.

The main shortcoming with this approach, and which makes it inapplicable to the fingerprint problem, is that if X consists of many components *subject to measurement error*, hash coding will not work.

24.4   Useful Information in a Variable

Chernoff introduced some relevant measures of the useful information in a variable. The specific choices need not be discussed here, but some characteristics should be mentioned in connection with a few examples.

Example 1
Sex

If half of the subjects are of each sex, and the determination of sex is accurate, then knowledge of the sex contributes one useful *bit* of information. This is in the sense that if the files were divided according to sex, then the knowledge of the sex would reduce the search by a factor of 2. If, however, the probability of error in determining sex is 0.02, the listing of sex would contribute only $1/4$ of a bit of useful information according to one of the measures. This substantial loss of useful information is an important indication of the potential danger of using *ad hoc* approaches without the benefit of theory.   ⬛

Example 2
Normal Errors

Suppose that X is an attribute measured by $Z = X + Y$ where Y is the measurement error. First we compare two observations on Z for the same X.

Let X be normally distributed with mean 0 and variance 1, i.e. $\mathcal{L}(X) = N(0,1)$. Let $Z_1 = X + Y_1$ and $Z_2 = X + Y_2$ where X, Y_1 and Y_2 are independently distributed with means 0 and variances 1, σ^2 and σ^2. Then

$$\mathcal{L}(Z_1 - Z_2) = N(0, 2\sigma^2) \qquad (24.4.1)$$

indicating how close two measurements on the same subject will differ.

Now let us compare two observation Z for different subjects. Here $Z_1 = X_1 + Y_1$ and $Z_2 = X_2 + Y_2$ for independent X_1, X_2, Y_1 and Y_2, and

$$\mathcal{L}(Z_1 - Z_2) = N(0, 2 + 2\sigma^2) \qquad (24.4.2)$$

It is clear that for small σ Z carries a great deal of information for determining matches.   ⬛

24.5 Allocation of Storage Space

If the measurement of X in the example of the preceeding section does not carry much useful information, we may not wish to devote much storage space for it. In particular, suppose that we wish to allocate only one bit of storage space in Example 2. Then symmetry considerations suggest the use of the sign of Z as the stored indicator of Z. In other words, we use $Z^* = 1$ for $Z < 0$, and $Z^* = 0$ for $Z \geq 0$.

Suppose now that we have two different and independent variables, X_1 and X_2 which are normally distributed with means 0 and variances 1, and subject to independent measurement errors Y_1 and Y_2 with variance σ^2. If we want to devote two bits of storage space to the resulting observations Z_1 and Z_2, it would be sensible to use the signs of Z_1 and Z_2.

Going one step further in conservatism, suppose we wish to use only one bit of storage space for the vector $\mathbf{Z} = (Z_1, Z_2)$. How should we divide the two dimensional Euclidean space $\mathbf{R}^2$ into two parts where $Z^* = 1$ and 0 respectively so as to maximize the useful information in the *reduced* variable Z^*?

My first reaction to this question, conditioned by experience as a statistician, was to conjecture that with two observations, we should be able to squeeze out more useful information than with one observation. However, several attempts at improvement failed, and I changed my conjecture to the reverse, *i.e.* that one could not improve over discarding Z_2 and letting $Z^* = 1$ or 0 depending on the sign of Z_1.

24.6 The Isoperimetric Problem

Stated formally, the optimization problem of the last section involves two, observations, $\mathbf{Z}^{(1)} = \mathbf{X} + \mathbf{Y}^{(1)}$ and $\mathbf{Z}^{(2)} = \mathbf{X} + \mathbf{Y}^{(2)}$ where $\mathbf{X} = (X_1, X_2)$, $\mathbf{Y}^{(1)} = (Y_1^{(1)}, Y_2^{(1)})$ and $\mathbf{Y}^{(2)} = (Y_1^{(2)}, Y_2^{(2)})$. The components of $\mathbf{X}, \mathbf{Y}^{(1)}$, and $\mathbf{Y}^{(2)}$ are all independent and normally distributed with means 0 and variances 1, σ^2 and σ^2 respectively. For various measures of useful information, the optimality condition for small σ^2 requires

$$P(\mathbf{X} \in R) = 1/2 \qquad (24.6.1)$$

as a first order effect and that $P(\mathbf{Z}^{(1)} \in R, \mathbf{Z}^{(2)} \ni R)$ be minimal as a second order effect.

Since

$$P(\mathbf{Z}^{(1)} \in R, \mathbf{Z}^{(2)} \ni R) \approx 2\sqrt{\pi}\sigma \int_{\partial R} \phi(x_1)\phi(x_2)\sqrt{dx_1^2 + dx_2^2} \quad (24.6.2)$$

where ∂R is the boundary of R, and ϕ is the standard normal density, our optimization problem is that of minimizing

$$L = \int_{\partial R} \phi(x_1)\phi(x2)\sqrt{dx_1^2 + dx_2^2} \quad (24.6.3)$$

subject to the condition

$$A = \int_R \phi(x_1)\phi(x_2)dx_1dx_2 = 1/2. \quad (24.6.4)$$

This is a variation of the standard isoperimetric problem where both area and perimeter are modified by the bivariate normal density.

An attempt, to prove that the region R which minimizes L subject to a given value of A between 0 and 1 consists of a half space, required the derivation of the inequality generalized by Cacoullos and Chen. Later, a more basic, and as yet unpublished, derivation was obtained by Beckner and Jerison, using an inequality due to Borell.

As indicated earlier, this solution of the isoperimetric problem implies that the asymptotic (as $\sigma \to 0$) solution of our storage problem is equivalent to discarding Z_2 and reporting the sign of Z_1.

24.7   Extensions

Suppose now that we are allowed more storage space for recording $\mathbf{Z}$. It seems perfectly reasonable to use two bits of storage space by reporting the signs of Z_1 and Z_2. But what if we are allowed still more storage space? For three bits of storage space we have to divide R into 8 regions of equal probability, so that the integrated density along the boundaries is minimal.

We might even consider using a fractional number of bits of storage space by dividing R into 3 regions of equal probability with minimal integrated density along the boundaries. So let us consider how to minimize L, the integrated density along the boundaries of n regions of equal probability for n greater than 2. We introduce several conjectures of increasing complexity which are supported on the grounds of the symmetry inherent in the standard bivariate normal distribution.

Conjecture 1

n rays.

Let the n regions consist of those divided by the equally spaced rays from the origin. Here

$$L = n \int_0^\infty \phi(0)\phi(x)dx = n\phi(0)/2 = 0.1997n. \qquad (24.7.1)$$

Conjecture 2

$(n-1)$ circles.

If we have n regions separated by $n-1$ circles centered at the origin, we note that the probability outside a circle of radius r is

$$P(X_1^2 + X_2^2 \geq r^2) = e^{-r^2/2} \qquad (24.7.2)$$

and the integrated density along the circle is

$$2\pi r \frac{1}{2\pi} e^{-r^2/2}. \qquad (24.7.3)$$

Thus, if the radii are $r_1, r_2, \ldots, r_{n-1}$, we need

$$e^{-r_i^2/2} = 1 - \frac{i}{n} \qquad (24.7.4)$$

and

$$L = \sum_{i=1}^{n-1} (1 - \frac{i}{n})\sqrt{-2\log(1 - \frac{i}{n})}$$

$$L = \frac{1}{n}\sum_{j=1}^{n-1} \sqrt{2\log(n/j)}$$

$$\approx 0.443n$$

Calculations show that n rays is preferable to $(n-1)$ circles for all values of n.

Conjecture 3

Circle with m of n regions inside.

In more detail, this third conjecture involves m partial rays from the origin to a circle of radius r centered at the origin, and $n - m$ partial rays from the circle to infinity outside. Two exceptional versions of this conjecture are when $m = 1$ and $n - 1$. If $m = 1$ there is no partial ray inside the circle. If $m = n - 1$ there is no partial ray outside the circle. For this conjecture we have

$$e^{-r^2/2} = 1 - m/n \qquad (24.7.5)$$

and

$$L = re^{-r^2/2} + m1(m1)\phi(0)(\Phi(r)-1/2)+(n-m)1(m < n-1)\phi(0)(1-\Phi(r)) \qquad (24.7.6)$$

where Φ is the standard normal c.d.f. and $1(\cdot)$ is the characteristic function.

Calculations for low values of n indicate that the best choice of m varies with n and does better than Conjecture 1 for $n4$. In fact, for $1 < m < n - 1$, we may write

$$L = A(r) + nB(r) \qquad (24.7.7)$$

where

$$A(r) = re^{-r^2/2} \qquad (24.7.8)$$

and

$$B(r) = \phi(0)\{(1 - e^{-r^2/2})(\Phi(r) - 1/2) + e^{-r^2/2}(1 - \Phi(r))\}. \qquad (24.7.9)$$

Since $B(r)$ achieves its minimum at $r = 0.913$ where $exp(-r^2/2) = 0.659$, $A(r) = 0.602$ and $B(r) = 0.0909$, it follows that for large n we would like to have $m \approx 0.341n$, and we would get

$$L \approx 0.602 + 0.0909n \qquad (24.7.10)$$

which crosses $n\phi(0)/2$ at $n = 5.5$. However, for $n = 5$, a circle with $m = 1$ does slightly better than 5 rays.

A more general conjecture which depends on symmetry is the following one.

Conjecture 4
Several circles with several regions in each ring.

More precisely, we take s circles with centers at the origin and radii $r_1 < r_2 < \ldots < r_s$. We select positive integers $m_1, m_2, \ldots, m_{s+1}$ and have m_1 regions in the interior of the first circle, m_i regions in the annulus between the $(i - 1)$-st and the i-th circles for $2 \leq i \leq s$, and finally m_{s+1} regions outside the i-th circle.

Taking $r_0 = 0$ and $r_{s+1} = \infty$, we have, for $1 \leq i \leq s$,

$$e^{-r_i^2/2} = \frac{m_{i+1} + m_{i+2} + \ldots + m_{s+1}}{n}$$

$$L = \sum_{i=1}^{s} r_i e^{-r_i^2/2} + \phi(0) \sum_{i=1}^{s+1} m_i 1(m_i 1)[\Phi(r_i) - \Phi(r_{i-1})].$$

If all $m_i 1$,

$$L = A(\mathbf{r}) + nB(\mathbf{r}) \qquad (24.7.11)$$

where

$$A(\mathbf{r}) = \sum r_i e^{-r_i^2/2} \qquad (24.7.12)$$

and

$$B(\mathbf{r}) = \sum_{i=1}^{s+1} [\phi(r_{i-1}) - \phi(r_i)][\Phi(r_i) - \Phi(r_{i-1})]. \qquad (24.7.13)$$

Minimizing $B(\mathbf{r})$ for several values of s, we obtain the following Table 24.1. This table, presenting approximations, ignoring the constraints that the m_i are integers not necessarily exceeding one, suggests that the break even choices of n, when one should shift from one value of s to the next, are approximately those given by Table 24.2.

In conclusion, we have not presented a formal proof that Conjecture 4 resolves the extension of the variation of the isoperimetric problem to the case of n regions for $n2$. We have not addressed the more general problem where the observed data are multivariate of dimension greater than 2. However, we now have some appreciation of what there is to gain in increased efficiency of information storage for the bivariate case.

References

1. Beckner, W. and Jerison, D., Gaussian Symmetry and an Optimal Information Problem. Unpublished Draft, 1–17.

2. Brascamp, H. J. and Lieb, E. H. (1976). On extensions of the Brunn-Minkowski and Prekopa-Leindler theorems including inequalities for log concave functions, and with an application to the diffusion equation. *Journal of Functional Analysis* **22**, 366–389.

3. Cacoullos, T. (1982). On upper and lower bounds for the variance of a function of a random variable. *Annals of Probability* **10**, 799–809.

4. Cacoullos, T. (1989).   Dual Poincare-type inequalities via the Cramer-Rao and the Cauchy-Schwarz inequalities and related characterizations. In *Statistical Data Analysis and Inferences* pp. 239–250. North-Holland, Amsterdam.

5. Chen, L. H. Y. (1982). An inequality for the multivariate normal distribution. *Journal of Multivariate Analysis* **12**, 306–315.

6. Chen, L. H. Y. (1985). Poincare-type inequalities via stochastic integrals. *Zeitschrift für Wahrscheinlichkeitstheorie und Verwandte Gebiete* **69**, 251–277.

7. Chernoff, H. (1980).   The identification of an element of a large population in the presence of noise. *Annals of Statistics* **8**, 1179–1197.

8. Chernoff, H. (1981). A note on an inequality involving the normal distribution. *Annals of Probability* **9**, 533–535.

TABLE 24.1

	$s = 0$	$s = 1$	$s = 2$	$s = 3$	$s = 4$	$s = 8$
$A(r)$	0.0000	0.6018	1.1036	1.5756	2.0364	3.8370
$B(r)$	0.1994	0.0909	0.0590	0.o437	0.0347	0.0191
$r1$		0.9133	0.6501	0.5221	0.4432	0.2923
$r2$			1.2340	0.9314	0.7706	0.4919
$r3$				1.4247	1.1116	0.6721
$r4$					1.5583	0.8500
$r5$						1.0367
$r6$						1.2453
$r7$						1.4998
$r8$						1.8682
m_1/n		0.3410	0.1905	0.1274	0.0935	0.0418
m_2/n		0.6590	0.3425	0.2245	0.1633	0.0721
m_3/n			0.4670	0.2856	0.2040	0.0882
m_4/n				0.3625	0.2422	0.1010
m_5/n					0.2970	0.1125
m_6/n						0.1238
m_7/n						0.1358
m_8/n						0.1501
m_9/n						0.1747

TABLE 24.2
Break even values of n for shifting s

s	$0 - 1$	$1 - 2$	$2 - 3$	$3 - 4$	$4 - 8$
n_s	5.5	15.7	30.8	51.2	114.9

25

On Finding A Single Positive Unit In Group Testing

Milton Sobel
University of California, Santa Barbara, CA

ABSTRACT For the group-testing situation with the total number N of units being infinite, a simple non-optimal procedure is proposed and is shown to be highly efficient compared to the optimal procedure.

Keywords and phrases Group testing, DH procedure, efficiency, robustness, Bayesian procedure

25.1   Introduction

In group testing, the total number N of units may be finite or infinite, but it is usually assumed that the units all have a common probability $p > 0$ of being positive (previously called defective or unsatisfactory) and $q = 1 - p$ of being negative (previously called nondefective or satisfactory). Any number x of units can be included in a single test but there are only two possible outcomes:

- all x units are negative,
- at least 1 of the x limits is positive.

We do not know which units are positive or how many are positive unless $x = 1$. This positive/negative terminology gives us a wider scope in applications since the positive units may be the ones we would rather not find or they may be the ones we are trying to find. The problem in this paper is to find a single positive unit, assuming $N = \infty$, so that (with $p > 0$) the probability is unity that it exists. For the two settings we consider: (1) q is known and (2) q is given by the Uniform(0,1) prior, the first one has already appeared in the literature, but not the second. We also wish to define a very simple (nonyoptimal) procedure (called DH for doubling and halving), and see how it compares with the optimal procedure in each of these two settings. The efficiency of a procedure is defined by comparing its expected number of tests required to that of the optimal procedure. The DH procedure does quite well in the first setting, reaching 94.22% efficiency (for some q values), but in the second setting, it does much better with a (Bayesian) efficiency of 99.6%. This is a remarkable result for a procedure so easy to implement.

25.2   Description of Procedures, Numerical Results

For the first setting with ($q < 1$ known) a procedure (denoted by R_1') was conjectured in Kumar and Sobel (1971) to be optimal and this conjecture was proved for $N = \infty$ in Hwang (1974) and then generalized for a finite N problem in Garey and Hwang (1973) and Hwang (1974). In Table 25.1 below we give the expectation and variance (of the number of tests required) for the optimal procedure for $q = .50(.05).95, .99$. The DH procedure (also given in Table 25.1) is defined as follows:

Procedure DH: Start, by testing 1 unit and keep on doubling the size of the next test as long as the results are negative. Then, test half of the positive set. Then, select the half that is positive either by direct test or by inference and continue to use halving until a positive unit is found.

The Procedure DH is easy to implement and does not use q for carrying it out, only for evaluating its performance. At $q = .618034 = (\sqrt{5}-1)/2$ the efficiency of the DH procedure reaches a maximum of over 94%.

For the second setting with $q < 1$ unknown, we propose a Bayesian procedure denoted by $R_{1B}^{(1)}$. It also starts by testing 1 unit and we use the posterior estimate $\hat{q}$ after each negative test result together with the value of x from Kumar and Sobel (1971, p. 825), assuming that $\hat{q}$ is the true value of q. The same table, extended to $x = 100$, appears in Sobel and Groll (1959, Table VII). The reduction part of the procedure $R_{1B}^{(1)}$ is based on the current values of m (size of the positive set) and s (the number of units shown to be negative); the current value of u (the number of units shown to be positive) is of course zero. Hence, except for the fact that $u = 0$ we have the same (F-situation) as in the procedure $R^{(1)}$ of Sobel and Groll (1966) and we use the tables and diagrams of that paper to complete the problem. The result is the nested procedure $R_{1B}^{(1)}$.

Obviously once we have a positive set, there is no need to do any "mixing" of binomial units with subsets of a positive set, since we only want to find one positive unit. Hence, we restrict our attention to nested procedures. Since the procedure R_1' in Kumar and Sobel (1971) and $R^{(1)}$ in Sobel and Groll (1966) are both optimal, we claim that the new procedure $R_{1B}^{(1)}$ is also optimal in the Bayesian sense, i.e., that it minimizes the Bayes average of the expected number of tests required to find a single positive unit. Since the first part of the procedure $R_{1B}^{(1)}$ is open-ended, it was necessary to cut it off after 20 attempts to find a positive subset; at this point, the probability is .999982 that a positive set would have been found and the contribution to the expected total number of tests is only in the fourth decimal place (as seen by the last column of Table 25.2).

Table 25.2 gives numerical results only for procedure $R_{1B}^{(1)}$; comparisons with procedure R_{DH} will be made later after some formulas are derived for the latter procedure in the next section. In Table 25.1, the symbol T_1 (resp., T) denotes the number of tests from the first part (resp., from both parts), of procedure R_{DH}. The optimal procedure R_1' described in Kumar and Sobel (1971) takes a common test size x_1 until a positive set (of size x_1) is found. After that, it follows the F-algorithm described in Sobel and Groll (1959) and Kumar and Sobel (1971) with boundary condition $F(1) = 0$. For $q = .99$, $x_1 = 69$ by Sobel and Groll (1959, Table VII). For $m = 69$, $x_2 = 32$; for $m = 32$ and $m = 37$, $x_2 = 16$. All other values can be obtained front Table IVA or Figure 4 of Sobel and Groll (1959). In Table 25.2 the symbol T_j denotes the test on which the first positive set was found, $\hat{q}_{1j}$ is the posterior estimate of q just before T_j, and $\hat{q}_{2j}$ is the same just after T_j. It is important to note that the results in the last column get close to 3 but still remain under 3; for an explanation, see Section 25.4 below. The entry in the j-th row of the last column of Table 25.2 is formed by the accumulation

$$(\text{col. 10, entry } j) - \sum_{\alpha=1}^{j} [\alpha + (\text{col. 9, entry } \alpha)](\text{col. 5, entry } \alpha)$$

$$(25.2.1)$$

and the limiting result in col. 10 is the desired $E\{T|R_{1B}^{(1)}\}$. Its exact value (less than 3) is not known.

One property that was seen to hold in Sobel and Groll (1966) was that for any positive set of size $m = 2^\alpha + \beta$ with $0 \le \beta < 2^\alpha$ and $u = 0$, the maximum size (col. 8 in Table 25.2) of the next test set is $2^{\alpha-1}$ for $2^\alpha \le m \le 3.2^{\alpha-1}$ and it is $m - 2^\alpha$ for the remaining values of m. Thus, for $\mu = 0$ and $m = 10, 11, 12, 13, 14$ and 15, the maximum test size for the next test is $4, 4, 4, 5, 6$ and 7, respectively. This property was used in the second part of Procedure $R_{1B}^{(1)}$ since it is inherent in procedure $R^{(1)}$ in Sobel and Groll (1966). To illustrate how this was used in calculations for Table 25.2, consider the row $j = 11$ where $m = 208$, $\alpha = 7$, $\beta = 80$. Since $m > 3 \cdot 2^6 = 192$, the maximum x-value for the next test is

TABLE 25.1
Comparison of procedure R_{DH} with the
optimal procedure R_1' for known q-values

q	E{} and σ^2{}	Procedure R_{DH}		Optimal procedure R_1'			Effiency (in %) of R_{DH} w.r.t. R_1' (Ratio of Expectations)	Remarks
		$T_1|R_{DH}$	$T|R_{DH}$	Common test size X_1	$T_1|R_1'$			
.50	Exp. var.	1.632843 0.513786	2.265686 2.055143	1	2.000000 2.000000		88.27%	
.55	Exp. Var.	1.731727 0.590715	2.463454 2.362860	1	2.222222 2.716049		90.21%	
.60	Exp. Var.	1.844464 0.678141	2.688928 2.712565	1	2.500000		92.97%	
$(\sqrt{5}-1)/2$ = .6180	Exp. Var.	1.889277 0.712768	2.778555 2.851072	1,2	2.618034 Discontinuity*		94.22%	Maximum efficiency at the golden mean
.65	Exp. Var.	1.975211 0.778899	2.950422 3.115595	2	2.731602 1.266843		92.58%	
.70	Exp. Var.	2.130118 0.896981	3.260236 3.587922	2	2.960784 1.883891		90.82&	
.75	Exp. Var.	2.318856 1.038412	3.637713 4.153649	2	3.285714 2.938776		90.32%	
.80	Exp. Var.	2.557891 1.212765	4.115782 4.851060	3	3.639344 2.391830		88.42%	
.85	Exp. Var.	2.878578 1.436453	4.757156 5.745810	4	4.092078 2.284712		86.02%	
.90	Exp. Var.	3.352652 1.742551	5.705303 6.970205	7	4.725119 1.912258		82.82%	
.95	Exp. Var.	4.213894 2.216688	7.427787 .66753	14	5.761577 2.012064		77.57%	
.99	Exp. Var.	6.377677 2.977411	11.755353 11.909645	69	8.105007 3.092540		68.95%	

* For $X_1 = 1$ (res. 2) the variance is 4.236068 (resp. 1.000000). Discontinuities in the Var$(T|R_1')$ can occur whenever the common X_1 changes, but they do not occur for $E(T|R_1')$.

TABLE 25.2
New Bayesian procedure $R_{1B}^{(1)}$ for finding a single positive unit

n									
1	.500001	1	1	.500000	.500000	.333333	-	0	0.50000
2	.666667	2	3	.250000	.750000	.533333	1	1.000000	1.250000
3	.800000	3	6	.107143	.857143	.700000	1	1.533333	1.735715
4	.875000	5	11	.0595238	.916667	.807692	2	2.280000	2.109524
5	.923077	9	20	.0357143	.952381	.881119	3	3.134238	2.400033
6	.954545	15	35	.0198413	.972222	.928747	6	3.873657	2.595939
7	.972973	25	60	.0113843	.983607	.956284	9	4.602791	2.728029
8	.983871	43	103	$.0^2677806$	.990385	.974501	16	5.380601	2.818724
9	.990476	73	176	$.0^2396567$	.994351	.984912	32	6.161282	2.878848
10	.994382	123	299	$.0^2231638$	.996667	.991078	50	6.861302	2.917905
11	.996678	208	507	$.0^2136483$	.998032	.994720	78	7.651099	2.943361
12	.998035	352	859	$.0^3805713$	.998838	.996876	128	8.414807	2.959809
13	.998839	596	1455	$.0^3475978$	.999314	.998153	256	9.205564	2.970371
14	.999314	1010	2465	$.0^3281298$	.999595	.998909	498	9.970184	2.977114
15	.999595	1710	4175	$.0^3166051$	.999761	.999355	686	10.702564	2.981381
16	.999761	2895	7070	$.0^4980409$	.999859	.999619	1024	11.454449	2.984073
17	.999859	4902	11972	$.0^4579015$	.999916	.999775	2048	12.223782	2.985765
18	.999916	8251	20223	$.0^4340751$	.999950	.999867	4096	13.008516	2.986822
19	.999951	14146	34369	$.0^4203511$	.999970	.999921	5954	13.75731	2.987488
20	.999971	23901	58270	$.0^4119340$	.999982	.999954	8192	14.499925	2.987900

$m - 2^\alpha = 80$. For the contribution to the 2^d part (col. 9 of Table 25.2), the values of $x = 77, 78, 79$, and 80 give respectively 7.655462, 7.651099, 7.66280, and 7.663130. The minimum of these is at $x = 78$ and that was used in Table 25.2.

25.3 Some Formulas for Procedure R_{DH}

Using T_1 for the number of tests needed to find a positive set under Procedure R_{DH}, we note that (before using the prior)

$$E\{T_1|R_{DH}\}$$
$$= 1 - q + 2(q - q^3) + 3(q^3 - q^7) + \cdots$$
$$= \sum_{j=0}^{\infty} q^{(2^j - 1)}. \tag{25.3.1}$$

For the total number of tests under Procedure R_{DH} we have the simple relation

$$T = 2T_1 - 1. \tag{25.3.2}$$

For the variance we used the fact that

$$E\{T_1^2|R_{DH}\}$$
$$= 1(1 - q) - q + 2^2(q - q^3) + 3^2(q^3 - q^7) + \cdots$$
$$= \sum_{j=0}^{\infty} (2j + 1)q^{(2^j - 1)}, \tag{25.3.3}$$

that $\sigma^2(T_1) = ET_1^2 - [E(T_1)]^2$ can be obtained from (25.3.1) and (25.3.3) and hence by (25.3.2)

$$\sigma^2(T|R_{DH}) = 4\sigma^2(T_1|R_{DH}). \tag{25.3.4}$$

Unlike the optimal procedure $R_{1B}^{(1)}$, where we only approximated $E\{T|R_{1B}^{(1)}\}$, the procedure R_{DH} now gives us two exact results. If we integrate the right side of (25.3.1) with respect to the Uniform prior on $[0, 1]$, we obtain

$$E\{T_1|R - DH\} = 1 + 1/2 + 1/4 + 1/8 + \cdots = 2 \qquad (25.3.5)$$

and from (25.3.2) we obtain our first (exact) result for procedure R_{DH}

$$E\{T|R_{DH}\} = 3. \qquad (25.3.6)$$

For the variance of T, we start by using (25.3.2) on each coefficient in (25.3.1), obtaining

$$E\{T^2|R_{DH}\} = 1^2(1 - q) + 3^2(a - a^3) + 5^2(q^3 - q^7)$$
$$+7^2(q^7 - q^{15}) + \cdots \qquad (25.3.7)$$
$$= 1 + 8 \sum_{j=1}^{\infty} jq^{(2^j - 1)} \qquad (25.3.8)$$

and upon integration with respect to the Uniform prior

$$E\{T^2|R_{DH}\} = 1 + 8 \sum_{j=1}^{\infty} j/2^j. \qquad (25.3.9)$$

Introducing a generating function parameter t^{j-1} in the summation in (25.3.8) we obtain, after first integrating from $t = 0$ to $t = 1$ and then differentiating we respect to t (and setting $t = 1$)

$$E\{T^2|R_{DH}\} = 1 + 8 \left[\frac{d}{dt} \left\{ \frac{t}{2 - t} \right\} \right]_{t=1}$$
$$= 1 + 8 \left[\frac{2}{(2 - t)^2} \right]_{t=1} = 17. \qquad (25.3.10)$$

This gives us our second (exact) result by using (25.3.6) for procedure R_{DH}

$$\sigma^2(T|R_{DH}) = 17 - 3^2 = 8. \qquad (25.3.11)$$

25.4   The Greedy Procedure R_G

At the end of Section 25.2, we described the maximum test size and how it was used in procedure $R_{1B}^{(1)}$. It can be noted that this maximum was used in 16 of the 20 rows of Table 25.2 (m is in Col. 3 and x is in Col. 8). Suppose we used the maximum test size throughout the second part of our procedure (without changing the first part as in $R_{1B}^{(1)}$), what results are obtained; denote this new procedure as R_G. Columns 1 through 6 of Table 25.2 are unchanged. Moreover since the value of x (in Col. 8) changes only for rows 5, 6, 10 and 11, the first four rows are exactly the same. Starting with row 5, the entries for rows 5, 10, 15 and 20 in Col. 10 become

1	Col. 10 $= E\{T|R_G\}$
5	2.400251
10	2.918631
15	2.982125
20	2.988644

Thus, we note that this gives an even better bound for the optimal procedure $R_{1B}^{(1)}$ and if we compute its efficiency we obtain 99.98%. As with the optimal procedure $R_{1B}^{(1)}$, this procedure R_G also does not give an exact precise value for an answer.

25.5   Conclusions

We now see why the value 3 was important to note in Table 25.2. If Procedure $R_{1B}^{(1)}$ is (Bayesian) optimal, then $E\{T|R_{1B}^{(1)}\}$ has to be less than the exact answer 3 for procedure $R - DH$, i.e., the value 3 is an upper bound for the optimal Bayes procedure $R_{1B}^{(1)}$. Comparing these two expectations, we obtain the spectacular result

$$\text{Efficiency } (R_{DH}) \geq 2.988/3 = 99.6\%, \qquad (25.5.1)$$

a remarkably good result for a procedure that is easy to implement and not fully adaptive. The procedure R_G has an even higher efficiency (99.98%) but does not yield simple expressions to approximate $E\{T\}$ for the optimal procedure.

25.6 Changing the Prior with Procedure R_{DH}

In an attempt to obtain an explicit formula for $E\{T\}$, we restrict our attention to procedure R_{DH} and consider the beta distribution $f(q) = Cq^{\alpha-1}(1-q)^{\beta-1}$ as a generalization of the Uniform(0,1) prior. By the same methods as used above we have

$$E\{T|R_{DH}\}$$

$$= C\sum_{j=1}^{\infty}(2j-1)\int_0^1 P_j\{X \geq 1\}q^{\alpha-1}(1-q)^{\beta-1}dq \qquad (25.6.1)$$

$$= C\sum_{j=1}^{\infty}(2j-1)\int_0^1 [q^{(2^{j-1}-1)} - q^{(2^j-1)}]q^{\alpha-1}(1-q^{\beta-1}dq \qquad (25.6.2)$$

$$= C\sum_{i=0}^{\beta-1}(-1)^i\binom{\beta-1}{i}\sum_{j=1}^{\infty}(2j-1)\left[\frac{1}{2^{j-1}+\alpha+i-1} - \frac{1}{2^j+\alpha+i-1}\right] \qquad (25.6.3)$$

$$= C\sum_{i=0}^{\beta-1}(-1)^i\binom{\beta-1}{i}\sum_{j=1}^{\infty}\left[\frac{2j-1}{2^{j-1}+\alpha+i-1} - \frac{(2j+1)}{2^j+\alpha+i-1}\right] \qquad (25.6.4)$$

$$= C\sum_{i=0}^{\beta-1}(-1)^i\binom{\beta-1}{i}\sum_{j=1}^{\infty}\frac{2}{2^j+\alpha+i-1}. \qquad (25.6.5)$$

The first double sum in (25.6.4) telescopes and sums to $C^{-1} = \frac{\Gamma(\alpha)\Gamma(\beta)}{\Gamma(\alpha+\beta)}$ and we obtain

$$E\{T|R_{DH}\}$$

$$= 1 + 2C\sum_{j=1}^{\infty}\sum_{i=0}^{\beta-1}(-1)^i\binom{\beta-1}{i}\frac{1}{2^j+\alpha+i-1} \qquad (25.6.6)$$

$$= 1 + 2C\sum_{j=1}^{\infty}\int_0^1 x^{2^j+\alpha-1}(1-x)^{\beta-1}dx \qquad (25.6.7)$$

$$= 1 + 2\frac{\Gamma(\alpha+\beta)}{\Gamma(\alpha)}\sum_{j=1}^{\infty}\frac{\Gamma(\alpha+2^j)}{\Gamma(\alpha+\beta+2^j)}. \qquad (25.6.8)$$

Computations based on (25.6.8) show that if we increase α (i.e., make the positive units scarcer) then the value of $E\{T\}$ increases, but if we keep the expectation $\alpha/(\alpha + \beta)$ fixed and increase the variance by replacing (α, β) by say, $(2\alpha, 2\beta)$, then the value of $E\{T\}$ decreases. The following short table illustrates this pattern.

Known Prior Density	Exp.	Var.	$E\{T\|R_{DH}\}$
$\alpha = \beta = 1$, Uniform$(0,1)$	$1/2$	$1/2$	3.000000
$\alpha = \beta = 2$, $6q(1-q)$	$1/2$	$1/20$	2.586995
$\alpha = \beta = 3$, $30q^2(1-q)^2$	$1/2$	$1/28$	2.468845
$\alpha = 4$, $\beta = 1$, $4q^3$	$4/5$	$2/75$	4.862089
$\alpha = 8$, $\beta = 2$, $72q^7(1-q)$	$4/5$	$4/275$	4.127280
$\alpha = 9$, $\beta = 1$, $9q^8$	$9/10$	$9/1100$	7.145056
$\alpha = 18$, $\beta = 2$, $342q^{17}(1-q)$	$9/10$	$9/2100$	6.369355
$\alpha = 99$, $\beta = 1$, $99q^{98}$	$99/100$	$99/101\ 10^{-4}$	12.394409
$\alpha = 198$, $\beta = 2$, $39402q^{197}(1-q)$	$99/100$	$99/201\ 10^{-4}$	12.522018

25.7 Robustness of Procedure R_{DH} for q Known

Suppose you assumed (or were told) that $q = .80$, but the correct value is .90. Using the optimal procedure R_1' for q known, your common test group size would be $x_1 = 3$ and the probability of a positive subset is $1 - (.9)^3 = .271$, so that $E\{T\} = 1/.271 = 3.6900137$. For the 2^d part of the procedure [using the table in Kumar and Sobel (1971)], you need an average of

$$E\{T_2|R_1'\} = \left[\frac{2 + q - 2q^3}{1 - q^3}\right]_{q=.9} = 5.321033, \qquad (25.7.1)$$

so that the total $E\{T|R_1'\} = 9.011070$.

For the procedure R_{DH} from Table 25.1 with $q = .90$ we obtain 3.52652 for $E\{T_1\}$ and 5.705303 for $E\{T\}$. Note that the value of q is not used in implementing the procedure, only for its evaluation. Hence, with an incorrect value of q you could be using an optimal procedure and operating at 63.3% efficiency. In other words, the procedure R_{DH} is robust against wrong information about the value of q.

Suppose now that $q = .90$ again but you assume (or are told) that $q = .99$. Then the test group size under the optimal procedure is $x_1 = 69$, the probability of a positive set is $1 - (.9)^{69} = .999304$ and $E\{T_1|R_1'\} = 1.000697$. For the second part of the procedure, [using Kumar and Sobel (1971)], you need an average

$$E\{T_2|R_1'\} = \left[7 + \frac{q^{59}}{1 - q^{69}}\right]_{q=.9} = 7.001998, \qquad (25.7.2)$$

so that the total $E\{T|R'_1\} = 8.002695$.

For procedure R_{DH}, $E\{T\}$ is the same as before, namely, 5.705303. Here, with the optimal procedure R'_1 you operate with 71.3% efficiency. Thus, from the viewpoint of robustness against wrong information about the q-value, the procedure R_{DH} is more robust "in both directions" than procedure R'_1 for q known.

Since the optimal Bayes solution $R^{(1)}_{1B}$ is similar to R_{DH} we don't expect the same results when using a prior for q and the posterior estimates of q after every test, i.e., we expect the optimal Bayes solution $R^{(1)}_{1B}$ to also be robust in both direction.

Acknowledgements Thanks are due to Professor Anton Boneh (Technion, Haifa, Israel) for recently pointing out to me that he studied (in a private communication) a modified and somewhat generalized form of Procedure R_{DH} above for the goal of finding *all* the defectives in a group-testing model with N given items (with p and $q = 1 - p$ unknown); he calls it the "DOD Policy" for Double or Divide. The results in his study also show robustness and proximity to optimality as in the present paper, where we look for only one single defective (or positive) unit.

References

1. Garey, M. and Hwang, F. K. (1973). Isolating a single defective using group-testing. *Preprint*.

2. Hwang, F. K. (1974). On finding a single defective in binomial group testing. *Journal of the American Statistical Association* **69**, 151–153.

3. Hwang, F. K. (1980). Optimal group-testing procedures in identifying a single defective from a finite population. *Bulletin of the Institute of Mathematics, Academia Sinica* **8**, 129–140.

4. Kumar, S. and Sobel, M. (1971). Finding a single defective in binomial group-testing. *Journal of the American Statistical Association* **66**, 824–828.

5. Sobel, M. and Groll, P. A. (1959). Group-testing to eliminate efficiently all defectives in a binomial sample. *Bell System Technical Journal* **38**, 1179–1252.

6. Sobel, M. and Groll, P. A. (1966). Binomial group-testing with an unknown proportion of defectives. *Technometrics* **8**, 631–656.

26

Testing Hypotheses on Variances in the Presence of Correlations

A. M. Mathai and P. G. Moschopoulos
McGill University, Montreal, Quebec, Canada
The University of Texas at El Paso, El Paso, TX

ABSTRACT The classical technique of testing the hypothesis of equality of variances in a bivariate normal population, when the population correlation is nonzero, is to construct linear functions of the variables so that the correlation between these linear functions is zero under the null hypothesis. Then the problem reduces to testing the hypothesis that the correlation of these linear functions is zero. In this paper we show that an extension of the technique to a p-variate normal when $p \geq 3$ is not possible. Thus we consider the likelihood ratio principle for the problem. We also consider the likelihood ratio test for several variations of the hypothesis. Both the exact null and nonnull moments and distributions of these test statistics are discussed in this paper.

Keywords and phrases Likelihood ratio, bivariate normal, multivariate normal, Lauricella function, Type-2 beta integral, gamma function

26.1 Bivariate Normal Population

Let X, with $X' = (x_1, \ldots, x_p)$, have a p-variate nonsingular normal distribution, that is, $X \sim N_p(\mu, \Sigma)$, $\Sigma > 0$. For $p = 2$ let the covariance matrix

$$\Sigma = \begin{bmatrix} \sigma_1^2 & \sigma_1 \sigma_2 \rho \\ \sigma_1 \sigma_2 \rho & \sigma_2^2 \end{bmatrix}.$$

The hypothesis of interest is $H_0 : \sigma_1^2 = \sigma_2^2$ where $\rho \neq 0$. For convenience let $x_1 = x$ and $x_2 = y$. An ingeneous technique used by Pitman (1939) is to consider the linear functions

$$u = x + y \quad \text{and} \quad v = x - y$$

Then the covariance between u and v is

$$\text{Cov}(u, v) = \text{Var}(x) - \text{Var}(y) = \sigma_1^2 - \sigma_2^2$$

which is zero under H_0 and thus the correlation between u and v is zero under H_0. Then, the hypothesis of equality of the two variances is equivalent to the hypothesis that the correlation between u and v is zero. Testing can be achieved by using the well known distribution of the sample correlation between u and v.

As it has been shown by Morgan (1939), for $p = 2$ the likelihood ratio principle also leads to u and v above. The maximum likelihood estimate for the population mean vector μ is the corresponding sample mean vector. Let $L(\hat{\mu})$ be the likelihood function at this estimate $\hat{\mu}$.

$$L(\hat{\mu}) = (2\pi)^{-Np/2} \mid \Sigma \mid^{-N/2} e^{-\frac{1}{2}\text{tr}(\Sigma^{-1}S)},$$

where S is the sample sum of products matrix

$$S = (s_{ij}), \quad s_{ij} = \Sigma_{k=1}^{N}(x_{ik} - \bar{x}_i)(x_{jk} - \bar{x}_j)$$

with x_{ik} denoting the k-th observation on the i-th component of X. For the general parameter space it is known that

$$\max(L) = \frac{(2\pi)^{-Np/2}e^{-Np/2}}{\mid S/N \mid^{N/2}}.$$

For $p = 2$, under H_0 let L be L_0. Then,

$$\Sigma = \Sigma_0 = \sigma^2 R = \sigma^2 \begin{bmatrix} 1 & \rho \\ \rho & 1 \end{bmatrix} \quad \text{and} \quad \mid \Sigma_0 \mid = (\sigma^2) \mid R \mid = (\sigma^2)^2(1 - \rho^2)$$

$$\frac{\partial \ln L_0}{\partial \sigma^2} = 0 \implies \hat{\sigma}_0^2 = \frac{\text{tr}(\hat{R}^{-1}S)}{Np} = \frac{\text{tr}(\hat{R}^{-1}S)}{2N} \qquad (26.1.1)$$

where $\hat{R}$ is the estimate of R. Taking the derivative of $\ln L_0$ with respect to ρ, equating to zero and solving for $\hat{\sigma}_0^2$ and $\hat{\rho}$ we have

$$\hat{\rho} = \frac{2s_{12}}{s_{11} + s_{22}} \quad \text{and} \quad \hat{\sigma}_o^2 = \frac{s_{11} + s_{22}}{2N}.$$

Substituting these back the λ-criterion is given by, denoting $\lambda^{2/N} = z$,

$$z = \lambda^{2/N} = \frac{s_{11}s_{22} - s_{12}^2}{(s_{11} + s_{22})^2 - 4s_{12}^2}.$$

Here we reject H_0 for small values of z.

But note that if we had used the Pitman approach then we would have started with $u = x_1 + x_2$ and $v = x_1 - x_2$. Let g denote the sample correlation between u and v. Then

$$g = \frac{\sum_{i=1}^{N}(u_i - \bar{u})(v_i - \bar{v})}{\sqrt{\sum_{i=1}^{N}(u_i - \bar{u})^2 \sum_{i=1}^{N}(v_i - \bar{v})^2}}$$

$$= \frac{\sum_{i=1}^{N}[(x_{1i} - \bar{x}_1) + (x_{2i} - \bar{x}_2)][(x_{1i} - \bar{x}_1) - (x_{2i} - \bar{x}_2)]}{\sqrt{\sum_{i=1}^{N}[(x_{1i} - \bar{x}_1) + (x_{2i} - \bar{x}_2)]^2 \sum_{i=1}^{N}[(x_{1i} - \bar{x}_1) - (x_{2i} - \bar{x}_2)]^2}}$$

$$= \frac{s_{11} - s_{22}}{\sqrt{(s_{11} + s_{22} + 2s_{12})(s_{11} + s_{22} - 2s_{12})}} = \frac{s_{11} - s_{22}}{\sqrt{(s_{11} + s_{22})^2 - 4s_{12}^2}}.$$

Consider

$$1 - g^2 = 1 - \frac{(s_{11} - s_{22})^2}{[(s_{11} + s_{22})^2 - 4s_{12}^2]} = \frac{4[s_{11}s_{22} - s_{12}^2]}{[(s_{11} + s_{22})^2 - 4s_{12}^2]}$$

$$= 4z = 4\lambda^{2/N}.$$

Thus both procedures lead to the same t-test, based on g; see also Cacoullos (2000) for the corresponding F-test.

It is not possible to extend the Pitman-Morgan procedure to the case $p = 3$. Consider the linear functions

$$u_1 = a_1 x_1 + a_2 x_2 + a_3 x_3 \text{ and } u_2 = b_1 x_1 + b_2 x_2 + b_3 x_3$$

where $X' = (x_1, x_2, x_3)$ and $X \sim N_3(\mu, \Sigma), \Sigma > 0$. The covariance between u_1 and u_2 is then

$$\begin{aligned}
Cov(u_1, u_2) = {} & a_1 b_1 Var(x_1) + a_2 b_2 Var(x_2) + a_3 b_3 Var(x_3) \\
& + a_1 b_2 Cov(x_1, x_2) + a_1 b_3 Cov(x_1, x_3) \\
& + a_2 b_1 Cov(x_1, x_2) + a_2 b_3 Cov(x_2, x_3) \\
& + a_3 b_1 Cov(x_1, x_3) + a_3 b_2 Cov(x_2, x_3).
\end{aligned}$$

Let $H_0 : \sigma_1^2 = \sigma_2^2 = \sigma_3^2$. Even under H_0, since the various correlations are arbitrary, if $Cov(u_1, u_2) = 0$, the following equations are to be satisfied.

$$a_1 b_1 + a_2 b_2 + a_3 b_3 = 0$$

$$a_1 b_2 + a_2 b_1 = 0$$

$$a_1 b_3 + a_3 b_1 = 0$$

$$a_2 b_3 + a_3 b_2 = 0.$$

The last three equations are in matrix notation

$$(a_1, a_2, a_3) \begin{bmatrix} b_2 & b_3 & 0 \\ b_1 & 0 & b_3 \\ 0 & b_1 & b_2 \end{bmatrix} = (0, 0, 0).$$

For a non-null solution for (a_1, a_2, a_3) we must have the determinant

$$\begin{vmatrix} b_2 & b_3 & 0 \\ b_1 & 0 & b_3 \\ 0 & b_1 & b_2 \end{vmatrix} = 0.$$

Then at least one of b_1, b_2, b_3 is zero. Thus at most two variables can enter into u_1 and u_2 and hence the technique cannot work for $p \geq 3$.

26.2   Modifying the Hypothesis

In the general p-variate normal the difficulties arise due to the fact that the population correlations are unequal. If the population correlations are equal, that is, $\rho_{ij} = \rho$ for all $i \neq j$, then the problem becomes simpler. Let us modify the hypothesis to the following form :

$$H_0 : \sigma_1^2 = \cdots = \sigma_p^2 = \sigma^2, \quad \rho_{ij} = \rho \text{ for all } i \text{ and } j, \ i \neq j$$

where σ^2 and ρ are unknown. Then it is not difficult to show that under H_0 we can construct two linear functions which are uncorrelated. For example for $p = 3$, consider

$$u = \sum_1^p x_i \text{ and } v = \sum_1^{p-1} x_i - (p-1)x_p.$$

The correlation between u and v is zero under the above H_0 and then the sample correlation between u and v can be used as a test statistic. Here the likelihood ratio procedure also works. Under H_0 we have :

$$\Sigma = \sigma^2 R, \quad R = \begin{bmatrix} 1 & \rho & \cdots & \rho \\ \rho & 1 & \cdots & \rho \\ \vdots & \vdots & \ddots & \vdots \\ \rho & \cdots & \cdots & 1 \end{bmatrix}$$

$$| \Sigma | = (\sigma^2)^p (1 - \rho)^{p-1} [1 + (p - 1)\rho]$$

and

$$\Sigma^{-1} = \frac{1}{\sigma^2 (1 - \rho)[1 + (p - 1)\rho]}$$

$$\times \begin{bmatrix} 1 + (p - 2)\rho & -\rho & \cdots & -\rho \\ -\rho & 1 + (p - 2)\rho & \cdots & -\rho \\ \vdots & \vdots & \ddots & \vdots \\ -\rho & -\rho & \cdots & 1 + (p - 2)\rho \end{bmatrix}.$$

The maximum likelihood estimate of σ^2 is as given in (26.1.1), that is,

$$\hat{\sigma}^2 = \frac{\mathrm{tr}(\hat{R}^{-1}S)}{Np}. \tag{26.2.1}$$

Let

$$\Sigma^{-1} = A = \begin{bmatrix} a & b & \cdots & b \\ b & a & \cdots & b \\ \vdots & \vdots & \ddots & \vdots \\ b & b & \cdots & a \end{bmatrix}.$$

Then

$$|A| = (a - b)^{p-1} [a + (p - 1)b].$$

Differentiating the log-likelihood with respect to a and b and equating to zero we have :

$$\frac{N(p - 1)}{\hat{a} - \hat{b}} + \frac{N}{\hat{a} + (p - 1)\hat{b}} - \mathrm{tr}(S) = 0, \tag{26.2.2}$$

$$\frac{-N(p - 1)}{\hat{a} - \hat{b}} + \frac{N(p - 1)}{\hat{a} + (p - 1)\hat{b}} - \sum_{i \neq j} s_{ij} = 0. \tag{26.2.3}$$

where N is the sample size. From (26.2.2) and (26.2.3) we get

$$\hat{a} + (p - 1)\hat{b} = \frac{Np}{\mathrm{tr}(S) + \sum_{i \neq j} s_{ij}} \tag{26.2.4}$$

and

$$\hat{a} - \hat{b} = \frac{Np(p-1)}{(p-1)\text{tr}(S) - \sum_{i \neq j}(s_{ij})}.$$

(26.2.5)

Therefore,

$$|\hat{\Sigma}_0|^{-1} = |\hat{A}| = \frac{N^p(p-1)^{p-1}p^p}{[(p-1)\text{tr}(S) - \sum_{i \neq j} s_{ij}]^{(p-1)}[\text{tr}(S) + \sum_{i \neq j} s_{ij}]}.$$

(26.2.6)

Then from (26.2.6), observing that the maximum of L in the whole parameter space is proportional to $|S|$ we have

$$u = \frac{\lambda^{2/N}}{p^p(p-1)^{p-1}} = \frac{|S|}{[(p-1)\text{tr}(S) - \sum_{i \neq j} s_{ij}]^{p-1}[\text{tr}(S) + \sum_{i \neq j} s_{ij}]}.$$

(26.2.7)

We note that the modified hypothesis agrees with the hypothesis of equality of variances and equality of covariances considered by Wilks (1946). The method used by Wilks to derive the null moments is quite lengthy. Here we will consider a simple alternative method of deriving the null and nonnull moments of u. The nonnull moments as well as the following procedures do not seem to be available in the literature.

26.3   Nonnull Moments

Let

$$u = \frac{|S|}{[(p-1)\text{tr}(S) - \sum_{i \neq j} s_{ij}]^{p-1}[\text{tr}(S) + \sum_{i \neq j} s_{ij}]}.$$

$$E\left(u^h\right) = \int\limits_{S>0} \frac{|S|^{\frac{n}{2}+h-\frac{p+1}{2}} e^{-\frac{1}{2}\text{tr}(\Sigma^{-1}S)}}{[(p-1)\text{tr}(S) - \sum_{i \neq j} s_{ij}]^{(p-1)h}[\text{tr}(S) + \sum_{i \neq j} s_{ij}]^h \, 2^{\frac{np}{2}} \, \Gamma_p\left(\frac{n}{2}\right) |\Sigma|^{\frac{n}{2}}} \, dS$$

for a general Σ, $n = N - 1$, where, for example,

$$\Gamma_p(\alpha) = \pi^{\frac{p(p-1)}{4}} \Gamma(\alpha)\Gamma\left(\alpha - \frac{1}{2}\right) \cdots \Gamma\left(\alpha - \frac{p-1}{2}\right), \, \Re(\alpha) > \frac{p-1}{2},$$

where $\Re(\cdot)$ denotes the real part of $(\cdot)$. Replace $\frac{S}{2}$ by S which will get rid of all factors containing 2. Replace two of the factors by equivalent integrals.

$$\frac{1}{\left[(p-1)\text{tr}(S) - \Sigma_{i \neq j} \ s_{ij}\right]^{(p-1)h}}$$
$$= \frac{1}{\Gamma[(p-1)h]} \int_{x>0} x^{(p-1)h-1} \ e^{-x\left[(p-1)\text{tr}(S) - \Sigma_{i \neq j} \ s_{ij}\right]} \ \mathrm{d}x,$$

for $\Re(h) > 0$.

$$\frac{1}{\left[\text{tr}(S) + \Sigma_{i \neq j} \ s_{ij}\right]^h}$$
$$= \frac{1}{\Gamma(h)} \int_{y>0} y^{h-1} \ e^{-y\left[\text{tr}(S) + \Sigma_{i \neq j} \ s_{ij}\right]} \ \mathrm{d}y, \ \text{ for } \Re(h) > 0.$$

The exponent in $E\left(u^h\right)$ reduces to the following, excluding $-1/2$:

$$\text{tr}\left(\Sigma^{-1}S\right) + x\left[(p-1)\text{tr}(S) - \Sigma_{i \neq j} \ s_{ij}\right] + y\left[\text{tr}(S) + \Sigma_{i \neq j} \ s_{ij}\right]$$

$$= \ \text{tr}\left(\Sigma^{-1}S\right) + [y + (p-1)x]\text{tr}(S) + (y - x)\Sigma_{i \neq j} \ s_{ij}$$

$$= \ \text{tr}\left(\Sigma^{-1}S\right) + \text{tr}(AS) = \text{tr}\left(\left(\Sigma^{-1} + A\right)S\right),$$

where

$$A = \begin{bmatrix} y + (p-1)x & y - x & \cdots & y - x \\ y - x & y + (p-1)x & & \vdots \\ \vdots & & \ddots & y - x \\ y - x & \cdots & y - x & y + (p-1)x \end{bmatrix}.$$

Integral over S, along with $\Gamma_p\left(\frac{n}{2}\right)|\Sigma|^{\frac{n}{2}}$ yields,

$$\frac{\left|\Sigma^{-1} + A\right|^{-\left(\frac{n}{2} + h\right)} \Gamma_p\left(\frac{n}{2} + h\right)}{\Gamma_p\left(\frac{n}{2}\right)|\Sigma|^{\frac{n}{2}}} = |\Sigma|^h \ \frac{\Gamma_p\left(\frac{n}{2} + h\right)}{\Gamma_p\left(\frac{n}{2}\right)} \ |I + A\Sigma|^{-\left(\frac{n}{2} + h\right)}.$$

$$(26.3.1)$$

Now we examine the factor $|I + A\Sigma|^{-\left(\frac{n}{2}+h\right)}$. One can write

$$A = px\,I + (y-x)\begin{bmatrix} 1 & \cdots & 1 \\ \vdots & & \\ 1 & \cdots & 1 \end{bmatrix}$$

$$= px\,I + (y-x)J, \quad J = \begin{bmatrix} 1 & \cdots & 1 \\ \vdots & & \\ 1 & \cdots & 1 \end{bmatrix}.$$

Note that since J is symmetric there exists an orthonormal matrix Q which will reduce J to its canonical form. The eigenvalues of J are $p, 0, \ldots, 0$. Hence

$$Q'JQ = \begin{bmatrix} p & 0 & \cdots & 0 \\ 0 & 0 & \cdots & 0 \\ \vdots & & \ddots & \vdots \\ 0 & 0 & \cdots & 0 \end{bmatrix}.$$

Then

$$px I + (y-x)J = pxQQ' + (y-x)QQ'JQQ'$$

and

$$|I + A\Sigma| = \left| I + \begin{bmatrix} py & O \\ O & O \end{bmatrix} V + \begin{bmatrix} O & O \\ O & pxI \end{bmatrix} V \right|, \quad V = Q'\Sigma Q.$$

Then

$$\begin{bmatrix} py & O \\ O & O \end{bmatrix} V = \begin{bmatrix} py & O \\ O & O \end{bmatrix} \begin{bmatrix} v_{11} & V_{12} \\ V_{21} & V_{22} \end{bmatrix} = \begin{bmatrix} v_{11}py & pyV_{12} \\ O & O \end{bmatrix}$$

and

$$\begin{bmatrix} O & O \\ O & pxI \end{bmatrix} \begin{bmatrix} v_{11} & V_{12} \\ V_{21} & V_{22} \end{bmatrix} = \begin{bmatrix} O & O \\ pxV_{21} & pxV_{22} \end{bmatrix}.$$

Now,

$$\left| I + \begin{bmatrix} py & O \\ O & O \end{bmatrix} V \right| = 1 + v_{11}py.$$

Therefore

$$|I + A\Sigma| = (1 + v_{11}py)\left| I + px\left(V_{22} - \frac{py}{1 + v_{11}py} V_{21}V_{12} \right) \right|.$$

Note that by expanding the following determinant in two different ways we have

$$\begin{vmatrix} 1 & \frac{(px)(py)}{1+v_{11}py} V_{12} \\ V_{21} & I + pxV_{22} \end{vmatrix}$$

$$= 1 \left| I + pxV_{22} - \frac{(px)(py)}{1 + v_{11}py} V_{21}V_{12} \right|$$

$$\equiv |I + pxV_{22}| \left[1 - \frac{(px)(py)}{1 + v_{11}py} V_{12} \left(I + pxV_{22} \right)^{-1} V_{21} \right].$$

Hence

$$|I + A\Sigma| = |I + pxV_{22}| \left\{ 1 + y \left[pv_{11} - p^2 x V_{12} \left(I + pxV_{22} \right)^{-1} V_{21} \right] \right\}.$$

Now we can integrate out y. That is,

$$\frac{1}{\Gamma(h)} \int_{y=0}^{\infty} y^{h-1} \left\{ 1 + y \left[pv_{11} - p^2 x V_{12} \left(I + pxV_{22} \right)^{-1} V_{21} \right] \right\}^{-\left(\frac{n}{2}+h\right)} dy$$

$$= \frac{\Gamma\left(\frac{n}{2}\right)}{\Gamma\left(\frac{n}{2} + h\right)} (pv_{11})^{-h} \left[1 - px \frac{V_{12}}{v_{11}} \left(I + pxV_{22} \right)^{-1} V_{21} \right]^{-h}. \quad (26.3.2)$$

Substituting

$$|I + pxV_{22}| \left[1 - px \frac{V_{12}}{v_{11}} \left(I + pxV_{22} \right)^{-1} V_{21} \right]$$

$$= \left| I + px \left(V_{22} - V_{21}v_{11}^{-1}V_{12} \right) \right|.$$

The integral over x, denoted by J_x, is of the following form:

$$J_x = \frac{1}{\Gamma[(p-1)h]} \int_{x=0}^{\infty} x^{(p-1)h-1}$$

$$\times |I + pxV_{22}|^{-\frac{n}{2}} \left| I + px \left(V_{22} - V_{21}v_{11}^{-1}V_{12} \right) \right|^{-h} dx.$$

Let the eigenvalues of pV_{22} be $\lambda_1, \ldots, \lambda_{p-1}$ and that of $p(V_{22}-V_{21}v_{11}^{-1}V_{12})$ be $\mu_1, \ldots, \mu_{p-1}$ where $\lambda_j, \mu_j > 0$, $j = 1, \ldots, p-1$ since pV_{22} and $p(V_{22} - V_{21}v_{11}^{-1}V_{12})$ are symmetric positive definite matrices. Then the integral reduces to the form

$$J_x = \frac{1}{\Gamma[(p-1)h]} \int_{x=0}^{\infty} x^{(p-1)h-1} \left[(1 + \lambda_1 x) \cdots (1 + \lambda_{p-1} x) \right]^{-\frac{n}{2}}$$

$$\times \left[(1 + \mu_1 x) \cdots (1 + \mu_{p-1} x) \right]^{-h} dx \quad (26.3.3)$$

Put

$$x = \frac{z}{1 - z} \rightarrow dx = \frac{dz}{(1 - z)^2} \text{ and } 0 < z < 1.$$

Then

$$J_x = \frac{1}{\Gamma[(p - 1)h]} \int_{z=0}^{1} z^{(p-1)h-1} (1 - z)^{(p-1)\frac{n}{2}-1}$$

$$\times \left[(1 - \lambda_1' z) \cdots \left(1 - \lambda_{p-1}' z\right)\right]^{-\frac{n}{2}} \left[(1 - \mu_1' z) \cdots \left(1 - \mu_{p-1}' z\right)\right]^{-h} dz,$$

$$\lambda_j' = 1 - \lambda_j, \ \mu_j' = 1 - \mu_j. \tag{26.3.4}$$

In order to write (26.3.4) as a Lauricella function we need a condition to be satisfied by the coefficient of z in the various factors. Take any large enough number a such that

$$a > \max\{\lambda_1, \ldots, \lambda_{p-1}, \ \mu_1, \ldots, \mu_{p-1}\}.$$

Replace x by $\frac{x}{a}$ in (26.3.3). Then (26.3.4) becomes

$$J_x = \frac{1}{\Gamma[(p - 1)h]a^{(p-1)h}} \int_{z=0}^{1} z^{(p-1)h-1} (1 - z)^{(p-1)\frac{n}{2}-1}$$

$$\times \left[(1 - \lambda_1^* z) \cdots \left(1 - \lambda_{p-1}^* z\right)\right]^{-\frac{n}{2}} \left[(1 - \mu_1^* z) \cdots \left(1 - \mu_{p-1}^* z\right)\right]^{-h} dz,$$

$$\lambda_j^* = 1 - \frac{\lambda_j}{a}, \ \mu_j^* = 1 - \frac{\mu_j}{a}, \ 0 < \lambda_j^* < 1, \ 0 < \mu_j^* < 1,$$

$$j = 1, \ldots, p - 1. \tag{26.3.5}$$

Then (26.3.5) can be written in terms of a Lauricella function F_D. From formula (4.8.14) of Mathai (1993) we have

$$J_x = \frac{1}{a^{(p-1)h}} \frac{\Gamma\left[(p - 1)\frac{n}{2}\right]}{\Gamma\left[(p - 1)\left(\frac{n}{2} + h\right)\right]} F_D\left((p - 1)h, \frac{n}{2}, \ldots, \frac{n}{2}, h, \ldots, h;\right.$$

$$\left.(p - 1)\left(\frac{n}{2} + h\right); \lambda_1^*, \ldots, \lambda_{p-1}^*, \mu_1^*, \ldots, \mu_{p-1}^*\right). \tag{26.3.6}$$

Now collecting all the factors we have the nonnull moment of u,

$$E\left(u^h\right) = |\Sigma|^h \frac{\Gamma_p\left(\frac{n}{2} + h\right)}{\Gamma_p\left(\frac{n}{2}\right)} \frac{\Gamma\left(\frac{n}{2}\right)}{\Gamma\left(\frac{n}{2} + h\right)} \frac{\Gamma\left[(p - 1)\frac{n}{2}\right]}{\Gamma\left[(p - 1)\left(\frac{n}{2} + h\right)\right]}$$

$$\times \frac{1}{(pv_{11})^h \, a^{(p-1)h}}$$

$$\times F_D\left((p-1)h, \ \frac{n}{2}, \ldots, \ \frac{n}{2}, \ h, \ldots, \ h; \ (p-1)\left(\frac{n}{2}+h\right); \right.$$

$$\left. \lambda_1^*, \ldots, \ \lambda_{p-1}^*, \ \mu_1^*, \ldots, \ \mu_{p-1}^* \right)$$

for $\Re(h) > \frac{p-n-1}{2}$, where λ_j^*, μ_j^* are defined in (26.3.5), F_D is given in (26.3.4) and v_{11} is available from (26.3.1).

26.4 Null Case

In the null case

$$|\Sigma| = (\sigma^2)^p (1-\rho)^{p-1}[1+(p-1)\rho]$$

and $I + A\Sigma$ is a matrix with the diagonal elements

$$1 + \sigma^2[1+(p-1)\rho]y + \sigma^2(p-1)(1-\rho)x$$

and the non-diagonal elements equal to $-\sigma^2(1-\rho)x + \sigma^2[1+(p-1)\rho]y$. Then the determinant is given by

$$|I + A\Sigma| = \{1+\sigma^2 p(1-\rho)x\}^{p-1}\{1+\sigma^2 p[1+(p-1)\rho]y\}.$$

Now, the integral over x, evaluating with the help of a type-2 beta integral, yields

$$\frac{1}{\Gamma[(p-1)h]} \int_0^\infty x^{(p-1)h-1}\{1+\sigma^2 p(1-\rho)x\}^{-(p-1)(\frac{n}{2}+h)}\mathrm{d}x$$

$$= [p(1-\rho)\sigma^2]^{-(p-1)h} \ \frac{\Gamma[(p-1)\frac{n}{2}]}{\Gamma[(p-1)(\frac{n}{2}+h)]} \ \text{ for } \Re(h) > -\frac{n}{2}.$$

Integrating y with the help of a type-2 beta integral yields

$$\frac{1}{\Gamma(h)} \int_0^\infty y^{h-1}\{1+p\sigma^2[1+(p-1)\rho]y\}^{-(\frac{n}{2}+h)}\mathrm{d}y$$

$$- \{p\sigma^2[1+(p-1)\rho]\}^{-h} \ \frac{\Gamma(\frac{n}{2})}{\Gamma(\frac{n}{2}+h)} \ \text{ for } \Re(h) > -\frac{n}{2}.$$

Combining all the factors and observing that $|\Sigma|^h$ is cancelled we have

$$E[(p^p u)^h | H_0] = \frac{\Gamma_p(\frac{n}{2}+h)}{\Gamma_p(\frac{n}{2})} \ \frac{\Gamma[(p-1)\frac{n}{2}]}{\Gamma[(p-1)(\frac{n}{2}+h)]} \ \frac{\Gamma(\frac{n}{2})}{\Gamma(\frac{n}{2}+h)}$$

$$\text{for } \Re(h) > -\frac{n}{2} + \frac{p-1}{2}.$$

Expanding $\Gamma[(p-1)(\frac{n}{2})]$ and $\Gamma[(p-1)(\frac{n}{2}+h)]$ with the help of the multiplication formula for gamma functions, namely,

$$\Gamma(mz) = (2\pi)^{\frac{1-m}{2}} m^{mz-\frac{1}{2}} \Gamma(z)\Gamma\left(z+\frac{1}{m}\right)\cdots\Gamma\left(z+\frac{m-1}{m}\right),$$

$$m = 1, 2, \cdots, \qquad (26.4.1)$$

and opening up $\Gamma_p(.)$ we have,

$$E[v^h|H_0] = c\,\frac{\Gamma(\frac{n-1}{2}+h)\Gamma(\frac{n-2}{2}+h)\cdots\Gamma(\frac{n}{2}-\frac{p-1}{2}+h)}{\Gamma(\frac{n}{2}+h)\Gamma(\frac{n}{2}+\frac{1}{p-1}+h)\cdots\Gamma(\frac{n}{2}+\frac{p-2}{p-1}+h)}\,(26.4.2)$$

where

$$v = (p-1)^{p-1}p^p u = \lambda^{2/N}$$

and c is a normalizing constant such that $E(v^h) = 1$ when $h = 0$. From the gamma structure in (4.2) it is easy to observe that the density of v in particular cases can be easily written down. For example when

$$p = 2, \ E(v^h) = c\,\frac{\Gamma(\frac{n}{2}-\frac{1}{2}+h)}{\Gamma(\frac{n}{2}+h)} \Rightarrow v \sim \text{type-1 beta}\ \left(\frac{n-1}{2},\frac{1}{2}\right).$$

For $p = 3$,

$$E(v^h) = c\,\frac{\Gamma(\frac{n-2}{2}+h)\Gamma(\frac{n-2}{2}+\frac{1}{2}+h)}{\Gamma(\frac{n}{2}+h)\Gamma(\frac{n}{2}+\frac{1}{2}+h)}.$$

Combining the gammas with the help of the multiplication formula for gamma functions of (4.1) with $m = 2$, we have

$$E(v^{\frac{1}{2}})^h = c_1\,\frac{\Gamma(n-2+h)}{\Gamma(n+h)},$$

where c_1 is a normalizing constant. Then $v^{\frac{1}{2}}$ is a type-1 beta with the parameters $(n-2, 2)$. For $p = 4$, two gammas in the numerator of (4.2) differ by an integer and hence the density is available in terms of a psi function. For $p = 5$, combining the gammas with the help of the multiplication formula we have

$$E(v^h) = c_2\,\frac{\Gamma(n-2+2h)\Gamma(n-4+2h)}{\Gamma(n+2h)\Gamma(n-\frac{1}{2}+2h)}.$$

Thus the density of $v^{\frac{1}{2}}$ can be evaluated in terms of a psi function since the numerator gammas differ by an integer. For $p \geq 6$ the density can be evaluated in terms of series involving psi and generalized zeta functions. For details see Mathai (1993).

From the gamma structure in (26.4.2) we can easily derive some approximations. To this end, replace h by $nh = \frac{n}{2}(2h)$ or consider the h-th moment of v^n. Then expand each gamma by using the asymptotic expansion of a gamma function, namely,

$$\Gamma(z + a) \approx \sqrt{2\pi} z^{z+a-\frac{1}{2}} e^{-z}, \qquad (26.4.3)$$

where $|z| \to \infty$ and a is bounded. In (26.4.2), with h replaced by nh, take $z = n(1 + 2h)/2$ and the remaining part in each gamma as a. Now, expanding and simplifying the gammas we have

$$E(v^n)^h \approx (1 + 2h)^{-\frac{1}{4}(p^2 + p - 4)}.$$

Consider the variable $w = -\ln(v^n)$. Then the moment generating function of w, taking h as the parameter, namely,

$$E(e^{hw}) = E\left(e^{h(-\ln v^n)}\right)$$

$$= E(v^n)^{-h} \approx (1 - 2h)^{-\frac{1}{4}(p^2 + p - 4)}.$$

Since the right side is the moment generating function of a chi-square with $\nu = \frac{1}{2}(p^2 + p - 4)$ degrees of freedom we have

$$w \approx \chi_\nu^2. \qquad (26.4.4)$$

If better approximations are needed then we can use an extended version of (26.4.3) which involves generalized Bernoulli polynomials. Then taking successive terms we can obtain w as a linear function of chi-square variables with different degrees of freedom where the leading term is that in (26.4.4). The approximation in (26.4.4) also agrees with the large sample approximation for $-2 \ln \lambda$.

26.5   The Conditional Hypothesis

We consider now the hypothesis

$$H_0 : \sigma_1^2 = \sigma_2^2 = \ldots = \sigma_p^2$$

given that $\rho_{ij} = \rho$. We call this the conditional hypothesis. Differentiation with respect to $\sigma_i^2, i = 1, \ldots, p$ yields the following likelihood equations :

$$N + c(p, \rho)\left\{ -[1 + (p-2)\rho]\frac{s_{11}}{\sigma_1^2} + \rho\left[\frac{s_{12}}{\sigma_1\sigma_2} + \cdots + \frac{s_{1p}}{\sigma_1\sigma_p}\right] \right\} = 0$$

$$\vdots$$

$$N + c(p, \rho) \left\{ -[1 + (p-2)\rho] \frac{s_{pp}}{\sigma_p^2} + \rho \left[\frac{s_{p1}}{\sigma_p \sigma_1} + \cdots + \frac{s_{p\ p-1}}{\sigma_p \sigma_{p-1}} \right] \right\} = 0$$

where

$$c(p, \rho) = \frac{1}{(1 - \rho)[1 + (p-1)\rho]}. \qquad (26.5.1)$$

Taking the sum of the above equations we have :

$$Np + c(p, \rho) \left\{ -[1 + (p-2)\rho] \left(\frac{s_{11}}{\sigma_1^2} + \cdots + \frac{s_{pp}}{\sigma_p^2} \right) + 2\rho \sum_{i<j} \frac{s_{ij}}{\sigma_i \sigma_j} \right\} = 0.$$

$$(26.5.2)$$

Differentiating with respect to ρ we have:

$$Np(p-1)\rho - \left[(p-2) \left(\frac{s_{11}}{\sigma_1^2} + \cdots + \frac{s_{pp}}{\sigma_p^2} \right) - 2 \sum_{i<j} \frac{s_{ij}}{\sigma_i \sigma_j} \right]$$

$$+ \frac{[(p-2) - 2(p-1)\rho]}{(1-\rho)[1 + (p-1)\rho]}$$

$$\times \left\{ (1 + (p-2)\rho) \left(\frac{s_{11}}{\sigma_1^2} + \cdots + \frac{s_{pp}}{\sigma_p^2} \right) - 2\rho \sum_{i<j} \frac{s_{ij}}{\sigma_i \sigma_j} \right\} = 0.$$

$$(26.5.3)$$

From (26.5.2) and (26.5.3) we have

$$\rho = \frac{p-2}{p-1} - \frac{1}{Np(p-1)} \left\{ (p-2) \left(\frac{s_{11}}{\sigma_1^2} + \cdots + \frac{s_{pp}}{\sigma_p^2} \right) - 2 \sum_{i<j} \frac{s_{ij}}{\sigma_i \sigma_j} \right\}.$$

$$(26.5.4)$$

The equations (26.5.2), (26.5.3) and (26.5.4) are satisfied for

$$\hat{\sigma}_i^2 = \frac{s_{ii}}{N}$$

and

$$\hat{\rho} = \frac{2}{p(p-1)} \sum_{i<j} r_{ij}, \quad r_{ij} = \frac{s_{ij}}{\sqrt{s_{ii}\, s_{jj}}}.$$

In the general space

$$\left|\hat{\Sigma}\right| = \frac{s_{11}\cdots s_{pp}}{N^p} \frac{[p(p-1) - 2\sum_{i<j} r_{ij}]^{p-1}\,[p + 2\sum_{i<j} r_{ij}]}{p^p\,(p-1)^{p-1}}.$$

Under the null hypothesis, we had evaluated the estimate of $|\Sigma|$ earlier. That is,

$$\left|\hat{\Sigma}_0\right| = \left[(p-1)\,\mathrm{tr}(S) - 2\sum_{i<j} s_{ij}\right]^{p-1}$$

$$\times \left[\mathrm{tr}(S) + 2\sum_{i<j} s_{ij}\right] / \left[(Np)^p (p-1)^{p-1}\right].$$

Let

$$\hat{R} = \begin{bmatrix} 1 & r_{12} & \cdots & r_{1p} \\ r_{12} & 1 & \cdots & r_{2p} \\ \vdots & & & \\ r_{p1} & r_{p2} & \cdots & 1 \end{bmatrix}, \quad D = \mathrm{diag}\left(\sqrt{s_{11}}, \ldots, \sqrt{s_{pp}}\right)$$

$$J = \begin{bmatrix} 1 & \cdots & 1 \\ \vdots & & \\ 1 & \cdots & 1 \end{bmatrix}, \quad S = D\hat{R}D.$$

Then

$$p(p-1) - 2\sum_{i<j} r_{ij} = p^2 - \mathrm{tr}(J\hat{R}) \Rightarrow$$

$$p + 2\sum_{i<j} r_{ij} = \mathrm{tr}(J\hat{R}).$$

$$p\,\mathrm{tr}(S) = p\,\mathrm{tr}(D^2)$$

$$(p-1)\,\mathrm{tr}(S) - 2\sum_{i<j} s_{ij} = p\,\mathrm{tr}(D^2) - \mathrm{tr}[(DJD)\hat{R}].$$

$$\mathrm{tr}(S) + 2\sum_{i<j} s_{ij} = \mathrm{tr}[(DJD)\hat{R}].$$

Let

$$u = N^p\,p^p\,(p-1)^{p-1}\,\lambda^{\frac{2}{N}} \tag{26.5.5}$$

where λ is the likelihood ratio criterion. Then

$$u = \frac{s_{11} \cdots s_{pp} \, [p^2 - \mathrm{tr}(J\hat{R})]^{p-1}[\mathrm{tr}(J\hat{R})]}{[p \, \mathrm{tr}(D^2) - \mathrm{tr}(DJD)\hat{R}]^{p-1}[\mathrm{tr}[(DJD)\hat{R}]}. \qquad (26.5.6)$$

In order to compute the h-th moment one can integrate over the density of S. For convenience replace $\frac{S}{2}$ by S and take the integral. Then

$$E(u^h | H_0) = \int_{S>0} \frac{(s_{11} \cdots s_{pp})^h [p^2 - \mathrm{tr}(J\hat{R})]^{(p-1)h}[\mathrm{tr}(J\hat{R})]^h}{[p \, \mathrm{tr}(D^2) - \mathrm{tr}\{(DJD)\hat{R}\}]^{(p-1)h}[\mathrm{tr}\{(DJD)\hat{R}\}]^h}$$

$$\times \frac{|S|^{\frac{n}{2} - \frac{p+1}{2}} \, e^{-\mathrm{tr}\Sigma_0^{-1}S}}{|\Sigma_0|^{\frac{n}{2}} \, \Gamma_p(\frac{n}{2})} \, dS$$

where

$$|S|^{\frac{n}{2} - \frac{p+1}{2}} = |D^2|^{\frac{n}{2} - \frac{p+1}{2}} \, |\hat{R}|^{\frac{n}{2} - \frac{p+1}{2}} = (s_{11} \cdots s_{pp})^{\frac{n}{2} - \frac{p+1}{2}} \, |\hat{R}|^{\frac{n}{2} - \frac{p+1}{2}}.$$

Then for large values of N we have $-2 \ln \lambda \approx \chi_{p-1}^2$ where λ is available from (26.5.5) and (26.5.6). The exact nonnull moments and the exact distributions seem to be difficult to evaluate.

References

1. Cacoullos, T. (2000). The F-test of homoscedasticity for correlated normal variables. *Statistics & Probability Letters*, to appear.

2. Mathai, A. M. (1993). *A Handbook of Generalized Special Functions for Statistical and Physical Sciences.* Oxford University Press, Oxford.

3. Morgan, W. A. (1939). A test for the significance of the difference between the two variances in a sample from a normal bivariate population. *Biometrika* **31**, 13–19.

4. Pitman, E. J. G. (1939). A note on normal correlation. *Biometrika* **31**, 9–12.

5. Wilks, S. S. (1946). Sample criteria for testing equality of means, equality of variances, and equality of covariances in a normal multivariate distribution. *Annals of Mathematical Statistics* **17**, 257–281.

27

Estimating the Smallest Scale Parameter: Universal Domination Results

Stavros Kourouklis
University of Patras, Patras, Greece

ABSTRACT In this work universal domination results are obtained for the problem of estimating the smallest of two scale parameters in independent populations with monotone likelihood ratios. In particular, for estimating the smallest of two normal variances several estimators universally dominating the minimum mean squared error estimator based on the corresponding sample variance are derived.

Keywords and phrases Decision theory, universal domination, ordered restricted inference, scale parameter, monotone likelihood ratio

27.1 Introduction

Wald's (1950) theory of statistical decisions is based on the assumption of a particular and fully specified loss. In practice, however, it is often difficult to specify the loss function exactly. Besides, if an estimator T_2 dominates another estimator T_1 with respect to a loss L_1 there is no guarantee in general that T_2 will still be better than T_1 under a different loss L_2. Hwang (1985) proposed the criterion of universal domination (u-domination) as a means of studying how robust the superiority of one estimator over another is with respect to the loss function. This criterion was further developed by Brown and Hwang (1989), whereas, earlier, similar concepts were considered in a less formal manner by Cohen and Sackrowitz (1970), Brown (1971) and Rukhin (1987).

For a scale parameter $\sigma > 0$ Hwang's (1985) criterion is defined as follows. If

$$EL\left(|T_2/\sigma - 1|\right) \leq EL\left(|T_1/\sigma - 1|\right) \tag{27.1.1}$$

for every nondecreasing function $L(\cdot)$ and every $\sigma > 0$, and for a particular $L(\cdot)$ the risk functions are not identical then T_2 u-dominates T_1 and the latter is said to be u-inadmissible. If there does not exist estimator that u-dominates T_1, then T_1 is called u-admissible. Hwang (1985) showed that u-domination is equivalent to stochastic domination which in our setup means that (27.1.1) holds iff

$$P\left(|T_2/\sigma - 1| \le c\right) \ge P\left(|T_1/\sigma - 1| \le c\right) \ \forall\, c > 0, \ \ \sigma > 0 \qquad (27.1.2)$$

and for some $\sigma > 0$ the distribution functions are not identical.

Recently Kushary (1998) dealt with u-admissibility and u-inadmissibility of equivariant estimators of the scale parameter in the one parameter gamma and the two parameter normal and exponential distributions. Subsequently, Kourouklis (1999) extended these results to a scale family of distributions with monotone likelihood ratio and to two normal or two exponential distributions for estimating the ratio of the respective scale parameters.

In this article we focus on ordered scale parameters. We refer to the book by Robertson, Wright, and Dykstra (1988) for an excellent account on order restricted inference. When σ_1, σ_2 are the variances of two normal populations and $\sigma_1 \le \sigma_2$, Kushary and Cohen (1989) demonstrated the u-inadmissibility of the standard estimator of σ_1, i.e. the minimum mean squared error estimator of σ_1 based only on the sample variance from the first population, in the equal sample sizes case by exhibiting a u-dominating estimator which uses both sample variances.

In Section 27.2 we assume that $\sigma_1 \le \sigma_2$ are scale parameters of two general independent populations with monotone likelihood ratios and derive sufficient conditions for the u-inadmissibility of an arbitrary scale equivariant estimator of σ_1. In particular, when $\sigma_1 \le \sigma_2$ are normal variances we establish u-inadmissibility of the standard estimator of σ_1 for general sample sizes as well as u-inadmissibility of Kushary's and Cohen's (1989) estimator by deriving u-dominating estimators also based on both sample variances. Then using also the sample means we derive additional estimators of σ_1 which u-dominate all the above ones. Analogous results hold when $\sigma_1 \le \sigma_2$ are the scale parameters of two independent exponential populations but for the sake of brevity they will not be presented here. Also, the results can easily be extended to more than two populations.

27.2   Main Results

We start with some auxiliary results. Let X be a statistic such that X/σ has density $f(x)I(x > 0)$ where $\sigma > 0$ is an unknown scale parameter. We assume that the family of distributions of X has the monotone likelihood ratio property, i.e.,

$$\frac{f(c_1 x)}{f(c_2 x)} \text{ is strictly increasing in x} > 0 \text{ for every } 0 < c_1 < c_2. \quad (27.2.1)$$

Our main results are based on the following lemmas.

LEMMA 27.2.1   *[Iliopoulos and Kourouklis (2000)]*
The function $f(x)$ is continuous on $(0, \infty)$, $xf(x)$ is strictly increasing and then strictly decreasing, and

$$\lim_{x \to 0} xf(x) = \lim_{x \to \infty} xf(x) = 0.$$

Lemma 27.2.1 ensures that $xf(x)$ has a unique mode, which we denote by a_0.

LEMMA 27.2.2   *[Kourouklis (1999)]*
Let $G(a) = P\left(\left|\dfrac{aX}{\sigma} - 1\right| \le c\right), a > 0, c > 0$. Then we have the following.
(i) $G(a)$ is strictly decreasing on $(2/a_0, \infty)$ for every $c > 0$.
(ii) If, in addition,

$$(x + y)f(x + y) > (x - y)f(x - y) \text{ whenever } a_0 \ge x > y > 0 \quad (27.2.2)$$

then $G(a)$ is strictly decreasing on $(1/a_0, \infty)$ for every $c > 0$.

PROOF (i) Let $F(\cdot)$ be the distribution function of X/σ. Then $G(a) = F\left(\dfrac{1+c}{a}\right) - F\left(\dfrac{1-c}{a}\right)$. Since X takes on positive values, for $c \ge 1, G(a) = F\left(\dfrac{1+c}{a}\right)$ which is obviously strictly decreasing in a. For $0 < c < 1$ we obtain $G'(a) = -\dfrac{1+c}{a^2}f\left(\dfrac{1+c}{a}\right) + \dfrac{1-c}{a^2}f\left(\dfrac{1-c}{a}\right)$ which is negative provided

$$\frac{1+c}{a}f\left(\frac{1+c}{a}\right) > \frac{1-c}{a}f\left(\frac{1}{a}\frac{c}{}\right). \quad (27.2.3)$$

Now for $a > 2/a_0$ we have $\dfrac{1-c}{a} < \dfrac{1+c}{a} < a_0$ and thus (27.2.3) holds by Lemma 27.2.1.
(ii) Applying (27.2.2) with $x = \dfrac{1}{a}, y = \dfrac{c}{a}, 0 < c < 1, a > 1/a_0$, we see that (27.2.3) holds, and hence the result follows.   ∎

LEMMA 27.2.3

If $f_1(x)$ and $f_2(x)$ are positive, unimodal functions on $(0, \infty)$ with modes x_1 and x_2 respectively (i.e., $f_i(x)$ is strictly increasing for $0 < x < x_i$ and strictly decreasing for $x > x_i$, $i = 1, 2$) and

$$\frac{f_2(x)}{f_1(x)} \text{ is strictly increasing in } x > 0 \qquad (27.2.4)$$

then $x_2 \geq x_1$.

PROOF Assume $x_2 < x_1$. Then on (x_2, x_1) $f_2(x)$ is decreasing and $f_1(x)$ is increasing, so that $\dfrac{f_2(x)}{f_1(x)}$ is decreasing on (x_2, x_1) which contradicts (27.2.4). ∎

Let now S_i, $i = 1, 2$, be independent statistics such that S_i/σ_i has density $g_i(x)I(x > 0)$, where σ_1 and σ_2 are unknown positive parameters with $\sigma_1 \leq \sigma_2$. We further assume that $g_i(x)$ satisfies (27.2.1). We will study two classes of estimators of σ_1, namely $C = \{aS_1 : a > 0\}$ and $D = \{\phi(V)S_1 : \phi \text{ is a positive function }\}$, where $V = S_2/S_1$. The first one is the class of scale equivariant estimators of σ_1 based only on S_1. This class does not make use of the full data and, in effect, ignores the information $\sigma_1 \leq \sigma_2$. The second one is the class of equivariant estimators under the group of transformations $(S_1, S_2) \rightarrow (aS_1, aS_2)$, $a > 0$. Clearly, $C \subset D$.

For $v > 0$ let $g_1(x|v; \sigma_1, \sigma_2)$ be the density of the conditional distribution of S_1/σ_1 given $V = v$, i.e.

$$g_1(x|v; \sigma_1, \sigma_2) \propto xg_1(x)g_2(\frac{\sigma_1}{\sigma_2}xv).$$

When $\sigma_1 = \sigma_2$ we simply write $g_1(x|v)$. We observe that since the $g_i(x)$'s satisfy (27.2.1) so does $g_1(x|v; \sigma_1, \sigma_2)$. We denote by $m(v; \sigma_1, \sigma_2)$ the mode of $xg_1(x|v; \sigma_1, \sigma_2)$ but when $\sigma_1 = \sigma_2$ we just write $m(v)$ (which does not depend on σ_i, $i = 1, 2$).

THEOREM 27.2.4

Let $\delta = \phi(V)S_1$ be an estimator in the class D. Then we have the following.
(i) δ is u-dominated by $\delta_1 = \min\{\phi(V), 2/m(V)\}S_1$ provided $\delta_1 \neq \delta$ with positive probability.
(ii) If for every $v > 0$ and $\sigma_1 \leq \sigma_2$ $g_1(x|v; \sigma_1, \sigma_2)$ satisfies (27.2.2) then both δ and δ_1 are u-dominated by $\delta_2 = \min\{\phi(V), 1/m(V)\}S_1$ provided $\delta_2 \neq \delta$ with positive probability.

PROOF (i) By Lemma 27.2.2(i), for every $v > 0$ the conditional probability

$$P\left(\left|a\frac{S_1}{\sigma_1} - 1\right| \le c | V = v\right) \text{ is decreasing for } a > \frac{2}{m(v; \sigma_1, \sigma_2)}. \text{ Further-}$$

more,

$$\frac{xg_1(x|v; \sigma_1, \sigma_2)}{xg_1(x|v)} \propto \frac{g_2(\frac{\sigma_1}{\sigma_2}xv)}{g_2(xv)}$$

is strictly increasing in $x > 0$ (unless $\sigma_1 = \sigma_2$) since g_2 satisfies (27.2.1) and $\sigma_1 \le \sigma_2$. Thus, Lemma 27.2.3 implies $m(v; \sigma_1, \sigma_2) \ge m(v)$ and consequently

$$P\left(\left|\frac{2}{m(v)}\frac{S_1}{\sigma_1} - 1\right| \le c | V = v\right) > P\left(\left|\phi(v)\frac{S_1}{\sigma_1} - 1\right| \le c | V = v\right) \tag{27.2.5}$$

if $\phi(v) > \frac{2}{m(v)}$. Upon taking expectations on both sides of (27.2.5) we conclude that

$$P\left(\left|\frac{\delta_1}{\sigma_1} - 1\right| \le c\right) > P\left(\left|\frac{\delta}{\sigma_1} - 1\right| \le c\right)$$

provided

$$P\left(\phi(V) > \frac{2}{m(V)}\right) > 0.$$

By the equivalence of u-domination and stochastic domination (see (27.1.1) and (27.1.2)) the result follows.

(ii) For the u-domination of δ repeat the argument of part (i) using Lemma 27.2.2(ii). Since $\delta_1 = \phi_1(V)S_1$ with $\phi_1(v) = \min\{\phi(v), 2/m(v)\}$, it is u-dominated by $\delta_2 = \min\{\phi_1(v), 1/m(v)\}S_1 = \delta_1$. ∎

REMARK Theorem 27.2.4 can be applied, in particular, to any estimator $\delta = aS_1$ in the class C. ∎

We now consider independent random samples $X_1, \ldots, X_{n_1} \sim N(\mu_1, \sigma_1)$ and $Y_1, \ldots, Y_{n_2} \sim N(\mu_2, \sigma_2)$, $n_i \ge 2$, where all the parameters are unknown and $\sigma_1 \le \sigma_2$. Set $S_1 = \sum_{i=1}^{n_1}(X_i - \overline{X})^2 \sim \sigma_1 \chi^2_{n_1-1}$ and $S_2 = \sum_{i=1}^{n_2}(Y_i - \overline{Y})^2 \sim \sigma_2 \chi^2_{n_2-1}$ and note that (27.2.1) is satisfied for the densities of S_i/σ_i, $i = 1, 2$. Also, in this case $g_1(x|v; \sigma_1, \sigma_2)$ is the density of gamma distribution $G\left(\frac{n_1 + n_2 - 2}{2}, \frac{2}{1 + (\sigma_1/\sigma_2)v}\right)$ which, as shown in Kushary (1998, Lemma 2.1), satisfies (27.2.2) with $a_0 = m(v; \sigma_1, \sigma_2) =$

$\dfrac{n_1 + n_2 - 2}{1 + (\sigma_1/\sigma_2)v}$. The best estimator in the class C with respect to squared error loss is

$$\delta_0 = \frac{1}{n_1 + 1} S_1. \tag{27.2.6}$$

Then the following result follows directly from Theorem 27.2.4.

THEOREM 27.2.5
(i) For $n_2 > n_1 + 4$, δ_0 is u-dominated by $\delta_1 = \min\left\{\delta_0, \dfrac{2(S_1 + S_2)}{n_1 + n_2 - 2}\right\}$.
(ii) For $n_2 > 3$, δ_0 is u-dominated by

$$\delta_2 = \min\left\{\delta_0, \frac{S_1 + S_2}{n_1 + n_2 - 2}\right\}. \tag{27.2.7}$$

REMARK $\delta_3 = \dfrac{S_1 + S_2}{n_1 + n_2 - 2}$ is the umvu estimator of σ_1 when $\sigma_1 = \sigma_2$, so that δ_2 chooses between δ_0 and δ_3 depending on which is the smallest of the two. ∎

In the case that the samples sizes are equal, i.e. $n_1 = n_2$, Kushary and Cohen (1989) have shown that δ_0 is u-dominated by

$$\delta_{KC} = \min\left\{\delta_0, \frac{2(S_1 + S_2)}{3(n_1 + 1)}\right\} \tag{27.2.8}$$

provided $n_1 > 6$. The next theorem demonstrates that δ_{KC} is u-inadmissible too.

THEOREM 27.2.6
Assume that $n_1 = n_2 > 7$. Then the estimator δ_{KC} in (27.2.8) is u-dominated by the estimator δ_2 in (27.2.7).

PROOF Writing $\delta_{KC} = \phi(V)S$, where $\phi(V) = \min\left\{\dfrac{1}{n_1 + 1}, \dfrac{2(1 + V)}{3(n_1 + 1)}\right\}$ and applying Theorem 27.2.4 we have that δ_{KC} is u-dominated by

$$\delta_4 = \min\left\{\frac{1}{n_1 + 1}, \frac{2(1 + V)}{3(n_1 + 1)}, \frac{1 + V}{2(n_1 - 1)}\right\} S_1.$$

Now for $n_1 > 7$, $\dfrac{2(1 + V)}{3(n_1 + 1)} > \dfrac{1 + V}{2(n_1 - 1)}$ and thus

$$\delta_4 = \min\left\{\frac{1}{n_1 + 1}, \frac{1 + V}{2(n_1 - 1)}\right\} S_1 = \delta_1.$$

∎

In the above normal setup all the u-dominating estimators are based only on the sample variances. We will next provide u-dominating estimators which, in addition, use the sample means. To this end , for $W_1 = n_1 \overline{X}^2 / S_1$ and $Z_1 = n_2 \overline{Y}^2 / S_1$, we consider the broader class $D_1 = \{\phi(V, W_1, Z_1) S_1 : \phi \text{ is a positive function}\}$ of equivariant estimators of σ_1 under the group of transformations

$$(\overline{X}, \overline{Y}, S_1, S_2) \rightarrow (\pm a\overline{X}, \pm a\overline{Y}, a^2 S_1, a^2 S_2), a > 0.$$

THEOREM 27.2.7
An estimator $\delta = \phi(V, W_1, Z_1) S_1$ in the class D_1 is u-dominated by

$$\delta^* = \min\left\{\delta, \frac{S_1 + n_1 \overline{X}^2 + S_2 + n_2 \overline{Y}^2}{n_1 + n_2}\right\}$$

(which is also in D_1) provided $\delta \neq \delta^*$ with positive probability.

PROOF Let L and K be independent Poisson random variables with parameters $n_1 \mu_1^2 / 2\sigma_1$ and $n_2 \mu_2^2 / 2\sigma_2$ respectively. We may and will assume that L and K are also independent of $(\overline{X}, S_1, \overline{Y}, S_2)$. Then it is easy to show that the conditional density of S_1 / σ_1 given $V = v, W_1 = w_1, Z_1 = z_1, L = l, K = k$ is gamma

$$G\left(\frac{n_1 + n_2 + 2l + 2k}{2}, \frac{2}{1 + w_1 + (\sigma_1/\sigma_2)(v + z_1)}\right)$$

which satisfies (27.2.2) with $a_0 = \dfrac{n_1 + n_2 + 2l + 2k}{1 + w_1 + (\sigma_1/\sigma_2)(v + z_1)}$ (see Kushary (1998, Lemma 2.1)). Hence, by Lemma 27.2.2(ii),

$$P\left(\left|a\frac{S_1}{\sigma_1} - 1\right| \leq c \middle| V = v, W_1 = w_1, Z_1 = z_1, L = l, K = k\right)$$

is strictly decreasing for $a > \dfrac{1 + w_1 + (\sigma_1/\sigma_2)(v + z_1)}{n_1 + n_2 + 2l + 2k}$. It follows that

$$P\left(\left|\frac{1 + w_1 + v + z_1}{n_1 + n_2}\frac{S_1}{\sigma_1} - 1\right| \leq c \middle| V = v, W_1 = w_1, \right.$$

$$\left. Z_1 = z_1, L = l, K = k\right)$$

$$> P\left(\left|\phi(v, w_1, z_1)\frac{S_1}{\sigma_1} - 1\right| \leq c \middle| V = v, W_1 = w_1, \right.$$

$$\left. Z_1 = z_1, L = l, K = k\right) \qquad (27.2.9)$$

if $\phi(v, w_1, z_1) > \dfrac{1 + w_1 + v + z_1}{n_1 + n_2}$. Upon taking expectations on both

sides of (27.2.9) we obtain that $P\left(\left|\dfrac{\delta^*}{\sigma_1} - 1\right| \le c\right) > P\left(\left|\dfrac{\delta}{\sigma_1} - 1\right| \le c\right)$

provided

$P\left(\phi(V, W_1, Z_1) > \dfrac{1 + W_1 + V + Z_1}{n_1 + n_2}\right) > 0$. By virtue of (27.1.2) the

proof is now complete. ∎

COROLLARY 27.2.8

(i) The estimator $\delta_0 = \dfrac{S_1}{n_1 + 1}$ *in (27.2.6) is u-dominated by*

$$\delta^* = \min\left\{\frac{S_1}{n_1 + 1}, \frac{S_1 + n_1\overline{X}^2 + S_2 + n_2\overline{Y}^2}{n_1 + n_2}\right\}.$$

(ii) The estimator δ_2 *in (27.2.7) is u-dominated by*

$$\delta_1^* = \min\left\{\frac{S_1}{n_1 + 1}, \frac{S_1 + S_2}{n_1 + n_2 - 2}, \frac{S_1 + n_1\overline{X}^2 + S_2 + n_2\overline{Y}^2}{n_1 + n_2}\right\}.$$

(iii) For $n_1 = n_2$ *and* $n_1 > 3$ *the estimator* δ_{KC} *in (27.2.8) is u-dominated by*

$$\delta_2^* = \min\left\{\frac{S_1}{n_1 + 1}, \frac{2(S_1 + S_2)}{3(n_1 + 1)}, \frac{S_1 + n_1\overline{X}^2 + S_2 + n_1\overline{Y}^2}{2n_1}\right\}.$$

PROOF Immediate from Theorem 27.2.7. ∎

REMARK We recognize that the estimator $\dfrac{S_1 + n_1\overline{X}^2 + S_2 + n_2\overline{Y}^2}{n_1 + n_2}$ that

appears in Theorem 27.2.7 is the umvue of σ_1 when $\sigma_1 = \sigma_2$ and

$\mu_1 = \mu_2 = 0$. ∎

REMARK The estimators δ_1^* and δ_2^* also u-dominate δ_0. ∎

References

1. Brown, L. D. (1971). Admissible estimators, recurrent diffusions, and insoluble boundary value problems. *Annals of Mathematical Statistics* **42**, 855–903.

2. Brown, L. D. and Hwang, J. T. (1989). Universal domination and stochastic domination: U-admissibility and U-inadmissibility of the least squares estimator. *Annals of Statistics* **17**, 252–267.

3. Cohen, A. and Kushary, D. (1998). Universal admissibility of maximum likelihood estimators in constrained spaces. *Statistics & Decisions* **16**, 131–146.

4. Cohen, A. and Sackrowitz, H. (1970). Estimation of the last mean of a monotone sequence. *Annals of Mathematical Statistics* **41**, 2021–2034.

5. Hwang, J. T. (1985). Universal domination and stochastic domination: Estimation simultaneously under a broad class of loss functions. *Annals of Statistics* **13**, 295–314.

6. Iliopoulos, G. and Kourouklis, S. (2000). Interval estimation for the ratio of scale parameters and for ordered scale parameters. *Statistics & Decisions*, to appear.

7. Kourouklis, S. (1999). On universal admissibility of scale parameter estimators. Unpublished manuscript.

8. Kushary, D. (1998). A note on universal admissibility of scale parameter estimators. *Statistics & Probability Letters* **38**, 59–67.

9. Kushary, D. and Cohen, A. (1989). Estimating ordered location and scale parameters. *Statistics & Decisions* **7**, 201–213.

10. Robertson, T., Wright, F. T., and Dykstra, R. L. (1988). *Order Restricted Statistical Inference*. John Wiley & Sons, New York.

11. Rukhin, A. L. (1987). Universal Bayes estimators. *Annals of Statistics* **6**, 1345–1351.

12. Wald, A. (1950). *Statistical Decision Functions*. John Wiley & Sons, New York.

28

On Sensitivity of Exponential Rate of Convergence for the Maximum Likelihood Estimator

James C. Fu
University of Manitoba, Winnipeg, Manitoba, Canada

ABSTRACT Maximum likelihood estimator derived from a parametric statistical model $M_O = \{R, F_\theta, \theta \in \Theta\}$ is often favored by statisticians as estimating and collecting the information for the unknown parameter θ. It has many large sample optimal properties. One of these optimal properties is that the maximum likelihood estimator has an optimal exponential rate of convergence to the parameter θ when the underlying distribution Q is a member of the model M_O. Motivated by robust estimation, this article mainly studies the sensitivity of exponential rate of convergence for the maximum likelihood estimator derived from the parametric model M_O when the true underlying distribution Q departs slightly from the model M_O. Applications of the results to exponential-family and t-family of distributions are studied. As a byproduct, it shows that the maximum likelihood estimator derived from the t-distribution with 6 degrees of freedom is the best robust estimator in the sense of local exponential rate with respect to all t-distributions.

Keywords and phrases t-distribution, robust estimation, maximum likelihood estimate, rate of convergence, Bahadur bound, exponential family, M-estimator, Cauchy distribution

28.1 Introduction

Let $M_O = \{R, F_\theta, \theta \in \Theta\}$ be a parametric statistical model, where $R = (-\infty, \infty)$ is the sample space, Θ is parameter space, an open subset of R, and for every $\theta \in \Theta$, F_θ is a probability distribution defined on R. Let $s = (x_1, \cdots, x_n)$ be a sample of n independent identically distributed ($i.i.d.$) observations. The maximum likelihood estimator (mle)

$\hat{\theta}_n(s)$ derived from the parametric model M_O is often favored by many statisticians for estimating and collecting the information of the unknown parameter θ. If the true underlying distribution $Q = F_\theta$ is a member of the statistical model M_O (the model is correctly specified), then the mle has many large sample optimal properties, for example consistency, asymptotic normality, and exponential rate of convergence. The consistency of the mle has been studied by Cramér (1946), Wald (1949), LeCam (1953, 1970), Bahadur (1967), and Pfanzagl (1973).

Under certain regularity conditions, the mle is asymptotically efficient in the sense that it is asymptotically normally distributed with variance achieving the Cramér-Rao lower bound; *i.e.*,

$$\sqrt{n}(\hat{\theta}_n(s) - \theta) \xrightarrow{L} \mathcal{N}(0, I^{-1}(\theta)), \qquad (28.1.1)$$

where $\xrightarrow{L}$ stands for convergence in distribution, and $I(\theta)$ is the Fisher information of the underlying distribution F_θ [see, for instance, Cramér (1946) and Rao (1963)].

Recently, Bahadur (1967, 1971, 1980), Fu (1973, 1975, 1982), Rukhin (1983), Steinebach (1978), Jureckova (1981), Kester (1981), Kester and Kallenberg (1986), and Rubin and Rukhin (1983) study the exponential rates of convergence for consistent estimators. They show that for any consistent estimator $T_n(s)$,

$$\liminf_{n \to \infty} \frac{1}{n} \log P(|T_n - \theta| \geq \varepsilon | F_\theta) \geq -B(\theta, \varepsilon) \qquad (28.1.2)$$

and

$$\liminf_{\varepsilon \to 0} \liminf_{n \to \infty} \frac{1}{n\varepsilon^2} \log P(|T_n - \theta| \geq \varepsilon | F_\theta) \geq -I(\theta)/2, \qquad (28.1.3)$$

where $B(\theta, \varepsilon)$ is given by

$$B(\theta, \varepsilon) = \inf_{\theta'}\{K(F_{\theta'}, F_\theta) : |\theta' - \theta| > \varepsilon\}, \qquad (28.1.4)$$

and

$$K(F_{\theta'}, F_\theta) = \begin{cases} \int_{-\infty}^{\infty} (\log \frac{dF_{\theta'}}{dF_\theta}) dF_{\theta'}, & \text{if } F_{\theta'} << F_\theta \\ \infty, & \text{otherwise}, \end{cases} \qquad (28.1.5)$$

is the Kullback-Leibler information of $F_{\theta'}$ with respect to F_θ. The positive constant $B(\theta, \varepsilon)$ is usually referred as the Bahadur bound. The estimator which achieves the lower bound of (28.1.3) is called asymptotically efficient in the Bahadur sense. It is well-known that for fixed ε, the mle achieves the Bahadur bound if and only if the underlying distribution is a member of an exponential family of distributions. The sufficient part was proved by Kester (1981) and the necessary part was proved by Cheng and Fu (1986). For $\varepsilon \to 0$, under very general conditions, the mle is *always locally optimal in the Bahadur sense that its exponential rate achieves the lower bound of inequality (28.1.3)*.

All these large sample optimal properties of the mle require the assumption that the true underlying distribution Q is a member of the parametric model M_O. When this basic assumption is false $(Q \notin M_O)$, what will happen to the mle $\hat{\theta}_n$ derived from the wrongly specified model M_O? The following questions are often asked:

(i) If the true underlying distribution Q does not belong to M_O, under what conditions does the mle $\hat{\theta}_n$ derived from the wrongly specified model M_O still converge under Q? If it does converge, what value does it converge to? Does this converge exponentially?

(ii) How sensitive is the exponential rate of convergence of the mle when the underlying distribution Q departs slightly from the wrongly specified model M_O?

The main goal of this article is to answer some of the above questions. It shows that if the underlying distribution Q is not too far away from the wrongly specified model M_O then the mle $\hat{\theta}_n(s)$ derived from the model M_O still converges to a value θ^* whose corresponding distribution $F_{\theta^*} \in M_O$ is the closest to the underlying distribution Q in terms of the Kullback-Leibler information. Further if the moment generating function of the score function of the specified model M_O exists under Q $(Q \notin M_O)$, then not only the mle $\hat{\theta}_n(s)$ converges to θ^* but also converges exponentially. This article is organized in the following way. Section 28.2 studies the main results mentioned above. In Section 28.3, exponential-family and t-family of distributions are used to illustrate the main results.

28.2 Main Results

For each $F_\theta \in M_O$, denote $f(x|\theta)$ as its density function, $l(\theta|x) = \log f(x|\theta)$, and $l^{(i)}(\theta|x) = (d/d\theta)^i l(\theta|x)$, $i = 1, \cdots, m$. For $i = 1$, the function $l^{(1)}(\theta|x)$ is referred as the score function. Given data $s = (x_1, \cdots, x_n)$, let $l_n(\theta|s) = \sum_{i=1}^{n} \log f(x_i|\theta)$ be the log-likelihood function. For simplicity, if there is no special specification, from here on the mle $\hat{\theta}_n(s)$ derived from the model M_O is assumed to be the unique solution of the likelihood equation

$$l_n^{(1)}(\theta|s) = 0. \tag{28.2.1}$$

Let $q(x)$ be the density function of the underlying distribution Q. Further, throughout this article, the underlying distribution Q is assumed to be absolutely continuous with respect to (wrt) every $F_\theta \in M_O$ $(Q \ll F_\theta)$ and the Kullback-Leibler information of Q wrt F_θ

$$K(Q, F_\theta) = \int_{-\infty}^{\infty} (\log \frac{q(x)}{f(x|\theta)}) q(x) dx < \infty, \qquad (28.2.2)$$

satisfies the following conditions:

(i) the Kullback-Leibler information $K(Q, F_\theta)$ is a convex function on Θ.

(ii) θ^* is the unique solution of the equation

$$K^{(1)}(Q, F_\theta) = \frac{d}{d\theta} K(Q, F_\theta) = 0. \qquad (28.2.3)$$

LEMMA 28.2.1
Given Q, if the Kullback-Leibler information of Q wrt F_θ, $K(Q, F_\theta)$, satisfies (2.2) then

$$\frac{1}{n} \sum_{i=1}^{n} l^{(1)}(\theta|x_i) \xrightarrow{P} -K^{(1)}(Q, F_\theta), \qquad (28.2.4)$$

in probability under Q as $n \to \infty$.

PROOF For given $\theta \in \Theta$, it follows form the condition (i) that

$$E_Q l^{(1)}(\theta|x) = \frac{d}{d\theta} E_Q l(\theta|x)$$

$$= \frac{d}{d\theta}(- \int_{-\infty}^{\infty} (\log \frac{q(x)}{f(x|\theta)}) q(x) dx)$$

$$= -K^{(1)}(Q, F_\theta). \qquad (28.2.5)$$

The result of (28.2.5) is an immediate consequence of (28.2.2) and the weak law of large numbers. This completes the proof. ∎

THEOREM 28.2.2
If the maximum likelihood estimator $\hat{\theta}_n(s)$ is the unique solution of the likelihood equation (28.2.1) and the Kullback-Leibler information $K(Q, F_\theta)$ satisfies conditions (i) and (ii), then

$$\hat{\theta}_n(s) \xrightarrow{P} \theta^*, \quad under \ Q, \ as \ n \to \infty, \qquad (28.2.6)$$

where $\theta^\star$ is given by (ii) and

$$K(Q, F_{\theta^\star}) = \inf_{\theta \in \Theta} K(Q, F_\theta). \tag{28.2.7}$$

PROOF Since the Kullback-Leibler information $K(Q, F_\theta)$ satisfies conditions (i) and (ii), it follows that

$$K^{(1)}(Q, F_\theta) \begin{cases} > 0, & \text{if } \theta > \theta^\star, \\ = 0, & \text{if } \theta = \theta^\star, \\ < 0, & \text{if } \theta < \theta^\star. \end{cases} \tag{28.2.8}$$

Noting that the mle $\hat{\theta}_n(s)$ is the unique solution of the likelihood equation (28.2.1). It follows that for every $\varepsilon > 0$,

$$P(\hat{\theta}_n(s) - \theta^\star > \varepsilon | Q) = P(l_n^{(1)}(\theta^\star + \varepsilon | s) > 0 | Q)$$
$$= P(\frac{1}{n} \sum_{i=1}^n l^{(1)}(\theta^\star + \varepsilon | x_i) > 0 | Q) \tag{28.2.9}$$

and

$$P(\hat{\theta}_n(s) - \theta^\star < -\varepsilon | Q) = P(l_n^{(1)}(\theta^\star - \varepsilon | s) < 0 | Q)$$
$$= P(\frac{1}{n} \sum_{i=1}^n l^{(1)}(\theta^\star - \varepsilon | x_i) < 0 | Q).$$
$$\tag{28.2.10}$$

For every $\varepsilon > 0$, there exists a $\delta > 0$ such that

$$K^{(1)}(Q, F_{\theta^\star + \varepsilon}) > \delta > 0$$

and

$$P(\hat{\theta}_n(s) - \theta^\star > \varepsilon | Q)$$
$$= P(\frac{1}{n} \sum_{i=1}^n l^{(1)}(\theta^\star + \varepsilon | x_i) > 0 | Q)$$
$$= P(\frac{1}{n} \sum_{i=1}^n l^{(1)}(\theta^\star + \varepsilon | x_i) + K^{(1)}(Q, F_{\theta^\star + \varepsilon}) > K^{(1)}(Q, F_{\theta^\star + \varepsilon}) | Q)$$
$$\leq P(\frac{1}{n} \sum_{i=1}^n l^{(1)}(\theta^\star + \varepsilon | x_i) + K^{(1)}(Q, F_{\theta^\star + \varepsilon}) > \delta | Q). \tag{28.2.11}$$

Lemma 28.2.1 and inequality (28.2.11) yield that

$$P(\hat{\theta}_n(s) - \theta^\star > \varepsilon|Q) \to 0, \quad \text{as } n \to \infty. \tag{28.2.12}$$

By the same token, it follows that

$$P(\hat{\theta}_n(s) - \theta^\star < -\varepsilon|Q) \to 0, \quad \text{as } n \to \infty. \tag{28.2.13}$$

Since ε is an arbitrary constant, the result that $\hat{\theta}_n(s)$ converges to $\theta^\star$ is an immediate consequence of the inequalities (28.2.12) and (28.2.13). Furthermore, the result (28.2.7) is a direct conclusion of condition (ii) and inequality (28.2.8). This completes the proof. ∎

If Q is not too far away from the specified model M_O in the sense that, for every $\theta \in \Theta$, the moment generating function of the score function $l^{(1)}(\theta|x)$ *wrt* Q,

$$\eta(t|\theta, Q) = E_Q \exp\{tl^{(1)}(\theta|x)\} < \infty, \tag{28.2.14}$$

exists in an interval $t \in (-\delta, \delta)$, then the following theorem shows that the mle $\hat{\theta}_n(s)$ converges to $\theta^\star$ exponentially.

THEOREM 28.2.3
If the condition (28.2.14) holds for every $\theta \in \Theta$ and $\hat{\theta}_n(s)$ is the unique solution of (28.2.1), the $\hat{\theta}_n(s)$ converges to $\theta^\star$ exponentially; i.e., for every $\varepsilon > 0$, there exists a positive constant β such that

$$\lim_{n \to \infty} \frac{1}{n} \log P(|\hat{\theta}_n - \theta^\star| > \varepsilon|Q) = -\beta, \tag{28.2.15}$$

where $\theta^\star$ satisfies (ii) and (28.2.7), and the exponential rate β is given by

$$\beta = -\log \max(\rho(\varepsilon), \rho(-\varepsilon)), \quad \text{and} \quad \rho(\pm\varepsilon) = \inf_{t \gtrless 0} \eta(t|\theta^\star \pm \varepsilon, Q). \tag{28.2.16}$$

PROOF Without loss of generality, we assume $\theta^\star \pm \varepsilon \in \Theta$. Since $\hat{\theta}_n(s)$ is a unique solution of (28.2.1), it follows from the definition of $\theta^\star$ that

$$P(\hat{\theta}_n(s) > \theta^\star + \varepsilon|Q) = P(l_n^{(1)}(\theta^\star + \varepsilon|s) > 0|Q)$$
$$= P(\sum_{i=1}^{n} l^{(1)}(\theta^\star + \varepsilon|x_i) > 0|Q) \tag{28.2.17}$$

and

$$P(\hat{\theta}_n(s) < \theta^\star - \varepsilon | Q) = P(l_n^{(1)}(\theta^\star - \varepsilon | s) < 0 | Q)$$

$$= P(\sum_{i=1}^{n} l^{(1)}(\theta^\star - \varepsilon | x_i) < 0 | Q). \quad (28.2.18)$$

The results follow immediately from Chernoff's theorem [see Chernoff (1952), Fu (1975)]. This completes the proof. $\blacksquare$

Let $\tilde{\theta}_n(s)$ be an M-estimator with respect to $\Psi(x, \theta)$. If $\exp\{-\Psi(x, \theta)\}$ is integrable and $\tilde{\theta}_n(s)$ is the unique solution of the equation

$$\sum_{i=1}^{n} \Psi^{(1)}(x_i, \theta) = \sum_{i=1}^{n} \frac{d}{d\theta} \Psi(x_i, \theta) = 0, \quad (28.2.19)$$

then the M-estimator $\tilde{\theta}_n(s)$ can be viewed as the maximum likelihood estimator derived from the model $M_O = \{R, F_\theta, \theta \in \Theta\}$ where F_θ has a density function given by

$$f(x, \theta) = C(\theta) \exp\{-\Psi(x, \theta)\} \quad (28.2.20)$$

with

$$C^{-1}(\theta) = \int_{-\infty}^{\infty} \exp\{-\Psi(x, \theta)\} dx.$$

Hence, all the above results for the mle can be extended to M-estimator. It follows from Theorem 28.2.2 that

$$\tilde{\theta}_n(s) \xrightarrow{p} \theta^\star, \quad (28.2.21)$$

as $n \to \infty$ under Q, where $\theta^\star$ is the unique solution of

$$\frac{C^{(1)}(\theta)}{C(\theta)} - \int_{-\infty}^{\infty} \Psi^{(1)}(x, \theta) dQ = 0. \quad (28.2.22)$$

Further, if

$$\xi(t | \theta, Q) = E_Q \exp\{-t\Psi^{(1)}(x, \theta)\} < \infty \quad (28.2.23)$$

exists in a neighborhood of zero, $t \subset (-\delta, \delta)$, then it follows from Theorem 28.2.2 that $\tilde{\theta}_n(s)$ converges to $\theta^\star$ exponentially with an exponential rate given by

$$\beta = -\log \max(\rho(\varepsilon), \rho(-\varepsilon)) \quad (28.2.24)$$

where

$$\rho(\pm\varepsilon) = \inf_{\substack{t > \\ t < 0}} e^{tC^{(1)}(\theta^\star \pm \varepsilon)/C(\theta^\star \pm \varepsilon)} \xi(t | \theta^\star \pm \varepsilon, Q). \quad (28.2.25)$$

It is worth mentioning that the least square estimator $\tilde{\theta}_n(s)$ for θ can also be viewed as mle *wrt* the statistical model $M_O = \{X, N_\theta, \ \theta \in R\}$, where N_θ is a normal distribution with location parameter θ. Under very mild conditions, it can be shown that the least square estimator $\tilde{\theta}_n(s)$ converges to $\theta^\star$ and is asymptotically normally distributed, but it does not converge to $\theta^\star$ exponentially when the tail probability of the underlying distribution Q tends to zero with a rate $1/x^\alpha$ $(3 < \alpha < \infty)$, as $x \to \infty$. More discussions about these will be given in the next section.

The condition that the mle is the unique solution of the likelihood equation (28.2.1) is often required in most literature of studying the large sample properties of maximum likelihood estimator. The main reason for requiring this condition, technically speaking, is that the uniqueness will make all the proofs much simpler and mathematically tractable. With some modifications of the above proofs, all the main results can be proved under the much weaker condition that the expected score function $l^{(1)}(x|\theta)$ has a unique solution, *i.e.*, mathematically it can be stated that for every $\theta_0 \in \Theta$, the equation

$$\int_{-\infty}^{\infty} l^{(1)}(x|\theta)dF_{\theta_0} \overset{set}{=} 0, \qquad (28.2.26)$$

has a unique solution in θ.

For example, in the case of Cauchy distribution with location parameter θ, the likelihood equation (2.1) has $(2n-1)$ roots (real and complex roots) but the equation

$$\int_{-\infty}^{\infty} l^{(1)}(x|\theta)dF_{\theta_0} = -\frac{2(\theta - \theta_0)}{(\theta - \theta_0)^2 + 4} \overset{set}{=} 0 \qquad (28.2.27)$$

has unique root at $\theta = \theta_0$. If the underlying distribution Q is a Cauchy distribution, Perlman (1983) and Reeds (1985) proved that all the real roots of the likelihood equation, except the mle (the global maximum), tend to $\pm\infty$ with probability one and the mle converges to θ_0. Bai and Fu (1986) proved a stronger result that the mle, in fact, converges to θ_0 exponentially and is also locally efficient in the Bahadur sense of Eq. (28.1.3).

28.3   Some Applications

Motivated by robust estimation, this section mainly studies the exponential rate of the mle when the underlying distribution Q departs slightly from the specified model. The statistical models considered in the following examples are respectively exponential families and families of t-distributions with location parameter θ.

28.3.1 Exponential Model

Let us consider the following exponential model

$$M_O = \{F_\theta : dF_\theta = g(x)\exp\{x\theta - C(\theta)\}dx, \theta \in R\}. \qquad (28.3.1)$$

The Kullback-Leibler information of the true underlying distribution Q with respect to a distribution F_θ in the exponential model M_O can be written as

$$K(Q, F_\theta) = \int_R \log \frac{dQ}{dF_\theta} dQ = K(Q, F_0) + C(\theta) - C(0) - \theta E_Q X. \quad (28.3.2)$$

The mle $\hat{\theta}_n(s)$ is a unique solution of the likelihood equation

$$\bar{x}_n - C^{(1)}(\theta) = 0, \qquad (28.3.3)$$

and it can be written as

$$\hat{\theta}_n(s) = h(\bar{x}_n), \qquad (28.3.4)$$

where $h(\cdot)$ is the inverse function of $C^{(1)}(\cdot)$. It is straightforward that $K^{(1)}(Q, F_\theta)$ has a unique maximum at $\theta^\star$ where $\theta^\star$ is a unique solution of

$$E_Q X - C^{(1)}(\theta) = 0, \qquad (28.3.5)$$

and it can be written as

$$\theta^\star = h(E_Q X). \qquad (28.3.6)$$

A straightforward application of our results, it yields that if

$$E_Q \exp\{t(X - C^{(1)}(\theta))\} < \infty \qquad (28.3.7)$$

exists for $t \in (-\delta, \delta)$, then the mle $\hat{\theta}_n(s)$ converges to $\theta^\star$ exponentially. This yields a stronger result than the result of McCulloch (1988).

In view of the mathematical form of the mle $\hat{\theta}_n(s)$ derived from (28.3.3), it follows that

$$P(\hat{\theta}_n(s) - \theta^\star > \varepsilon | Q) = P(\bar{X}_n > C^{(1)}(\theta^\star + \varepsilon) | Q), \qquad (28.3.8)$$

hence, the mle $\hat{\theta}_n(s)$ converges to $\theta^\star$ exponentially if, and only if, the tail probability of the underlying distribution Q tends to zero exponentially which is equivalent to that (28.3.7) exists in a neighborhood of $t = 0$. If the tail probability of the underlying distribution Q tends to zero with a rate of $1/x^\alpha$ ($\alpha > 3$), then the generating function (28.3.7) does not exist. If this is the case, then the mle $\hat{\theta}_n(s)$ remains convergence to θ^* in probability, and is asymptotically normally distributed. However, $\hat{\theta}_n(s)$ does not converge to θ^* exponentially. Summing up these evidences, we could conclude that the mle $\hat{\theta}_n(s)$ derived from the exponential model (28.3.1) is nonrobust in the sense of exponential rate of convergence with respect to the usual metric topology. More importantly, it shows that the criterion of selecting the optimal consistent estimator based on the exponential rate is more refined than the criterion based on the variance of the asymptotic distribution.

28.3.2   Families of t-distributions with Location Parameter

For each $d = 1, 2, \cdots, \infty$, define

$$M_d = \{dF_{\theta,d}(x) : \text{all } t\text{-distributions with } d \text{ degrees of freedom and}$$
$$\text{location parameter } \theta \in R\}, \tag{28.3.9}$$

where $dF_{\theta,d}$ has the form

$$dF_{\theta,d}(x) = \frac{\Gamma(\frac{d+1}{2})}{\sqrt{2\pi}\Gamma(\frac{d}{2})}(1 + \frac{(x - \theta)^2}{d})^{-\frac{d+1}{2}} dx. \tag{28.3.10}$$

Further, let

$$M = \{dF_{\theta,d} : \text{all } t\text{-distributions with } d = 1, 2, \cdots, \infty, \text{ and } \theta \in R\} \tag{28.3.11}$$

be the family of all t-distributions with location parameter θ. The statistical model M includes all the models of M_d, $d = 1, \cdots, \infty$. The model M covers Cauchy ($d = 1$) and Normal ($d = \infty$) distributions, which are of great interest both in theory and practice.

Let $\hat{\theta}_d(d = 1, \cdots, \infty)$ be the mle derived from the model M_d. Write $\hat{\theta}_\infty = \hat{\theta}_N$ and $\hat{\theta}_1 = \hat{\theta}_C$ as mles derived from Normal and Cauchy models, respectively. In this example, our main interest is to study the exponential rate of convergence of $\hat{\theta}_d$ when the true underlying distribution Q does not belong to the model M_d but it is a member of the larger model M. This assumption is motivated by the robust estimation for the location parameter.

The exponential rate of convergence of the mle $\hat{\theta}_d$ is deeply associated with the expectation of the score function $l^{(1)}(\theta|x, d) = (\partial/\partial\theta) \log dF_{\theta,d}$ of specified model. Define

$$S(\theta|d)$$

$$= E(l^{(1)}(\theta|x,d)|F_{0,d})$$

$$= \int_{-\infty}^{\infty} \frac{\Gamma(\frac{d+1}{2})}{\sqrt{\pi d}\Gamma(\frac{d}{2})} \frac{(d+1)(x-\theta)}{(d+(x-\theta)^2))(1+x^2/d)^{(d+1)/2}} dx \quad (28.3.12)$$

as the expected score function $l^{(1)}(\theta|x,d)$ with respect to the underlying distribution $Q = F_{0,d}$. For the two extreme cases of Cauchy and Normal distributions, we have, respectively,

$$S(\theta|1) = \int_{-\infty}^{\infty} \frac{2(x-\theta)}{\pi(1+(x-\theta)^2)(1+x^2)} dx = -\frac{2\theta}{\theta^2+4}, \quad (28.3.13)$$

and

$$S(\theta|\infty) = \int_{-\infty}^{\infty} (x-\theta)\frac{1}{\sqrt{2\pi}}e^{-1/2x^2} dx = -\theta. \quad (28.3.14)$$

Note that for every d, $S(\theta|d)$ is a continuous function in θ and its graph crosses the θ-axis only once at $\theta = 0$. Further, for any degree of freedom d ($d < \infty$), the function $S(\theta|d)$ is bounded and $S(\theta|d) \to 0$, as $|\theta| \to \infty$. For the Normal distribution ($d = \infty$), the expected score function is $S(\theta|\infty) = -\theta$, which is continuous but unbounded. By numerically integrating (28.3.12), it yields the following Figure 28.1 which illustrates the behaviors of the expected score functions $S(\theta|d)$, for $d = 1, 6, 15, \infty$. In addition, it follows from the law of large numbers

$$\frac{1}{n}l_n^{(1)}(\theta) = \frac{1}{n}\sum_{i=1}^{n} l^{(1)}(\theta|x_i,d) \xrightarrow{p} S(\theta|d), \quad (28.3.15)$$

as $n \to \infty$, under $Q = F_{0,d}$.

For the Cauchy distribution, the empirical score function

$$\frac{1}{n}l_n^{(1)}(\theta) = \frac{1}{n}\sum_{i=1}^{n} \frac{2(x_i-\theta)}{1+(x_i-\theta)^2} \quad (28.3.16)$$

based on observed data $(x_1, \cdots, x_n)$ is also plotted against the expected score function $S(\theta|1)$ in Figure 28.1. Perlman (1983) and Reeds (1985) proved that, for Cauchy distribution, all the roots of likelihood equation except the mle (the global maximum) tend to $\pm\infty$ with probability one, and the mle converges to θ_0 ($\theta_0 = 0$ in Figure 28.1). This convergence behavior is also observed in Figure 28.1 for all t-distributions with degrees of freedom $d = 1, 2, \cdots$.

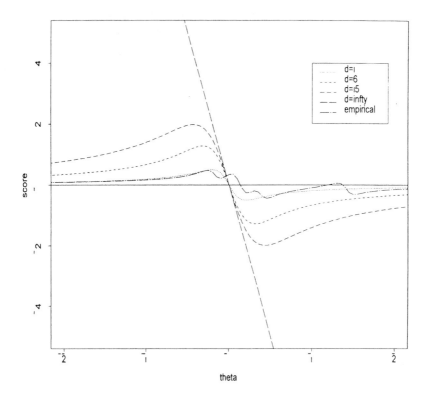

FIGURE 28.1
The expected scores $S(\theta|d)$ for $d = 1, 6, 15, \infty$, and the empirical
scores $\frac{1}{n}l_n^{(1)}(\theta)$ based on Eq. (28.3.16) for $n = 5$ under model M_1

THEOREM 28.3.1

Let M be the family of t-distributions defined by equations (28.3.9) and (28.3.10). Then the following results hold:

(i)   *For any $Q = F_{\theta,d} \in M$, and $d \geq 3$, the mle $\hat{\theta}_N = \bar{x}_n$ derived from the normal model M_∞ is consistent and asymptotically normally distributed,*

(ii) *the estimator $\hat{\theta}_N$ does not converge to $\theta(\theta = \theta^\star)$ exponentially for any $d = 1, 2, \cdots$, except for $d = \infty$ (nonrobust estimator),*

(iii) *the mle $\hat{\theta}_C$ derived from the Cauchy model converges to θ exponentially for every d including $d = \infty$ (robust estimator).*

PROOF For the maximum likelihood estimator $\hat{\theta}_N = \bar{x}_n$ derived from the Normal model, it follows from Theorem 28.2.2 that $\theta^\star = \theta$ and $\hat{\theta}_N$ converges to θ as $n \to \infty$ for any $d \geq 3$. The asymptotic normality of $\hat{\theta}_N$ follows directly from the central limit theorem. This completes (i). Note that for any $d = 1, 2, \cdots$, (except $d = \infty$)

$$\int_{-\infty}^{\infty} e^{-(x-\theta)t} [1 + \frac{(x - \theta')^2}{d}]^{-\frac{d+1}{2}} dx = \infty \qquad (28.3.17)$$

does not exist for t in any neighborhood of zero. Hence $\hat{\theta}_N$ does not converge to θ exponentially. This completes the proof of (ii). Contrary to the notion of (28.3.17), for any θ and θ', and $d = 1, 2, \cdots, \infty$, there exists $\delta > 0$, such that

$$\int_{-\infty}^{\infty} \exp\{-\frac{t(x - \theta)}{1 + (x - \theta)^2}\} dF_{\theta',d} < \infty \qquad (28.3.18)$$

for all $t \in (-\delta, \delta)$. It follows from Theorem 28.2.2 that the mle derived from the Cauchy model M_1 converges to θ exponentially for any underlying distribution $Q = F_{\theta,d}$ inside the model M. This completes the proof. ∎

REMARK In view of this theorem, the mle $\hat{\theta}_N$ has asymptotic variance σ_d^2/n $(d \geq 3)$ under the t-distribution, where σ_d^2 is the variance of the t-distribution with d degrees of freedom but the mle $\hat{\theta}_n$ does not converge to θ exponentially. This substantiates the criticisms of asymptotic normal theory applying to large sample estimation problems by many statisticians, for example Weiss and Wolfowitz (1966, 1967). It also shows that the asymptotic normal theory could not distinguish the rates of convergence among the consistent estimators. In general, roughly speaking, $\hat{\theta}_N$ is a slow consistent estimator *wrt* heavy tail underlying distribution. ∎

For any d, the mle $\hat{\theta}_d$ derived from the model M_d $(d = 1, 2, \cdots,$ and $d \neq \infty)$, the following general results hold.

THEOREM 28.3.2
For any $d = 1, 2, \cdots$, and $d \neq \infty$, the mle $\hat{\theta}_d$ converges to θ under $Q = F_{\theta,d'}$, $d' = 1, \cdots, \infty$ as $n \to \infty$, having an exponential rate given by

$$\beta_{d,d'}(\varepsilon) = -\log \rho_{d,d'}, \qquad (28.3.19)$$

where

$$\rho_{d,d'} = \max(\rho_{d,d'}(\varepsilon), \rho_{d,d'}(-\varepsilon)), \qquad (28.3.20)$$

$$\rho_{d,d'}(\pm\varepsilon) = \inf_{t \gtrless 0} \eta_d(t|\theta \pm \varepsilon, F_{\theta,d'}), \qquad (28.3.21)$$

and

$$\eta_d(t|\theta \pm \varepsilon, F_{\theta,d'}) = \int_{-\infty}^{\infty} (\exp\{-t\frac{(d+1)(x - \theta \pm \varepsilon)}{d + (x - \theta \pm \varepsilon)^2}\}) dF_{\theta,d'}(x). \quad (28.3.22)$$

PROOF For every $d = 1, 2, \cdots$, the moment generating function $\eta_d(t|\theta \pm \varepsilon, F_{\theta,d'})$ exists in a neighborhood of $t = 0$ for every $d' = 1, 2, \cdots, \infty$, hence the result follows immediately from Theorem 28.2.2. ∎

28.4   Discussion

For ε small, the exponential rate $\beta_{d,d'}(\varepsilon)$ of the mle $\hat{\theta}_d(s)$, with respect to the underlying distribution $Q = F_{\theta,d'}$, admits the expansion [see Fu (1975, 1982)]

$$\beta_{d,d'}(\varepsilon) = \alpha_{d,d'}\frac{\varepsilon^2}{2} + o(\varepsilon^2), \qquad (28.4.1)$$

where $\alpha_{d,d'}$ is called the local exponential rate. The local exponential rate $\alpha_{d,1}$ of $\hat{\theta}_d$ with respect to the Cauchy underlying distribution is a decreasing function of degree of freedom d having $\alpha_{1,1} = I_C$ (Fisher information of Cauchy distribution) and $\alpha_{\infty,1} = 0$. The local exponential rate $\alpha_{d,\infty}$ of $\hat{\theta}_d$ with respect to the Normal underlying distribution is an increasing function of d having $\alpha_{1,\infty} > 0$, and $\alpha_{\infty,\infty} = I_N$ (Fisher information of Normal). Hence the following inequalities

$$\max(\alpha_{1,1}, \alpha_{\infty,\infty}) \geq \max(\alpha_{d,1}, \alpha_{d,\infty})$$

$$\geq \min(\alpha_{d,1}, \alpha_{d,\infty}) \geq \min(\alpha_{1,1}, \alpha_{\infty,\infty}) \quad (28.4.2)$$

hold for every $d = 1, 2, \cdots, \infty$. Further, for every fixed d, $\alpha_{d,d'}$ is a convex function of d' and has a maximum at $d = d'$, and

$$\min_{d'} \alpha_{d,d'} = \min(\alpha_{d,1}, \alpha_{d,\infty}). \tag{28.4.3}$$

The local exponential rate $\alpha_{d,d'}$ can be used as a criterion for robust estimation. A consistent estimator is called a robust estimator with respect to the model M if its local exponential rate is greater than zero for every distribution in M.

Note that Theorem 28.3.1 yields

$$\alpha_{\infty,d'} \equiv 0, \quad \text{for every } d' = 1, 2, \cdots \text{ except for } d' = \infty \tag{28.4.4}$$

and

$$\alpha_{d,d'} > 0, \quad \text{for every } d = 1, 2, \cdots (d \neq \infty), \text{ and every } d' = 1, 2, \cdots, \infty \tag{28.4.5}$$

Hence the mle $\hat{\theta}_N$ derived from the normal model is a non-robust estimator, and on the contrary, the mle's $\hat{\theta}_d$ derived from M_d, $d < \infty$ are robust estimators with respect to M in the sense of exponential rate of convergence.

The mle $\hat{\theta}_{d^*}$ derived from the model M_{d^*} is called the best *robust estimator* among all the mle's $\hat{\theta}_d$, $d = 1, \cdots, \infty$, with respect to the model M (maximum solution), if

$$\alpha_{d^*,1} = \alpha_{d^*,\infty} = \max_d \min_{d'} \alpha_{d,d'} = \max_d \min(\alpha_{d,1}, \alpha_{d,\infty}). \tag{28.4.6}$$

Unfortunately, for given d and d', the exact analytical expression of the local exponential rate $\alpha_{d,d'}$ could not be obtained. Hence d^* could not be obtained analytically. By using Theorem 28.3.2 and numerical integration of evaluating the local exponential rate $\alpha_{d,d'} \sim \beta_{d,d'}(\varepsilon)/\varepsilon^2$ at very small $\varepsilon (\varepsilon = 0.0001)$, it yields that d^* is approximately "6". In words, among all the maximum likelihood estimators $\hat{\theta}_d$, $d = 1, 2, \cdots, \infty$, the mle $\hat{\theta}_6$ is the best robust estimator with respect to the family of all t-distributions in the sense of local exponential rate.

Acknowledgments. This work was supported in part by the National Sciences and Engineering Research Council of Canada under the grant A-9216.

References

1. Bahadur, R. R. (1967). Rates of convergence of estimates and test statistics. *Annals of Mathematical Statistics* **38**, 303–324.

2. Bahadur, R. R. (1971). Some limit theorems in statistics. *Regional Conference Series in Applied Mathematics*, Society for Industry and Applied Mathematics, Philadelphia.

3. Bahadur, R. R. (1980). On large deviations of maximum likelihood and related estimates. *Technical Report*. Department of Statistics, University of Chicago.

4. Bai, Z. D. and Fu, J. C. (1986). Likelihood principle and maximum likelihood estimator of location parameter for Cauchy distribution. *Canadian Journal of Statistics* **15**, 137–146.

5. Cheng, P. and Fu, J. C. (1986). On a fundamental optimality of the maximum likelihood estimator. *Statistics & Probability Letters* **4**, 173–178.

6. Chernoff, H. (1952). A measure of asymptotic efficiency for tests of a hypothesis based on sum of observations. *Annals of Mathematical Statistics* **23**, 493–502.

7. Cramér, H. (1946). *Mathematical Methods of Statistics*, Princeton University Press, Princeton, New Jersey.

8. Fu, J. C. (1973). On a theorem of Bahadur on the rate of convergence of point estimators. *Annals of Statistics* **1**, 745–749.

9. Fu, J. C. (1975). The rate of convergence of consistent point estimators. *Annals of Statistics* **3**, 234–240.

10. Fu, J. C. (1982). Large sample point estimation: A large deviation approach. *Annals of Statistics* **10**, 762–771.

11. Jureckova, J. (1981). Tail-behavior of location estimators. *Annals of Statistics* **9**, 578–585.

12. Kester, A. (1981). Large deviation optimality of MLE's in exponential families. *Technical Report No. 160*, Free University of Amsterdam.

13. Kester, A. and Kallenberg, W. C. M. (1986). Large deviation of estimators. *Annals of Statistics* **14**, 648–664.

14. LeCam, L. M. (1953). On some asymptotic properties of maximum likelihood estimates and related Bayes estimates. *Proceedings of the Berkeley Symposium* **1**, 277–328.

15. LeCam, L. M. (1970). On the assumptions used to prove asymptotic normality of maximum likelihood estimates. *Annals of Mathematical Statistics* **41**, 802-828.

16. McCulloch, R. E. (1988). Information function in exponential families. *The American Statistician* **42**, 73–75.

17. Perlman, M. D. (1983). The limiting behavior of multiple roots

of the likelihood equation. *Recent Advances in Statistics*, pp. 339–370, Academic Press, New York.

18. Pfanzagl, J. (1973). Asymptotic optimum estimation and test procedures. *Proceedings of Prague Symposium on Asymptotic Statistics*, 201–272, Charles University, Prague.

19. Rao, C. R. (1963). Criteria of estimation in large samples. *Sankhyā* **25**, 189–206.

20. Reeds, J. A. (1985). Asymptotic number of roots of Cauchy location likelihood equations. *Annals of Statistics* **13**, 775–784.

21. Rubin, H. and Rukhin, A. L. (1983). Convergence rates of large deviation probabilities for point estimators. *Statistics & Probability Letters* **1**, 197–202.

22. Rukhin, A. L. (1983). Convergence rates of estimators of finite parameter: how small can error probabilities be. *Annals of Statistics* **11**, 202-206.

23. Steinebach, J. (1978). Convergence rates of large deviation probabilities in the multidimensional case. *Annals of Probability* **2**, 751–759.

24. Wald, A. (1949). A note on the consistency of maximum likelihood estimates. *Annals of Mathematical Statistics* **20**, 595–601.

25. Weiss, L. and Wolfowitz, J. (1966). Generalized maximum likelihood estimators. *Teoriya Vyeroyatnostey* **11**, 68–93.

26. Weiss, L. and Wolfowitz, J. (1967). Maximum probability estimators. *Annals of the Institute of Statistical Mathematics* **19**, 193–206.

29

A Closer Look at Weighted Likelihood in the Context of Mixtures

Marianthi Markatou
Columbia University, New York, NY

ABSTRACT The performance of the weighted likelihood methodology in the context of mixtures is studied in detail. Specifically, we study the behavior of the method under two types of misspecification. Those are (1) probability model misspecification, which refers to both, component misspecification and probability distribution misspecification, and (2) variance structure misspecification. We contrast the behavior of the weighted likelihood estimates with that of Huber-type M-estimates and maximum likelihood estimates. We present simulation results which exemplify the role of the starting values in the convergence of the weighted likelihood algorithm. We then discuss the relationship of these results with the problem of model selection.

Keywords and phrases Estimating equations, model selection, mixtures, robustness, weighted likelihood

29.1 Introduction

The problem of parametric statistical inference in the presence of a finite mixture of distributions has been studied extensively. An excellent survey of it was given by Redner and Walker (1984). Moreover, the recent book by Lindsay (1995) gives a detailed account of problems that can be formulated as mixture model problems.

It is also a common statistical practice to study the robustness of a statistical procedure by constructing a class of alternative mixture models. For example, if the central model is $N(\mu, \sigma^2)$ and we are interested in estimating the parameter μ, we might assess the robustness of the estimator by measuring its performance, that is its bias and mean squared error, on data that arise from the model $(1 - \epsilon)N(\mu, \sigma^2) + \epsilon N(\mu, \tau^2)$

where $\tau^2 > \sigma^2, 0 < \epsilon < 1$.

One of the limitations of the mixture model theory is that it seems that very little is known about the robustness aspect of this methodology. To quote Lindsay (1995) "It is natural and desirable to ask the question: what are the consequences of slight errors in the misspecification of the model? How stable are my parameters under contamination? When a mixture model is being specified, and I specify a mixture of two normals with different means, what are the consequences if the mixture is actually of 3 normals? Or of two t-distributions?"

A number of authors have tried to answer, at least a few of the above questions, by trying to adopt the usual M-estimation methods in the context of mixtures. Some of these attempts include a robustification of the EM algorithm [De Veaux and Krieger (1990)], the introduction of a measure of typicality [Campbell (1984)] which reflects the contribution of the mth observation to the mean of the kth component, and the adoption of multivariate M-estimation techniques for multivariate normal mixtures [McLachlan and Basford (1988)]. Additional work includes Aitkin and Tunnicliffe (1980), Clarke and Heathcote (1994), Gray (1994), and Windham (1996).

A direction that has seen some development when the problem of robustness is of interest, is to employ minimum distance ideas for parameter estimation. This approach has been taken by Woodward *et al.* (1984), Woodward, Whitney, and Eslinger (1995), and Cutler and Cordero-Brana (1996). The problem with these procedures is that they are computationally complex. Moreover, there is no natural definition of the concept of outliers.

In this paper we take a closer look at the issue of robustness as it relates to the mixture model. In particular, we address the effects of misspecification of the variance structure and mixture model misspecification on the parameter estimates of the mixture model. We study the performance of the weighted likelihood estimating equations (WLEE) methodology in this setting and compare it with that of the usual likelihood methodology and with the performance of classical M-estimators as they are adapted in the mixture model setting by McLachlan and Basford (1988).

The paper is organized as follows. Section 29.2 briefly discusses the WLEE methodology in the context of mixtures. Section 29.3 presents simulation results. Section 29.4 discusses the implications of the WLEE as they relate to the model selection topic. Finally, Section 29.5 offers conclusions.

29.2 Background

Let $X_1, X_2, \ldots, X_n$ be a random sample of completely unclassified observations from a finite mixture distribution with g components. The probability density function of an observation X is of the form $m(x; \phi) = \sum_{i=1}^{g} p_i f(x; \theta_i)$, where $f(x; \theta_i)$ is the probability density or mass function of the ith subpopulation and θ_i is the component parameter that describes the specific attributes of the ith component population. Any parameters that describe unknown characteristics common to the entire population are assumed to be absent for the time being. However, if those parameters are of interest their corresponding estimating equations can be easily incorporated into the estimating scheme. The vector $\phi = (p_1, p_2, \ldots, p_g, \theta_1, \ldots, \theta_g)^T$ of all unknown parameters belongs to a parameter space Ω subject to $\sum_{i=1}^{g} p_i = 1, p_i \geq 0, i = 1, 2, \ldots, g$, and θ_i belongs to Θ.

In this context, the weighted likelihood estimating equations are

$$\sum_{j=1}^{n} w(\delta(x_j)) \nabla_\phi [lnm(x_j; \phi)] = 0 \qquad (29.2.1)$$

where $\delta(x_j)$ is the Pearson residual evaluated at x_j and $w(.)$ is a weight function that downweights observations which have large residuals. If $m(x; \phi)$ is a probability mass function the Pearson residual is defined as $\delta(t) = (d(t) - m(t; \phi))/m(t; \phi)$, [Lindsay (1994)], where $d(t)$ is the proportion of observations in the sample with value equal to t. If the components are continuous, the Pearson residual is

$$\delta(x_j) = \{f^*(x_j)/m^*(x_j; \phi)\} - 1,$$

where $f^*(.)$ is a kernel density estimator and $m^*(.; \phi)$ is the mixture model smoothed with the same kernel used to obtain the density estimator. Note that the range of the Pearson residuals is the interval $[-1, \infty)$. A value of $\delta(x_i)$ close to zero indicates agreement between the data and the hypothesized model in the neighborhood of the observation x_i. On the other hand, a large value of the residual indicates a discrepancy between the data and the model in the form of an excess of observations relative to the model prediction. Thus, we call an observation an outlier if it has a large Pearson residual.

The weight functions $w(.)$ are unimodal in that they decline smoothly as the residual δ departs from 0 towards -1 or $+\infty$ and take the maximal value of 1 when the residual δ is 0. An observation that is consistent with the assumed model receives a weight of approximately 1. If it is inconsistent with the model it receives a small weight; a weight of approximately 0 indicates that the observation is highly inconsistent with the model. The final fitted weights indicate which of the data points were downweighted in the final solution relative to the maximum likelihood estimator.

The motivation for this weight construction comes from minimum disparity estimation. In the discrete case, given any strictly convex function G(.), a nonnegative disparity measure between the model m_ϕ and the data d is given by $\rho_G(d, m_\phi) = \sum_t m_\phi(t) G(\delta(t))$. The value of ϕ that minimizes the disparity is the minimum disparity estimator corresponding to G [Lindsay (1994)]. In the continuous case , the disparity between the data and the model is defined as $\rho_G(f^*, m_\phi^*) = \int G(\delta(x)) m_\phi^*(x) dx$, [Basu and Lindsay (1994)]. When G is selected appropriately a large class of important distances can be developed. For example, if $G(\delta) = (\delta + 1) ln(\delta + 1)$ a version of the Kullback-Leibler divergence is generated. We will focus attention on the robust disparity measure generated by $G(\delta) = 2\delta^2/(\delta + 2)$ which corresponds to the symmetric chisquared distance. The symmetric chisquared distance is defined as

$$\sum_t \frac{[d(t) - m(t; \phi)]^2}{(1/2)d(t) + (1/2)m(t; \phi)}$$

in the discrete case, and

$$\int \frac{[f^*(x) - m^*(x; \phi)]^2}{(1/2)f^*(x) + (1/2)m^*(x; \phi)} dx$$

in the continous case. The corresponding weight function is unimodal, twice differentiable at $\delta = 0$, with $w'(0) = 0$ and $w''(0) < 0$, where prime denotes differentiation with respect to δ. The weight function relates to the function G via the equation $(0.5[(\delta + 1)G'(\delta) - G(\delta)] + 1)/(\delta + 1)$, and G' is the derivative of G with respect to δ. An additional advantage of the chisquared weight function is that in the continuous case we can construct a *parallel* disparity with G specified as above. This disparity could be useful as it can form the basis for the construction of a goodness of fit test for model. How this can be done will be discussed in Section 29.4. Finally, we would like to note that Markatou, Basu, and Lindsay (1997) suggested a general scheme for weight construction which can also be adopted here.

From equation (29.2.1) the estimating equations for θ_i are

$$\sum_{j=1}^n \sum_{i=1}^g \frac{w(\delta(x_j))}{m(x_j; \phi)} p_i f(x_j; \theta_i) u(x_j; \theta_i) = 0 \qquad (29.2.2)$$

where $u(x_j; \theta_i) = \nabla_{\theta_i} ln f(x_j; \theta_i)$.

For the vector $p = (p_1, p_2, \ldots, p_g)^T$ the equations are

$$p_i = p_i \sum_{j=1}^{n} \frac{w(\delta(x_j))}{\sum_{j=1}^{n} w(\delta(x_j))} \frac{f(x_j; \theta_i)}{m(x_j; \phi)}. \tag{29.2.3}$$

The solutions of the system (29.2.2), (29.2.3) are the weighted likelihood estimating equations (WLEE) estimators of the parameter vector $\phi = (p_1, p_2, \ldots, p_g, \theta_1, \theta_2, \ldots, \theta_g)^T$.

A key element to the success of this methodology is the appropriate construction of the Pearson residuals in the continuous case, which requires selection of a kernel $k(x; t, h)$ and a degree of smoothing h in order to create f^* and m^*. The choice of kernel does not seem critical and hence we recommend choosing it based on convenience [Markatou (2000)]. The problem that arises next is that one would like to do the smoothing so that the robustness properties are homogeneous throughout the space of model parameters. This requirement suggests that the smoothing parameter should be proportional to the variance of X under the true model. When the model is a finite mixture of $N(\mu_i, \sigma^2)$ a natural kernel is the normal density with variance h^2. The variance h^2 serves as the bandwidth parameter and determines the robustness properties of the WLEE estimator. Markatou (2000), proposes to select the smoothing parameter h^2 at the $(i+1)$st step of the iteration as a multiple of the estimate of model variance at the ith step of the iteration, that is, $h_{(i+1)}^2 = c\hat{\sigma}_{(i)}^2$, where $c > 0$. To complete the bandwidth selection we need to have a rule to determine c. Markatou (2000) suggests obtaining c by solving the equation

$$-\frac{A_2}{2} \left\{ \frac{(1+c)^{3/2}}{c(3+c)^{1/2}} - 1 \right\} = \gamma_0,$$

where $\gamma_0 > 0$ is the number of observations to be downweighted and it is selected by the user, $A_2 = w''(0)$.

In the multivariate case the smoothing matrix is selected as $H = c\hat{\Sigma}$, where $\hat{\Sigma}$ is the estimated covariance and the constant c is obtained by solving the equation

$$\frac{A_2}{2} \left\{ \left[\frac{(1+c)^{3/2}}{c(3+c)^{1/2}} \right]^m - 1 \right\} = \gamma_0$$

where $m = dim(\phi)$, [Markatou (2000)]. This equation assumes that $\Sigma = diag(\sigma_i^2)$. Note that, by selecting the smoothing parameter as a function of the scale has the additional advantage of obtaining invariant estimators under scale transformations.

When the model is a finite mixture of normal components with parameters (μ_i, σ_i^2), and the kernel is normal with variance h^2, then select

$h^2 = c \cdot (\sum_{i=1}^{g} \hat{p}_i \hat{\sigma}_i^2), \hat{p}_i, \hat{\sigma}_i^2$ are estimates of the proportion and variance of the *ith* subpopulation and c is determined as before. In the multivariate case we select $H = c \sum_{i=1}^{g} \hat{p}_i \hat{\Sigma}_i$ [Markatou (2000)].

29.3   Simulation Experiments and Results

We now describe in detail a simulation study that was designed to offer insight in a number of issues associated with the performance of the weighted likelihood methodology in the context of mixtures.

First, we offer some general comments about the performance of the WLEE.

The weighted likelihood estimating equations need not have a unique solution. Verification of the conditions of Markatou, Basu, and Lindsay (1998) for the component densities of a mixture model together with the additional assumption of parameter identifiability guarantees the existence of a root in the neighborhood of the true parameter value. Under these conditions, the consistency and asymptotic normality of the estimates is ensured. The asymptotic distribution of the weighted likelihood estimators is normal with mean 0 and covariance matrix the inverse of the Fisher information. This result states that the estimates are first order efficient. As such, their influence function is the same with the influence function of the corresponding MLE. Therefore, in order to study the robustness properties of these estimates we need to use tools other than the influence function.

When one has an equation with multiple roots one might wish to know if the presence of a root represents some underlying structure. This is particularly relevant in the context of mixtures since the existence of different roots could indicate multiple potential mixture model fits due to the presence of more components than originally specified. The weighted likelihood equations tend to have multiple roots when one attempts to fit a two component model to data from a three component density. When the mixing proportion of the third component is sufficiently small the weighted likelihood equations mostly exhibit one root that corresponds to the model comprised from the well represented components. As the mixing proportion of the third component increases the weighted likelihood equations have multiple roots that correspond to two component mixture models that fit different portions of the data. In this case, the robustness of the procedure is defined by its ability to detect possible data substructures in the form of multiple roots.

When an additional component that is not included in the model specification is close to a component of the specified mixture model, robustness, defined as resistance of the statistical procedure to deviations from the stipulated model, may be more relevant. The weighted likelihood methodology downweights the data from the additional component, and the power of discrimination goes down as the method tries to distinguish between fitting a model with a larger variance that encompasses all data points or a more robust fit.

This empirical property corresponds to a stability of the estimating equation and its roots under contamination that can be investigated using an approach similar to that of Lindsay (1994). Let $\{\xi_j; j = 1, 2, \ldots\}$ be a sequence of elements in the sample space. Let $\hat{F}_j(x) = (1-\epsilon)\hat{F}(x) + \epsilon\Delta_{\xi_j}(x)$ be the contaminated distribution and $f_j^*(x) = \int k(x, t, h)d\hat{F}_j(x)$ the corresponding kernel smoothed data. Then $\{\xi_j\}$ is an outlier sequence for $m_\phi(x)$ and data $\hat{F}$ if $\delta(\xi_j) = \frac{f_j^*(\xi_j)}{m_\phi^*(\xi_j)} - 1$ converge to infinity and at the same time $m_\phi^*(\xi_j)$ converges to zero.

The weighted likelihood score along this sequence is $\int w(x; \hat{F}_j, M_\phi) u(x; \phi)d\hat{F}_j$ and the estimating equations will ignore the outliers if the limit of the weighted likelihood sequence does not depend on the outliers. Verification of the conditions given in Markatou, Basu, and Lindsay (1998) shows that the weighted likelihood equations satisfy the relationship

$$\int w(x; \hat{F}_j, M_\phi)u(x; \phi)d\hat{F}_j \rightarrow \int w(x; \hat{F}_\epsilon, M_\phi)u(x; \phi)d\hat{F}_\epsilon$$

where $\hat{F}_j(x) = (1 - \epsilon)\hat{F}(x) + \epsilon\Delta_{\xi_j}(x)$ is a contaminated distribution, $\{\xi_j; j = 1, 2, \ldots\}$ is a sequence of elements of the sample space and $\hat{F}_\epsilon(x) = (1 - \epsilon)\hat{F}(x)$ is a subdistribution, $0 < \epsilon < 1$.

The stability of the set of roots in identifying important structures in the data is of paramount importance as it could indicate the presence of additional components than those originally specified in the model.

Because the estimating equations do not necessarily have a unique solution, we need to use multiple random starts to find all reasonable solutions with 'high probability'. Markatou (2000), suggests to use as starting values the method of moment estimates (MME) of the parameters of the mixture [Lindsay (1989), Furman and Lindsay (1994), and Lindsay and Basak (1993)]. These MME are calculated on subsamples drawn randomly from the data. There were several reasons that led us to select the method of moment estimates as starting values. The estimates are unique solutions of the moment equations, they can be computed quickly using a simple bisection algorithm, and they are consistent estimates of the true parameter values.

We used $B = 50$ bootstrap subsamples for our univariate case studies and $B = 100$ bootstrap subsamples for the bivariate case. However, $B = 50$ is sufficient for the bivariate case as well, in that the important data substructures can be identified. The subsample size was equal to the number of parameters to be estimated, consistent with the recommendation of Markatou (2000). To terminate the algorithm we use the stopping rule recommended in Markatou (2000) with $\alpha = 0.001$.

To calculate the maximum likelihood estimates we used the EM algorithm. It is often suggested to stop the iteration when $|\ell_{i+1} - \ell_i| \leq tol$, where tol is a small specified constant and ℓ_{i+1}, ℓ_i are the values of the log-likelihood at steps $(i+1)$st and ith. We used the implementation of the EM algorithm given in McLachlan and Basford (1988, pages 213–216). The stopping rule bounds the difference between the log-likelihood at the ith step and the log-likelihood at the (i-10)th step by the number 10^{-4} multiplied by the log-likelihood at the (i-10)th step of the iteration. If this bound is unreachable, the algorithm stops at the 150th iteration. Bohning *et al.* (1994) discuss the use of the EM-algorithm with a different stopping rule. Their rule is based on the use of Aitken's acceleration, which is a device to exploit the regularity of the convergence process. In complex practical problems this may be a better stopping rule to use. For additional discussion about algorithms in the context of mixtures see also Bohning (1999).

In what follows we describe the simulation experiments and discuss our results.

29.3.1   Normal Mixtures with Equal Component Variance

In this section we describe the simulation experiments performed. We first concentrate in the case of normal mixtures with equal component variance.

The nominal model was $p_1 N(\mu_1, \sigma^2) + p_2 N(\mu_2, \sigma^2)$, $p_1 + p_2 = 1$. The data were generated from the model $p_1 N(0, 1) + p_2 N(8, 1) + p_3 N(12, 1)$, where $p_1 = p_2$ and $p_3 = 0, 0.04, 0.10, 0.16, 0.20, 0.26$.

In our calculations we used the chisquared weight function. It is defined as $w(\delta) = 1 - \delta^2/(\delta+2)^2$ [see Markatou, Basu, and Lindsay (1998)]. To compute the Pearson residuals we used normal kernel with $h^2 = c\sigma^2$ with two different levels of c, either 0.050 or 0.010, which correspond to downweighting on average 5 and 14 observations respectively. We considered samples of size 100. At each sample model, we simulated 100 data sets. All samples were generated in S-plus and all programs were written in FORTRAN77. All calculations were carried out on a DEC5000/50 station.

For each data set we used the method of moment estimates calculated on the entire sample to produce starting values; we also carried out a bootstrap root search. Given a sample we took $B = 50$ bootstrap subsamples of size 4 and use them to construct our starting values.

TABLE 29.1
WLEE estimators with starting values MMEs computed using the
entire sample. The sample model is $p_1 N(0, 1) + p_2 N(8, 1) + p_3 N(12, 1)$,
with $p_1 = p_2$. The sample size is 100 and the number of Monte
Carlo replications is 100

% of mixing p_3	Estimates	Parameter Estimates			
		p_1	μ_1	μ_2	σ
0%	MLE	0.4998	8.0096	0.0281	0.9916
	WLEE ($c = 0.050$)	0.4998	8.0145	0.0267	0.9582
	WLEE ($c = 0.010$)	0.4994	8.0124	0.0236	0.9121
4%	MLE	0.5200	8.5102	0.2898	1.3834
	WLEE ($c = 0.050$)	0.5009	8.0258	0.0068	0.9685
	WLEE ($c = 0.010$)	0.4997	8.0253	0.0070	0.9129
10%	MLE	0.5511	8.7098	-0.0265	1.6377
	WLEE ($c = 0.050$)	0.5027	8.0156	-0.0238	0.9699
	WLEE ($c = 0.010$)	0.5024	7.9969	-0.0246	0.9143
16%	MLE	0.5813	9.0233	0.0028	1.7025
	WLEE ($c = 0.050$)	0.5105	8.0705	0.0137	1.0123
	WLEE ($c = 0.010$)	0.5061	8.0165	0.0203	0.9255
20%	MLE	0.6010	9.3029	-0.0256	1.7220
	WLEE ($c = 0.050$)	0.5206	8.2703	-0.0175	1.1383
	WLEE ($c = 0.010$)	0.5028	8.0355	-0.0195	0.9561
26%	MLE	0.6307	9.6287	-0.0243	1.7072
	WLEE ($c = 0.050$)	0.6021	9.3780	-0.0101	1.7127
	WLEE ($c = 0.010$)	0.5603	8.8517	-0.0158	1.4084

Table 29.1 presents the WLEE estimators under the various three
component sampling models when the MME calculated on the entire
sample are used as starting values. For p_3 up to 0.16 and for $c =
0.050, 0.010$ the WLEE algorithm converges to a root similar to the model
$p_1 N(0, 1) + p_2 N(8, 1)$ ignoring the third component. When $p_3 > 0.16$
and depending on the value of c we observe an increase in the bias of
the WLEE estimators with a final convergence to a root similar to that
suggested by the maximum likelihood. This root estimates fairly accurately the location of the $N(0, 1)$ component while the second location
is shifted to the right.

When bootstrap search is used several roots which correspond to different possibilities for fitting different two component mixture models
emerge. Depending on the mixing proportion of the third normal com-

TABLE 29.2
Frequency of identified roots. The sample model is
$p_1 N(0,1) + p_2 N(8,1) + p_3 N(12,1), p_1 = p_2$. **Model I is**
$p_1 N(0,1) + p_2 N(8,1)$, **II is** $p_1 N(8,1) + p_2 N(12,1)$, **III is**
$p_1 N(0,1) + p_2 N(12,1)$ **and model IV stands for an MLE-like fit. The**
+ sign indicates that the model was suggested, the - sign indicates
absence of it. The frequency is over 100 samples if size 100 with 50
bootstrap searches in each sample

Suggested Models				% of Mixing p_3				
I	II	III	IV	4%	10%	16%	20%	26%
+	+	+	+				4	54
+	-	-	-	49	17	3	1	
-	-	-	+			1		
+	+	-	-	49	67	70	29	
+	-	+	-	2	7	2		
-	+	+	-			2		
-	+	-	+			2	6	13
+	+	+	-		9	18	34	3
+	+	-	+			2	26	28
+	-	+	+					1
-	+	+	+					1

ponent some root patterns are favored over others.

Table 29.2 presents the different root patterns found in our simulation for $c = 0.050$. When the mixing proportion of the third component is 0.04 in approximately half of the Monte Carlo samples a mixture of $N(0,1)$ and $N(8,1)$ was suggested. The remaining samples suggested a mixture of $N(0,1)$ and $N(8,1)$ as well as a mixture between $N(8,1)$ and $N(12,1)$. In only two samples a mixture between $N(0,1)$ and $N(12,1)$ was observed. For up to $p_3 = 0.10$ the above described types of roots are present in a large number of samples. Most of the samples exhibit roots that stay close to fitting a mixture between $N(0,1)$ and $N(8,1)$, a mixture between $N(8,1)$ and $N(12,1)$ and a mixture between $N(0,1)$ and $N(12,1)$, over the 50 bootstrap starting values. In all samples and for all values of c the mixture model $p_1 N(0,1) + p_2 N(8,1)$ was suggested.

TABLE 29.3
Bootstrap estimates of the parameters of the mixture
$p_1 N(0, 1) + p_2 N(8, 1), p_1 + p_2 = 1.$ **The sampling model is**
$p_1 N(0, 1) + p_2 N(8, 1) + p_3 N(12, 1), p_1 = p_2.$ **The number of Monte Carlo**
replications is 100, the sample size is 100 and the number of
bootstrap samples is 50

% of mixing p_3	Estimates	$\hat{p}_1$	$\tilde{\mu}_1$	$\tilde{\mu}_2$	$\hat{\sigma}$
0%	WLEE (0.050)	0.4998	8.0145	0.0267	0.9582
	WLEE (0.010)	0.4994	8.0124	0.0236	0.9121
4%	WLEE (0.050)	0.5009	8.0258	0.0068	0.9685
	WLEE (0.010)	0.4997	8.0253	0.0070	0.9129
10%	WLEE (0.050)	0.5027	8.0156	-0.0238	0.9699
	WLEE (0.010)	0.5024	7.9969	-0.0246	0.9143
16%	WLEE (0.050)	0.5105	8.0705	0.0137	1.0123
	WLEE (0.010)	0.5061	8.0165	0.0203	0.9255
20%	WLEE (0.050)	0.5041	8.0045	-0.0122	0.9902
	WLEE (0.010)	0.5028	8.0355	-0.0195	0.9561
26%	WLEE (0.050)	0.5024	7.9919	-0.0283	0.9688
	WLEE (0.010)	0.5019	8.0048	-0.0283	0.9367

Table 29.3 shows the bootstrap estimates for proportion, location, and scale of the mixture model comprised from the N(0, 1) and N(8, 1) components for various values of p_3. The estimates presented are averages over the 100 simulated samples and the 50 bootstrap random starts. A comparison between tables (1.1), (1.2), and (1.3) illustrates the behaviour, in terms of convergence, of the random starts. When the algorithm is started at the MMEs computed using the entire sample it converges, for low values of p_3, to a mixture model wich includes only the well represented components in the sample. For larger values of p_3 it converges to an MLE-like root. On the other hand, when random starts are used different root configurations, and thus multiple model fits, are present.

The algorithm converged very fast. Our finding was that the mean, median and the standard deviation of the number of iterations depend on the mixing proportion of the third component and they increase with it. For example, when the MMEs calculated on the entire sample are used as starting values in the weighted likelihood algorithm with $p_3 = 0.10$ and smoothing constant $c = 0.050$ the average number of iterations is 8.78, the median is 9 and the standard deviation is 1.508 over 100 samples. When the smoothing constant is 0.010 the mean number of iterations is 8.26, the median 8 and the standard deviation 1.260. However, when the mixing proportion $p_3 = 0.20$ and for $c = 0.050$ the average number of iterations is 14.54, the median is 14 and the standard deviation is 6.090. When $c = 0.010$ the corresponding numbers are 15.2, 14, 4.575. We also found that in our large simulation runs, each individual B=50 search averaged about 8 minutes in real time. Thus in no case were the calculations excessive.

When the components are not well separated the number of iterations needed to converge to a root increases. As an example, when the sampling model is a mixture of equal proportions of N(0,1) and N(2,1) the mean number of iterations for the weighted likelihood algorithm with $c = 0.050$ was 14.52, the median was 10 and the standard deviation 16.455. The EM algorithm had a mean of 18.24, median of 16 and standard deviation 7.425 iterations.

29.3.2   Normal Mixtures with Unequal Component Variance

In this section we describe the simulation experiments and results obtained for the case of normal mixtures with unequal component variances. Moreover, we study the case of multivariate normal mixtures with equal covariance matrices and compare our results with those obtained from applying a Huber-type M-estimation procedure.

The sampling model is $p_1 N(0, 1) + p_2 N(8, 4) + p_3 N(16, 1)$ where $p_1 =$

p_2 and p_3 is 0.04, 0.10, 0.14, 0.16, 0.20, 0.26. The nominal model is assumed to be $p_1 N(\mu_1, \sigma_1^2) + p_2 N(\mu_2, \sigma_2^2)$, $p_1 + p_2 = 1$. The starting values are the MME calculated on the entire sample as well as bootstrap sample based estimates. Notice that the starting values have been computed under the assumption that the normal components have the same variance. However, this variance misspecification in the starting values does not affect the final estimates. An interesting question that remains yet unanswered is how can one construct method of moments estimates when the component submodels have different variances. The values of the smoothing constant in the normal kernel were again 0.050, 0.010. When the normal kernel is used for up to $p_3 = 0.14$ the weighted likelihood algorithm started at the MMEs calculated on the entire sample, converges to the root closer to the mixture that is comprised from the N(0, 1) and N(8, 4) models. For $p_3 > 0.14$ the algorithm suggests multiple mixture model fits. This type of behavior is different from the one observed in the case of equal component variance where the MMEs computed on the entire sample converge to an MLE-like root when the third mixing proportion is increased. Convergence here is slightly slower than in the case of equal component variances. For example, when $p_3 = 0.04$ and $c = 0.05$ the mean number of iterations is 6.62, the median is 6.50 and the standard deviation is 2.058. When $p_3 = 0.10$ and $c = 0.050$ the mean, median and standard deviation are 9.15, 7 and 7.48 respectively. When bootstrap is involved we always find the interesting roots. When the percentage of mixing of the third component becomes high we can detect multiple roots as it is expected.

We also carried out a simulation study where the sampling model was a three component mixture, with equal proportions from the normal populations with means $(0,0)^T$ and $(4,4)^T$ and a varying proportion from a third component. The third component is a normal with mean $(1 + \xi - 5, 1 - (\xi - 5))^T$. This third component is thought of as an outlying component. All normal populations have the same identity covariance matrix. We used values of $\xi = 1, 3, 4$. The mixing proportion of this third component is 0.04, 0.10, 0.14, 0.16, 0.20, 0.26. As starting values we use the method of moment estimates computed on the entire sample. We also use bootstrap root search. Normal kernel smoothing was used to compute the Pearson residuals with bandwidth parameter $H_{(i+1)} = c\hat{\Sigma}_{(i)}$, where $\hat{\Sigma}_{(i)}$ is the estimated covariance at the ith step of the iteration. The value of $c = 0.3$ corresponds to mean downweighting of 1.599 or approximately 2% of the sample size. We generated 100 samples of size 100 over the range of B.

We first examine the behaviour with a fixed starting value and hence a single solution. When the mixing proportion of the outlying normal component is less that 0.16 and the MME computed on the entire sample were used as starting values the WLEE algorithm converged in every case to the root that corresponds to fitting a mixture of normals with means $(0,0)^T$ and $(4,4)^T$ respectively, ignoring the third component. When the mixing proportion of the third component is 0.16 in 89 out of 100 samples a root corresponding to fitting the mixture of $N((0,0),I)$ and $N((4,4),I)$ was identified. In the remaining 11 samples the WLEE algorithm converged to a root fitting one normal component with mean near $(-3,5)^T$ and a second component with mean similar to the mean of the remaining data. When the mixing proportion of the normal component with mean $(-3,5)^T$ is 0.26, different data samples resulted in solutions which corresponded to fitting one normal component among means $(0,0)^T, (4,4)^T$ and $(-3,5)^T$ respectively and a second component with mean similar to the one from the average of the remaining data. This behaviour is different from the one observed in the univariate case in which the WLEE algorithm started with MMEs computed on the entire sample converge to one type of root.

We next examine the value of the bootstrap root search. When bootstrap is implemented depending on the mixing proportion of the normal component with mean $(-3,5)^T$ all interesting root types are present, as it is expected. Therefore, the algorithm has the ability to unearth different data substructures when indeed those are present. On the other hand, when data are generated from a mixture of equal proportions of $N((0,0),I)$ and $N((4,4),I)$ observations then for all bootstrap starting values and all samples one and only one root is present and that is the root that corresponds to fitting the model from which the data were generated.

When the mean of the third component is $(0,2)^T$, one of the root types, the MLE-like root is found in most cases for smoothing parameter equal to 0.3. For values of the smoothing parameter less than 0.3 a root corresponding closer to fitting a mixture of the normal components with means $(0,0)^T$ and $(4,4)^T$ is identified.

For the model $p_1 N((0,0)^T, I) + p_2 N((4,4)^T, I) + p_3 N((-3,5)^T, I)$, $p_1 = p_2$ and p_3 as above we also calculated Huber's type M-estimators, as proposed by McLachlan and Basford (1988, Chapter 2, Section 2.8). These estimators replace the normal scores with a ψ function. We use Huber's psi-function defined as

$$\psi(s) = \begin{cases} s & \text{if } |s| \leq 3.0034 \\ 3.0034 sign(s) & \text{otherwise} \end{cases}$$

where the constant 3.0034 is calculated according to the formula $k_1(p) = [p\{1-(2/9)p +((2/9)p)^{1/2}1.645\}^3]^{1/2}$ recommended by Campbell (1984). Their modified EM algorithm was used with the MLE as starting values. The algorithm was terminated if the difference of the estimators between two consecutive iterations was less than or equal to 10^{-5}.

TABLE 29.4
Huber's estimators for the mean parameters and covariance
matrix Σ. The sampling model is
$p_1 N((0,0), I) + p_2 N((4,4), I) + p_3 N((-3,5), I), p_1 = p_2$. **The sample size**
is 100 and the number of Monte Carlo replications is 100

% of mixing proportion				Parameter Estimates	
p_3	$\hat{p}_1$	$\hat{\mu}_1$	$\hat{\mu}_2$	$\hat{\Sigma}$	
0%	0.5001	-0.007	3.9807	0.5278	0.0060
		-0.003	3.9993	0.0060	0.5574
4%	0.5182	0.0185	3.9256	0.5847	0.0415
		0.0423	3.9806	0.0415	0.6038
10%	0.5534	0.0153	3.6423	0.7140	0.1785
		0.1807	4.0399	0.1785	0.7500
14%	0.5671	-0.007	3.6019	0.7423	0.1574
		0.2413	4.0557	0.1574	0.7595
16%	0.5823	0.0914	3.3172	0.8404	0.3116
		0.3420	4.0614	0.3116	0.9127
20%	0.6273	0.4023	2.2049	1.2717	0.7085
		0.7222	4.2533	0.7085	1.3497
26%	0.6458	0.8686	0.1058	2.7084	1.1726
		1.2912	4.5714	1.1726	2.2481

Table 29.4 presents the modified Huber's estimates for various values of the mixing parameter p_3. From $p_3 = 0$, it is clear that the method does not consistently estimate the normal covariance Σ, but it does recover the correct shape up to a constant. However, the shape is distorted as p_3 increases. Notice that the location parameters are estimated fairly accurately for low values of p_3. As the value of p_3 increases the bias of the estimators increases. Moreover, the method does not have the diagnostic value of the weighted likelihood, as it treats the observations coming from the third component as contamination.

29.3.3   Other Models

We performed also simulations using mixtures of t-distributions with 2 and 3 degrees of freedom. The data were rescaled to have variance 1. We generated samples of size 100 and randomly selected 50 data points to which the constant 2.56 was added. The model fitted was a mixture of normals with different means and variances.

In order to interpret the results it is instructive to bear in mind what a scientist might want from an analysis of a mixture problem. The scientist might require good estimates of the mixing proportions and good estimates of the central locations of the subpopulations, so as to be able to describe these populations. The actual scale estimates are probably of scientific interest only in describing the extend to which the populations overlap.

Table 29.5 presents the estimates of the mixture parameters when the MMEs computed using the entire sample initialized the algorithm. The WLEE estimates of location have mean squared errors that are about ten times smaller than the mean squared error of the corresponding MMEs. It is no longer clear what the correct value for the scale parameters is, so it is hard to define bias. Notice however, that the ratio of the two scale estimates is approximately one. When bootstrap was implemented only one root was identified. McLachlan and Peel (1998) discuss a robust approach to clustering using mixtures of multivariate t-distributions. This t-mixture model can be fitted using the EMMIX program of McLachlan *et al.* (1998).

TABLE 29.5
Means of MME, MLE, and WLEE estimates and their mean
squared errors. The fitted model is $p_1 N(\mu_1, \sigma_1^2) + p_2 N(\mu_2, \sigma_2^2)$. The
notation $t_3^{(r)}$ means that the sample is rescaled to have variance 1.
Also $t_3^{(r)}(2.56)$ means that the second location is 2.56. In
parenthesis the mean squared error is reported. The sample size is
100 and the number of Monte Carlo replications is 100

Sampling Model: $0.5t_3^{(r)} + 0.5t_3^{(r)}(2.56)$

Estimates	$\hat{\pi}$	$\hat{\mu}_1$	$\hat{\mu}_2$	$\hat{\sigma}_1$	$\hat{\sigma}_2$
MME	0.5222	2.4763	-0.0149	0.9423	0.9423
	(0.0172)	(0.1494)	(0.1330)	(0.0471)	(0.0471)
MLE	0.4934	2.6132	-0.0178	0.7614	0.8068
	(0.0044)	(0.0310)	(0.0408)	(0.0900)	(0.0850)
WLE(0.300)	0.5067	2.5785	-0.0380	0.6997	0.6971
	(0.0020)	(0.0254)	(0.0132)	(0.1147)	(0.1147)
WLE(0.100)	0.5023	2.5749	-0.0203	0.6540	0.6613
	(0.0011)	(0.0127)	(0.0121)	(0.1324)	(0.1308)
WLE(0.050)	0.5025	2.5664	-0.0121	0.6387	0.6458
	(0.0012)	(0.0127)	(0.0122)	(0.1431)	(0.1483)
WLE(0.010)	0.5019	2.5499	0.0112	0.6017	0.6078
	(0.0015)	(0.0127)	(0.0127)	(0.1719)	(0.1717)

29.4   Model Selection

A natural question associated with the existence of multiple roots is that of selecting the model that describes adequately the true distribution. This adequate description of the distribution is viewed as a goodness of fit problem, and the parallel disparity measure between the data as expressed by $f^*(x)$ and the model $m_\phi^*(x)$ provides a natural avenue to address this problem.

In what follows we will briefly describe our framework.

We assume there exists a true probability distribution τ that has generated the data. The class $M = \{m_\beta : \beta \in \Theta\}$ is a tentative family of probability densities which we will call the *model*. The true model τ may or may not belong to this class.

The goodness of fit approach that we take here treats the question of model adequacy, which is equivalent to the question of whether or not τ belongs in M, as a null hypothesis. To select therefore among the different models suggested by the WLEE methodology we use

$$\rho(f^*, m_\phi^*) = \int G(\delta(x)) m_\phi^*(x) dx,$$

where G is a strictly convex, thrice differentiable function, such that $G(0) = 0$ and $G'(0) = 0$.

Assume now that the true probability model τ that generated the data belongs to the class M and that the null hypothesis $H_0 : \tau = m_{\phi_0}$ is simple, that is ϕ_0 is completely specified. In this case, a reasonable test statistic is

$$T_1 = 2n[G''(0)]^{-1}\rho(f^*, m_{\phi_0}^*),$$

where $G''(0) > 0$, because the function G is strictly convex.

However, in practice, the parameters of the hypothesized model are not known and need to be estimated from the data. Therefore, the null hypothesis becomes $H_0 : \tau \in M$, while the alternative states that τ does not belong in the class M. In this case, we propose to use as a test statistic the quantity

$$T_2 = 2n[G''(0)]^{-1}\rho(f^*, m_{\hat{\phi}}^*)$$

where $\hat{\phi}$ is an estimate of the parameter ϕ.

The development of the asymptotic distributions of these test statistics under the null hypothesis and alternatives is work in progress and the subject of a different paper. Here, we would like to point out that evaluation of the statistic T_2 at the different roots suggested by the weighted likelihood methodology provides a way of selecting among the different proposed models.

29.5 Conclusions

In this paper we take a closer look at the performance of weighted likelihood in the context of mixture models. It is shown that the weighted likelihood methodology produces robust and first order efficient estimators for the model parameters. When the number of components in the true model is higher than the number of components specified in the hypothesized model, the weighted likelihood equations have multiple roots. In this case, the number of roots can be used as a heuristic to identify multiple potential mixture model fits.

The existence of multiple roots naturally leads to the model selection question. We have proposed two test statistics that may be used to address this problem.

Acknowledgements This work was supported by NSF grant DMS-9973569. The author would like to thank a referee for constructive suggestions that improve the structure of the paper.

References

1. Aitkin, M. and Tunnicliffe, W. G. (1980). Mixture models, outliers, and the EM algorithm. *Technometrics* **22**, 325–331.

2. Basu, A. and Lindsay, B. G. (1994). Minimum disparity estimation for continuous models: Efficiency, distribution and robustness. *Annals of Institute of Statistical Mathematics* **46**, 683–705.

3. Bohning, D. (1999). *Computer-Assisted Analysis of Mixtures and Applications*. Chapman and Hall, London, England.

4. Bohning, D., Dietz, E., Schaub, R., Schlattmann, P., and Lindsay, B. G. (1994). The distribution of the likelihood ratio for mixtures of densities from the one-parameter exponential family. *Annals of the Institute of Statistical Mathematics* **46**, 373–388.

5. Campbell, N. A. (1984). Mixture models and atypical values. *Mathematical Geology* **16**, 465–477.

6. Clarke, B. R. and Heathcote, C. R. (1994). Robust estimation of k-component univariate normal mixtures. *Annals of the Institute of Statistical Mathematics* **46**, 83–93.

7. Cutler, A. and Cordero-Brana, O. (1996). Minimum Hellinger distance estimation for finite mixture models. *Journal of the American Statistical Association* **91**, 1716–1723.

8. De Veaux, R. D. and Krieger, A. M. (1990). Robust estimation of a normal mixture. *Statistics & Probability Letters* **10**, 1–7.

9. Furman, D. W. and Lindsay, B. G. (1994). Measuring the relative effectiveness of moment estimators as starting values in maximizing likelihoods. *Computational Statistics and Data Analysis* **17**, 493–507.

10. Gray, G. (1994). Bias in misspecified mixtures. *Biometrics* **50**, 457–470.

11. Lindsay, B. G. (1989). Moment matrices: Applications in mixtures. *Annals of Statistics* **17**, 722–740.

12. Lindsay, B. G. (1994). Efficiency versus robustness: The case of minimum Hellinger distance and related methods. *Annals of Statistics* **22**, 1018–1114.

13. Lindsay, B. G. (1995). *Mixture Models: Theory, Geometry and Applications.* NSF-CBMS Regional conference series in *Probability & Statistics*, Vol. 5, Institute of Mathematical Statistics, Hayward, CA.

14. Lindsay, B. G. and Basak, P. (1993). Multivariate normal mixtures: A fast consistent method of moments. *Journal of the American Statistical Association* **88**, 468–476.

15. Markatou, M., Basu, A., and Lindsay, B. G. (1997). Weighted likelihood estimating equations: The discrete case with applications to logistic regression. *Journal of Statistical Planning and Inference* **57**, 215–232.

16. Markatou, M., Basu, A., and Lindsay, B. G. (1998). Weighted likelihood estimating equations with a bootstrap root search. *Journal of the American Statistical Association* **93**, 740–750.

17. Markatou, M. (1996). Robust statistical inference: Weighted likelihoods or usual M-estimation? *Communications in Statistics—Theory and Methods* **25**, 2597–2613.

18. Markatou, M. (2000). Mixture models, robustness and the weighted likelihood methodology. *Biometrics* (to appear).

19. McLachlan, G. J. and Basford, K. E. (1988). *Mixture Models: Inference and Applications to Clustering.* Marcel Dekker Inc., New York.

20. McLachlan, G. J. and Peel, D. (1998). Robust cluster analysis via mixtures of multivariate t-distributions. In *Lecture Notes in Computer Science*, Vol. 1451, (Eds., A. Amin, D. Dori, P. Pudil, and H. Freeman), pp. 658–666. Spinger-Verlag, Berlin.

21. McLachlan, G. J., Peel, D., Basford, K. E., and Adams, R. (1998). *EMMIX* program.
    http://www.maths.uq.edu.au/gjm/emmix/emmix.html.

22. Redner, R. A. and Walker, H. F. (1984). Mixture densities, maximum likelihood and the EM algorithm. *SIAM Review* **26**, 195–239.

23. Windham, M. P. (1996). Robustizing mixture analysis using model weighting. Preprint.

24. Woodward, W. A., Parr, W. C., Schucany, W. R., and Lindsey, H. (1984). A comparison of minimum distance and maximum likelihood estimation of the mixture proportion. *Journal of the American Statistical Association* **79**, 590–598.

25. Woodward, W. A., Whitney, P., and Eslinger, P. W. (1995). Minimum Hellinger distance estimation for mixture proportions. *Journal of Statistical Planning and Inference* **48**, 303–319.

30

On Nonparametric Function Estimation
with Infinite-Order Flat-Top Kernels

Dimitris N. Politis
University of California at San Diego, La Jolla, CA

ABSTRACT The problem of nonparametric estimation of a smooth, real-valued function of a vector argument is addressed. In particular, we focus on a family of infinite-order smoothing kernels that is characterized by the flatness near the origin of the Fourier transform of each member of the family; hence, the term 'flat-top' kernels. Smoothing with the proposed infinite-order flat-top kernels has optimal Mean Squared Error properties. We review some recent advances, as well as give two new results on density estimation in two cases of interest: (i) case of a smooth density over a finite domain, and (ii) case of infinite domain with some discontinuities.

Keywords and phrases Nonparametric estimation, flat-top kernels, density estimation, mean squared error

30.1 Introduction: A General Family of Flat-Top Kernels of Infinite Order

Let $f : R^d \to R$ be an unknown function to be estimated from data. In the typical nonparametric set-up, nothing is assumed about f except that it possesses a certain degree of smoothness. Usually, a preliminary estimator of f can be easily calculated that, however, lacks the required smoothness; e.g., in the case where f is a probability density. Often, the preliminary estimator is even inconsistent; e.g., in the case where f is a spectral density and the preliminary estimator is the periodogram. Rosenblatt (1991) discusses these two cases in an integrated framework.

In order to obtain an estimator (denoted by $\hat{f}$) with good properties, for example, large-sample consistency and smoothness, one can smooth the preliminary estimator by convolving it with a function $\Lambda : R^d \to R$ called the 'kernel', and satisfying $\int \Lambda(x)dx = 1$; unless otherwise noted, integrals will be over the whole of R^d. It is convenient to also define the Fourier transform of the kernel as $\lambda(s) = \int \Lambda(x)e^{i(s \cdot x)}dx$, where $s = (s_1, \ldots, s_d)$, $x = (x_1, \ldots, x_d) \in R^d$, $(s \cdot x) = \sum_k s_k x_k$ is the inner product between s and x.

Typically, as the sample size increases, the kernel $\Lambda(\cdot)$ becomes more and more concentrated near the origin. To achieve this behavior, we let $\Lambda(\cdot)$ and $\lambda(\cdot)$ depend on a real-valued, positive 'bandwidth' parameter h, that is, we assume that $\Lambda(x) = h^{-d}\Omega(x/h)$, and $\lambda(s) = \omega(hs)$, where $\Omega(\cdot)$ and $\omega(\cdot)$ are some fixed (not depending on h) bounded functions, satisfying $\omega(s) = \int \Omega(x)e^{i(s \cdot x)}dx$; the bandwidth h will be assumed to be a decreasing function of the sample size.

If Ω has finite moments up to qth order, and moments of order up to $q - 1$ equal to zero, then q is called the 'order' of the kernel Ω. If the unknown function f has r bounded continuous derivatives, it typically follows that

$$Bias(\hat{f}(x)) = E\hat{f}(x) - f(x) = c_{f,\Omega}(x)h^k + o(h^k), \qquad (30.1.1)$$

where $k = \min(q, r)$, and $c_f(x)$ is a bounded function depending on Ω, on f, and on f's derivatives. Note that existence and boundedness of derivatives up to order r includes existence and boundedness of mixed derivatives of total order r; cf. Rosenblatt (1991, p. 8).

This idea of choosing a kernel of order q in order to get the $Bias(\hat{f}(x))$ to be $O(h^k)$ dates back to Parzen (1962) and Bartlett (1963); the paper by Cacoullos (1966) seems to be the first contribution on the multivariate case. Some more recent references on 'higher-order' kernels include the following: Devroye (1987), Gasser, Müller, and Mammitzsch (1985), Granovsky and Müller (1991), Jones (1995), Jones and Foster (1993), Marron (1994), Marron and Wand (1992), Müller (1988), Nadaraya (1989), Silverman (1986), Scott (1992), and Wand and Jones (1993).

Note that the asymptotic order of the bias is limited by the order of the kernel if the true density is very smooth, i.e., if r is large. To avoid this limitation, one can define a 'superkernel' as a kernel whose order can be any positive integer; Devroye (1992) contains a detailed analysis of superkernels in the case of (univariate) probability density estimation. Thus, if f has r bounded continuous derivatives, a superkernel will result in an estimator with bias of order $O(h^r)$, no matter how large r may be; so, we might say that a superkernel is a kernel with 'infinite order'.

However, it might be more appropriate to say that a kernel has 'infinite order' if it results in an estimator with bias of order $O(h^r)$ no matter how large r may be *regardless of whether the kernel has finite moments*. It seems that the finite-moment assumption for Ω is just a technical one, and that existence of the Lebesgue integrals used to calculate the moments is *not* necessarily required in order that a kernel has favorable bias performance; rather, it seems that if the integrals defining the moments of Ω have a Cauchy principal value of zero then the favorable bias performance follows, and this is in turn ensured by setting ω to be constant over an open neighborhood of the origin.

A preliminary report on a specific type of such infinite order kernel in the univariate case (that corresponds to an ω of 'trapezoidal' shape) was given in Politis and Romano (1993). Consequently, in Politis and Romano (1996, 1999) a general family of multivariate flat-top kernels of infinite order was proposed, and the favorable large-sample bias (and Mean Squared Error) properties of the resulting estimators were shown in the cases of probability and spectral density estimation.

Presently, we will propose a slightly bigger, more general class of multivariate flat-top kernels of infinite order with similar optimality properties — as shown in Section 30.2. Similarly to the Politis and Romano (1993, 1996, 1999) papers, our arguments here will focus on rates of convergence without explicit calculation of the proportionality constants involved; the reader should be aware of the increasing concern in the literature regarding those constants, and their corresponding finite-sample effects — see, for instance, Marron and Wand (1992). Finally, in Section 30.3 we will address the interesting case where the unknown function f possesses the required smoothness only over a subset of the domain; in particular, we will investigate to what extend the performance of $\hat{f}(x)$ is affected by a discontinuity of f (or its derivatives) at points away from x.

The general family of multivariate flat-top kernels of infinite order can be defined as follows.

DEFINITION 30.1.1

Let C be a compact, convex subset of R^d that contains an open neighborhood of the origin; in other words, there is an $\epsilon > 0$ such that $\epsilon D \subset C \subset \epsilon^{-1}D$, where D is the Euclidean unit ball in R^d.

The kernel Ω_C is said to be a member of the general family of multivariate flat-top kernels of infinite order if

$$\Omega_C(x) = (2\pi)^{-d} \int \omega_C(s)e^{-i(s\cdot x)}ds,$$

where the Fourier transform $\omega_C(s)$ satisfies the following properties:

(i) $\omega_C(s) = 1$ for all $s \in C$;

(ii) $\int |\omega_C(s)|^2 ds < \infty$; and

(iii) $w_C(s) = w_C(-s)$, *for any* $s \in R^d$.

Property (i) guarantees the favorable bias properties and the 'infinite' order, while property (ii) ensures a finite variance of the resulting estimator $\hat{f}$; finally, property (iii) guarantees that Ω_C is real-valued.

In practically working with such a flat-top kernel, one must choose C. A typical choice for C is the unit ball in l_p, with some choice of p satisfying $1 \leq p \leq \infty$; see Politis and Romano (1999). In addition, it is natural to impose the condition that w_C be a continuous function with the property $|w_C(s)| \leq 1$, for any $s \in R^d$. Nevertheless, only properties (i), (ii), (iii) are required for our results.

30.2   Multivariate Density Estimation: A Review

Suppose $X_1, \ldots, X_N$ are independent, identically distributed (i.i.d.) random vectors taking values in R^d, and possessing a probability density function f; the assumption of independence is not crucial here. The arguments apply equally well if the observations are stationary and weakly dependent, where weak dependence can be quantified through the use of mixing coefficients — see, for example, Györfi *et al.* (1989).

The objective is to estimate $f(x)$ for some $x \in R^d$, assuming f possesses a certain degree of smoothness. In particular, it will be assumed that the characteristic function $\phi(s) = \int e^{i(s \cdot x)} f(x) dx$ tends to zero sufficiently fast as $||s||_p \to \infty$; here $s = (s_1, \ldots, s_d)$, $x = (x_1, \ldots, x_d) \in R^d$, $(s \cdot x) = \sum_k s_k x_k$ is the inner product between s and x, and $||\cdot||_p$ is the l_p norm, i.e., $||s||_p = (\sum_k |s_k|^p)^{1/p}$, if $1 \leq p < \infty$, and $||s||_\infty = \max_k |s_k|$.

We define the flat-top kernel smoothed estimator of $f(x)$, for some $x \in R^d$, by

$$\hat{f}(x) = \frac{1}{N} \sum_{k=1}^N \Lambda_C(x - X_k) = \frac{1}{(2\pi)^d} \int \lambda_C(s)\phi_N(s)e^{-i(s \cdot x)} ds, \quad (30.2.1)$$

where $\lambda_C(s) = \int \Lambda_C(x)e^{i(s \cdot x)} dx$, $\Lambda_C(x) = h^{-d}\Omega_C(x/h)$, for some chosen bandwidth $h > 0$, and Ω_C satisfies the properties of Definition 30.1.1; also note that $\phi_N(s)$ is the sample characteristic function defined by

$$\phi_N(s) = \frac{1}{N} \sum_{k=1}^N e^{i(s \cdot X_k)}.$$

Now it is well known [cf. Rosenblatt (1991, p. 7)] that if f is continuous at x, and $f(x) > 0$, then

$$Var(\hat{f}(x)) = \frac{1}{h^d N} f(x) \int \Omega^2(x) dx + O(1/N). \qquad (30.2.2)$$

Hence, the order of magnitude of the Mean Squared Error (MSE) of $\hat{f}(x)$ hinges on the order of magnitude of its bias. To quantify the bias (and resulting MSE) of $\hat{f}(x)$, we formulate three different conditions based on the rate of decay of ϕ that are in the same spirit as the conditions in Watson and Leadbetter (1963).

Condition C_1: For some $p \in [1, \infty]$, there is an $r > 0$, such that $\int ||s||_p^r |\phi(s)| ds < \infty$

Condition C_2: For some $p \in [1, \infty]$, there are positive constants B and K such that $|\phi(s)| \leq Be^{-K||s||_p}$ for all $s \in R^d$.

Condition C_3: For some $p \in [1, \infty]$, there is a positive constant B such that $|\phi(s)| = 0$, if $||s||_p \geq B$.

Note that if one of Conditions C_1 to C_3 holds for some $p \in [1, \infty]$, then, by the equivalence of l_p norms for R^d, that same Condition would hold for *any* $p \in [1, \infty]$, perhaps with a change in the constants B and K.

Conditions C_1 to C_3 can be interpreted as different conditions on the smoothness of the density $f(x)$ for $x \in R^d$; cf. Katznelson (1968), Butzer and Nessel (1971), Stein and Weiss (1971), and the references therein. Note that they are given in increasing order of strength, i.e., if Condition C_2 holds, then Condition C_1 holds as well, and if Condition C_3 holds, then Conditions C_1 and C_2 hold as well. Also note that if Condition C_1 holds, then f must necessarily have $[r]$ bounded, continuous derivatives over R^d, where $[\cdot]$ is the integer part; cf. Katznelson (1968, p. 123). Obviously, if Condition C_2 holds, then f has bounded, continuous derivatives of *any* order over R^d.

The following theorem quantifies the performance of the proposed family of flat-top estimators. It was first proved in Politis and Romano (1999) in the case where C is the l_p unit ball with $1 \leq p \leq \infty$; we restate it—without proof—below in this slightly more general case.

THEOREM 30.2.1 *[Politis and Romano (1999)]*
Assume that $N \to \infty$.
(a) Under Condition C_1, and letting $h \sim AN^{-1/(2r+d)}$, for some constant

A > 0, it follows that

$$\sup_{x \in R^d} MSE(\hat{f}(x)) = O(N^{-2r/(2r+d)}).$$

(b) Under Condition C_2, and letting $h \sim A/\log N$, where A is a constant such that $A < 2K$, it follows that

$$\sup_{x \in R^d} MSE(\hat{f}(x)) = O(\frac{\log^d N}{N}).$$

(c) Under Condition C_3, and letting h be some constant small enough such that $h \leq B^{-1}$, it follows that

$$\sup_{x \in R^d} MSE(\hat{f}(x)) = O(1/N).$$

REMARK The special case where $\omega_C(s) = 0$ for all $s \notin C$ has been considered by many authors in the literature, e.g. Parzen (1962), Davis (1977), and Ibragimov and Hasminksii (1982). Nevertheless, the choice $\omega_C(s) = 0$ for all $s \notin C$ is *not* recommendable in practice; see Politis and Romano (1999) for more details on such practical concerns, including choosing the bandwidth h in practice. ∎

REMARK A rather surprising observation is that smoothing with flat-top kernels does not seem to be plagued by the 'curse of dimensionality' in case the underlying density is ultra-smooth, possessing derivatives of all orders, i.e., under Condition C_2 (or C_3). For example, in Theorem 30.2.1b under Condition C_2, the MSE of estimation achieved by flat-top kernel smoothing is of order $O(\frac{\log^d N}{N})$, i.e., depending on the dimension d only through the slowly varying function $\log^d N$. A more extreme result obtains under Condition C_3: Theorem 30.2.1c shows that in that case the MSE of estimation becomes exactly $O(1/N)$ which is identical to the parametric rate of estimation, and does not depend on the dimension d at all. ∎

REMARK It is also noteworthy that, even in the univariate case $d = 1$, the MSE of estimation is identical to the $O(1/N)$ parametric rate of estimation under Condition C_3, and is very close to $O(1/N)$ under Condition C_2. In other words, if a practitioner is to decide between fitting a particular parametric model to the data vs. assuming that the unknown density has derivatives of all orders (i.e., Condition C_2) and using our proposed flat-top kernel smoothing, there is no real benefit (in terms of rate of convergence) in favor of the parametric model. As a matter of fact, the smoothness Condition C_2 may be viewed as defining a huge class of functions that includes all the usual parametric models; the proposed flat-top kernel smoothing can then proceed to estimate the unknown function with accuracy comparable to the accuracy of a parametric estimator. ∎

REMARK It is well-known in the literature [see, for example, Müller (1988) or Scott (1992)] that kernel density estimators corresponding to kernels of order bigger than two are not necessarily nonnegative functions; it goes without saying that the same applies for our estimators $\hat{f}$ that are obtained using kernels of infinite order. Nevertheless, the nonnegativity is not a serious issue as there is a natural fix-up, namely using the modified estimator $\hat{f}^+(x) = \max(\hat{f}(x), 0)$; see also Gajek (1986) and Hall and Murison (1992). Note that the estimator $\hat{f}^+(x)$ is not only nonnegative, but is more accurate as well, in the sense that $MSE(\hat{f}^+(x)) \leq MSE(\hat{f}(x))$, for all x; this fact follows from the obvious inequality $|\hat{f}^+(x) - f(x)| \leq |\hat{f}(x) - f(x)|$. In addition, if $f(x) > 0$, an application of Chebychev's inequality shows that $Prob\{\hat{f}(x) = \hat{f}^+(x)\} \to 1$ under the assumptions of our Theorem 30.2.1; on the other hand, if $f(x) = 0$, then the large-sample distribution of either $\sqrt{h^d N}\hat{f}^+(x)$, or $\sqrt{h^d N}\hat{f}(x)$, degenerates to a point mass at zero. ∎

REMARK By the formal analogy between probability spectral density estimation [see, for example, Rosenblatt (1991)] it should not be surprising that flat-top kernels might be applicable in a context of nonparametric spectral density estimation. In Politis and Romano (1995, 1996), kernels belonging to a subset of the family of flat-top kernels are employed for the purpose of spectral density estimation using data consisting of a realization of a stationary time series or a homogeneous random field. Incidentally, note a typo in the statement of Theorem 2 in Politis and Romano (1996): instead of $M_i \sim dc_i \log N_i$ it should read $m_i \sim dc_i \log N_i$. ∎

30.3 Further Issues on Density Estimation

In this section we will continue the discussion on probability density estimation based on i.i.d. data $X_1, \ldots, X_N$, and will investigate to what extend the performance of $\hat{f}(x)$ is affected by a discontinuity of f (or its derivatives) at points away from x.

30.3.1   Case of Smooth Density over a Finite Domain

To fix ideas, consider first the univariate case $d = 1$; it is well-known that, if the random variables $X_1, \ldots, X_N$ are bounded, that is, if the density f has domain the finite interval $[a, b]$ as opposed to R, then the characteristic function ϕ will not satisfy the smoothness Conditions C_1, C_2, or C_3. The situation is exemplified by the smoothest of such densities, namely the uniform density over the interval $[-\theta, \theta]$ whose characteristic function is given by $\phi(s) = \frac{\sin \theta s}{\theta s}$; cf. Rao (1973, p. 151).

In general, suppose $\bar{f}$ is a very smooth density function (e.g., satisfying one of the smoothness Conditions C_1, C_2, or C_3), and let $f(x) = c \, \bar{f}(x) 1_{[-\theta, \theta]}(x)$, where $1_{[-\theta, \theta]}(x)$ is the indicator function, and $c = 1/\int_{-\theta}^{\theta} \bar{f}(x) dx$. Then the characteristic function of f is given by $\phi(s) = 2\theta c \, \bar{\phi}(s) * \frac{\sin \theta s}{\theta s}$, where $\bar{\phi}$ is the characteristic function of $\bar{f}$, and $*$ denotes convolution. In other words, the term $\frac{\sin \theta s}{\theta s}$ seems unavoidable, and is due to the truncation of the random variables.

Nevertheless, it seems intuitive that for the ultra-smooth uniform density over the interval $[-\theta, \theta]$, smoothing should give good results; this is indeed true as the following discussion shows. First note that if $\theta = \pi$, then $\frac{\sin \theta s}{\theta s} = 0$ for all $s \in Z - \{0\}$; this observation naturally brings us to Fourier series on the circle defined by 'wrapping' the interval $[-\pi, \pi]$ around on a circle, or — in high dimensions — Fourier series on the d-dimensional torus.

So, without loss of generality (and possibly having to use a linear/affine transformation in pre-processing the data), assume that $X_1, \ldots, X_N$ are i.i.d. with probability density f defined on the torus $T = [-\pi, \pi]^d$. Now let the characteristic function $\phi(s) = \int_T e^{i(s \cdot x)} f(x) dx$, and the sample characteristic function $\phi_N(s) = \frac{1}{N} \sum_{k=1}^{N} e^{i(s \cdot X_k)}$. Recall the Fourier series formula

$$f(x) = (2\pi)^{-d} \sum_{s \in Z^d} e^{-i(s \cdot x)} \phi(s),$$

and define our estimator

$$\hat{f}(x) = (2\pi)^{-d} \sum_{s \in Z^d} \lambda_C(s) e^{-i(s \cdot x)} \phi_N(s),$$

where $\lambda_C(s)$ was defined in Section 30.2. As before, we define some smoothness conditions based on the characteristic function ϕ.

Condition K_1: For some $p \in [1, \infty]$, there is an $r > 0$, such that $\sum_{s \in Z^d} ||s||_p^r |\phi(s)| < \infty$

Condition K_2: For some $p \in [1, \infty]$, there are positive constants B and K such that $|\phi(s)| \leq B e^{-K||s||_p}$ for all $s \in Z^d$.

Condition K_3: For some $p \in [1, \infty]$, there is a positive constant B such that $|\phi(s)| = 0$, if $||s||_p \geq B$ (with $s \in Z^d$).

The following theorem quantifies the performance of the general family of flat-top estimators; its proof follows closely the proof in Politis and Romano (1999) and is omitted.

THEOREM 30.3.1

Assume that $N \to \infty$.
(a) Under Condition K_1, and letting $h \sim AN^{-1/(2r+d)}$, for some constant $A > 0$, it follows that

$$\sup_{x \in T} MSE(\hat{f}(x)) = O(N^{-2r/(2r+d)}).$$

(b) Under Condition K_2, and letting $h \sim A/\log N$, where A is a constant such that $A < 2K$, it follows that

$$\sup_{x \in T} MSE(\hat{f}(x)) = O(\frac{\log^d N}{N}).$$

(c) Under Condition K_3, and letting h be some constant small enough such that $h \leq B^{-1}$, it follows that

$$\sup_{x \in T} MSE(\hat{f}(x)) = O(1/N).$$

REMARK It is easy to see that the uniform density on T satisfies condition K_3, and thus smoothing with a flat-top kernel achieves the parametric $\sqrt{N}$-rate in this case (with no dependence on the dimensionality d) which is remarkable. Nevertheless, Conditions K_1, K_2, K_3 are quite stringent as they imply smoothness/differentiability of f over the whole torus T; this is equivalent to assuming that a periodic extension of f over R^d is smooth/differentiable over the whole of R^d. ∎

To fix ideas, we again return to the case $d - 1$, and note that Condition K_1 implies that f has $[r]$ bounded, continuous derivatives over T, where $[\cdot]$ is the positive part; this implies, in particular, that $f(-\pi) = f(\pi)$, $f'(-\pi) = f'(\pi)$, $f''(-\pi) = f''(\pi)$, and so forth up to the $[r]$-th derivative. If f is smooth/differentiable over $(-\pi, \pi)^d$ but not over the whole torus T then the Fourier series method is not appropriate; rather, a technique of extension of f over the whole of R^d might be useful as elaborated upon in the next subsection.

30.3.2   Case of Infinite Domain with Some Discontinuities

We now return to the set-up of Section 30.2 where $X_1, \ldots, X_N$ are i.i.d. random vectors taking values in R^d possessing a probability density function f. The objective is to estimate $f(x)$ for some x in the interior of I, where I is a compact rectangle in R^d over which f possesses a certain degree of smoothness. Again without loss of generality (and possibly having to use a linear/affine transformation in pre-processing the data), assume that $I = [-a, a] \times [-a, a] \times \cdots [-a, a]$ for some $a > 0$. Outside the rectangle I, f and its derivatives might have discontinuities, and f might even be zero (bringing us to the set-up of bounded random variables as in the previous subsection). We now define the following condition which is related to our previous Condition C_1.

Condition $C^*[r]$: For some positive integer r, f has r bounded, continuous derivatives over the closed region I.

We again define the flat-top kernel smoothed estimator of $f(x)$ by equation (30.2.1). It is intuitive that, if the tails of $\Omega_C(x)$ were negligible, the influence on $\hat{f}(x)$ of some X_k observations that are far away from x (and may even correspond to a region where f is not smooth) would be insignificant; this is indeed a true observation, and leads to the following result. To state it, we define the smaller rectangle $J = [-b, b] \times [-b, b] \times \cdots \times [-b, b]$ where $0 < b < a$; b should be thought to be close to a such that the point x of interest will also belong to J.

THEOREM 30.3.2
Assume that

$$\Omega_C(x) = O((1 + \max_i |x_i|)^{-q}), \qquad\qquad (30.3.1)$$

for some real number $q > d$. Let $p = Prob\{X_1 \in I\} = \int_I f(x)dx$ be strictly in $(0, 1)$.
Under Condition $C^*[r + 1]$, and letting $h \sim AN^{-1/(2r^*+d)}$, for some constant $A > 0$, it follows that

$$\sup_{x \in J} MSE(\hat{f}(x)) = O(N^{-2r^*/(2r^*+d)})$$

as $N \to \infty$, where $r^* = \min(r, q - d)$.

PROOF The order of $MSE(\hat{f}(x))$ again depends on the bias of $\hat{f}(x)$ since $Var(\hat{f}(x))$ is of order $1/(h^d N)$ as before. To estimate the bias, consider the following argument.

Let $\bar{f}$ be a probability density that has (at least) $r+1$ bounded derivatives over the whole of R^d, and such that $\bar{f}(x) = f(x)$ for all $x \in I$; this extension of f over the whole of R^d can be done in many ways—see e.g. Stein (1970). Re-order the X_is in such a way that $X_1, \ldots, X_K$ are in I, whereas $X_{K+1}, \ldots, X_N$ are in I^c, i.e., the complement of I.

Construct a new sample $Y_1, \ldots, Y_N$ with the property that $Y_i = X_i$ for $i = 1, \ldots, K$, and such that $Y_{K+1}, \ldots, Y_N$ are drawn i.i.d. from density $\bar{f}(x) 1_{I^c}(x)/(1-p)$, where $p = Prob\{X_1 \in I\} = \int_I f(x)dx$. It is apparent now that the sample $Y_1, \ldots, Y_N$ can be considered as a *bona fide* i.i.d. sample from density $\bar{f}(x)$ for $x \in R^d$.

Now recall that (together with the re-ordering) we have

$$\hat{f}(x) = \frac{1}{N} \sum_{k=1}^{N} \Lambda_C (x - X_k) = \frac{1}{N} \sum_{k=1}^{K} \Lambda_C (x - X_k) + (\frac{N-K}{N})u_1.$$

Note that, due to assumption (30.3.1), and to the fact that an observation outside I will be at a distance of at least $a - b$ (in an l_∞ sense) from the point $x \in J$, we have $u_1 = O(\frac{h^{q-d}}{(a-b)^q})$ almost surely.

Finally, observe that

$$\frac{1}{N} \sum_{k=1}^{K} \Lambda_C (x - X_k) = \frac{1}{N} \sum_{k=1}^{K} \Lambda_C (x - Y_k)$$

$$= \frac{1}{N} \sum_{k=1}^{N} \Lambda_C (x - Y_k) + (\frac{N-K}{N})u_2,$$

where again $u_2 = O(\frac{h^{q-d}}{(a-b)^q})$ almost surely.

To summarize:

$$\hat{f}(x) = \frac{1}{N} \sum_{k=1}^{N} \Lambda_C (x - Y_k) + (u_1 + u_2)(\frac{N-K}{N}). \tag{30.3.2}$$

Taking expectations in equation (30.3.2), and recalling that $E(\frac{N-K}{N}) = 1 - p$, and

$$E\left(\frac{1}{N} \sum_{k=1}^{N} \Lambda_C (x - Y_k)\right) = \bar{f}(x) + o(h^r),$$

and that $\bar{f}(x) = f(x)$ for all $x \in I$ (and thus for all $x \in J$ as well), the theorem is proven. ∎

REMARK Theorem 30.3.2 shows that, in the possible presence of discontinuities in f or its derivatives at points away from the target region I, it is important to use a flat-top kernel $\Omega_C(x)$ that is chosen to have small tails. For example, the case where $w_C(s) = 0$ for all $s \notin C$ that was considered by Parzen (1962), Davis (1977), and Ibragimov and

Hasminksii (1982) satisfies equation (30.3.1) with $q = 1$ and is *not* recommendable. By contrast, the simple kernel $\Lambda_c^{PROD}(x)$ of Politis and Romano (1999) satisfies (30.3.1) with $q = 2$ as long as $c > 1$. ∎

REMARK It is easy to construct flat-top kernels satisfying equation (30.3.1) with $q > 2$; all it takes is to make sure that $\omega_C(s)$ has a high degree of smoothness (e.g., high number of derivatives) for all s. To do this, one must pay special attention at the boundary of the region C since, inside C, $\omega_C(s)$ is infinitely differentiable with all derivatives being zero. We now give an explicit construction of a flat-top kernel satisfying (30.3.1) with an arbitrary exponent q in the case where C is the l_∞ unit ball. Let $\omega_1(s_1)$ be a member of the flat-top family in the case $d = 1$, and having the following properties: $\omega_1(s_1) = \omega_1(-s_1)$ for all $s_1 \in R$; $\omega_1(s_1) = 1$ for all $|s_1| \leq 1$; $\omega_1(s_1) = 0$ for all $|s_1| \geq c$ for some $c > 1$ ($c = 2$ is a useful practical choice); $\omega_1(s_1)$ is monotone decreasing for $1 < s_1 < c$; and $\omega_1(s_1)$ possesses n continuous derivatives for all $s_1 \in R$. Then, the d-dimensional flat-top kernel $\Omega_C(x)$ with Fourier transform equal to $\omega_C(s) = \prod_{i=1}^{d} \omega_1(s_i)$ satisfies (30.3.1) with exponent $q > n+1$. ∎

As a last remark, note that Condition $C^*[r+1]$ in Theorem 30.3.2 can be relaxed to $C^*[r]$ if $\Omega_C(x)$ is chosen to have r finite moments which in turn can be guaranteed by having $q > r + 1$ in equation (30.3.1). The following corollary should be compared to Theorem 30.2.1(a).

COROLLARY 30.3.3
Assume Condition $C^*[r]$, as well as equation (30.3.1) with some $q > r+1$ and $q \geq r + d$. Let $p = Prob\{X_1 \in I\} = \int_I f(x)dx$ be strictly in $(0,1)$. Letting $h \sim AN^{-1/(2r+d)}$, for some constant $A > 0$, it follows that

$$\sup_{x \in J} MSE(\hat{f}(x)) = O(N^{-2r/(2r+d)})$$

as $N \to \infty$.

Acknowledgement This research was partially supported by NSF grant DMS 97-03964. Many thanks are due to Prof. George Kyriazis of the University of Cyprus for many helpful discussions, and to Prof. E. Masry of the University of California at San Diego for pointing out a mistake in Politis and Romano (1996).

References

1. Bartlett, M. S. (1963). Statistical estimation of density functions. *Sankhyā, Series A* **25**, 245–254.

2. Butzer, P. and Nessel, R. (1971). *Fourier Analysis and Approximation.* Academic Press, New York.

3. Cacoullos, T. (1966). Estimation of a multivariate density. *Annals of the Institute of Statistical Mathematics* **18**, 178–189.

4. Davis, K. B. (1977). Mean integrated square error properties of density estimates. *Annals of Statistics* **5**, 530–535.

5. Devroye, L. (1987). *A Course in Density Estimation.* Birkhäuser, Boston.

6. Devroye, L. (1992). A note on the usefulness of superkernels in density estimation. *Annals of Statistics* **20**, 2037–2056.

7. Gajek, L. (1986). On improving density estimators which are not bona fide functions. *Annals of Statistics* **14**, 1612–1618.

8. Gasser, T., Müller, H. G., and Mammitzsch, V. (1985). Kernels for nonparametric curve estimation. *Journal of the Royal Statistical Society, Series B* **47**, 238–252.

9. Granovsky, B. L. and Müller, H. G. (1991). Optimal kernel methods: A unifying variational principle. *International Statistical Review* **59**, 373–388.

10. Györfi, L., Härdle, W., Sarda, P., and Vieu, P. (1989). *Nonparametric Curve Estimation from Time Series.* Lecture Notes in Statistics No. 60, Springer-Verlag, New York.

11. Hall, P. and Murison, R. D. (1992). Correcting the negativity of high-order kernel density estimators. *Journal of Multivariate Analysis* **47**, 103–122.

12. Ibragimov, I. A. and Hasminksii, R. Z. (1982). Estimation of distribution density belonging to a class of entire functions. *Theory of Probability and Its Applications* **27**, 551–562.

13. Jones, M. C. (1995). On higher order kernels. *Journal of Nonparametric Statistics* **5**, 215–221.

14. Jones, M. C. and Foster, P. J. (1993). Generalized jackknifing and higher order kernels. *Journal of Nonparametric Statistics* **3**, 81–94.

15. Katznelson, Y. (1968). *An Introduction to Harmonic Analysis.* Dover, New York.

16. Marron, J. S. (1994). Visual understanding of higher order kernels. *Journal Computational and Graphical Statistics* **3**, 447–458.

17. Marron, J. S. and Wand, M. P. (1992). Exact mean integrated squared error. *Annals of Statistics* **20**, 712–736.

18. Müller, H. G. (1988). *Nonparametric Regression Analysis of Longitudinal Data.* Springer-Verlag, Berlin.

19. Nadaraya, E. A. (1989). *Nonparametric Estimation of Probability Densities and Regression Curves.* Kluwer Academic Publishers, Dordrecht, The Netherlands.

20. Parzen, E. (1962). On estimation of a probability density function and its mode. *Annals of Mathematical Statistics* **33**, 1065–1076.

21. Politis, D. N. and Romano, J. P. (1993). On a family of smoothing kernels of infinite order. In *Computing Science and Statistics, Proceedings of the 25th Symposium on the Interface*, San Diego, California, April 14-17, 1993 (Eds., M. Tarter and M. Lock), pp. 141–145, The Interface Foundation of North America.

22. Politis, D. N. and Romano, J. P. (1995). Bias-corrected nonparametric spectral estimation. *Journal of Time Series Analysis* **16**, 67–104.

23. Politis, D. N. and Romano, J. P. (1996). On flat-top kernel spectral density estimators for homogeneous random fields. *Journal of Statistical Planning and Inference* **51**, 41–53.

24. Politis, D. N. and Romano, J. P. (1999). Multivariate density estimation with general flat-top kernels of infinite order. *Journal of Multivariate Analysis* **68**, 1–25.

25. Priestley, M. B. (1981), *Spectral Analysis and Time Series.* Academic Press, New York.

26. Rao, C. R. (1973). *Linear Statistical Inference and its Applications*, Second Edition. John Wiley & Sons, New York.

27. Rosenblatt, M. (1991). *Stochastic Curve Estimation.* NSF-CBMS Regional Conference Series Vol. 3, Institute of Mathematical Statistics, Hayward, CA.

28. Scott, D. W. (1992). *Multivariate Density Estimation: Theory, Practice, and Visualization.* John Wiley & Sons, New York.

29. Silverman, B. W. (1986). *Density Estimation for Statistics and Data Analysis.* Chapman and Hall, London.

30. Stein, E. M. (1970). *Singular Integrals and Differentiability Properties of Functions.* Princeton University Press, Princeton, New Jersey.

31. Stein, E. M. and Weiss, W. (1971). *Introduction to Fourier Analysis on Euclidean Spaces.* Princeton University Press, Princeton, New Jersey.

32. Wand, M. P. and Jones, M. C. (1993). Comparison of smoothing parameterizations in bivariate kernel density estimation. *Journal of the American Statistical Association* **88**, 520–528.

33. Watson, G. S. and Leadbetter, M. R. (1963). On the estimation of the probability density I. *Annals of Mathematical Statistics* **33**, 480–491.

31

Multipolishing Large Two-Way Tables

Kaye Basford, Stephan Morgenthaler and John W. Tukey
University of Queensland, Brisbane, Australia
Ecole Polytechnique Fédérale de Lausanne, Lausanne, Switzerland
Princeton University, Princeton, NJ

ABSTRACT This paper discusses tools for the more detailed analysis of large two-way tables. Because of its simple structure, the additive fit is often inappropriate for large tables. To go further, multipolishing, where the rows are fitted as function of a vector representing the columns, provides a possible approach. The paper compares this technique with ideas based on the singular value decomposition. The methods are illustrated with a plant breeding example.

Keywords and phrases Analysis of variance, nonadditivity, interactions, graphical displays of two-way tables, singular value decomposition, plant breeding

31.1 Introduction

Plant breeding experiments involve the assessment of the quality of genetically different lines. If a single attribute is measured in a variety of environments, a genotype by environment ($G \times E$) table is determined. The standard analysis of $G \times E$ tables decomposes the observed table (y_{ge}) ($g = 1, \ldots, G$, $e = 1, \ldots, E$) into subtables of simple and interpretable structure. Of course, this is but one of a multitude of classes of examples involving two-way tables. When the table is large, such as 30×40 or even 50×200, fitting the additive decomposition $m + u_g + v_e$, using one common effect, a vector $\boldsymbol{u}$ of individual genotype effects, and a vector $\boldsymbol{v}$ of individual environment effects will typically not satisfy the user. This class of fits, which describes the eth column $(y_{1e}, \ldots, y_{Ge})$ as

$$(m + v_e) + \boldsymbol{u}, \qquad (31.1.1)$$

is not rich enough to describe adequately the detailed aspects of the behavior of the response, which can be made visible because of the large amount of data. Note that we used boldfaced letters to denote vectors. This will be the convention throughout the paper. Adding the whole of the usual interaction to (31.1.1) is not helpful and would in any case not result in an interpretable description. We should ask ourselves, in what ways we can add to the simple additive model to arrive at a decomposition that provides more insights in typical cases. The types of decompositions we are led to use have been described by, among others, Tukey (1962, Section VII), Gollob (1968), Hoaglin, Mosteller and Tukey (1985, Ch. 3) and have been extensively discussed by Mandel (1995). Suppose the genotype factor is associated with the numerical row variable $(u_1, \ldots, u_G)$ and the environment factor is similarly associated with the column variable $(v_1, \ldots, v_E)$. A variable such as $\boldsymbol{u}$ comes in handy in the description of the columns of the table. We might, for example, consider a description of the eth column $(y_{1e}, \ldots, y_{Ge})$ in terms of a linear function of the form

$$p_e + q_e \, \boldsymbol{u} \, , \qquad (31.1.2)$$

and similarly for the rows of the table. Examples with such external variables occur frequently in physics and chemistry (and economics and demographics). Suppose, for example, that the rows correspond to differing amounts of a catalyst. A simple (conditional) model for each column vector of responses might then postulate a linear dependence on these amounts. In some other classes of examples, there is no useful external variable, in which case we propose to define one internally, based on the observations. Various reasonable choices present themselves, for example, the row and column effects from an additive fit to the table or the two principal eigenvectors from a singular value decomposition of the table.

In this paper we explore the use of such models and describe several possible plots for visualizing the resulting fit, reserving the use of one of more (histograms) rootograms of residuals at one or more stages of the fit for a later paper.

31.2   Bilinear Multipolishes

The additive fit to a two-way table (y_{ge}) with G rows and E columns is of the form $m + u_g + v_e$, where u_g are the row effects and v_e denotes the column effects. Finding least-squares estimates of m, u_g and v_e can be achieved by **polishing** or **row-and-column sweeping**. This is an algorithm one form of which works as follows:

[P1] Compute the row means and sweep them from the table:
$$\tilde{u}_g = \sum_{e=1}^{E} y_{ge}/E$$
$$\tilde{r}_{ge} = y_{ge} - \tilde{u}_g$$

[P2] Compute the column means and sweep them from the table, and from the vector of preliminary effects created in step [P1]:
$$m = \sum_{g=1}^{G} \tilde{u}_g/G$$
$$u_g = \tilde{u}_g - m$$
$$v_e = \sum_{g=1}^{G} \tilde{r}_{ge}/G$$
$$r_{ge} = \tilde{r}_{ge} - v_e.$$

This method is inspired by Eq. (31.1.1), which shows that for centered $\boldsymbol{u}$, that is $\bar{u} = 0$, we can find $m + v_e$ by taking the mean of the e-th column thus formed. The above algorithm will thus produce effects satisfying $\sum_g u_g = \sum_e v_e = 0$.

In a **multipolish**, it is not simply the mean that is being swept from a row or a column, but a more complex function involving an auxiliary variable. Let $\boldsymbol{u} = (u_g)$ be a centered variable for the rows and $\boldsymbol{v} = (v_e)$ a centered variable for the columns. Centering means that $\sum_g u_g = \sum_e v_e = 0$. Inspired by Eq. (31.1.2) one can then modify the above algorithm in the following way:

[**MP1**] Fit a linear function in $\boldsymbol{v}$ to each row and sweep it from the table:
$$\tilde{a}_g = \sum_{e=1}^{E} y_{ge}/E$$
$$\tilde{b}_g = \sum_{e=1}^{E} y_{ge}v_e / \sum_{e=1}^{E} v_e^2$$
$$\tilde{r}_{ge} = y_{ge} - \tilde{a}_g - \tilde{b}_g v_e;$$

[**MP2**] Sweep a linear function in $\boldsymbol{u}$ from each column of the table and from the vectors of preliminary effects created in step [MP1]:
$$m_{11} = \sum_{g=1}^{G} \tilde{a}_g/G$$
$$m_{21} = \sum_{g=1}^{G} \tilde{a}_g u_g / \sum_{g=1}^{G} u_g^2$$
$$a_g = \tilde{a}_g - m_{11} - m_{21} u_g$$
$$m_{12} = \sum_{g=1}^{G} \tilde{b}_g/G$$
$$m_{22} = \sum_{g=1}^{G} \tilde{b}_g u_g / \sum_{g=1}^{G} u_g^2$$
$$b_g = \tilde{b}_g - m_{12} - m_{22} u_g$$
$$p_e = \sum_{g=1}^{G} \tilde{r}_{ge}/G$$
$$q_e = \sum_{g=1}^{G} \tilde{r}_{ge} u_g / \sum_{g=1}^{G} u_g^2$$
$$r_{ge} = \tilde{r}_{ge} - p_e - q_e u_g$$

Of course, this algorithm produces effects that are not only centered, but are also orthogonal to the auxiliary variables. To express these conditions, Eq. (31.1.2) must be written as

$$(m_{11} + m_{12}v_e + p_e) + (m_{21} + m_{22}v_e + q_e)\, \boldsymbol{u} ,$$

where the symbols correspond to the ones used in the multipolish algorithm. The cycle [MP1], [MP2] also adds two terms for the effects swept from the rows, which shows that the fit to the e-th column is in fact equal to

$$(m_{11} + m_{12}v_e + p_e) + (m_{21} + m_{22}v_e + q_e)\, \boldsymbol{u} + \boldsymbol{a} + v_e\boldsymbol{b}. \qquad (31.2.1)$$

31.2.1   Choosing the Auxiliary Variables Equal to the Effects of the Additive Fit

As we have indicated, it is quite natural to use [P1], [P2] in order to create two variables $\boldsymbol{u}$ and $\boldsymbol{v}$, and then to apply [MP1], [MP2]. To describe this algorithm, it is helpful to put it into a matrix notation. Let $Y = (y_{ge})$ be the data matrix of dimension $G \times E$. The additive fit, obtained by [P1] and [P2], satisfies

- $m = \boldsymbol{1_G}^T Y \boldsymbol{1_E}/(EG)$
- $\boldsymbol{u} = Y\boldsymbol{1_E}/E - m\boldsymbol{1_G}$
- $\boldsymbol{v} = Y^T\boldsymbol{1_G}/G - m\boldsymbol{1_E}$

where $\boldsymbol{1_n}$ denotes a vector of length n whose components are all equal to 1. If we use these vectors as auxiliary variables, we obtain the following fit.

PROPOSITION 31.2.1
The bilinear multipolish fit to the table $Y \in \mathbb{R}^{G \times E}$ with the effects of the additive fit as auxiliary variables is equal to

$$m\boldsymbol{1_G}\boldsymbol{1_E}^T + \boldsymbol{u}\boldsymbol{1_E}^T + \boldsymbol{1_G}\boldsymbol{v}^T + m_{22}\boldsymbol{u}\boldsymbol{v}^T + \boldsymbol{b}\boldsymbol{v}^T + \boldsymbol{u}\boldsymbol{q}^T , \qquad (31.2.2)$$

where

- $m = \boldsymbol{1_G}^T Y \boldsymbol{1_E}/(EG)$
- $m_{22} = \boldsymbol{u}^T Y \boldsymbol{v}/((\boldsymbol{u}^T\boldsymbol{u})(\boldsymbol{v}^T\boldsymbol{v}))$
- $\boldsymbol{b} = Y\boldsymbol{v}/\boldsymbol{v}^T\boldsymbol{v} - \boldsymbol{1_G} - m_{22}\boldsymbol{u}$
- $\boldsymbol{q} = Y^T\boldsymbol{u}/\boldsymbol{u}^T\boldsymbol{u} - \boldsymbol{1_E} - m_{22}\boldsymbol{v}.$

PROOF The step [MP1] leads to

- $\tilde{\boldsymbol{a}} = Y\boldsymbol{1_E}/E = \boldsymbol{u} + m\boldsymbol{1_G}$
- $\tilde{\boldsymbol{b}} = Y\boldsymbol{v}/\boldsymbol{v}^T\boldsymbol{v}$
- $\tilde{R} = (\tilde{r}_{ge}) = Y - \tilde{\boldsymbol{a}}\boldsymbol{1_E}^T - \tilde{\boldsymbol{b}}\boldsymbol{v}^T$

The sweeps in [MP2] will then result in

- $m_{11} = m$
- $m_{21} = 1$
- $\boldsymbol{a} = \boldsymbol{0}_G$
- $m_{12} = \boldsymbol{1}_G{}^T Y \boldsymbol{v}/(G\boldsymbol{v}^T\boldsymbol{v}) = G(\boldsymbol{v}^T + m\boldsymbol{1}_E{}^T)\boldsymbol{v}/(G\boldsymbol{v}^T\boldsymbol{v}) = 1$
- $m_{22} = \boldsymbol{u}^T Y \boldsymbol{v}/((\boldsymbol{u}^T\boldsymbol{u})(\boldsymbol{v}^T\boldsymbol{v}))$
- $\boldsymbol{b} = Y\boldsymbol{v}/\boldsymbol{v}^T\boldsymbol{v} - \boldsymbol{1}_G - m_{22}\boldsymbol{u}$
- $\boldsymbol{p} = \tilde{R}^T \boldsymbol{1}_G/G = Y^T \boldsymbol{1}_G/G - \boldsymbol{1}_E \tilde{\boldsymbol{a}}^T \boldsymbol{1}_G/G - \boldsymbol{v}\tilde{\boldsymbol{b}}^T \boldsymbol{1}_G/G = (\boldsymbol{v}+m\boldsymbol{1}_E) - m\boldsymbol{1}_E - \boldsymbol{v} = \boldsymbol{0}_E$
- $\boldsymbol{q} = \tilde{R}^T \boldsymbol{u}/\boldsymbol{u}^T\boldsymbol{u} = Y^T\boldsymbol{u}/\boldsymbol{u}^T\boldsymbol{u} - \boldsymbol{1}_E - m_{22}\boldsymbol{v}$

∎

The first three terms in (31.2.2) are equal to the additive fit, adding the forth term leads to Tukey's (1949) ODOFFNA, whereas the last two terms are result of fitting a straight line in the auxiliary variable to each row and column.

31.2.2 Robust Alternatives

It is easy and straightforward to create a robust polish and multipolish. Instead of using the least-square criterion when sweeping from the rows and columns, one is in fact free to use any other method. If one replaces the mean with the media in [P1] and in [P2], for example, the median polish is obtained [Tukey (1977, Ch. 11)]. There is, however, one important change in the algorithm. In general it does not converge in two steps, [P1]–[P2], but has to be interated [P1]–[P2]–[P1]–[P2]–[P1]– and so on until convergence or until the changes become small enough. Similarly, in [MP1] and [MP2], the least-squares fitting of a straight line can be replaced by a robust line such as the biweight line.

31.3 Matrix Approximations

In this section we present an alternative way of thinking about the fit obtained in (31.2.2), based on the following result about rank-one approximations.

LEMMA 31.3.1
Let $A \in \mathbb{R}^{G \times E}$ be an arbitrary matrix and let $\boldsymbol{v} \in \mathbb{R}^E$ be an arbitrary nonzero vector. The best least-squares approximation of the form $\boldsymbol{u}\boldsymbol{v}^T$

to the matrix A is obtained by choosing

$$\boldsymbol{u} = A\boldsymbol{v}/\boldsymbol{v}^T\boldsymbol{v} \propto A\boldsymbol{v}\,. \qquad (31.3.1)$$

The corresponding approximation is

$$A \approx \frac{A\boldsymbol{v}\boldsymbol{v}^T}{\boldsymbol{v}^T\boldsymbol{v}} \qquad (31.3.2)$$

and the sum of the squared elements of the difference between these matrices is $\mathrm{tr}(A^T A) - \boldsymbol{v}^T A^T A\boldsymbol{v}/\boldsymbol{v}^T\boldsymbol{v}$.

PROOF The sum of the squared entries of the difference $A - \boldsymbol{u}\boldsymbol{v}^T$ can be computed as

$$\mathrm{tr}\Big(\big(A - \boldsymbol{u}\boldsymbol{v}^T\big)^T \big(A - \boldsymbol{u}\boldsymbol{v}^T\big)\Big),$$

where tr denotes the trace. Using elementary properties of the trace, we can express this sum of squares as $\mathrm{tr}\big(A^T A\big) - 2\boldsymbol{u}^T A\boldsymbol{v} + (\boldsymbol{u}^T\boldsymbol{u})(\boldsymbol{v}^T\boldsymbol{v})$. This quadratic form in $\boldsymbol{u}$ is minimized when

$$2\boldsymbol{u}(\boldsymbol{v}^T\boldsymbol{v}) - 2A\boldsymbol{v} = \boldsymbol{0}_G\,.$$

∎

We could, of course, just as well have started with a given vector $\boldsymbol{u} \in \mathbb{R}^G$ and found the least-squares approximation based on $\boldsymbol{u}$ alone,

$$A \approx \frac{\boldsymbol{u}\boldsymbol{u}^T A}{\boldsymbol{u}^T\boldsymbol{u}}\,. \qquad (31.3.3)$$

One can also iterate, for example, starting with a vector $\boldsymbol{v_0}$. The first approximation is then based on $\boldsymbol{u_0} = A\boldsymbol{v_0}/\boldsymbol{v_0}^T\boldsymbol{v_0}$. If we use this vector to compute an approximation, it will make use of $\boldsymbol{v_1} = A^T\boldsymbol{u_0}/\boldsymbol{u_0}^T\boldsymbol{u_0}$. If iterating these two matrix multiplications converges, the limiting vectors $\boldsymbol{u_\infty}$ and $\boldsymbol{v_\infty}$ must be fixed points of the iteration. This implies

$$\boldsymbol{u_{\infty+1}} = A\boldsymbol{v_\infty}/\boldsymbol{v_\infty}^T\boldsymbol{v_\infty}$$

$$\boldsymbol{v_{\infty+1}} = A^T\boldsymbol{u_{\infty+1}}/\boldsymbol{u_{\infty+1}}^T\boldsymbol{u_{\infty+1}}$$

$$= (A^T A\boldsymbol{v_\infty})(\boldsymbol{v_\infty}^T\boldsymbol{v_\infty})/\boldsymbol{v_\infty}^T A^T A\boldsymbol{v_\infty}$$

$$= \boldsymbol{v_\infty}$$

Thus

$$(A^T A)\frac{\boldsymbol{v_\infty}}{(\boldsymbol{v_\infty}^T\boldsymbol{v_\infty})^{1/2}} = \frac{(\boldsymbol{v_\infty}^T A^T A\boldsymbol{v_\infty})}{(\boldsymbol{v_\infty}^T\boldsymbol{v_\infty})}\frac{\boldsymbol{v_\infty}}{(\boldsymbol{v_\infty}^T\boldsymbol{v_\infty})^{1/2}}\,.$$

The resulting approximation is of the form

$$A \approx d \, \frac{u_\infty}{\left(u_\infty{}^\mathrm{T} u_\infty\right)^{1/2}} \, \frac{v_\infty{}^\mathrm{T}}{\left(v_\infty{}^\mathrm{T} v_\infty\right)^{1/2}} \, ,$$

with $d = \left(v_\infty{}^T A^T A v_\infty\right)^{1/2}$. This is the well-known rank-one singular value approximation of the matrix A, with d being the largest singular value. This shows that the approximation given in the above lemma can be interpreted as a one-step version of the rank-one singular value approximation, starting from the vector v.

31.3.1 Approximations of Two-Way Tables

By applying the above lemma one can derive the fit described in (31.2.2). The first stage consists in centering the matrix,

$$Y = m \, \mathbf{1}_G \mathbf{1}_E{}^\mathrm{T} + \left(Y - m \, \mathbf{1}_G \mathbf{1}_E{}^\mathrm{T}\right).$$

Next, we apply the rank-one approximation based on the vector $v_0 = \mathbf{1}_E$ to the centered matrix. Eq. (31.3.1) gives

$$u = \left(Y - m \, \mathbf{1}_G \mathbf{1}_E{}^\mathrm{T}\right) \mathbf{1}_E / \mathbf{1}_E{}^T \mathbf{1}_E = Y \mathbf{1}_E / E - m \, \mathbf{1}_G,$$

which is equal to the vector of row effects (see Section 31.2.1). The approximation (31.3.2) is $u \mathbf{1}_E{}^T$. If we add this to the previous fit, we have

$$Y = m \, \mathbf{1}_G \mathbf{1}_E{}^\mathrm{T} + u \mathbf{1}_E{}^\mathrm{T} + \left(Y - m \, \mathbf{1}_G \mathbf{1}_E{}^\mathrm{T} - u \mathbf{1}_E{}^\mathrm{T}\right).$$

Next, one applies the rank-one approximation based on $u_0 = \mathbf{1}_G$ to the new remainder matrix. The lemma shows that

$$v = \left(Y - m \, \mathbf{1}_G \mathbf{1}_E{}^\mathrm{T} - u \mathbf{1}_E{}^\mathrm{T}\right)^T \mathbf{1}_G / G = Y \mathbf{1}_G / G - m \mathbf{1}_E$$

leads to an optimal approximation $\mathbf{1}_G v^T$. Adding this term into the fit gets us to the usual additive fit

$$Y = m \, \mathbf{1}_G \mathbf{1}_E{}^\mathrm{T} + u \mathbf{1}_E{}^\mathrm{T} + \mathbf{1}_G v^\mathrm{T} + \left(Y - m \, \mathbf{1}_G \mathbf{1}_E{}^\mathrm{T} - u \mathbf{1}_E{}^\mathrm{T} - \mathbf{1}_G v^\mathrm{T}\right).$$

At this stage, several directions in which we might proceed further present themselves. We could iterate the one-step procedure we have used up to now, which would lead us to a one-rank singular value approximation of the centered data matrix. Or, we can proceed with the method described in our lemma with the aim of extracting further terms for the model from the residual matrix. Of course, we need to change our starting vectors, but we have natural candidates in $\boldsymbol{u}$ and $\boldsymbol{v}$. Starting with the vector $\boldsymbol{v}$, we would then like to approximate the residuals from the additive fit. The corresponding term in the fit is equal to $\left(\mathrm{Y} - m\,\boldsymbol{1_G}\boldsymbol{1_E}^{\mathrm{T}} - \boldsymbol{u}\boldsymbol{1_E}^{\mathrm{T}} - \boldsymbol{1_G}\boldsymbol{v}^{\mathrm{T}}\right)\boldsymbol{v}\boldsymbol{v}^{\mathrm{T}}/\boldsymbol{v}^{\mathrm{T}}\boldsymbol{v} = \left(\mathrm{Y}\boldsymbol{v} - \boldsymbol{1_G}\boldsymbol{v}^{\mathrm{T}}\boldsymbol{v}\right)\boldsymbol{v}^{\mathrm{T}}/\boldsymbol{v}^{\mathrm{T}}\boldsymbol{v} = \mathrm{Y}\boldsymbol{v}\boldsymbol{v}^{\mathrm{T}}/\boldsymbol{v}^{\mathrm{T}}\boldsymbol{v} - \boldsymbol{1_G}\boldsymbol{v}^{\mathrm{T}}$, where we made us of the fact that $\boldsymbol{1_E}^{\mathrm{T}}\boldsymbol{v} = 0$. After this step, the fit has become

$$Y = m\,\boldsymbol{1_G}\boldsymbol{1_E}^{\mathrm{T}} + \boldsymbol{u}\boldsymbol{1_E}^{\mathrm{T}} + \mathrm{Y}\boldsymbol{v}\boldsymbol{v}^{\mathrm{T}}/\boldsymbol{v}^{\mathrm{T}}\boldsymbol{v} + \text{remainder matrix}.$$

Next, we compute the rank-one approximation based on $\boldsymbol{u}$ of the new remainder matrix. This term in the approximation is equal to

$$\boldsymbol{u}\boldsymbol{u}^{\mathrm{T}}\left(\mathrm{Y} - m\,\boldsymbol{1_G}\boldsymbol{1_E}^{\mathrm{T}} - \boldsymbol{u}\boldsymbol{1_E}^{\mathrm{T}} - \mathrm{Y}\boldsymbol{v}\boldsymbol{v}^{\mathrm{T}}/\boldsymbol{v}^{\mathrm{T}}\boldsymbol{v}\right)/\boldsymbol{u}^{\mathrm{T}}\boldsymbol{u}$$
$$= \boldsymbol{u}\boldsymbol{u}^{\mathrm{T}}\mathrm{Y}/\boldsymbol{u}^{\mathrm{T}}\boldsymbol{u} - \boldsymbol{u}\boldsymbol{1_E}^{\mathrm{T}} - \boldsymbol{u}\boldsymbol{u}^{\mathrm{T}}\mathrm{Y}\boldsymbol{v}\boldsymbol{v}^{\mathrm{T}}/((\boldsymbol{v}^{\mathrm{T}}\boldsymbol{v})(\boldsymbol{u}^{\mathrm{T}}\boldsymbol{u})).$$

At this stage, the fit is

$$Y = m\,\boldsymbol{1_G}\boldsymbol{1_E}^{\mathrm{T}} + \boldsymbol{u}\boldsymbol{u}^{\mathrm{T}}\mathrm{Y}/\boldsymbol{u}^{\mathrm{T}}\boldsymbol{u} + \mathrm{Y}\boldsymbol{v}\boldsymbol{v}^{\mathrm{T}}/\boldsymbol{v}^{\mathrm{T}}\boldsymbol{v}$$
$$- \boldsymbol{u}\boldsymbol{u}^{\mathrm{T}}\mathrm{Y}\boldsymbol{v}\boldsymbol{v}^{\mathrm{T}}/((\boldsymbol{v}^{\mathrm{T}}\boldsymbol{v})(\boldsymbol{u}^{\mathrm{T}}\boldsymbol{u})) + \mathrm{R}. \tag{31.3.4}$$

Since the fit in (31.3.4) is equal to (31.2.2), multipolishing (by least squares) with the vectors of effects $\boldsymbol{u}$ and $\boldsymbol{v}$ is equal to the one-step least-squares approximation of our lemma. Of course it would again be possible to iterate and to replace the terms we added to the additive model by the rank-one singular value approximation.

31.4   Displays

If we wish to display the fitted values (31.3.4), we can decompose them into rows. The fitted gth row is equal to

$$m\,\boldsymbol{1_E} + \frac{u_g \mathrm{Y}^{\mathrm{T}}\boldsymbol{u}}{\boldsymbol{u}^{\mathrm{T}}\boldsymbol{u}} + \boldsymbol{v}\left(\frac{\sum_e v_e\, y_{ge}}{\boldsymbol{v}^{\mathrm{T}}\boldsymbol{v}} - u_g\frac{\boldsymbol{u}^{\mathrm{T}}\mathrm{Y}\boldsymbol{v}}{(\boldsymbol{v}^{\mathrm{T}}\boldsymbol{v})(\boldsymbol{u}^{\mathrm{T}}\boldsymbol{u})}\right),$$

which we can plot against the vector $\boldsymbol{v}$. Doing this for $1 \leq g \leq G$ results in a set of straight lines if the additional condition $\mathrm{Y}^{\mathrm{T}}\boldsymbol{u} \propto \boldsymbol{v}$ is satisfied. Note that this holds for $\boldsymbol{u_\infty}$ and $\boldsymbol{v_\infty}$. However, replacing $\boldsymbol{u}$ and $\boldsymbol{v}$ by these two vectors would not lead to the analogue of (31.3.4). In general, this additional condition will not be true and the curves will be jagged lines instead of straight ones. One could separate the second from the other terms. In this case, what is left over is a family of straight lines intersecting at the origin and a family of jagged lines. Looking from the multipolish point of view (31.2.2) at these same terms, one notes immediately, that the jagged part is due to the vectors b and q. So, an alternative way of plotting the rows would show b as a function of u. In the resulting scatter plot, one would be able to spot unusual genotypes (rows).

If one wants to study the dependence of the the fitted eth column against u, one is led to

$$m\, 1_G + \frac{v_g Y v}{v^T v} + u\left(\frac{\sum_g u_g\, y_{ge}}{u^T u} - v_g \frac{u^T Y u}{(v^T v)(u^T u)}\right).$$

31.5 Example

The G×E data set under investigation comes from the International Maize and Wheat Improvement Centre (CIMMYT), an internationally funded, non-profit research organization aimed at improving the productivity, profitability, and sustainability of maize and wheat farmers in low-income countries. CIMMYT disseminates most of its elite germplasm through a system of international nurseries, where the term "nursery" means a collection of wheat lines assembled to meet a breeding objective (Fox, 1996). One of the major CIMMYT nurseries was the International Spring Wheat Yield Nursery (ISWYN), with the first, $ISWYN_1$, distributed in 1964 and the last, $ISWYN_{30}$, distributed in 1994 to co-operators around the world as pre-packaged seed, ready for sowing in an international multi-environment trial.

The one chosen to analyse here was $ISWYN_{11}$. It contains the yield (in tonnes per hectare) of 50 genotypes grown in each of 65 environments. The 50 genotypes comprised 49 entries that were chosen by the CIM-MYT breeders (from their lines and nominations from different parts of the world) and a local check. The 65 environments comprised various locations around the world, with co-operators sometimes using the same location under different conditions or time of planting (and hence labeling it as a different environment). Over the full thirty ISWYNs, a total of 605 lines (generally in subsets of 50) were tested in 407 environments (of which 193 were used in three or more ISWYNs).

CIMMYT orientates breeding on the concept of mega-environments, defined as broad (not necessarily continuous and frequently transconti-nental) areas, characterised by similar biotic and abiotic stresses, crop-

TABLE 31.1
This table shows part of a 50×65 genotype by environment dataset. The rightmost column and the last row show the additive effects as fitted by least squares

	E1	E2	E3	E4	E5	E6	E7	E8
G1	5.1	1.3	4.0	5.5	3.9	2.1	2.5	3.0
G2	3.1	0.7	3.5	5.8	4.8	2.4	2.2	2.9
G3	3.5	1.6	3.8	3.7	4.4	2.2	2.0	2.3
G4	4.6	1.5	3.4	5.2	2.8	2.2	2.8	2.4
G5	5.9	1.3	4.1	5.7	3.3	1.0	2.5	3.0
G6	4.7	1.3	3.3	6.3	5.3	4.1	2.4	2.4
G7	5.0	0.9	2.1	5.7	2.4	3.0	2.3	2.6
G8	4.9	1.1	4.2	6.5	2.5	3.1	2.2	2.4
G9	4.6	1.0	4.2	6.6	3.0	1.2	3.0	3.1
G10	5.4	1.1	3.1	6.8	3.4	2.4	2.3	2.6
G11	4.6	1.5	4.3	6.3	3.8	2.3	2.7	2.9
G12	6.0	1.6	4.1	6.3	5.5	3.1	2.8	3.1
G13	5.6	1.3	2.2	6.5	2.8	2.8	2.4	2.7
G14	4.6	1.5	3.4	4.1	2.9	2.0	2.1	2.4
G15	4.3	0.9	3.5	5.1	1.9	2.7	1.8	2.2
G16	5.7	1.5	3.5	6.1	4.0	4.0	2.6	2.2
G17	6.4	1.4	3.7	5.4	5.2	2.2	2.4	3.2
E effect	1.11	−2.55	−0.23	1.69	−0.17	−1.09	−1.40	−1.26

	E9	E10	E11	E12	E13	E14	E15	G effect
G1	3.0	4.7	6.4	4.8	5.2	4.1	3.8	0.15
G2	2.5	3.6	6.4	4.6	4.9	5.5	2.9	−0.47
G3	2.6	4.1	6.6	4.9	5.2	4.1	3.5	−0.15
G4	3.9	3.6	6.2	4.4	5.0	3.4	3.1	−0.28
G5	3.7	3.8	5.8	4.9	5.1	3.7	3.8	0.06
G6	3.6	4.7	6.9	5.1	5.5	3.3	4.1	0.12
G7	4.6	4.3	7.0	5.3	5.6	3.5	4.1	0.23
G8	3.5	4.7	6.7	5.2	5.7	4.4	3.9	0.03
G9	3.1	2.0	6.1	4.9	4.9	3.3	3.9	−0.14
G10	3.9	4.6	6.3	5.6	5.7	3.9	4.2	0.04
G11	3.9	5.4	7.0	5.4	5.7	4.1	4.5	0.23
G12	3.3	4.8	6.3	5.5	5.4	4.4	5.5	0.30
G13	3.7	3.1	8.0	5.6	6.2	4.1	3.9	0.16
G14	2.7	3.3	3.4	3.9	3.3	2.4	3.1	−0.67
G15	2.9	4.6	7.3	5.6	6.2	4.2	4.5	−0.04
G16	3.2	4.6	6.9	5.2	5.5	3.7	3.8	0.28
G17	3.4	4.1	6.1	4.1	6.0	5.1	4.7	0.38
E effect	−0.69	0.46	2.62	1.27	1.72	−0.03	−0.09	3.81

ping system requirements and consumer preferences. Germplasm may accommodate major stresses through the mega-environment for which it was developed, but perhaps not all the significant secondary stresses. Hence it was expected that there would be considerably less interaction and that main effects would not be sufficient to explain the variation in the data.

The four following figures show the genotypes in dependence of the environments and vice versa. The jagged part of the curves is quite pronounced for some genotypes. Apart from a few exceptions, the plot of the genotypes is quite encouraging in the sense that the good ones are relatively good across all environments. Plotting the environments as a function of genotypes would show flat lines, if none of the genotypes were well adapted to some class of environments. This is largely the case in this example, but note that some of lines rise as one goes across the genotypes.

Fig. 31.5 shows the individual effects of the genotypes. On the horizontal axis are the effects from an additive model, whereas on the vertical axis are the components of the vector b, that is the individual slopes, when regression the responses of one genotype one the environments. The genotypes G19, G28, and G45 are highlighted. The first one has the smallest u_g and also a very small b_g. The others have values of u_g slightly above average and are opposites with regard to b_g. Fig. 31.6 shows that this has the expected effects, namely a slower rise of the fitted response for G45 and a faster one for G28. The genotype G19 is almost everywhere worst.

31.6 Concluding Remarks

We have discussed a special case of a method for fitting large two-way tables. With relatively few degrees of freedom, namely $1+(G-2)+(E-2)$ additional parameters, it allows the user to explain important parts of the interaction. A big advantage of the proposed method is the relative ease of interpretation. We have seen that one can display the results and use the fitted effects to identify remarkable environments or genotypes.

Acknowledgements. The research by Stephen Morgenthaler has been supported by a grant from the Swiss National Science Foundation.

References

1. Fox, P. N. (1996). *The CIMMYT Wheat Program's International Multi-environment Trials, Plant Adaptation and Crop Improvement* (Eds., M. Cooper and G. L. Hammer), pp. 139–164. CAB International, Oxford, England.

2. Gollob, H. F. (1968). A statistical model which combines features of factor analytic and analysis of variance techniques. *Psychometrika* **33**, 73–116.

3. Hoaglin, D. C., Mosteller, F. and Tukey, J. W. (1985). *Exploring Data Tables, Trends, and Shapes.* John Wiley & Sons, New York.

4. Mandel, J. (1995). *Analysis of Two-Way Layouts.* Chapman and Hall, London.

5. Tukey, J. W. (1949). One degree of freedom for non-additivity. *Biometrics* **5**, 232–242.

6. Tukey, J. W. (1962). The future of data analysis. *Annals of Mathematical Statistics* **33**, 1–67.

7. Tukey, J. W. (1977). *Exploratory Data Analysis.* Addison-Wesley, Reading, Massachussetts.

8. Tukey, J. W. (1977). *The Collected Works of John W. Tukey: Volume III, Philosophy and Principles of Data Analysis from 1949 to 1964.* Wadsworth & Brooks/Cole, Monterey, California.

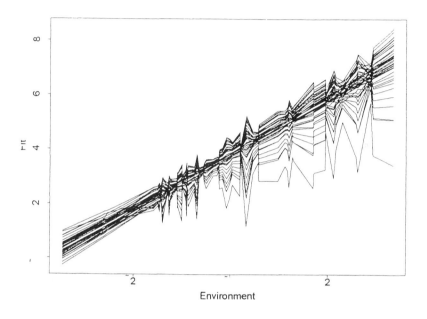

FIGURE 31.1
Each curve corresponds to a genotype and shows the fitted value
as a function of the environment effects

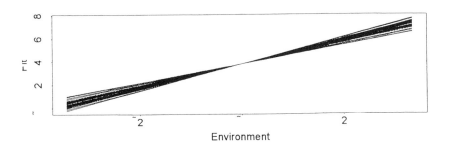

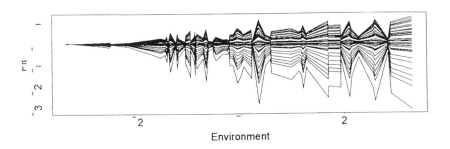

FIGURE 31.2
The top panel shows the G-linear part of the fit, whereas the
bottom contains the rest of the fit

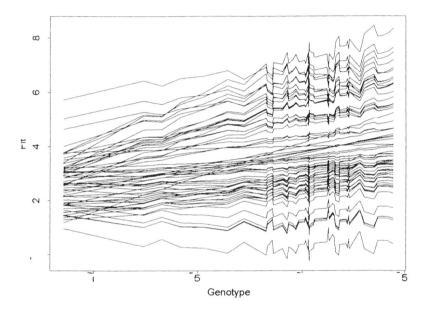

FIGURE 31.3
Each curve corresponds to an environment and shows the fitted
value as a function of the genotype effects

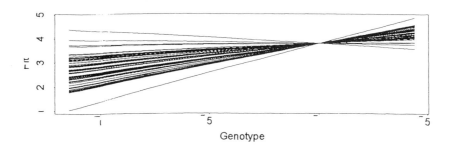

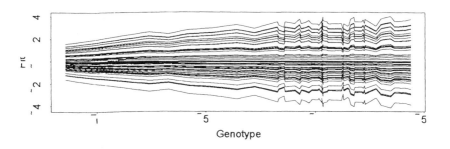

FIGURE 31.4
The top panel shows the linear part of the fit, whereas the bottom
contains the jagged part

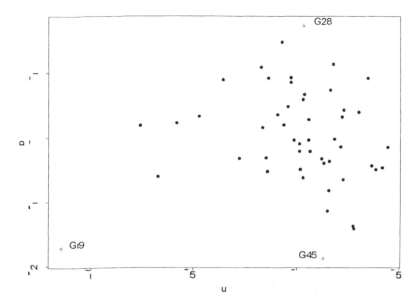

FIGURE 31.5
The individual slopes *b* are shown against the effects *u* of the
purely additive fit

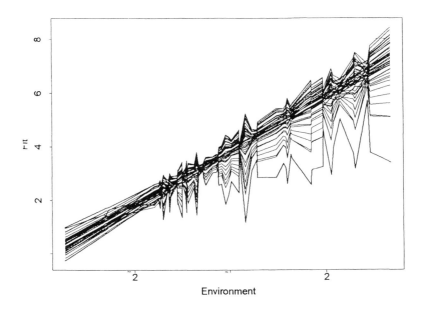

FIGURE 31.6
This is the same as Fig. 31.1, with the three selected genotypes
highlighted

32

On Distances and Measures of
Information: A Case of Diversity

Takis Papaioannou
University of Piraeus, Greece

ABSTRACT We critically review the state of the art. The concept of statistical information, measures of information, their properties and their use, impact, and role in inference are discussed. New properties are presented along with recent applications and interconnections. The existing large number of measures of information or divergence, the apparent failure of unification, i.e., coming up with the measure for all cases, settings, concepts and properties and the multitude of applications ranging from the theory of estimation to Bayesian analysis lead to the conclusion that measures of information will remain a case of diversity. The concept of information will continue to be intangible.

Keywords and phrases Measure of information, measure of divergence, maximum entropy principle, measure of dependence

---

32.1 Introduction

Distances or divergences and measures of information (m.o.i.s) are of fundamental importance in statistics. They have been in existence and use for many years, but still their role in statistics is marginal, mysterious or even magical. They deal with the concept of information, statistical information, they are functionals of distributions or densities and, at other occasions, they express the distance or divergence between two distributions or probability measures or even the affinity or similarity between several distributions.

There are many m.o.i.s, places, and frameworks in which they appear. Attempts have been made by the author, his coworkers, and other researchers to present a unified theory of statistical information and the result may be considered a partial success.

While information is a basic and fundamental concept in statistics, there is no universal agreement on how to define and measure it in a unique way. In statistics we have Fisher information, Kullback–Leibler information, entropy information, Lindley information, and Jaynes's maximum entropy principle. We also have the Akaike information criterion. There is prior and posterior information, weight of evidence, mutual information, uncertainty etc. Since statistical information cannot be defined mathematically, several statements have been used to delimit the concept. For details and related references see the review articles by Kendall (1973), Csiszar (1977), Papaioannou (1985), Aczel (1986), Soofi (1994), and Pardo (1999) as well as Kullback (1959), Papaioannou and Kempthorne (1970), and Ferentinos and Papaioannou (1981).

All these concepts and m.o.i.s are interrelated, one comes from or may be considered a special case of the other, etc. and statisticians disagree on the geometric figure representing these informations and their interrelationships. Even review papers differ among themselves depending on the author's perspective. Soofi (1994) presents information–theoretic statistics in terms of a *pyramid* whose top vertex is Shannon information and the vertices of its base are Lindley, Kullback and Jaynes informations. We disagree with this representation. Jaynes did not introduce a new m.o.i but he put forward the maximum entropy principle. We believe statistical information theory should be represented by a *triangle* whose top vertex is Fisher information and base vertices are Shannon and Kullback. Soofi considers Fisher's information as a byproduct of the Kullback-Leibler information. Fisher's information is a historically leading concept in statistics and has played and is playing an important and major role in developing statistics. Entropy is a specialized concept more appropriate to communication theory.

In this paper we review the state of the art in this area. In Section 32.2 we give a summary of m.o.i.s presenting mainly but briefly new measures and commenting on their interrelationships and interpretations. The properties, mainly statistically oriented properties, are discussed in Section 32.3. New properties are collected which give ideas for further research. The role of m.o.i.s in statistical inference is discussed in Section 32.4 with applications in Section 32.5.

Due to the vastness of the subject we cite the most pertinent papers and the reader is referred to the articles given in these references.

32.2 Measuring Information – Measures of Information

Measures of information have been classified in three categories: (i) Fisher-type, (ii) divergence-type and (iii) entropy-type. We shall follow the notation of Papaioannou (1985). Sometimes measures in the first two classes are called parametric and nonparametric respectively.

The representative measure in the first class is the classical m.o.i introduced by Fisher and given by

$$I_X^F(\theta) = \int f(x,\theta) \left(\frac{\partial \ln f(x,\theta)}{\partial \theta} \right)^2 dx$$

based on a gpdf $f(x,\theta)$, where θ is a univariate parameter.

The representative measure of the second class (divergence-type) is the Kullback-Leibler given by

$$I_X^{KL}(f_1, f_1) = \int f_1(x) \ln[f_1(x)/f_2(x)]dx$$

and based on two gpdfs f_1, f_2. It is the difference between Kerridge's inaccuracy and Shannon's entropy and it is also called relative or cross-entropy.

An important measure is this class which produced a high degree of unification is Csiszar's measure or φ-divergence

$$I_X^C(f_1, f_2) = \int \varphi(f_1/f_2)f_2 dx,$$

where φ is a real valued convex function on $[0, \infty)$ with $0\varphi(0/0) = 0$, $0\varphi(a/0) = a\varphi_\infty$, $\varphi_\infty = \lim[\varphi(u)/u]$ as $u \to \infty$.

For additional measures in these two classes see Papaioannou (1985) and Pardo (1999).

In the last twenty years, additional divergence-type measures of information have been introduced in the literature. One class of these measures are functions (s-power functions) of the Hellinger distance of order r between f_1 and f_2 or f_1 and $(f_1 + f_2)/2$ and, in a sense, are generalizations of Renyi's distance between these two distributions. They are usually called unified (r, s)-divergences. Special values or limiting cases of the parameters or indices of these measures yield some of the previous and simpler measures, obtaining in this way a unification in the study of divergence-type measures of information. These measures cannot be obtained from Csizar's φ-divergence by a special choice of φ [cf. Taneja (1989)]. Other m.o.i's are the λ divergence and λ-mutual information based on Fererri's hypoentropy

$$(1 + \lambda^{-1}) \ln(1 + \lambda^{-1}) - (1/\lambda) \int (1 + \lambda f) \ln(1 + \lambda f) dx$$

[cf. Morales *et al.* (1993), etc.].

A group of Spanish researches led by L. Pardo introduced a further generalization of φ-divergence, the so-called (h, φ)-divergence, aiming at obtaining a higher degree of unification (i.e., to include more m.o.i.s than the Csiszar or (r, s)-divergences). They took a nonnegative real valued C^2 function $h(\cdot)$ with $h(0) = 0$ and evaluated it at the φ-divergence. This new measure shares some of the properties and results of statistical information but has not received wide application because of its complexity and the fact that the user has to select not only φ but h as well [cf. Morales *et al.* (1994) and Salicru *et al.* (1994)].

Another general measure, which is very closely related with the m.o.i.s in this class (Renyi), has been studied in detail and received eminence particularly for multinomial populations and for reasons of goodness of fit, is the Cressie and Read power divergence statistic

$$I_X^{CR}(P, Q) = 2[\lambda(\lambda + 1)]^{-1} \sum P_i[(P_i/Q_i)^\lambda - 1], \qquad \lambda \in \mathbb{R}$$

[cf. Read and Cressie (1988)].

The representative measure for the third class is Shannon's entropy or differential entropy

$$H[f] = -\Sigma p_i \ln p_i, \qquad H[f] = -\int f(x) \ln f(x) dx$$

for discrete and continuous distributions, respectively. Some authors use $H[f]$ in statistics as a measure of dispersion or concentration. A distribution with large $H[f]$ is less concentrated, thus it is more difficult to predict an outcome of X, than a distribution with small $H[f]$. Entropy itself, among other things, has been used in statistics to measure information in a realization of a r.v. X with gpdf f. Large values of $H[f]$ correspond to small amounts of information contained in an observed value of X, so information in X is given by $- H[f]$ [cf. Ebrahimi and Soofi (1990)].

There are many entropy-type m.o.i.s which have been produced by entropy characterizations, i.e., by finding the mathematical function of the p_is satisfying variations or relaxations or generalizations of Shannon's entropy axioms [see, for example, Aczel (1980)]. We have not followed these developments since we believe that, these m.o.i.s do not have much relevance in statistical inference.

Another well-known index is the Gini–Simpson index Σp_i^2, also called information energy, associated with distribution $(p_1, \ldots, p_k)$ and having interesting statistical applications [cf. Pardo and Taneja (1990)].

Pardo and his coworkers introduced another divergence, the generalized R-divergence based on the concavity property of the (h, φ)-entropy. For details see Pardo *et al.* (1993). A further generalization can be obtained using weights [cf. Pardo (1999)].

The previous categorization is not absolute in the sense that a measure may originate from one class but, depending on the distribution(s) involved or the mathematical operator applied upon it, it may end up into another. For example, Fisher's measure is a limiting case of the Kullback-Leibler measure as one distribution converges to the other parametrically.

There is a fourth class of measures which are difficult to classify distinctly in the previous three classes since they originate from the Bayesian perspective. In this perspective there are many distributions involved. E.g., with the obvious notation, $\xi(\theta)$, $p(x|\lambda)$, $p(x)$, $\xi(\theta|x)$ and their number increases as we discriminate between two models or consider Bayes hierachical models, i.e., models in which the parameters of the prior distributions are r.v.s themselves depending on other parameters (hyperparameters) and so on [Goel (1983)]. M.o.i.s can be constructed in different ways using the previous Fisher, divergence and entropy type m.o.i.s and the distributions involved. The most representative m.o.i. in this category is Lindley's (1956) m.o.i

$$I_X^L(\xi) = H(\xi) - E[H(\xi|X)]$$

which is always ≥ 0 as being the divergence between $p(x, \theta) = p(x|\theta)\xi(\theta)$ and $p(x)\xi(\theta)$.

Recently a new m.o.i. almost identical with Fisher information has made its presence in statistical literature. It is Fisher information with differentiation taken with respect to the argument x of f rather than the parameter θ. It is usually called Fisher information number

$$I_X^F = E[(\partial \ln f(X)/\partial x]^2.$$

This number coincides with Fisher's m.o.i. when θ is a location parameter and there remains an open question to explore its role and interelationship with statistical information theory (total variation) [see, for example, Thelen (1989), Papathanasiou (1993) and Kagan and Landsman (1997)].

A standard feature of most of the newly introduced m.o.i.s is that they have a close relationship with the older (classical) ones. They coincide with them for special choices of their parameters or of the functions h and φ. Their usefulness lies in the idea of unification but is limited from the practical point of view barring some statistical applications to be discussed below. Some of the new m.o.i.s are outside the scope of statistical information theory since the do not satisfy some of its basic properties.

32.3   Properties of Measures of Information

As stated before there are many properties of m.o.i.s. For a list of these properties see Papaioannou (1985). We simply mention by name five basic ones: nonnegativity, additivity, maximal information, invariance under sufficient transformations and appearance in Cramer-Rao inequalities.

Statisticians believe that more data means more information. This is justified by the additivity, nonnegativity and maximal information properties.

As research progresses new interesting properties are being added up or other neglected ones call to be investigated. The previous list may be completed as follows:

1. Let (Z, δ) be the pair of variables associated with a randomly censored experiment, where we observe X or Y (the censoring variable) and $Z = \min(X, Y)$, $\delta = I(X < Y)$. Then we should have
   (i) $E[\text{Information}(X)] \geq E[\text{Information}(Z, \delta)]$, for every X, Y (a kind of maximal information property).
   (ii) $E[\text{Information}(Z_1, \delta_1)] \leq E[\text{Information}(Z_2, \delta_2)]$ for every X with $Y_1 \leq_{st} Y_2$, [information decreases as censoring increases or equivalently $I(X, Y_1) \leq I(X, Y_2)$] [Hollander, Proschan, and Sconing (1987)].

2. If $f(x, \theta)$, θ real, has a monotone likelihood ratio in x then

$$\theta_1 < \theta_2 < \theta_3 \rightarrow I(F_{\theta_1}, f_{\theta_2}) < I(f_{\theta_1}, f_{\theta_3})$$

[Ali and Silvey (1974)].

3. Additivity for hierarchical distributions: if $f_1 \subset f_2 \subset f_3$ then

$$I(f_1, f_3) = I(f_1, f_2) + I(f_2, f_3).$$

4. Lindley's concavity and monotonicity in n: Let $X_1, X_2, \ldots$ be i.i.d. r.v.s; then $I_{X_n, X_{n+1}}$ is concave in n and increasing in n [Lindley (1956)].

5. If $f_1 = p(x, \theta)$ and $f_2 = p(x)\xi(\theta)$, where p and ξ are gpdfs, then $I(f_1, f_2)$ is concave in $\xi(\theta)$ [Lindley (1956)].

6. Convexity of $I(f_1, f_2)$ in either f_1 or f_2 [Csiszar (1975)].

7. Relationship with Type II error rate: $\log_2 \beta_n = -nI(f_1, f_2)$ for large n, where β_n is the Type II error rate in testing f_1 versus f_2.

8. Markov chain inequalities:
   (i) if $X_1, X_2, X_3)$ is a Markov chain, then

$$I(f_3(x_3), f_{3|1}(x_3|x_1)) \leq I(f_2(x_2), f_{2|1}(x_2|x_1))$$

[Goel (1983)],
(ii) if (X_1, X_2, X_3, X_4) is a Markov chain then

$$I(f_1(x_1), f_4(x_4)) \leq I(f_2(x_2), f_3(x_3))$$

[Csiszar (1977)].

9. The information generating function: a function whose derivatives produce m.o.i.s [Guiasu and Reischer (1985)].

Do existing measures or which m.o.i.s satisfy the above or which of the above properties and what are their implications to the foundations and inference? This remains to be investigated.

32.4 Measures of Information and Inference

Measures of information have been and are extensively used in statistical inference, mainly in testing statistical hypotheses developing primarily tests of significance based on asymptotic theory.

In the theory of point estimation, Fisher's m.o.i. plays a dominant role since it appears in Cramer-Rao type inequalities. Of course there are some limitations since the Cramer-Rao lower bound is applicable for regular families of distributions. We made some attempts to measure information about θ contained in the data for broader families of distributions including non regular ones but the results were of limited scope since the new parametric measures lacked simplicity and were limiting cases of divergence-type m.o.i.s [see Ferentinos and Papaioannou (1981)]. This led us to conjecture that Fisher's m.o.i. is unique up to a constant.

Some years ago we thought to use m.o.i.s as loss functions and produce information optimum (I-optimum) estimators, i.e., estimators minimizing the risk

$$E(I(f_\theta, f_{\hat{\theta}}))$$

for all estimators $\hat{\theta}$ belonging to a certain class uniformly in θ. We obtained results analogous to the results of decision theory on point estimation, but these results relied heavily on the convexity of $I(f_\theta, f_{\theta^*})$ in θ^* for all θ and we did not pursue the topic further. It is well known that, for exponential families of distributions, maximum likelihood estimators $\hat{\theta}_n$ of θ converge almost surely to θ^*, the value which minimizes over Θ the Kullback-Leibler divergence $I^{KL}(f_0, f(x, \theta))$ between f_0, the true distribution of the data, and $f(x, \theta)$ [cf. McCulloch (1988)]. Now divergence-type m.o.i.s have been used in the general theory of estimation as loss functions to introduce minimum distance/divegrence estimatiors and obtain generalizations of classical results such as consistency, asymptotic normality, etc. [see Papaioannou (1971), Morales, Pardo and Vajda (1995), Barron and Hengartner (1998), Cutler and Cordero-Brana (1996), and Hartigan (1998)].

A promising application of m.o.i.s in Bayesian estimation is to use divergence-type measures to determine the prior needed in Bayesian analysis so that the Bayes estimator is as close possible in terms of divergence with the maximum likelihood estimator.

In testing statistical hypotheses, m.o.i.s have been used to develop tests of significance. The idea is to consider a sample estimator of the m.o.i and derive its asymptotic distribution (usually normal or χ^2).

Initially sample estimates of entropy both finite and differential, were obtained and their asymptotic distributions were established [see, for example, Nayak (1985)]. The asymptotic distribution of $\hat{H}$ is normal. Then sample estimates of entropy and divergence-type m.o.i.s were examined. In all cases the sample statistics were evaluated exactly or approximately. The asymptotic distribution of $I(f_1, f_2)$ was considered in two sampling cases: one random sample from f_1 with f_2 known and two independent samples one from each f_i. Also two model cases were considered: f_1 and/or f_2 to be either multinomial or general parametric distributions. In all cases the asymptotic results lead to tests of significance for goodness of fit, tests of equality of divergence, tests of independence, tests of homogeneity in contingency tables under random or stratified random sampling with proportional allocation and independence among strata, tests of homogeneity of variances, tests of partial homogeneity as, for example, in two multivariate normal populations, etc. [cf. Pardo *et al.* (1993), Morales *et al.* (1994), Salicru *et al.* (1994), and Pardo *et al.* (1995)]. Analogous results for estimates of Fisher information do not exist.

The results may representatively be summarized as follows. We shall use Cziszar's φ divergence as the most general and widely accepted unifying m.o.i and multinomial or product multinomial populations.

1. For a single multinomial population $\mathrm{Mult}(N, p_1, \ldots, p_k)$

$$\sqrt{N}[I^C(\hat{P}, Q) - I^C(P, Q)]/\hat{\sigma}_c \xrightarrow{L} N(0, 1)$$

$$2N[I^C(\hat{P}, Q) - \varphi(1)]/\varphi''(1) \xrightarrow{L} \chi^2_{k-1} \quad \text{if } P = Q$$

where $\hat{P}$ is the sample estimate of $P = (p_1, \ldots, p_k)$, $Q = (q_1, \ldots, q_k)$ is a known distribution and for $\hat{\sigma}_c^2$, see Zografos, Ferentinos and Papaioannou (1990).

2. In rxc multinomial tables with $\Pi = (\pi_{ij})$ and $P = (p_{ij})$ the cell and sample probabilities and $\Pi_0 = (\pi_{i.}\pi_{.j})$, $P_0 = (p_{i.}p_{.j})$

$$\sqrt{N}[I^C(P, P_0) - I^C(\Pi, \Pi_0)] \xrightarrow{L} N(0, \sigma^2)$$

$$2N[I^C(P, P_0) - I^C(\Pi, \Pi_0)] \xrightarrow{L} \text{Linear combination of } \chi_1^2$$

$$2N[I^C(P, P_0) - \varphi(1)]/\varphi''(1) \xrightarrow{L} \chi_{(r-1)(c-1)}^2 | H_0 : \pi_{ij} = \pi_i \pi_{.j},$$

provided that φ satisfies some smoothness conditions and for σ^2 [cf. Zografos (1993)].

3. For two independent multinomial populations $M(N, p_1, \ldots, p_k)$, $M(M, q_1, \ldots, q_k)$

$$\sqrt{N + M}[I^C(\hat{P}, \hat{Q}) - I^C(P, Q)] \xrightarrow{L} N(0, \sigma^2)$$

$$2MN(M + N)^{-1}[I^C(\hat{P}, \hat{Q}) - \varphi(1)]/\varphi''(1) \xrightarrow{L} \chi_{k-1}^2 | P = Q,$$

$i = 1, \ldots, k$, [cf. Zografos (1998)].

Similar results exist for general parametric distributions and for other divergence-type m.o.i.s or dissimilarities and can be found in Salicru *et al.* (1994). They are based on the old results by Kupperman (1957) for the Kullback-Leibler divergence and have wide applications. Moreover similar asymptotic results hold for the estimated Rao distance or geodesic or Riemannian distance between two parametric distributions which is induced from the Fisher information matrix and is a generalization of Mahalanobis distance [cf. Jensen (1993)]. Optimality of goodness of fit tests based on divergences in terms of Pitman and Bahadur efficiencies has also been investigated [cf. Read and Cressie (1988), Zografos, Ferentinos, and Papaioannou (1990), and Pardo (1999)].

32.5 Applications

As it is evident from the above discussion, the applications of m.o.i.s are numerous. We shall only mention a few topics such as Bayesian information theory, differential geometry and information, observed Fisher information and EM algorithm, maximum entropy principle, measures of dependence, etc. and refer the reader to the papers given below and the references cited therein. In addition we mention selectively by title two areas where we have used m.o.i.s to measure information and to introduce and study new statistical models respectively: (i) random censoring, quantal random censoring with random or not recording of uncensored observations and total information [cf. Tsairidis *et al.* (1996,

2000)] and (ii) association and symmetry/asymmetry models for contingency tables based on f-divergence which are the closest to independence or symmetry [cf. Kateri and Papaioannou (1994, 1997)].

32.6 Conclusions

To seek a measure satisfying *all* (reasonable) properties is unrealistic. Statisticians have not agreed on the set of reasonable properties. M.o.i.s will continue to play an important but diversified role in statistics and the pattern of research is expected to continue as in the last fifteen years with special emphasis however in Bayesian applications. M.o.i.s provide tools to prove results or give models for inference but the case to yield a superior methodology than the classical one remains to be seen. Information remains an intangible concept.

Acknowledgement Most of this work was done while the author was at the University of Ioannina, Greece.

References

1. Aczel, J. (1986). Characterizing information measures: Approaching the end of an era. In *Uncertainty in Knowledge-Based Systems*. Lecture Notes in Computer Science (Eds., B. Bouchon and R. R. Yager), pp. 359–384. Springer-Verlag, New York.

2. Ali, S. M. and Silvey, S. D. (1966). A general class of coefficients of divergence of one distribution from another. *Journal of the Royal Statistical Society, Series B* **28**, 131–142.

3. Barron, A. and Hengartner, N. (1998). Information theory and super efficiency. *Annals of Statistics* **26**, 1800–1825.

4. Csiszar, I. (1977). Information measures: A critical review. *Transactions of the 7th Prague Conference on Information Theory, Statistical Decision Functions and Random Processes*, pp. 73–86, Prague, 1974, Academia.

5. Cutler, A. and Cordero-Brana, O. I. (1996). Minimun Hellinger distance estimation for finite mixture models. *Journal of the American Statistical Association* **91**, 1716–1723.

6. Ebrahimi, N. and Soofi, E. S. (1990). Relative information loss under Type II censored experimental data. *Biometrika* **77**, 429–

435.

7. Ferentinos, K. and Papaioannou, T. (1981). New parametric measures of information. *Information and Control* **51**, 193–208.

8. Guiasu, S. and Reischer, C. (1985). The relative information generating function. *Information and Sciences* **35**, 235–241.

9. Hartigan, J. A. (1998). The maximum likelihood prior. *Annals of Statistics* **26**, 2083–2103.

10. Hollander, M., Proschan, F., and Sconing, J. (1987). Measuring information in the right-censored models. *Naval Research Logistics* **34**, 669–687.

11. Jensen, U. (1993). Derivation and calculation of Rao distances: a review. *Technical Report* 71/1993. Institut fur Statistik und Okonometrie der Universitat Kiel.

12. Kagan, A. and Landsman, Z. (1997). Statistical meaning of Carlen's superadditivity of the Fisher information. *Statistics & Probability Letters* **32**, 175–179.

13. Kateri, M. and Papaioannou, T. (1994). f-divergence association models. *International Journal of Mathematical and Statistical Sciences* **3**, 179–203.

14. Kateri, M. and Papaioannou, T. (1997). Symmetry and asymmetry models for rectangular contingency tables. *Journal of the American Statistical Association* **92**, 1124–1131.

15. Kendall, M. G. (1973). Entropy, probability and information. *International Statistical Review* **41**, 59–68.

16. Kullback, S. (1959). *Information Theory and Statistics*. John Wiley & Sons, New York.

17. Kupperman, M. (1957). Further Applications of Information Theory to Multivariate Analysis and Statistical Inference. *Ph.D. dissertation*, George Washington University, Washington, D.C.

18. Lindley, D. V. (1956). On a measure of the information provided by an experiment. *Annals of Mathematical Statistics* **27**, 986–1005.

19. Morales, D., Pardo, L., Salicru, M., and Menendez, M. L. (1993). The λ-divergence and the λ-mutual information: Estimation in stratified sampling. *Journal of Computational and Applied Mathematics* **47**, 1–10.

20. Morales, D., Pardo, L., Salicru, M., and Menendez, M. L. (1994). Asymptotic properties of divergence statistics in stratified random sampling and its application to test statistical hypotheses. *Journal of Statistical Planning and Inference* **38**, 201–223.

21. Morales, D., Pardo, L., and Vajda, I. (1995). Asymptotic divergence of estimates of discrete distributions. *Journal of Statistical Planning Inference* **48**, 347–369.

22. Nayak, T. K. (1985). On diversity measures based on entropy functions. *Communications in Statistics — Theory and Methods* **14**, 203–215.

23. Papaioannou, T. (1971). A criterion of estimation based on measures of information, Unpublished manuscript, University of Georgia. Abstract, *Annals of Mathematical Statistics* **42**, 2179.

24. Papaioannou, T. (1985). Measures of information. In *Encyclopedia of Statistical Sciences* **5** (Eds., S. Kotz and N. L. Johnson), pp. 391–397. John Wiley & Sons, New York.

25. Papaioannou, T. and Kempthorne, O. (1971). On Statistical Information Theory and Related Measures of Information. *Technical Report* No. ARL. 71-0059, Aerospace Research Laboratories, Wright-Patterson A.F.B., Ohio.

26. Papathanasiou, V. (1993). Some characteristic properties of the Fisher information matrix via Cacoullos-type inequalities. *Journal of Multivariate Analysis*, **44**, 256-265.

27. Pardo, L. (1999). Generalized divergence measures: statistical applications. In *Encyclopedia of Microcomputers*, pp. 163–191. Marcel-Dekker, New York.

28. Pardo, L., Morales, D., Salicru, M., and Menendez, M. L. (1993). The φ-divergence statistic in bivariate multinomial populations including stratification. *Metrika* **40**, 223–235.

29. Pardo, L., Salicru, M., Menendez, M. L., and Morales, D. (1995). A test for homogeneity of variances based on Shannonss entropy. *Metron* **53**, 135–146.

30. Pardo, L. and Taneja, I. J. (1991). Information energy and its applications. *Advances in Electronics and Electron Physics* **90**, 165–241.

31. Read, T. R. C. and Cressie, N. A. C. (1988). *Goodness-up-Fit Statistics for Discrete Multivariate Data.* Springer-Verlag, New York.

32. Salicru, M., Morales, D., Menendez, M. L., and Pardo, L. (1994). On the applications of divergence type measures in testing statistical hypotheses. *Journal of Multivariate Analysis* **51**, 372–391.

33. Soofi, E. S. (1994). Capturing the intangible concept of information. *Journal of the American Statistical Association* **89**, 1243–1254.

34. Taneja, I. J. (1989). On generalized information measures and their applications. *Advances in Electronics and Electron Physics* **76**, 327–413.

35. Thelen, B. J. (1989). Fisher information and dichotomies in equivalence/contiguity. *Annals of Statistics* **17**, 1664–1690.

36. Tsairidis, Ch., Ferentinos, K., and Papaioannou, T. (1996). Information and random censoring. *Information Sciences* **92**, 159–174.

37. Tsairidis, Ch., Zografos, K., Ferentinos, K., and Papaioannou, T. (2000). Information in quantal response data and random censoring. *Annals of the Institute of Statistical Mathematics* (to appear).

38. Zografos, K. (1993). Asymptotic properties of φ-divergence statistic and its applications in contingency tables. *International Journal of Mathematical and Statistical Sciences* **2**, 5–21.

39. Zografos, K. (1998). f-dissimilarity of several distributions in testing statistical hypotheses. *Annals of the Institute of Statistical Mathematics* **50**, 295–310.

40. Zografos, K., Ferentinos, K., and Papaioannou, T. (1990). φ-diverg-ence statistics: sampling properties and multinomial goodness of fit and divergence tests. *Communications in Statistics — Theory and Methods* **19**, 1785–1802.

33

Representation Formulae for Probabilities of Correct Classification

Wolf-Dieter Richter
University of Rostock, Rostock, Germany

ABSTRACT Representation formulae for probabilities of correct classification are derived for a minimum-distance type decision rule. These formulae are essentially based upon the two-dimensional Gaussian distribution and allow new derivations of recent results in Krause and Richter (1999).

Keywords and phrases Generalized minimum-distance rule, repeated measurements, two-dimensional decision space, two-dimensional representation formulae, doubly noncentral F-distribution, linear model approach

33.1 Introduction

Let an individual having a Gaussian distributed feature variable belong to one of two distinct populations. Assume that we are given measurements from a sample of individuals giving rise to independent and identically distributed Gaussian feature variables. Several methods of allocating the individual or the whole sample to one of the two populations have been studied in the literature. For an introduction to this area see, e.g., Anderson (1984) or McLachlan (1992). A certain subclass of classification rules is given by the so-called distance rules. Because of the great variety of distances existing in statistics, there are several approaches to distance based classification rules, as only to mention Cacoullos and Koutras (1985), Cacoullos (1992), and Cacoullos and Koutras (1996). When choosing a method for classifying an individual or a sample of individuals, one has to distinguish between the cases of known or unknown moments. Certain sample-distance based classification rules, however, work without assumptions concerning the second

517

order moments and probabilities of correct classification can be described explicitly in terms of these moments. This advantage has been exploited to some extent recently in Krause and Richter (1999). The method developed there combines a geometric sample measure representation formula for the multivariate Gaussian measure with a certain non classic linear model approach due to Krause and Richter (1994). This linear model type approach will be modified in the present paper to derive new representation formulae for probabilities of correct classification which are based upon the two-dimensional Gaussian law. Further transformation of these formulae yields expressions in terms of the doubly noncentral F-distribution as has been derived recently in another way in Krause and Richter (1999).

Let n_1, n_2 and n_3 observations belong to the three populations Π_1, Π_2, and Π_3, respectively. We suppose that Π_1 and Π_2 are distinguishable with respect to their expectations and that population Π_3 can be understood as a copy of one of Π_1 or Π_2. The overall sample vector

$$Y_{(n)} = \left(Y_{(n_1)}^{1T}, Y_{(n_2)}^{2T}, Y_{(n_3)}^{3T} \right)^T \quad , \quad n = n_1 + n_2 + n_3,$$

consisting of the repeated measurement vectors $Y_{(n_i)}^i = (Y_{i1}, \ldots, Y_{in_i})^T$ from the three populations $\Pi_i, i = 1, 2, 3$ will be assumed to define a Gaussian statistical structure

$$S_n = (R^n, B^n, \{\Phi_{\mu, \Sigma}, \mu \in M, \Sigma \in \Theta\}).$$

Here, M denotes the range of $EY_{(n)}$ and will be called the model space, although it is not a linear vector space. It satisfies the representation

$$M = \{\mu \in R^n : \mu = \mu_1 1^{+00} + \mu_2 1^{0+0} + \mu_3 1^{00+},$$

$$\mu_3 \in \{\mu_1, \mu_2\}, (\mu_1, \mu_2) \in R^2, \mu_1 \neq \mu_2\},$$

where

$$1^{+00} = (1_{n_1}^T O_{n_2+n_3}^T)^T, 1^{0+0} = (0_{n_1}^T 1_{n_2}^T 0_{n_3}^T)^T, 1^{00+} = (0_{n_1+n_2}^T 1_{n_3}^T)^T,$$

$$1_{n_i} = (1, \ldots, 1)^T \in R^{n_i}, 0_m = (0, \ldots, 0)^T \in R^m.$$

Further,

$$\Theta = \left\{ \Sigma = \begin{pmatrix} \sigma_1^2 I_{n_1} & & \\ & \sigma_2^2 I_{n_2} & \\ & & \sigma_3^2 I_{n_3} \end{pmatrix}, (\sigma_1^2, \sigma_2^2) \in R^+ \times R^+, \sigma_3^2 \in \{\sigma_1^2, \sigma_2^2\} \right\}$$

is a set of block diagonal matrices where I_{n_i} denotes a $n_i \times n_i$ unit matrix.

The problem of interest here is, on the basis of the overall sample vector $Y_{(n)}$, to decide between the hypotheses

$$H_{1/3} : \mu_3 = \mu_1 \text{ and } H_{2/3} : \mu_3 = \mu_2.$$

33.2 Vector Algebraic Preliminaries

Put

$$1^{+0+} = 1^{+00} + 1^{00+}, 1^{0++} = 1^{0+0} + 1^{00+}, 1^{+++} = 1^{+00} + 1^{0+0} + 1^{00+}$$

and denote by

$$M_{1/3} = L(1^{+0+}, 1^{0+0}) \text{ and } M_{2/3} = L(1^{0++}, 1^{+00})$$

subspaces of the sample space spanned up by the vectors standing within the brackets. These spaces can be understood as hypotheses spaces or restricted model spaces under the hypotheses $H_{1/3}$ or $H_{2/3}$, respectively. Second basis representations for these spaces are

$$M_{1/3} = L(1^{+++}, 1^{0+0}), \quad M_{2/3} = L(1^{+++}, 1^{+00}).$$

Note that $M_{1/3}$ and $M_{2/3}$ are not orthogonal to each other,

$$M_{1/3} \cap M_{2/3} = L(1^{+++})$$

and

$$1^{0+0} \perp 1^{+00}.$$

While the dimensions of the hypotheses spaces satisfy the equations

$$\dim M_{i/3} = 2, \ i = 1, 2$$

the dimension of

$$\widetilde{M} = L(1^{+00}, 1^{0+0}, 1^{00+}),$$

i.e. the smallest subspace of the sample space containing both $M_{1/3}$ and $M_{2/3}$, equals three. A second basis representation for this so called extended model space is

$$\widetilde{M} = L(1^{+++}, 1^{+00}, 1^{0+0}).$$

The spaces $L(1^{+++})$ and $L(1^{+00}, 1^{0+0})$ are linearly independent but not orthogonal. A third basis representation for $\widetilde{M}$ is

$$\widetilde{M} = L(1^{+++}, 1^{-0+}, 1^{0-+}),$$

where

$$1^{-0+} = -\frac{1}{n_1}1^{+00} + \frac{1}{n_3}1^{00+}, 1^{0-+} = -\frac{1}{n_2}1^{0+0} + \frac{1}{n_3}1^{00+}.$$

The spaces $L(1^{+++})$ and $L(1^{-0+}, 1^{0-+})$ are orthogonal but

$$(1^{-0+}, 1^{0-+}) = \frac{1}{n_3}. \tag{33.2.1}$$

Since

$$\Pi_{1^{-0+}}\mu = (\mu_3 - \mu_1)1^{-0+} \text{ and } \Pi_{1^{0-+}}\mu = (\mu_3 - \mu_2)1^{0-+},$$

the two-dimensional space

$$W = L(1^{-0+}, 1^{0-+})$$

will be called effect space or decision space. These notations correspond to the circumstances that changes of the differences $(\mu_3 - \mu_i), i = 1, 2$ are immediately reflected in the space W and decisions concerning the magnitude of these differences should be based upon considerations within this space. This can be taken as motivation to define the decision rules

$$d_c | R^n \longrightarrow \{1, 2\}, \ c > 0$$

for deciding between the hypotheses $H_{1/3}$ and $H_{2/3}$ as

$$d_c(y_{(n)}) = 2 - I\{\|\Pi_{1^{-0+}}y_{(n)}\| < c\|\Pi_{1^{0-+}}y_{(n)}\|\} \tag{33.2.2}$$

for arbitrary $c > 0$. Here, $I(A)$ denotes the indicator of the random event A. Notice that

$$d_c(Y_{(n)}) = 1$$

holds iff

$$\|\Pi_{1^{-0+}}\Pi_W Y_{(n)}\| < c\|\Pi_{1^{0-+}}\Pi_W Y_{(n)}\|.$$

Recognize further that

$$d_1(Y_{(n)}) = 2 - I\{\|Y_{(n)} - \Pi_{M_{1/3}}Y_{(n)}\| < \|Y_{(n)} - \Pi_{M_{2/3}}Y_{(n)}\|\}.$$

Hence the geometrically motivated decision rule d_c will be called, throughout the present paper, a generalized minimum-distance classification rule.

Let us now consider the orthogonal projection of μ onto the effect space W, $\Pi_W \mu$.

Note that $\mu \in M \subset \widetilde{M} = L(1^{+++}, W)$ and $W \perp 1^{+++}$. Hence, $\Pi_W \mu = \mu - \Pi_{1^{+++}}\mu$, i.e.

$$\Pi_W \mu = a(\mu_1, \mu_2, \mu_3)1^{+00} + b(\mu_1, \mu_2, \mu_3)1^{0+0} + c(\mu_1, \mu_2, \mu_3)1^{00+} \tag{33.2.3}$$

with

$$n \cdot a(x, y, z) = n_2(x - y) + n_3(x - z),$$
$$n \cdot b(x, y, z) = n_1(y - x) + n_3(y - z),$$
$$n \cdot c(x, y, z) = n_1(z - x) + n_2(z - y).$$

It is easily seen from this $\widetilde{M}$–basis representation of $\Pi_W \mu$ that the co-efficients must depend on each other. We shall therefore try to reduce the number of parameters included in the model. To this end we start with a first reparametrisation. This reparametrisation is based upon a certain partial orthogonalisation.

LEMMA 33.2.1

$$\Pi_W \mu = -n_1 a(\mu_1, \mu_2, \mu_3) 1^{-+0} - n_2 b(\mu_1, \mu_2, \mu_3) 1^{0-+},$$

$$\Pi_{1+++} \mu = \frac{n_1 \mu_1 + n_2 \mu_2 + n_3 \mu_3}{n_1 + n_2 + n_3} 1^{+++}.$$

The proof of the second assertion is obvious. The first assertion follows immediately from the following lemma. Notice that, e.g., the dimension-depending new parameter

$$m(\mu_1, \mu_2, \mu_2) = \frac{1}{n} \sum_{i=1}^{3} \mu_i n_i$$

does not allow an immediate interpretation in the original problem.

LEMMA 33.2.2
The $\widetilde{M}$-vector
$$u = a1^{+00} + b1^{0+0} + c1^{00+} \tag{33.2.4}$$

belongs to the subspace W and allows the representation

$$u = \varphi 1^{-0+} + \psi 1^{0-+} \tag{33.2.5}$$

for some $(\varphi, \psi) \in R^2$ if and only if

$$an_1 + bn_2 + cn_3 = 0. \tag{33.2.6}$$

In this case,
$$\varphi = -n_1 a, \quad \psi = -n_2 b. \tag{33.2.7}$$

PROOF Replacing the two vectors in (33.2.5) by their definitions yields

$$u = -\frac{\varphi}{n_1} 1^{+00} - \frac{\psi}{n_2} 1^{0+0} + \frac{\varphi + \psi}{n_3} 1^{00+}.$$

Equating coefficients from the latter formula with corresponding coefficients from (33.2.4) gives (33.2.7) and

$$\frac{\varphi + \psi}{n_3} = -\frac{n_1 a + n_2 b}{n_3}.$$

The latter quantity coincides with the coefficient c if and only if condition (33.2.6) is fulfilled. ∎

Let us define by

$$\overline{Y}_{i\cdot} = \frac{1}{n_i} \sum_{j=1}^{n_i} Y_{ij}$$

the mean in the i-th population, $i = 1, 2, 3$ and by $m(\overline{Y}_{1\cdot}, \overline{Y}_{2\cdot}, \overline{Y}_{3\cdot})$ the overall mean. It follows then that

$$\Pi_{1+++} Y = m(\overline{Y}_{1\cdot}, \overline{Y}_{2\cdot}, \overline{Y}_{3\cdot}) 1^{+++}$$

and

$$\Pi_{\widetilde{M}} Y = \overline{Y}_{1\cdot} 1^{+00} + \overline{Y}_{2\cdot} 1^{0+0} + \overline{Y}_{3\cdot} 1^{00+}.$$

Hence, an $\widetilde{M}$–basis representation for $\Pi_W Y$, i.e. for $\Pi_{\widetilde{M}} Y - \Pi_{1+++} Y$ is

$$\Pi_W Y = a(\overline{Y}_{1\cdot}, \overline{Y}_{2\cdot}, \overline{Y}_{3\cdot}) 1^{+00} + b(\overline{Y}_{1\cdot}, \overline{Y}_{2\cdot}, \overline{Y}_{3\cdot}) 1^{0+0} + c(\overline{Y}_{1\cdot}, \overline{Y}_{2\cdot}, \overline{Y}_{3\cdot}) 1^{00+}.$$

The last three equations define $\widetilde{M}$–basis representations of the least squares estimations for the quantities $\Pi_{1+++} \mu$, $\Pi_{\widetilde{M}} \mu$ and $\Pi_W \mu$, respectively.

COROLLARY 33.2.3
Using the above defined functions a and b, a W-basis representation formula for $\Pi_W Y$ is given by

$$\Pi_W Y = -n_1 a(\overline{Y}_{1\cdot}, \overline{Y}_{2\cdot}, \overline{Y}_{3\cdot}) 1^{-0+} - n_2 b(\overline{Y}_{1\cdot}, \overline{Y}_{2\cdot}, \overline{Y}_{3\cdot}) 1^{-0+}.$$

Note that
$$W = L(b_1, b_2)$$

with
$$b_1 = 1^{-0+} \text{ and } b_2 = \frac{1}{n_1 + n_3} 1^{+0+} - \frac{1}{n_2} 1^{0+0} \tag{33.2.8}$$

defines an orthogonal basis representation for W where

$$b_2 = 1^{0-+} - \Pi_{1-0+} 1^{0-+}.$$

From (33.2.1) and

$$\|1^{-0+}\|^2 = \frac{n_1 + n_3}{n_1 n_3} \tag{33.2.9}$$

we get

$$\Pi_{1^{-0+}} 1^{0-+} = \frac{n_1}{n_1 + n_3} 1^{-0+}.$$

Hence,

$$b_2 = -\frac{1}{n_2} 1^{0+0} + \frac{1}{n_3} 1^{00+} + \frac{1}{n_1 + n_3} 1^{+00} - \frac{n_1}{n_3(n_1 + n_3)} 1^{00+}.$$

This yields the second assertion in (33.2.8). The following lemma presents a reparametrisation which is based upon orthogonalisation.

LEMMA 33.2.4
The quantity $\Pi_W \mu$ can be written as

$$\Pi_W \mu = \frac{n_1 n_3}{n_1 + n_3} (\mu_3 - \mu_1) b_1 + d(\mu_1, \mu_2, \mu_3) b_2,$$

whereby the new parameter d satisfies the two representation formulae

$$nd(\mu_1, \mu_2, \mu_3) = n_1 n_2 (\mu_1 - \mu_2) + n_2 n_3 (\mu_3 - \mu_2), \tag{33.2.10}$$

and

$$nd(\mu_1, \mu_2, \mu_3) = n_2 (n_1 + n_3) \left(m^{(1/3)}(\mu_1, \mu_3) - \mu_2 \right) \tag{33.2.11}$$

with

$$m^{(1/3)}(x, y) = \frac{n_1 x + n_3 y}{n_1 + n_3}.$$

PROOF Making use of equations (33.2.3) and (33.2.7) above as well as (33.2.12) below one can see that

$$u = \Pi_W \mu = \Pi_W (\mu_1 1^{+00} + \mu_2 1^{0+0} + \mu_3 1^{00+})$$

allows the representation

$$u = \vartheta b_1 + \nu b_2$$

with

$$\vartheta = -n_1 a(\mu_1, \mu_2, \mu_3) - \frac{n_1 n_2}{n_1 + n_3} b(\mu_1, \mu_2, \mu_3)$$

$$= \frac{n_1 n_2 (\mu_2 - \mu_1) + n_1 n_3 (\mu_3 - \mu_1)}{n}$$

$$+ \frac{n_1^2 n_2 (\mu_1 - \mu_2) + n_1 n_2 n_3 (\mu_3 - \mu_2)}{n(n_1 + n_3)}$$

and

$$\nu = \psi = -n_2 b(\mu_1, \mu_2, \mu_3) = \frac{n_2 n_1(\mu_1 - \mu_2) + n_2 n_3(\mu_3 - \mu_2)}{n}.$$

Hence

$$n(n_1 + n_3)\vartheta = (n_1 + n_3)[n_1 n_2(\mu_2 - \mu_1) + n_1 n_3(\mu_3 - \mu_1)]$$
$$+ n_1^2 n_2(\mu_1 - \mu_2) + n_1 n_2 n_3(\mu_3 - \mu_2)$$
$$= n_1 n_3 n(\mu_3 - \mu_1)$$

and

$$n\nu = n_2(n_1 + n_3)\left[\frac{n_1\mu_1 + n_3\mu_3}{n_1 + n_3} - \mu_2\right],$$

which proves the assertions of the lemma. ∎

Observe that since the new parameters d and $m^{(1/3)}(\mu_1, \mu_3)$ depend on the sample sizes they do not allow immediate interpretations with respect to the original problem.

LEMMA 33.2.5
The $\widetilde{M}$-vector u from (33.2.4) belongs to the subspace W and allows the representation

$$u = \vartheta b_1 + \nu b_2 \qquad (33.2.12)$$

for certain values of $(\vartheta, \nu) \in R^2$ if and only if the condition (33.2.6) is satisfied. In this case we have, with (φ, ψ) from (33.2.7),

$$\vartheta = \varphi + \frac{n_1}{n_1 + n_3}\psi, \quad \nu = \psi.$$

PROOF Equating coefficients, (33.2.12) follows from

$$a1^{+00} + b1^{0+0} + c1^{00+} = u = \vartheta b_1 + \nu b_2$$
$$= \vartheta\left(-\frac{1}{n_1}1^{+00} + \frac{1}{n_3}1^{00+}\right) + \nu\left(\frac{1}{n_1 + n_3}1^{+0+} - \frac{1}{n_2}1^{0+0}\right)$$
$$= \left(-\frac{\vartheta}{n_1} + \frac{\nu}{n_1 + n_3}\right)1^{+00} - \frac{\nu}{n_2}1^{0+0} + \left(\frac{\vartheta}{n_3} + \frac{\nu}{n_1 + n_3}\right)1^{00+}.$$

In the same way it follows that

$$c = \frac{\vartheta}{n_3} + \frac{\nu}{n_1 + n_3} = -\frac{n_1}{n_3}a - \frac{n_1 n_2}{n_3(n_1 + n_3)}b - \frac{n_2}{n_1 + n_3}b,$$

which is equivalent to (33.2.6). ∎

By definition of b_1,

$$\Pi_{b_1} Y = \frac{n_1 n_3}{n_1 + n_3} (\overline{Y}_{3\cdot} - \overline{Y}_{1\cdot}) b_1.$$

Using the notation

$$\overline{Y^{(1/3)}_{\cdot\cdot}} = \frac{1}{n_1 + n_3} \sum_{i \in \{1,3\}} \sum_{j=1}^{n_i} Y_{ij} = m^{(1/3)}(\overline{Y}_{1\cdot}, \overline{Y}_{3\cdot})$$

for the pooled mean of the values from the union of the first and third subsamples $Y^1_{(n_1)}$ and $Y^3_{(n_3)}$, we get

$$\Pi_{b_2} Y = d(\overline{Y}_{1\cdot}, \overline{Y}_{2\cdot}, \overline{Y}_{3\cdot}) b_2$$
$$= \frac{1}{n}(n_1 n_2 (\overline{Y}_{1\cdot} - \overline{Y}_{2\cdot}) + n_2 n_3 (\overline{Y}_{3\cdot} - \overline{Y}_{2\cdot})) b_2$$
$$= \frac{n_2}{n}(n_1 + n_3)\left(\overline{Y^{(1/3)}_{\cdot\cdot}} - \overline{Y}_{2\cdot}\right) b_2.$$

Hence we arrive at least squares estimates for $\Pi_{b_1}\mu$ and $\Pi_{b_2}\mu$.
With

$$\|b_1\| = \sqrt{\frac{n_1 + n_3}{n_1 \cdot n_3}} \text{ and } \|b_2\| = \sqrt{\frac{n_1 + n_2 + n_3}{n_2(n_1 + n_3)}} \qquad (33.2.13)$$

we get the following representation formula for the least squares estimate of $\Pi_W \mu$ with respect to the normalized orthogonal basis $\{B_1, B_2\}$ where $B_i = b_i / \|b_i\|$:

$$\Pi_W Y = \sqrt{\frac{n_1 n_3}{n_1 + n_3}} (\overline{Y}_{3\cdot} - \overline{Y}_{1\cdot}) B_1 + \sqrt{\frac{n_2(n_1 + n_3)}{n_1 + n_2 + n_3}} \left(\overline{Y^{(1/3)}_{\cdot\cdot}} - \overline{Y}_{2\cdot}\right) B_2 .$$
$$(33.2.14)$$

33.3   Distributional Results

33.3.1   Representation Formulae Based upon the Two-Dimensional Gaussian Law

The random variables $\overline{Y}_{3\cdot} - \overline{Y}_{1\cdot}$ and $\overline{Y^{(1/3)}_{\cdot\cdot}} - \overline{Y}_{2\cdot}$ play an essential role in the basic formula (33.2.14). They are coefficients of the projections of $Y_{(n)}$ onto the normalized orthogonal basis vectors B_1 and B_2 of the decision space W and are therefore uncorrelated. They are linear combinations of the components of the Gaussian random vector $Y_{(n)}$ and are therefore jointly Gaussian distributed and consequently independent from each other.

Let $\eta_1, \eta_2, \ldots$ denote independent standard Gaussian distributed random variables. The symbol $X_i \sim N_i(a_i, \lambda_i^2)$ will be used to indicate that X_i follows the same distribution law as $a_i + \lambda_i \eta_i$. Thus

$$\sqrt{\frac{n_1 n_3}{n_1 + n_3}} (\overline{Y}_{3.} - \overline{Y}_{1.}) \sim N_1 \left(\sqrt{\frac{n_1 n_3}{n_1 + n_3}} (\mu_3 - \mu_1), \frac{n_1 \sigma_3^2 + n_3 \sigma_1^2}{n_1 + n_3} \right).$$

$$(33.3.1)$$

The first and second order moments of this distribution will be denoted by ξ_1 and δ_1^2, respectively. In a similar way as above, one can show that the random variables $\overline{Y}^{(1/3)}$ und $\overline{Y}_{2.}$ are independent. Using the notations

$$\overline{Y}_{2.} \sim N_3 \left(\mu_2, \frac{\sigma_2^2}{n_2} \right)$$

and

$$\overline{Y}_{..}^{(1/3)} \sim N_4 \left(m^{(1/3)}(\mu_1, \mu_3), \frac{n_1 \sigma_1^2 + n_3 \sigma_3^3}{(n_1 + n_3)^2} \right),$$

we get

$$\sqrt{\frac{n_2(n_1 + n_3)}{n_1 + n_2 + n_3}} (\overline{Y}_{..}^{(1/3)} - \overline{Y}_{2.}) \sim N_2(\xi_2, \delta_2^2) \qquad (33.3.2)$$

with

$$\xi_2 = \sqrt{\frac{n_2(n_1 + n_3)}{n_1 + n_2 + n_3}} (m^{(1/3)}(\mu_1, \mu_3) - \mu_2)$$

$$= \frac{n_1 n_2(\mu_1 - \mu_2) + n_3 n_2(\mu_3 - \mu_2)}{\sqrt{(n_1 + n_2 + n_3) n_2(n_1 + n_3)}}$$

and

$$\delta_2^2 = \frac{(n_1 + n_3)^2 \sigma_2^2 + n_2(n_1 \sigma_1^2 + n_3 \sigma_3^2)}{(n_1 + n_2 + n_3)(n_1 + n_3)}.$$

Recall that

$$d_c(Y_{(n)}) = 1$$

holds if and only if

$$N_1 B_1 + N_2 B_2 \in \{z \in W : \|\Pi_{1-0+} z\| < c \|\Pi_{1^0-+} z\|\}. \qquad (33.3.3)$$

The following lemma concerns reducing dimension and norming.

LEMMA 33.3.1

The relation

$$N_1 B_1 + N_2 B_2 \in \{z \in W : \|\Pi_{1-0+} z\| < c \, \|\Pi_{10-+} z\|\}$$

is true iff

$$\left(\frac{N_1}{\delta_1}, \frac{N_2}{\delta_2}\right)^T \in \left\{(t_1, t_2)^T \in R^2 : \frac{|t_1|}{|t_1 + \kappa t_2|} < \zeta \cdot c\right\},$$

where

$$\kappa = \sqrt{\frac{n_3[(n_1 + n_3)^2 \sigma_2^2 + n_2(n_1 \sigma_1^2 + n_3 \sigma_3^2)]}{n_1 n_2 (n_1 \sigma_3^2 + n_3 \sigma_1^2)}} \tag{33.3.4}$$

and

$$\zeta = \sqrt{\frac{n_1 n_2}{(n_1 + n_3)(n_2 + n_3)}} = \frac{1}{\sqrt{(1 + n_3/n_1)(1 + n_3/n_2)}}. \tag{33.3.5}$$

PROOF Put $A = \{z \in W : \|\Pi_{1-0+} z\| / \|\Pi_{10-+} z\| < c\}$. Then

$$A = \left\{z_1 B_1 + z_2 B_2 : \frac{\|\Pi_{1-0+} (z_1 B_1 + z_2 B_2)\|}{\|\Pi_{10-+} (z_1 B_1 + z_2 B_2)\|} < c, \ (z_1, z_2)^T \in R^2\right\}.$$

With

$$(B_1, 1^{0-+}) = \sqrt{\frac{n_1}{n_3(n_1 + n_3)}}$$

and

$$(B_2, 1^{0-+}) = \sqrt{\frac{n_1 + n_2 + n_3}{n_2(n_1 + n_3)}}$$

it follows that

$$A = \{z_1 B_1 + z_2 B_2 : |z_1| / |z_1 + \chi \cdot z_2| < \zeta \cdot c, (z_1, z_2)^T \in R^2\},$$

where

$$\chi = \sqrt{n_3(n_1 + n_2 + n_3)} / \sqrt{n_1 n_2}.$$

Hence,

$$N_1 B_1 + N_2 B_2 \in A$$

iff

$$(N_1, N_2)^T \in \{(z_1, z_2)^T \in R^2 : |z_1| / |z_1 + \chi z_2| < \zeta \cdot c\}.$$

The assertion of the lemma follows now with $\kappa = (\delta_2 / \delta_1) \chi$. ∎

Put

$$\nu_1 = \frac{\xi_1}{\delta_1} = \sqrt{\frac{n_1 n_3}{n_1 \sigma_3^2 + n_3 \sigma_1^2}}(\mu_3 - \mu_1) \tag{33.3.6}$$

and

$$\nu_2 = \frac{\xi_2}{\delta_2} = \frac{\sqrt{n_2}(n_1 + n_3)[m^{(1/3)}(\mu_1, \mu_3) - \mu_2]}{\sqrt{(n_1 + n_3)^2 \sigma_2^2 + n_2(n_1 \sigma_1^2 + n_3 \sigma_3^2)}}$$

$$= \frac{n_1(\mu_1 - \mu_2) + n_3(\mu_3 - \mu_2)}{\sqrt{(n_1 + n_3)^2 \sigma_2^2 + n_2(n_1 \sigma_1^2 + n_3 \sigma_3^2)}}. \tag{33.3.7}$$

If the hypothesis $H_{1/3}$ is true, then $m^{(1/3)}(\mu_1, \mu_3) = \mu_1$,

$$\nu_1 = 0 \tag{33.3.8}$$

and

$$\nu_2 = \sqrt{n_2}\frac{\mu_1 - \mu_2}{\sigma_2}\sqrt{\frac{1 + n_3/n_1}{1 + n_3/n_1 + (n_2\sigma_1^2)/(n_1\sigma_2^2)}}$$

$$= \frac{\sqrt{n_1 + n_3}(\mu_1 - \mu_2)}{\sqrt{(n_1 + n_3)\sigma_2^2 + n_2\sigma_1^2}}. \tag{33.3.9}$$

If the hypothesis $H_{1/3}$ is true, then

$$\kappa = \sqrt{\frac{n_3}{n_1} + \frac{(n_1 + n_3)n_3\sigma_2^2}{n_1 n_2 \sigma_1^2}}, \tag{33.3.10}$$

but if $H_{2/3}$ is true, then

$$\kappa = \sqrt{\frac{n_3\sigma_2^2(n_2 n_3 + (n_1 + n_3)^2) + n_1 n_2 n_3 \sigma_1^2}{n_1 n_2(n_1 \sigma_2^2 + n_3 \sigma_1^2)}}. \tag{33.3.11}$$

If $H_{1/3}$ is true, then the random event of correct classification

$$CC_1(c) = \{d_c(Y_{(n)}) = 1\},$$

in view of (33.3.3) and Lemma 33.3.1, obtains the representation

$$CC_1(c) = \left\{\frac{|N_1(0,1)|}{|N_1(0,1) + \kappa N_2(\nu_2, 1)|} < \zeta \cdot c\right\} \tag{33.3.12}$$

with independent random variables N_1 and N_2, ν_2 from (33.3.9), κ from (33.3.10) and ζ from (33.3.5). This proves the following theorem.

THEOREM 33.3.2

If the hypothesis $H_{1/3}$ is true, then $P_1(CC_1(c))$, the probability of correct classification into the population Π_1, satisfies the representation formula

$$P_1(CC_1(c)) = \Phi_{(0,\nu_2)^T,\begin{pmatrix} 1 & 0 \\ 0 & 1 \end{pmatrix}}(CC_1^*(c)) \qquad (33.3.13)$$

for all $c > 0$, with

$$CC_1^*(c) = \left\{ (t_1, t_2)^T \in R^2 : \frac{|t_1|}{|t_1 + \kappa t_2|} < \zeta \cdot c \right\}$$

and ν_2, κ, and ζ as in (33.3.9), (33.3.10), and (33.3.5),respectively.

If the hypothesis $H_{2/3}$ is true, then the random event of correct classification into the population Π_2,

$$CC_2(c) = \{d_c(Y_{(m)}) = 2\},$$

satisfies for all $c > 0$ the representation

$$CC_2(c) = \left\{ \frac{|N_1(\nu_1, 1)|}{|N_1(\nu_1, 1) + \kappa N_2(\nu_2, 1)|} > \zeta \cdot c \right\}, \qquad (33.3.14)$$

where κ and ζ are to be chosen according to (33.3.11) and (33.3.5), respectively,

$$\nu_1 = \frac{\mu_2 - \mu_1}{\sqrt{\sigma_2^2/n_3 + \sigma_1^2/n_1}} = \frac{\sqrt{n_1 n_3}(\mu_2 - \mu_1)}{\sqrt{n_1 \sigma_2^2 + n_3 \sigma_1^2}} \qquad (33.3.15)$$

and, since

$$m^{1/3}(\mu_1, \mu_2) - \mu_2 = \frac{\mu_1 - \mu_2}{1 + n_3/n_1},$$

ν_2 is defined as

$$\nu_2 = \sqrt{n_2} \frac{\mu_1 - \mu_2}{\sigma_2} \Bigg/ \sqrt{\left(1 + \frac{n_3}{n_1}\right)^2 + \frac{n_2}{n_1}\frac{\sigma_1^2}{\sigma_2^2} + \frac{n_2 n_3}{n_1^2}}$$

$$= \frac{n_1(\mu_1 - \mu_2)}{\sqrt{\sigma_2^2(n_2 n_3 + (n_1 + n_3)^2) + \sigma_1^2 n_1 n_2}}. \qquad (33.3.16)$$

The following theorem is an immediate consequence.

THEOREM 33.3.3
*If $H_{2/3}$ is true then the probability of correct classification into the pop-
ulation Π_2, $P_2(CC_2(c))$, satisfies the representation formula*

$$P_2(CC_2(c)) = \Phi_{(\nu_1,\nu_2)^T, \left(\begin{array}{cc} 1 & 0 \\ 0 & 1 \end{array} \right)}(CC_2^*(c)) \tag{33.3.17}$$

for all $c > 0$, with

$$CC_2^*(c) = \left\{ (t_1, t_2)^T \in R^2 : \frac{|t_1|}{|t_1 + \kappa t_2|} > \zeta \cdot c \right\}$$

*and ν_1, ν_2, κ, ζ as in (33.3.15), (33.3.16), (33.3.11), and (33.3.5), re-
spectively.*

The representation formulae for $P_1(CC_1(c))$ in Theorem 33.3.2 and for
$P_2(CC_2(c))$ in Theorem 33.3.3 do not only reflect different quantitative
situations but they are also of different qualitative nature. The most
obvious difference between $\nu_1 = 0$ in (33.3.8) and $\nu_1 \neq 0$ in (33.3.15) can
be easily detected. The following theorem presents a second representation formula for $P_2(CC_2(c))$ which corresponds in quality to that for
$P_1(CC_1(c))$ in Theorem 33.3.2. Its proof repeats that of Theorem 33.3.2
and will be suppressed therefore here.

THEOREM 33.3.4
*If $H_{2/3}$ is true then the probability of correct classification into the pop-
ulation Π_2 satisfies for all $c > 0$ the representation*

$$P_2(CC_2(c)) = \Phi_{(0,\tilde{\nu}_2)^T, \left(\begin{array}{cc} 1 & 0 \\ 0 & 1 \end{array} \right)}(\widetilde{CC}_2(c)), \tag{33.3.18}$$

where

$$\widetilde{CC}_2(c) = \left\{ (t_1, t_2)^T \in R^2 : \frac{|t_1|}{|t_1 + \tilde{\kappa} t_2|} < \frac{\zeta}{c} \right\}$$

with ζ as in (33.3.5),

$$\tilde{\kappa} = \sqrt{\frac{n_3}{n_2} + \frac{n_3(n_2 + n_3)}{n_1 n_2} \frac{\sigma_1^2}{\sigma_2^2}} \tag{33.3.19}$$

and

$$\tilde{\nu}_2 = \frac{\sqrt{n_2 + n_3}(\mu_2 - \mu_1)}{\sqrt{(n_2 + n_3)\sigma_1^2 + n_1\sigma_2^2}}. \tag{33.3.20}$$

The next theorem follows by analogy.

THEOREM 33.3.5

If $H_{1/3}$ is true then it holds for all $c > 0$

$$P_1(CC_1(c)) = \Phi_{(\tilde{\nu}_1, \tilde{\nu}_2)^T, \begin{pmatrix} 1 & 0 \\ 0 & 1 \end{pmatrix}} (\widetilde{CC_1}(c)), \qquad (33.3.21)$$

where

$$\widetilde{CC_1}(c) = \left\{ (t_1, t_2)^T \in R^2 : \frac{|t_1|}{|t_1 + \tilde{\kappa} t_2|} > \frac{\zeta}{c} \right\}$$

with ζ as in (33.3.5),

$$\tilde{\kappa} = \sqrt{\frac{n_3 \sigma_1^2(n_1 n_3 + (n_2 + n_3)^2) + n_1 n_2 n_3 \sigma_2^2}{n_1 n_2 (n_1 \sigma_1^2 + n_3 \sigma_2^2)}} \qquad (33.3.22)$$

and

$$\tilde{\nu}_1 = \frac{\sqrt{n_2 n_3}(\mu_1 - \mu_2)}{\sqrt{n_2 \sigma_1^2 + n_3 \sigma_2^2}} \qquad (33.3.23)$$

as well as

$$\tilde{\nu}_2 = \frac{n_2(\mu_2 - \mu_1)}{\sqrt{\sigma_1^2(n_1 n_3 + (n_2 + n_3)^2) + n_1 n_2 \sigma_2^2}}. \qquad (33.3.24)$$

33.3.2 Representation Formulae Based upon the Doubly Noncentral F-Distribution

It was shown in John (1961) and Moran (1975) that the probabilities of correct classification can be expressed in terms of the doubly noncentral F-distribution if expectations are unknown but covariance matrices are known and equal and the linear discriminant function is used for classifying an individual into one of the populations Π_1 and Π_2. It has been recently proved, in Krause and Richter (1999), that the probabilities of correct classification can be also expressed in terms of the doubly noncentral F-distribution if both expectations and covariance matrices are unknown but a certain generalized minimum-distance rule is used for making the decision. Here, a result will be derived which is equivalent in content to the latter one but diffcrent from it in form. The method of proving this result developed here differs from that in Krause and Richter (1999) in using basically a two-dimensional representation formula from the preceding section whereas the proof of the corresponding result in Krause and Richter (1999) starts from a sample space measure representation formula.

The aim of what follows is to determine the Gaussian measure of $CC_1^*(c)$ in accordance with Theorem 33.3.2. To this end we describe the boundary of the set of points satisfying the inequality

$$\frac{|t_1|}{|t_1 + \kappa t_2|} < \zeta c \qquad (33.3.25)$$

with the help of the straight lines

$$g_j : t_2 = -\frac{1}{\kappa}\left(1 + \frac{(-1)^j}{\zeta c}\right) t_1, j = 1, 2. \qquad (33.3.26)$$

It turns out that the set of solutions of (33.3.25) includes the t_2-axis in its inner part. The straight line g_2 belongs for all values of ζc to the union of the second and third quadrants in a cartesian coordinate system, i.e. it belongs to the union of the sets $\{t_1 < 0, t_2 > 0\}$ and $\{t_1 > 0, t_2 > 0\}$.

The straight line g_1 belongs to the same set if $\zeta c > 1$ but to the union of the first and fourth quadrants if $0 < \zeta c < 1$. The straight lines g_1 and g_2 intersect within the set (33.3.25) under an angle α satisfying

$$\alpha = \pi - \arctan\left(-\frac{1}{\kappa}\left(1 + \frac{1}{\zeta c}\right)\right)$$
$$+ (-1)^{I\{\zeta c < 1\}} \cdot \arctan\left(-\frac{1}{\kappa}\left(1 - \frac{1}{\zeta c}\right)\right). \qquad (33.3.27)$$

Notice that the set of points $(t_1, t_2)^T$ corresponding to (33.3.25) represents a cone. That is why there exists a vector $(t_{10}, t_{20})^T \in R^2$ and a positive real number $d = d(\zeta c)$ such that a vector $(t_1, t_2)^T \in R^2$ satisfies condition (33.3.25) if and only if it satisfies the condition

$$\frac{\|(t_1, t_2)^T - \Pi_{(t_{10}, t_{20})^T}(t_1, t_2)^T\|^2}{\|\Pi_{(t_{10}, t_{20})^T}(t_1, t_2)^T\|^2} < d^2. \qquad (33.3.28)$$

The latter condition is equivalent to

$$\frac{(t_1 t_{20} - t_2 t_{10})^2}{(t_1 t_{10} + t_2 t_{20})^2} < d^2$$

or

$$(q - t_2/t_1)^2 < d^2(1 + q t_2/t_1)^2 \qquad (33.3.29)$$

where

$$q = t_{20}/t_{10}.$$

Let us determine now a solution (q, d). Recall that the boundary of (33.3.25) can be described by the equations (33.3.26). The temporary assumption that we have in (33.3.25) and (33.3.29) equalities instead of inequalities leads us to the equation systems

$$\frac{q-d}{dq+1} = -\frac{1}{\kappa} + \frac{1}{\kappa\zeta c} \quad , \quad \frac{q+d}{1-dq} = -\frac{1}{\kappa} - \frac{1}{\kappa\zeta c} \qquad (33.3.30)$$

and

$$\frac{q-d}{dq+1} = -\frac{1}{\kappa} - \frac{1}{\kappa\zeta c} \quad , \quad \frac{q+d}{1-dq} = -\frac{1}{\kappa} + \frac{1}{\kappa\zeta c}. \qquad (33.3.31)$$

The solution of (33.3.30) is given by

$$d = \frac{1}{2}\left(\kappa\zeta c + \frac{\zeta c}{\kappa} - \frac{1}{\kappa\zeta c}\right) + \frac{1}{2}\sqrt{\left(\kappa\zeta c + \frac{\zeta c}{\kappa} - \frac{1}{\kappa\zeta c}\right)^2 + 4}, \qquad (33.3.32)$$

$$q = \left[d - \frac{1}{\kappa} + \frac{1}{\kappa\zeta c}\right] / \left[1 + \frac{d}{\kappa} - \frac{d}{\kappa\zeta c}\right]. \qquad (33.3.33)$$

The solution of (33.3.31) is given by

$$d = -\frac{1}{2}\left(\kappa\zeta c + \frac{\zeta c}{\kappa} - \frac{1}{\kappa\zeta c}\right) + \frac{1}{2}\sqrt{\left(\kappa\zeta c + \frac{\zeta c}{\kappa} - \frac{1}{\kappa\zeta c}\right)^2 + 4}, \qquad (33.3.34)$$

$$q = \left[d - \frac{1}{\kappa} - \frac{1}{\kappa\zeta c}\right] / \left[1 + \frac{d}{\kappa} + \frac{d}{\kappa\zeta c}\right]. \qquad (33.3.35)$$

Consequently, the representation (33.3.25) for the cone under consideration is equivalent to the representation (33.3.29) if the quantities d and q are chosen there either according to (33.3.32) and (33.3.33) or according to (33.3.34) and (33.3.35), respectively. Hence, the following lemma has been proved.

LEMMA 33.3.6
The probability of correct classification into Π_1 satisfies the representation

$$P_1(CC_1(c))$$

$$= \Phi_{(0,v_2)^T,\begin{pmatrix}1 & 0\\0 & 1\end{pmatrix}}\left(\left\{(t_1,t_2)^T \in R^2 : \frac{\|(t_1,t_2)^T - \Pi_{(t_{10},t_{20})^T}(t_1,t_2)^T\|^2}{\|\Pi_{(t_{10},t_{20})^T}(t_1,t_2)^T\|^2} < d^2\right\}\right).$$

Here, v_2 is chosen as in (33.3.9), $(t_{10},t_{20})^T \in R^2$ is an arbitrary vector satisfying $t_{20}/t_{10} = q$ and d and q are to be chosen according to either (33.3.32) and (33.3.33) or (33.3.34) and (33.3.35).

It follows from the definition of the doubly noncentral F–distribution with $(1,1)$ degrees of freedom and from the invariance of the two-dimensional standard Gaussian measure with respect to orthogonal transformations that $P_1(CC_1(c))$ can be expressed as a suitable value of the cumulative distribution function $F_{1,1,\Delta_1^2,\Delta_2^2}$. The noncentrality parameters of this distribution are

$$\Delta_2^2 = \left\|\Pi_{(t_{10},t_{20})^T}(0,\nu_2)^T\right\|^2 = \frac{t_{20}^2 \nu_2^2}{t_{10}^2 + t_{20}^2}$$

and

$$\Delta_1^2 = \left\|(0,\nu_2)^T - \Pi_{(t_{10},t_{20})^T}(0,\nu_2)^T\right\|^2 = \frac{t_{10}^2 \nu_2^2}{t_{10}^2 + t_{20}^2}.$$

As a result, the following theorem has been proved.

THEOREM 33.3.7
If $H_{1/3}$ is true, then

$$P_1(CC_1(c)) = F_{1,1,\Delta_1^2,\Delta_2^2}(d^2)$$

with ν_2, d, q as in Lemma 33.3.6 and

$$\Delta_1^2 = \frac{\nu_2^{\,2}}{1+q^2}, \quad \Delta_2^2 = \frac{\nu_2^2}{1+1/q^2}.$$

Notice that a corresponding result in Krause and Richter (1999) is different in form and it is not obvious how to transform one of the results into the other in a direct way.

References

1. Anderson, T. W. (1984). *Multivariate Statistical Analysis.* John Wiley & Sons, New York.

2. Cacoullos, T. (1992). Two LDF characterizations of the normal as a spherical distribution. *Journal of Multivariate Analysis* **40**, 205–212.

3. Cacoullos, T. and Koutras, M. (1985). Minimum distance discrimination for spherical distributions. In *Statistical Theory and Data Analysis* (Ed., K. Matusita), pp. 91–102, North-Holland, Amsterdam.

4. Cacoullos, T. and Koutras, M. (1996). On the performance of minimum-distance classification rules for Kotz-type elliptical distributions. In *Advances in the Theory and Practice of Statistics: A Volume in Honor of Samuel Kotz* (Eds., N. L. Johnson and N. Balakrishnan), pp. 209–224. John Wiley & Sons, New York.

5. Hills, M. (1966). Allocation rules and their error rates. *Journal of the Royal Statistical Society, Series B* **28**, 1–31.

6. John, S. (1961). Errors in discrimination. *Annals of Mathematical Statistics* **32**, 1125–1144.

7. Krause, D. and Richter, W.-D. (1994). Geometric approach to evaluating probabilities of correct classification into two Gaussian or spherical categories. In *Information Systems and Data Analysis* (Eds., H.-H. Bock, W. Lenski, and M. Richter), pp. 242–250. Springer-Verlag, Berlin.

8. Krause, D. and Richter, W.-D. (1999). Exact probabilities of correct classifications for repeated measurements from elliptically contoured distributions. Submitted for publication.

9. McLachlan, G. (1992). *Discriminant Analysis and Statistical Pattern Recognition*. John Wiley & Sons, New York.

10. Moran, M. A. (1975). On the expectation of errors of allocation associated with a linear discrimination function. *Biometrika* **62**, 141–148.

11. Schaafsma, W. and van Vark, G. (1977). Classification and discrimination problems with applications, Part I. *Statistica Neerlandica* **31**, 25–45.

34

Estimation of Cycling Effect on Reliability

Vilijandas Bagdonavičius and Mikhail Nikulin
Vilnius University, Lithuania
University Bordeaux 2, Bordeaux, France

ABSTRACT Accelerated life models including cycling effect are introduced. These models are modifications of the additive accumulation of damages and generalized proportional hazards models. Semiparametric estimation procedures are proposed.

Keywords and phrases Accelerated life models, additive accumulation of damages, cycling stress, generalized proportional hazards, semiparametric estimation

34.1 Models

If an item is functioning under a periodic stress, the number of cycles has an effect on reliability. The greater the number of stress cycles, the shorter is the life of items. The purpose of the paper is to formulate models including the effect of cycling and to give semiparametric estimation procedures for reliability characteristics corresponding to the usual stress conditions using experiments under accelerated stresses with greater numbers of cycles and possibly greater amplitudes then those of the usual stress.

Consider modification of the AAD [additive accumulation of damages, Bagdonavičius (1978)] model. Suppose that stresses are differentiable periodic time functions $x(\cdot) : [0, \infty) \to B \in \mathbf{R}$ and let $T_{x(\cdot)}$ be the time-to-failure under the stress $x(\cdot)$.

Under the AAD model the stress changes locally only the scale: the survival function $S_{x(\cdot)}(t) = \mathbf{P}\{T_{x(\cdot)} > t\}$ has the form

$$S_{x(\cdot)}(t) = G\left\{ \int_0^t r\{x(u)\}\, du \right\}.$$

When cycling effect is present, this model is not appropriate. Denote by Δx the amplitude, by Δt the period and by

$$\bar{x} = \frac{1}{\Delta t} \int_0^{\Delta t} x(u)du$$

the mean value of $x(\cdot)$. The number of cycles in the interval $[0, t]$ is

$$n(t) = \int_0^t | \, d\mathbf{1}\{x'(u) > 0\} \, | \, .$$

Consider [cf. Schabe and Viertl (1995)] the following generalization of the AAD model:

$$S_{x(\cdot)}(t) = G\left\{ r_1\{\bar{x}\} \, t + r_2\{\Delta x\} \int_0^t | \, d\mathbf{1}\{x'(u) > 0\} \, | \right\}. \quad (34.1.1)$$

The second term includes the effect of cycling: if the number of cycles and the amplitude are greater then the probability to survive until the moment t is smaller (the function G is decreasing).

Consider also a modification of the generalized proportional hazards (GPH) model [Bagdonavičius and Nikulin (1999b)]. Let

$$\alpha_{x(\cdot)}(t) = \lim_{h \downarrow 0} \frac{1}{h} \mathbf{P}\{T_{x(\cdot)} \in (t, t+h] \mid T_{x(\cdot)} > t\} = -\frac{S'_{x(\cdot)}(t)}{S_{x(\cdot)}(t)}, \ t \geq 0,$$

be the hazard rate function under the stress $x(\cdot)$. Denote by

$$A_{x(\cdot)}(t) = \int_0^t \alpha_{x(\cdot)}(u)du = -ln\{S_{x(\cdot)}(t)\}, \ t \geq 0,$$

the accumulated hazard rate under $x(\cdot)$.

The GPH model holds on E if for all $x(\cdot) \in E$

$$\alpha_{x(\cdot)}(t) = r\{x(t)\} \, q\{A_{x(\cdot)}(t)\} \, \alpha_0(t).$$

The particular cases of the GPH model are the proportional hazards [PH, Cox (1972)] model ($q(u) \equiv 1$) and the AAD model ($\alpha_0(t) \equiv \alpha_0 = const$).

Examples of other particular cases are:

$$\alpha_{x(\cdot)}(t) = r\{x(t)\}(1 + A_{x(\cdot)}(t))^{\gamma+1}\alpha_0(t),$$

$$\alpha_{x(\cdot)}(t) = r(x(t)) \, e^{\gamma A_{x(\cdot)}(t)} \, \alpha_0(t),$$

$$\alpha_{x(\cdot)}(t) = r(x(t)) \, \frac{1}{1 + \gamma A_{x(\cdot)}(t)} \, \alpha_0(t).$$

In terms of survival functions, the GPH model is written in the form:

$$S_{x(\cdot)}(t) = G\left\{\int_0^t r\{x(u)\}\, dH(S_0(u))\right\},$$

where the survival function G is such that it's inverse function H has the form

$$H(u) = \int_0^{-\ln u} \frac{dv}{q(v)},$$

and the baseline function S_0 doesn't depend on the stress. Generalization of the GPH model, including the effect of cycling, has the form

$$S_{x(\cdot)}(t) = G\left\{r_1\{\bar{x}\}\, H(S_0(t)) + r_2\{\Delta x\}\int_0^t |\, d\mathbf{1}\{x'(u) > 0\}\,|\right\}.$$

$$(34.1.2)$$

34.2 Semiparametric Estimation

Semiparametric estimation for the AAD model was considered by Basu and Ebrahimi (1982), Sethuraman and Singpurwalla (1982), Shaked and Singpurwalla (1983), Schmoyer (1986, 1991), Robins and Tsiatis (1992), Lin and Ying (1995), and Bagdonavičius and Nikulin (1999a,b and 2000).

Estimation for the PH model was developed by Cox (1972), Tsiatis (1981), and Andersen and Gill (1982).

Particular cases of the GPH model with parameterization of q was considered by Andersen *et al.* (1993) and Bagdonavičius and Nikulin (1999b), with completely known q by Dabrowska and Doksum (1988a,b), Cheng, Wei and Ying (1995), Murphy, Rossini and van der Vaart (1997), and Bagdonavičius and Nikulin (1997a,b,c).

34.2.1 The First Model

Suppose that the model (34.1.1) is considered, the function G is unknown and the functions r_1 and r_2 are parameterized. The most common parameterizations for this type of models are

$$r_j(u, \beta_j) = \exp\{\beta_{j0} + \beta_{j1}\varphi_j(u)\} \quad (j = 1, 2),$$

where φ_j are specified and $\beta_j = (\beta_{j0}, \beta_{j1})^T$. We can take $\beta_{10} = 0$ because the function G is unknown. Set $\beta = (\beta_1, \beta_2^T)^T$.

Let $x_1(\cdot), ..., x_k(\cdot)$ be periodic differentiable stresses with different cycling rates and possibly different amplitudes.

Suppose that k groups of items are observed. The ith group of n_i items is tested under the stress $x_i(\cdot)$ time t_i.

Denote by $N_i(\tau)$ the number of observed failures of the ith group in the interval $[0, \tau]$ and by $Y_i(\tau)$ the number of items at risk just before the moment τ, by $T_{i1} \leq \ldots \leq T_{im_i}$ the moments of failures of the ith group, $m_i = N_i(t_i)$.

The purpose is to estimate reliability of items under the usual stress $x_0(\cdot)$ with smaller cycling rate and possibly smaller amplitude than those of the stresses $x_i(\cdot)$.

The survival function under the stress x_i is

$$S_{x_i(\cdot)}(t) = G\left\{ r_1\{\bar{x}_i, \beta_1\} t + r_2\{\Delta x_i, \beta_2\} \int_0^t |\, d\mathbf{1}\{x_i'(u) > 0\}\,| \right\}.$$

In the particular case when $\bar{x}_i = \bar{x}_0$ for all i, we can take $r_1(u) \equiv 1$ because G is unknown.

The idea of estimation procedure is similar to the idea of the EM algorithm. At first a pseudoestimator (depending on β) of the survival function G is obtained. After the parameter β is estimated using the maximum likelihood method.

The random variables

$$R_{ij} = r_1\{\bar{x}_i, \beta_1\} T_{ij} + r_2\{\Delta x_i, \beta_2\} \int_0^{T_{ij}} |\, d\mathbf{1}\{x_i'(u) > 0\}\,|$$

can be considered as "observed" pseudofailures in the experiment where n items, $n = \sum_{i=1}^m n_i$, with the survival function G are tested and $n_i - m_i$ among them are censored at the moment $g_i(t_i, \beta)$, where

$$g_i(t, \beta) = r_1\{x_i(u), \beta_1\} t + r_2\{\Delta x_i, \beta_2\} \int_0^t |\, d\mathbf{1}\{x_i'(u) > 0\}\,|.$$

Denote by $h_i(t, \beta)$ the inverse function of g_i with β being fixed. Then

$$N^R(\tau, \beta) = \sum_{i=1}^k N_i(h_i(\tau, \beta))$$

is the number of "observed" failures in the interval $[0, \tau]$ and

$$Y^R(\tau, \beta) = \sum_{i=1}^k Y_i(h_i(\tau, \beta))$$

is the number of items at risk just before the moment τ.

The survival function G can be "estimated" by the Kaplan-Meier estimator [see Andersen et al. (1993)]: for all $s \leq \max_i\{g_i(t_i, \beta)\}$ we have

$$\tilde{G}(s, \beta) = \prod_{\tau \leq s}\left(1 - \frac{\Delta N^R(\tau, \beta)}{Y^R(\tau, \beta)}\right) = \prod_{\tau \leq s}\left(1 - \frac{\sum_{l=1}^k \Delta N_l(h_l(\tau, \beta))}{\sum_{l=1}^k Y_l(h_l(\tau, \beta))}\right),$$

where

$$\Delta N^R(\tau, \beta) = N^R(\tau, \beta) - N^R(\tau-, \beta).$$

If there is no *ex aequo* (ties) then the pseudoestimator $\tilde{G}$ can be written:

$$\tilde{G}(s, \beta) = \prod_{(i,j):T_{ij} \leq h_i(s,\beta)} \left(1 - \frac{1}{\sum_{l=1}^{k} Y_l(h_l(g_i(T_{ij}, \beta), \beta))} \right).$$

The likelihood function is defined as

$$L(\beta) = \prod_{i=1}^{k} \prod_{j=1}^{m_i} [\tilde{G}(g_i(T_{ij}-, \beta), \beta) - \tilde{G}(g_i(T_{ij}, \beta), \beta)] \tilde{G}^{n_i - m_i}(g_i(t_i, \beta), \beta),$$

where

$$\tilde{G}(u-, \beta) = \lim_{\varepsilon \downarrow 0} \tilde{G}(u - \varepsilon, \beta).$$

In parametric models the likelihood function is a product of densities in the observed failure points and survival functions in the censoring points. In our case the density $f = -G'$ is unknown and the factor which corresponds to a failure is replaced by a jump of the function $\tilde{G}$, therefore factors

$$\tilde{G}(g_i(T_{ij}-, \beta), \beta) - \tilde{G}(g_i(T_{ij}, \beta), \beta).$$

are included in $L(\beta)$.

If there are *ex aequo* then denote by $T_1^*(\beta) < \ldots < T_q^*(\beta)$ the different moments among $g_i(T_{ij}, \beta)$, d_j - the number of pseudofailures at the moment $T_j^*(\beta)$. Then for any $s \leq \max_i\{g_i(t_i, \beta)\}$

$$\tilde{G}(s, \beta) = \prod_{j:T_j^*(\beta) \leq s} \left(1 - \frac{d_j}{\sum_{l=1}^{m} Y_l(h_l(T_j^*, \beta)(\beta))} \right)$$

and

$$L(\beta) = \prod_{j=1}^{q} [\tilde{G}(T_{j-1}^*(\beta), \beta) - \tilde{G}(T_j^*(\beta), \beta)]^{d_j} \prod_{i=1}^{k} \tilde{G}^{n_i - m_i}(g_i(t_i, \beta), \beta).$$

The maximum likelihood estimator maximizes $L(\beta)$: $\hat{\beta} = Argmax_\beta L(\beta)$.

The survival function under the usual stress $x_0(\cdot)$ is estimated for any $s \leq \max_i\{g_i(t_i, \beta)\}$ by replacing unknown parameters β by their estimators $\hat{\beta}$ and the stresses $x(\cdot)$ by the usual stress $x_0(\cdot)$ in the formula (34.1.1):

$$\hat{S}_{x_0}(s) = \tilde{G}\left\{ r_1\{\bar{x}_0, \hat{\beta}_1\}t + r_2\{\Delta x_0, \hat{\beta}_2\} \int_0^t | d\mathbf{1}\{x_0'(u) > 0\} |, \hat{\beta} \right\}.$$

34.2.2 The Second Model

Consider the model (34.1.2) with pre-specified parameterization $r_i = r_i(u, \beta_i)$ and $q = q(u, \gamma)$ (or, equivalently, $G = G(t, \gamma) = G_\gamma(t)$) via the parameters $\theta = (\beta_1, \beta_2^T, \gamma)^T$, $\beta_2 = (\beta_{20}, \beta_{21})$ and an unknown baseline function $\alpha_0(t)$.

Suppose that the plan of experiments is as before. Put

$$X_{ij} = T_{ij} \wedge t_i, \quad \delta_{ij} = \mathbf{1}_{\{T_{ij} \le t_i\}}, \quad N_{ij}(t) = \mathbf{1}_{\{T_{ij} \le t, \delta_{ij}=1\}},$$

$$Y_{ij}(t) = \mathbf{1}_{\{X_{ij} \ge t\}}, \quad A_i = A_{x_i(\cdot)}.$$

Estimation is also a two-stage procedure. To obtain a pseudoestimator of the baseline function S_0 we note that the compensator $\Lambda_i(t)$ of the counting process $N_i(t)$ is:

$$\Lambda_i(t) = \int_0^t Y_i(u)\psi(A_i(u-))(r_1(\bar{x}_i)d\,H(S_0(u))$$

$$+ r_2(\Delta x_i) \mid d\mathbf{1}\{x_i'(u) > 0\} \mid),$$

where

$$\psi(u) = \alpha(H(e^{-u})), \quad \alpha = -G'/G.$$

It indicates a pseudoestimator (still depending on θ) for S_0:

$$\tilde{S}_0(t, \theta)$$
$$= G_\gamma \left\{ \int_0^t \frac{dN(t) - \sum_{i=1}^k r_2\{\Delta x_i, \beta_2\} Y_i(u)\,\psi_\gamma(\hat{A}_i(u-)))|d\mathbf{1}\{x_i'(u)>0\}|}{\sum_{i=1}^k r_1(\bar{x}_i, \beta_1) Y_i(u)\,\psi_\gamma(\hat{A}_i(u-))} \right\},$$

where $\hat{A}_i$ is the Nelson-Aalen estimator [see Andersen *et al.* (1993)] of A_i:

$$\hat{A}_i(t) = \int_0^t \frac{dN_i(u)}{Y_i(u)}.$$

The likelihood function

$$L(\theta) = \prod_{i=1}^k \prod_{j=1}^{m_i} [U_i(T_{ij}-, \theta) - U_i(T_{ij}, \theta)]\, U_i^{n_i-m_i}(t_i, \theta), \quad (34.2.1)$$

where

$$U_i(t, \theta) = G_\gamma\{r_1(\bar{x}_i, \beta_1)$$
$$\int_0^t \frac{dN(u) - \sum_{l=1}^k r_2\{\Delta x_l, \beta_2\} Y_l(u)\,\psi_\gamma(\hat{A}_l(u-))\,|d\mathbf{1}\{x_l'(u)>0\}|}{\sum_{l=1}^k r_1(\bar{x}_l, \beta_1)\, Y_l(u)\,\psi_\gamma(A_l(u-))}$$
$$+ r_2\{\Delta x_i, \beta_2\} \mid d\mathbf{1}\{x_i'(u) > 0\} \mid\}. \quad (34.2.2)$$

If k_i are small, the Nelson-Aalen estimators may become inefficient and the estimators $\hat{A}_i$ should be replaced by estimators which are functions of all data:these estimators are determined recurrently [cf. Bagdonavičius and Nikulin (1999b)]:

$$\tilde{A}_i(t,\theta) = \int_0^t \psi_\gamma(\tilde{A}_i(u-,\theta))(r_1(\bar{x}_i,\beta_1)\,dH_\gamma(\hat{S}_0(u,\theta))$$

$$+ r_2(\Delta x_i,\beta_2)\mid d\mathbf{1}\{x_i'(u) > 0\}\mid), \qquad (34.2.3)$$

where

$$H_\gamma(\hat{S}_0(u,\theta))$$

$$= \int_0^t \frac{dN(t) - \sum_{i=1}^k r_2\{\Delta x_i,\beta_2\}Y_i(u)\,\psi_\gamma(\tilde{A}_i(u-,\theta))\mid d\mathbf{1}\{x_i'(u) > 0\}\mid}{\sum_{i=1}^k r_1(\bar{x}_i,\beta_1)Y_i(u)\,\psi_\gamma(\tilde{A}_i(u-,\theta))}.$$

The likelihood function has the form (34.2.1), where the functions $U_i(t,\theta)$ are as in (34.2.2) with the only difference being that the Nelson-Aalen estimators $\hat{A}_i(t)$ are replaced by $\tilde{A}_i(t,\theta)$. Note that to find an initial estimator $\hat{\theta}^{(0)}$ you need initial estimators $\tilde{A}_i^{(0)}(t,\theta)$ and *vice versa*: to find initial estimators $\tilde{A}_i^{(0)}(t,\theta)$ you need an initial estimator $\hat{\theta}^{(0)}$. This problem can be solved by taking the initial estimator $\hat{\theta}^{(0)}$ obtained by the first method with Nelson-Aalen estimators. Then the initial estimators $\tilde{A}_i^{(0)}(t,\theta^{(0)})$ are obtained recurrently from the equations (34.2.3). The estimator $\hat{\theta}^{(1)}$ is obtained by the likelihood function, where $U_i(t,\theta)$ are obtained by the formula (34.2.2), replacing $\hat{A}_i(t)$ by $\tilde{A}_i^{(0)}(t,\theta^{(0)})$ and so on.

The estimator of the survival function $S_{x_0}(t)$, $t \leq \tau$, under the usual stress is

$$\hat{S}_{x(\cdot)}(t)$$

$$= G\left\{r_1\{\bar{x}_0,\hat{\beta}_1\}\,H_{\hat{\gamma}}(\hat{S}_0(t,\hat{\theta})) + r_2\{\Delta x_0,\hat{\beta}_2\}\int_0^t \mid d\mathbf{1}\{x_0'(u) > 0\}\mid\right\}.$$

The asymptotic normality of estimators can be proved by arguments similar to the ones used in Bagdonavičius and Nikulin (1999b) for GPH models.

References

1. Andersen, P. K. and Gill, R. D. (1982). Cox's regression model for counting processes: A large sample study. *Annals of Statistics* **10**, 1100–1120.

2. Andersen, P. K., Borgan, O., Gill, R. D., and Keiding, N. (1993). *Statistical Models Based on Counting Processes.* Springer-Verlag, New York.

3. Bagdonavičius, V. (1978). Testing the hyphothesis of the additive accumulation of damages. *Probability Theory and its Applications.* **23**, 403–408.

4. Bagdonavičius, V. and Nikulin, M. (1997a). Transfer functionals and semiparametric regression models. *Biometrika* **84**, 365–378.

5. Bagdonavičius, V. and Nikulin, M. (1997b). Asymptotic analysis of semiparametric models in survival analysis and accelerated life testing. *Statistics* **29**, 261–283.

6. Bagdonavičius, V. and Nikulin, M. (1997c). Analysis of general semiparametric models with random covariates. *Romanian Journal of Pure and Applied Mathematics* **42**, 351–369.

7. Bagdonavičius, V. and Nikulin, M. (1999a). On semiparametric estimation of reliability from accelerated data. In *Statistical and Probabilistic Models in Reliability* (Eds., D. C. Ionescu and N. Limnios), pp. 75–89. Birkhaüser, Boston.

8. Bagdonavičius, V. and Nikulin, M. (1999b). Generalized proportional hazards models: estimation. *Lifetime Data Analysis* **5**, 329–350.

9. Bagdonavičius, V. and Nikulin, M. (2000). On nonparametric estimation in accelerated experiments with step-stresses. *Statistics* **33**, 349–365.

10. Basu, A. P. and Ebrahimi, N. (1982). Nonparametric accelerated life testing. *IEEE Transactions on Reliability* **31**, 432–435.

11. Cheng, S. C., Wei, L. J., and Ying, Z. (1995). Analysis of transformation models with censored data. *Biometrika* **82**, 835–846.

12. Cox, D. R. (1972). Regression models and life tables. *Journal of the Royal Statistical Society, Series B* **34**, 187–220.

13. Cox, D. R. and Oakes, D. (1984). *Analysis of Survival Data*, Chapman and Hall, London.

14. Dabrowska, D. M. and Doksum, K. A. (1988a). Estimation and testing in a two-sample generalized odds-rate model. *Journal of*

the *American Statistical Association* **83** 744–749.

15. Dabrowska, D. M. and Doksum, K. A. (1988b). Partial likelihood in transformation model with censored data. *Scandinavian Journal of Statistics* **15**, 1–23.

16. Lin, D. Y. and Ying, Z. (1995). Semiparametric inference for the accelerated life model with time dependent covariates. *Journal of Statistical Planning and Inference* **44**, 47–63.

17. Murphy, S. A., Rossini, A. J., and van der Vaart, A. W. (1997). Maximum likelihood estimation in the proportional odds model. *Journal of the American Statistical Association* **92**, 968–976.

18. Robins, J. M. and Tsiatis, A. A. (1992). Semiparametric estimation of an accelerated failure time model with time dependent covariates. *Biometrika* **79**, 311–319.

19. Schabe, H. and Viertl, R. (1995). An axiomatic approach to models of accelerating life testing. *Engineering Fracture Mechanics* **30**, 203–217.

20. Schmoyer, R. (1986). An exact distribution-free analysis for accelerated life testing at several levels of a single stress. *Technometrics* **28**, 165–175.

21. Schmoyer, R. (1991). Nonparametric analysis for two-level single-stress accelerated life tests. *Technometrics* **33**, 175–186.

22. Sethuraman, J. and Singpurwalla, N. D. (1982). Testing of hypotheses for distributions in accelerated life tests. *Journal of the American Statistical Association* **77**, 204–208.

23. Shaked, M. and Singpurwalla, N. D. (1983). Inference for step-stress accelerated life tests. *Journal of Statistical Planning and Inference* **7**, 295–306.

24. Tsiatis, A. A. (1981). A large sample study of Cox's regression model. *Annals of Statistics* **9**, 93–108.

PART VI

Applications to Biology and Medicine

35

A New Test for Treatment vs. Control in an Ordered 2×3 Contingency Table

Arthur Cohen and H. B. Sackrowitz
Rutgers University, New Brunswick, NJ

ABSTRACT Suppose two treatments are to be compared where the responses to the treatments fall into a $2 \times k$ contingency table with ordered categories. Test the null hypothesis that the treatments have the same effect against the alternative that the second treatment is "better" than the first. "Better" means that the probability of falling into the higher of the ordered categories is larger for the second treatment than for the first treatment. (In technical terms what we define as better is stochastically larger.) For $k = 3$ we propose a new test that has very desirable properties. The small sample version of the new test is unbiased, conditionally unbiased, admissible, and has a monotone power function property. It is not easily carried out. The large sample version of the test is approximately unbiased, has desirable monotonicity properties, and is easy to carry out. The new test seems preferable to the likelihood ratio test and to the Wilcoxon-Mann-Whitney test.

Keywords and phrases Wilcoxon-Mann-Whitney test, likelihood ratio test, unbiased test, likelihood ratio order, stochastic order, independence

35.1   Introduction

A classical model in determining the effectiveness of a treatment is to administer a placebo to n_1 individuals and to administer the treatment to n_2 individuals. Suppose the responses are categorical and the categories are ordered. For example, the categories oftentimes are no improvement, some improvement, cured. For such a model one wishes to test the null hypothesis that treatment and placebo effects are the same against some one sided alternative that the treatment is "better" than the placebo.

There are various ways to formally define "better". One way, which is oftentimes convincing, is to say that the treatment is better than the placebo if the proportion of individuals for the treatment group in the higher categories of the ordering, is larger than the comparable proportion in the placebo group. This way of describing "better" is formally called stochastic ordering.

The situation above arises in connection with many $2 \times k$ ordered contingency tables. Each row of the table represents a treatment or a condition while the ordered k levels of the other factor represent some reaction or status that may very well be influenced and can be correlated with the treatment or condition. There are an abundance of such tables. In fact Moses, Emerson, and Hosseini (1984) identified 27 instances of $2 \times k$ tables in a survey of articles in vol. 306, 1982 of the *New England Journal of Medicine*. Many of these articles involve 2×3 tables. In some of the $2 \times k$ tables the row totals are not predetermined as in our opening model. Nevertheless, in the situation where only the total number in all cells is predetermined a comparable testing problem is to test a null of independence against a one sided alternative of stochastic ordering of conditional probabilities.

A very popular approach to this testing problem is to use nonparametric types of tests including the Wilcoxon-Mann-Whitney (WMW) rank test. See for example Rahlfs and Zimmerman (1993), Emerson and Moses (1985), Agresti and Finlay (1997), and Moses, Emerson, and Hosseini (1984). This latter paper is essentially reproduced in the book *Medical Uses of Statistics*, second edition (1992). Reference in the article is made to Stat Xact, a software program based on the WMW test. Grove (1980) derives and studies the likelihood ratio test (LRT) for testing independence vs. stochastic ordering of conditional probabilities assuming a $2 \times k$ ordered contingency table when the total number of observations is fixed (full multinomial model). Bhattacharya and Dykstra (1994) derive and study the LRT for testing equality of two multinomial distributions against stochastic ordering when there are two independent distributions (product multinomial model). More recently, Berger, Permutt, and Ivanova (1998) offers tests for a 2×3 table. Cohen and Sackrowitz (1998) offer a new testing methodology that can be applied in this situation when samples are not large. See also Cohen and Sackrowitz (1999b).

In this paper, for 2×3 ordered contingency tables (under either the full multinomial model or the product multinomial) we will propose a new test. This new test will be decidedly preferable to the WMW test and on an overall basis, preferable to the LRT. It is preferable to the tests proposed by Berger *et al.* and for large samples, preferable to the test methodology of Cohen and Sackrowitz (1998). The test recommended for the problem has very strong theoretical properties that the other tests do not have. In fact the theory exposes serious weaknesses of the WMW test implying that it should not be used. The LRT and Berger *et al.* tests are shown to have shortcomings. The simulation study bears out the theoretical support for the test we recommend.

In the next section we describe the test we recommend. Call it CST. One additional advantage of CST is that it is easy to implement for moderate samples sizes. We give a numerical example. In Section 35.3 we give a simulation study that compares the size and power of CST, LRT, and WMW. In Section 35.4 we indicate some theoretical properties of CST and the shortcomings of the other tests. All proofs are given in the Appendix.

35.2 New Test, Implementation and Example

Consider the following $2 \times k$ contingency table

X_{11}	X_{12}	$\cdots$	X_{1k}	n_1
X_{21}	X_{22}	$\cdots$	X_{2k}	n_2
T_1	T_2		T_k	n

The X_{ij}, $i = 1, 2; j = 1, 2, \ldots, k$ are cell frequencies with $\sum_{j=1}^{k} X_{ij} = n_i$ and $n_1 + n_2 = n$. The T_j, $j = 1, 2, \ldots, k$ are column totals. For the product multinomial model the n_i are fixed sample sizes and $\mathbf{X}_i = (X_{i1}, \ldots, X_{ik})$ has a multinomial distribution with probability vector $\mathbf{p}'_i = (p_{i1}, \ldots, p_{ik})$ such that $\sum_{j=1}^{n_i} p_{ij} = 1$. Test $H_0 : \mathbf{p}_1 = \mathbf{p}_2$ vs. $H_1 - H_0$ where

$$H_1 : \sum_{j=\ell}^{k} p_{1j} \le \sum_{j=\ell}^{k} p_{2j}, \qquad \text{for } \ell = 2, \ldots, k. \qquad (35.2.1)$$

This alternative is called stochastic ordering.

Now let $k = 3$. We first describe our test procedure when n_1 and n_2 have moderate to large samples sizes ($n_i \ge 15$). Let $Y_1 = X_{11}$, $Y_2 = X_{11} + X_{12}$, $\mu_{y_1} = n_1 T_1/n$, $\mu_{y_2} = n_1(T_1 + T_2)/n$, $\sigma_{y_1}^2 = n_1(n - n_1)T_1(n - T_1)/n^2(n-1)$, $\sigma_{y_2}^2 = n_1(n - n_1)(T_1 + T_2)(n - T_1 - T_2)/n^2(n-1)$, $\rho = \sqrt{T_1(n - T_1 - T_2)}/\sqrt{(n - T_1)(T_1 + T_2)}$, $B_1 = (Y_1 - \mu_{y_1})/\sigma_{y_1}$, $B_2 = (Y_2 - \mu_{y_2})/\sigma_{y_2}$, $M = (\sigma_{y_1} - \sigma_{y_2}\rho)/(\sigma_{y_2} - \sigma_{y_1}\rho)$, and let $R(\omega)$ be the decreasing function defined as

$$R(\omega) = \varphi(\omega)/\Phi(\omega), \qquad (35.2.2)$$

TABLE 35.1
Body weight class

	Light	Medium	Heavy
condition absent (p_1)	22	28	14
condition present (p_2)	6	16	18

where $\varphi(\omega)$ is the standard normal density function and $\Phi(\omega)$ is the standard normal cumulative distribution function. (Note $R(-\omega)$ is Mill's Ratio.) Our test procedure is carried out by determining a p-value. If the p-value is to be used for testing at a particular significance level, clearly we reject if the p-value is less than the given level.

Determine the p-value as follows:

Consider whether

$$MR\left(\frac{B_1 - B_2\rho}{\sqrt{1-\rho^2}}\right) \overset{<}{\underset{>}{=}} R\left(\frac{B_2 - B_1\rho}{\sqrt{1-\rho^2}}\right). \qquad (35.2.3)$$

If the left hand side of (35.2.3) > right hand side fix B_2 and find $B_1^* > B_1$ such that equality holds in (35.2.3). Find the p-value from tables of the bivariate normal distribution $BVN((0,0);(1,1,\rho))$ by finding $1-F(B_1^*, B_2)$ where F is the bivariate normal cumulative distribution function. If the left hand side of (35.2.3) < right hand side fix B_1 and find $B_2^* > B_2$ such that equality in (35.2.3) holds and proceed as before.

Note that tables of the bivariate normal $BVN((0,0);(1,1,\rho))$ appear in *Tables of the Bivariate Normal Distribution and Related Functions*, National Bureau of Standards (1959) and is also available in IMSL.

As an example we use some of the data in Zar (1996). Consider Table 35.1.

The data are the frequencies of occurrence among elderly women of a skeletal condition, tabulated by body weight. The null hypothesis is that the distribution for condition present is the same as for condition absent. We note the weight categories are ordered and the alternative hypothesis is that the condition present distribution is stochastically larger than the condition absent distribution.

Find $n_1 = 64$, $n_2 = 40$, $T_1 = 28$, $T_2 = 44$, $\mu_1 = 17.23$, $\mu_2 = 27.08$, $\sigma_{y_1} = 2.21$, $\sigma_{y_2} = 2.30$, $\rho = .405$, $M = .91$, $B_1 = 2.16$, $B_2 = 2.47$. Using (2.3) and tables of the standard normal distribution find $B_1^* = 2.44$. Use IMSL subroutines to find a p-value of .014.

REMARK For the full multinomial model $H_0 : p_{ij} = p_{i.}p_{.j}$ and $H_1 :$ $\sum_{j=\ell}^{k} p_{1j}/p_{1.} \leq \sum_{j=\ell}^{k} p_{2j}/p_{2.}$, where p_{ij} is the cell probability, $p_{i.} = \sum_{j=1}^{k} p_{ij}$, $p_{.j} = \sum_{i=1}^{2} p_{ij}$. The test procedure and all results are exactly the same. This is true since the test is performed conditionally given all marginal totals. Thus it doesn't matter if n_1 and n_2 are treated as fixed. ∎

When the sample sizes are not large the new test can be described but is more difficult to implement. Let $k = 3$. The new test is performed conditionally given n_1, T_1, T_2. Let $\psi(y_1, y_2)$ be the test function i.e. $\psi(y_1, y_2)$ is the probability of rejecting H_0 when (y_1, y_2) is observed. Then CST is as follows:

$$
\psi(y_1, y_2; C_1, C_2, \gamma_1, \gamma_2) = \begin{cases} 1 & \text{if } y_1 > C_1 \text{ or } y_2 > C_2 \\ \gamma_1 & \text{if } 0 \leq y_1 < C_1,\, y_2 = C_2 \\ \gamma_2 & \text{if } y_1 = C_1,\, 0 \leq y_2 < C_2 \\ \max(\gamma_1, \gamma_2) & \text{if } y_1 = C_1,\, y_2 = C_2 \\ 0 & \text{otherwise,} \end{cases}
$$

$$(35.2.4)$$

where $\gamma_1, \gamma_2, C_1, C_2$ are chosen so that the test has size α and satisfies

$$
E_{H_0}\big\{\big[(Y_2 - Y_1) - E_{H_0}(Y_2 - Y_1)\big]\psi(Y_1, Y_2)\big|n_1, T_1, T_2\big\} = 0. \,(35.2.5)
$$

$(E_{H_0}\{\cdot|n_1, T_1, T_2\}$ means conditional expected value under H_0.)

REMARK When sample sizes are not large considerable randomization may be required. In this case a method described in Berger and Sackrowitz (1997) can be used to improve the test described in (35.2.4). ∎

35.3 Simulation Study

We consider the product multinomial model.

In this section we offer simulated power functions for the LRT, the WMW rank test, and for CST. All tests in the simulation were carried out at the .05 level of significance. The LRT was carried out using Theorem 8.F.1 of Bhattacharya and Dykstra (1994). That is, the LRT statistic is calculated and its p-value is determined using the probability bound of Theorem 8.F.1. If the p-value $\leq .05$ the test rejects.

The WMW test was carried out using the limiting normal distribution of the normalized test statistic.

Table 35.2 contains simulated power for the LRT, WMW rank test, and the CST. The simulation was based on 10,000 iterations. The sample sizes for each multinomial was 25. From the table we note that samples sizes of 25 seem adequate for using the normal approximation associated with CST since the size of the test is close to .05. The size of the LRT will always be less than .05. We also note that for those alternatives considered, which represent a wide diversity of alternatives, CST is doing better than LRT in the majority of cases and CST is decidedly better than the WMW rank test.

Additional simulations for sample sizes $(n_1 = 15, n_2 = 20)$, $(n_1 = 25, n_2 = 15)$, $(n_1 = 50, n_2 = 40)$ were done and the same pattern as in the case $(n_1 = 25, n_2 = 25)$ was observed. Thus we feel that the simulation supports the theoretical findings to be discussed and leads us strongly to recommend the CST.

In regard to the WMW test we can refer to Hilton and Mehta (1993). In that paper it was noted that if the WMW test is performed unconditionally and without regard to randomization then its power will be poor. Furthermore the alternative hypothesis for which WMW is considered in that reference is not the stochastic order alternative we study. The alternative which these authors and Mehta, Patel, and Tsiatis (1984) study is the more stringent alternative called likelihood ratio order. See Cohen and Sackrowitz (1999a) for a discussion of various alternatives.

Two further comments regarding CST and WMW. In our simulation the power of CST is doing well relative to WMW even for likelihood ratio order alternatives. Second, our theory will demonstrate that WMW must have very poor conditional power (less than α) for many stochastic order alternatives. It follows that if the alternative is stochastic ordering WMW should not be used.

35.4 Theoretical Properties

The test CST was developed so that it would have the desirable properties of being unbiased and it would lie in a nontrivial complete class of tests. It has the further property of being conditionally unbiased given the marginal totals are held fixed. Still further it has a strictly increasing power function in certain directions in the alternative space.

We describe the theoretical properties in terms of the product multinomial model and remark that the properties hold for the full multinomial model. Recall that the product multinomial model entails $\mathbf{X}_i$, $i = 1, 2$ as independent multinomial vectors denoted by $M(n_i, k, \mathbf{p})$. Let the odds ratio θ_j be defined as $\theta_j = (p_{1j} p_{2k} / p_{2j} p_{1k})$ and let $\nu_j = \log \theta_j$. Let $T_j = X_{1j} + X_{2j}$. The joint density of $\mathbf{X}_1, \mathbf{X}_2$ is

TABLE 35.2
Simulated power functions

POWER

Parameter	CST	LRT	WMW
p_{11} p_{12} p_{13} p_{21} p_{22} p_{23}			
.33 .33 .33 .33 .33 .33	.0511	.0440	.0409
.2 .4 .4 .2 .4 .4	.0504	.0429	.0376
.25 .25 .25 .25 .25 .25	.0499	.0436	.0360
.33 .33 .33 .25 .25 .5	.2657	.2308	.2461
.4 .4 .2 .2 .2 .6	.8564	.8271	.8081
.25 .5 .25 .2 .2 .6	.6838	.7325	.5665
.33 .33 .33 .2 .33 .47	.2866	.2303	.2693
.4 .1 .5 .2 .2 .6	.4139	.3832	.2204
.2 .4 .4 .2 .2 .6	.2559	.3345	.2223
.6 .2 .2 .2 .6 .2	.7807	.8361	.5962
.5 .1 .4 .4 .2 .4	.1237	.1707	.0754
.45 .1 .45 .40 .15 .45	.0837	.0996	.0548
.4 .1 .5 .2 .1 .7	.4716	.3694	.3505
.4 .1 .5 .1 .1 .8	.8195	.7315	.7295
.4 .2 .4 .1 .2 .7	.8209	.7446	.7455
.5 .3 .2 .3 .3 .4	.4805	.3969	.4699
.5 .2 .3 .25 .2 .55	.6022	.5008	.5683

$$\frac{n_1!n_2!}{\prod_{i,j} x_{ij}!} \prod_{i,j} p_{ij}^{x_{ij}}, \quad \sum_{j=1}^{k} x_{ij} = n_i, \quad \sum_{j=1}^{k} p_{ij} = 1. \qquad (35.4.1)$$

Under $H_0 : \mathbf{p}_1 = \mathbf{p}_2$, it follows that $T_j, j = 1, 2, \ldots, k$ are sufficient and complete statistics. The conditional distribution of $\mathbf{X}_1, \mathbf{X}_2$ given $T_1, \ldots, T_k$ is the multivariate extended hypergeometric distribution. Its density can be expressed as

$$\left[\prod_{j=1}^{k} \binom{T_1}{x_{1j}} \prod_{j=1}^{k-1} \theta_j^{x_{1j}} \right] \Big/ \left[\sum_{\mathbf{x}_1 \in A} \prod_{j=1}^{k} \binom{T_j}{x_{1j}} \prod_{j=1}^{k-1} \theta_j^{x_{1j}} \right], \qquad (35.4.2)$$

where $A = \{\mathbf{x}_1^{(1)} : 0 \leq x_{1j} \leq T_j, j = 1, \ldots, k-1, x_{1k} = n_1 - \sum_{j=1}^{k-1} x_{1j} \leq t_k, \sum_{j=1}^{k-1} x_{1j} \leq n_1\}$. From (35.4.2) we note that the conditional distribution of $\mathbf{X}^{(1)}$ given $\mathbf{T}$ is a multivariate exponential family distribution of the form

$$f(\mathbf{x}^{(1)}; \boldsymbol{\nu}) = C_{\mathbf{T}}(\boldsymbol{\nu}) h(\mathbf{x}^{(1)}) e^{\mathbf{x}^{(1)'} \boldsymbol{\nu}}, \qquad (35.4.3)$$

with $\nu_j, j = 1, 2, \ldots, k-1$ as the natural parameters. (The ν_j are log odds ratios formed by the jth and k^{th} columns of the $2 \times k$ table.) Furthermore, under H_0, $\boldsymbol{\nu} = \mathbf{0}$. From Lemma 2.1 of Berger and Sackrowitz (1997) it follows that given a value of $\boldsymbol{\nu}$ with $\nu_1 > 0$, there exists a pair $(\mathbf{p}_1, \mathbf{p}_2)$ such that $\mathbf{p}_1 \leq_{st} \mathbf{p}_2$ ($\mathbf{p}_1$ is stochastically less or equal to $\mathbf{p}_2$) and $(\mathbf{p}_1, \mathbf{p}_2)$ yield the given value of $\boldsymbol{\nu}$. (For that same value of $\boldsymbol{\nu}$ there could exist $(\mathbf{p}_1, \mathbf{p}_2)$ such that $\mathbf{p}_1 \nleq_{st} \mathbf{p}_2$.) For purposes of determining a complete class, an admissible test, or an unbiased test for H_0 vs. H_1, we can find a complete class, an admissible test and an unbiased test for the conditional problem of testing $H_0^* : \boldsymbol{\nu} = \mathbf{0}$ vs. $H_1^* : \nu_1 > 0$. This was noted in Berger and Sackrowitz (1997). Furthermore from (35.4.3) and using arguments of Eaton (1970) it follows that a complete class of tests, $\mathcal{C}$, consists of test functions that are monotone nondecreasing as a function of x_{11} when $x_{12}, \ldots, x_{1(k-1)}$ are fixed. This too was noted in Berger and Sackrowitz (1997).

We now concentrate on the special case $k = 3$. We first offer a set of sufficient conditions for a test $\psi(x_{11}, x_{12})$ to be conditionally unbiased, hence unconditionally unbiased, and to lie in the nontrivial complete class noted earlier. Afterwards we will show that CST satisfies the conditions.

THEOREM 35.4.1
Sufficient conditions for a test $\psi(x_{11}, x_{12})$ to be conditionally unbiased of size α and to lie in C are

$$\psi \text{ is nondecreasing in } x_{11} \text{ for fixed } x_{11} + x_{12} \qquad (35.4.4)$$

$$\psi \text{ is nondecreasing in } x_{11} + x_{12} \text{ for fixed } x_{11} \qquad (35.4.5)$$

$$E_{\nu=0}\psi(X_{11}, X_{12}) = \alpha \qquad (35.4.6)$$

$$E_{\nu=0}[X_{12} - E_{\nu=0}X_{12}]\psi(X_{11}, X_{12}) = 0 \qquad (35.4.7)$$

The function

$$g(X_{12}) = E_{\nu=0}[\psi(X_{11}, X_{12})|X_{12}] \qquad (35.4.8)$$

decreases and then increases.

PROOF See the Appendix. ∎

REMARK The proof of Theorem 35.4.1 demonstrates a stronger result than conditional unbiasedness of the test. The test ψ satisfying properties (35.4.4)–(35.4.8) has a conditional (unconditional) power function that is increasing as a function of ν_1 for every fixed ν_2. ∎

Next we state

THEOREM 35.4.2
The test in (35.2.4) is conditionally unbiased and lies in C.

PROOF See the Appendix. ∎

In Section 35.2 we indicated how to implement the test for moderate or large samples. This implementation was based on some limiting results. The implementation was possible because of

THEOREM 35.4.3

The approximate p-value of the test procedure in (35.2.4) is as follows: Consider whether

$$MR\left(\frac{B_1 - B_2\rho}{\sqrt{1-\rho^2}}\right) \overset{<}{\underset{>}{=}} R\left(\frac{B_2 - B_1\rho}{\sqrt{1-\rho^2}}\right). \tag{35.4.9}$$

If the left hand side of (35.4.9) > right hand side fix B_2 and find $B_1^* > B_1$ such that equality holds in (35.4.9). Find the p-value from tables of the bivariate normal distribution $BVN((0,0);(1,1,\rho))$ by finding $1 - F(B_1^*, B_2)$. If the left hand side of (35.4.9) < right hand side fix B_1 and find $B_2^* > B_2$ such that equality (35.4.9) holds, and proceed similarly as before.

PROOF See the Appendix. ∎

We now make some comments regarding the LRT, the WMW test, and the tests proposed by Berger *et al.* The LRT is not performed conditionally given the marginal totals. The marginal totals form a sufficient and complete statistic under H_0, and therefore to achieve a similarity property the test should be performed conditionally. Hence the LRT test is not similar, meaning that its power properties on alternatives near the null hypothesis space cannot be good. Furthermore the LRT as well as the tests proposed by Berger *et al.* lack an intuitive monotonicity property called concordance monotonicity. See Cohen and Sackrowitz (1998) for a discussion of this property. Still further the asymptotic distribution of the test statistic upon which the LRT is based is difficult to determine. Bhattacharya and Dykstra (1994) offer a bound on the probability of rejecting the null hypothesis by the LRT. If one uses the bound the test is conservative, rejecting less, and resulting in small power.

In this section we pose the testing problem in terms of ν_1, ν_2. It can be shown that the power function of the WMW test increases as ν_2 varies from $-\infty$ to ∞ when ν_1 is zero. This implies that the power of the WMW test will be below α (the power when $\nu_1 = 0$, $\nu_2 = 0$) for negative values of ν_2 when ν_1 is close to zero. What's more the power gets worse as ν_2 becomes more negative. This is a very poor property and for this reason the WMW test cannot be recommended for this testing problem.

Appendix

PROOF OF THEOREM 35.4.1

A complete class of tests for H_0^* vs. H_1^* are weak $*$ limits of Bayes tests. A Bayes test is

$$\psi_B(x_{11}, x_{12}) = \begin{cases} 1 & \text{if } \int_{\Omega_1} C(\boldsymbol{\nu}) e^{x_{11}\nu_1 + x_{12}\nu_2} d\xi(\boldsymbol{\nu}) > K \\ \gamma & \text{if } \int_{\Omega_1} C(\boldsymbol{\nu}) e^{x_{11}\nu_1 + x_{12}\nu_2} d\xi(\boldsymbol{\nu}) = K \\ 0 & \text{otherwise,} \end{cases} \qquad (A.1)$$

where $\Omega_1 = \{\boldsymbol{\nu} : \nu_1 > 0\}$ and $\xi(\boldsymbol{\nu})$ is a prior distribution on Ω_1.

Clearly each Bayes test is nondecreasing in x_{11} for fixed x_{12}. It follows that this property must also hold for $\psi^*(x_{11}, x_{12})$ which is a weak $*$ limit of a sequence of Bayes tests. Therefore the class of tests $\mathcal{C}$, consisting of tests which are nondecreasing in x_{11} for fixed x_{12}, is a complete class.

Now it can be shown that if ψ satisfies (35.4.4) and (35.4.5) it is nondecreasing in x_{11} for fixed x_{12}. Hence $\psi \in \mathcal{C}$.

Next we prove conditional unbiasedness of size α. The size α property is condition (35.4.6). Let $\beta_\psi(\nu_1, \nu_2)$ denote the power function of the test ψ. The partial derivative of the conditional power function with respect to ν_1 is

$$\frac{\partial \beta}{\partial \nu_1} = E_\nu X_{11} \psi(\mathbf{X}) - E_\nu X_{11} E_\nu \psi(\mathbf{X}). \qquad (A.2)$$

Conditions (35.4.4) and (35.4.5) imply that ψ is monotone in the partial sums, $Y_1 = X_{11}$, $Y_2 = X_{11} + X_{12}$. The partial sums of these variables with the hypergeometric distribution of (35.4.2) are associated. See Cohen and Sackrowitz (1994). Hence (A.2) ≥ 0 for every (ν_1, ν_2).

To demonstrate conditional unbiasedness it suffices to show that the conditional power function when $\nu_1 = 0$, $\beta_\psi(0, \nu_2)$ is minimized at $\nu_2 = 0$. Note that the derivative of the function $\beta_\psi(0, \nu_2)$ with respect to ν_2 is

$$E_{\nu_1 = 0, \nu_2} X_{12} \psi(X_{11}, X_{12}) - E_{\nu_1 = 0, \nu_2} X_{12} E_{\nu_1 = 0, \nu_2} \psi(X_{11}, X_{12}). \quad (A.3)$$

Also note that

$$\beta_\psi(0, \nu_2) = E_{\nu_2}\{E_{\boldsymbol{\nu} = \mathbf{0}} \psi(X_{11}, X_{12}) | X_{12}\} = E_{\nu_2} g(X_{12}). \qquad (A.4)$$

Now we use condition (35.4.7) and (A.3) to conclude that $\beta_\psi(0, \nu_2)$ has an extremum at $\nu_2 = 0$. Furthermore condition (35.4.8), (A.5) and a sign change theorem for a one dimensional exponential family density imply that the extremum is a minimum and $\beta_\psi(0, \nu_2)$ is nondecreasing as ν_2 moves away from 0 in both directions. These properties of the conditional power function of ψ along with (A.2) ≥ 0 imply conditional and unconditional unbiasedness. ∎

PROOF OF THEOREM 35.4.2

We prove the theorem by showing there exists a test of the form (35.2.4) satisfying (35.4.4)–(35.4.8). Clearly any test of the form (35.2.4) satisfies (35.4.4) and (35.4.5). Next we show that any such test satisfies (35.4.8). Note that for fixed $X_{12} = Y_2 - Y_1 < C_2 - C_1$, the test rejects if $X_{11} > C_1$. However the conditional distribution of $X_{11}|X_{12}$ when $\boldsymbol{\nu} = 0$ is stochastically decreasing in X_{12}. This implies that $g(X_{12})$ increases as X_{12} varies from $C_2 - C_1$ to smaller values. Furthermore when $X_{12} > C_2 - C_1$, the test rejects when $X_{11} + X_{12} > C_2$. But the conditional distribution of $X_{11} + X_{12}|X_{12}$ is stochastically increasing as X_{12} increases. This implies that $g(X_{12})$ increases as X_{12} increases from $C_2 - C_1$ and establishes condition (35.4.8).

To show that (35.2.4) satisfies (35.4.6) and 35.4.7) we note first that tests in (35.2.4) that are non-randomized in one of the two directions can be written as

$$\psi(\mathbf{Y}; C_1, C_2, 0, \gamma_2) = \psi(\mathbf{Y}; C_1 + 1, C_2, 1, \gamma_2) \text{ and}$$

$$\psi(\mathbf{Y}; C_1, C_2, \gamma_1, 1) = \psi(\mathbf{Y}; C_1, C_2 - 1, \gamma_1, 0). \quad (A.5)$$

Now note that there exists a test $\psi_1^*(Y_1)$ defined by

$$\psi_1^*(Y_1) = \begin{cases} 1 & \text{if } Y_1 > C_1^* \\ \gamma_1^* & \text{if } Y_1 = C_1^* \\ 0 & \text{if } Y_1 < C_1^*, \end{cases} \quad (A.6)$$

such that $0 \leq \gamma_1^* < 1$ and $E_0\psi_1^*(Y_1) = \alpha$. Also there exists a test

$$\psi_2^*(Y_2) = \begin{cases} 1 & \text{if } Y_2 > C_2^* \\ \gamma_2^* & \text{if } Y_2 = C_2^* \\ 0 & \text{if } Y_2 < C_2^*, \end{cases} \quad (A.7)$$

such that $0 < \gamma_2^* \leq 1$ and $E_0\psi_2^*(Y_2) = \alpha$. In the above definitions of ψ_1^*, ψ_2^*, if either test is non-randomized then $\gamma_1^* = 0$ or $\gamma_2^* = 1$ respectively.

We next illustrate that there exists a collection of tests $\psi_\beta(\mathbf{Y}) = \psi(\mathbf{Y}; C_1(\beta), C_2(\beta), \gamma_1(\beta), \gamma_2(\beta))$, indexed by $\beta \in [0, 1]$ such that

$$\psi_\beta(\mathbf{Y}) \text{ is of the form (35.2.4)} \qquad (A.8)$$

$$\psi_0(\mathbf{Y}) = \psi_2^*(Y_2) \text{ and } \psi_1(\mathbf{Y}) = \psi_1^*(\mathbf{Y}) \qquad (A.9)$$

$$E_0\psi_\beta(\mathbf{Y}) = \alpha \text{ for all } \beta \in [0,1] \qquad (A.10)$$

$$E_0(X_{12} - E_0 X_{12})\psi_\beta = \text{cov}_0(X_{12}, \psi_\beta(\mathbf{Y}))$$
$$\text{is a continuous function of } \beta. \qquad (A.11)$$

To see this note

$$\psi_1^*(Y_1) = \psi(\mathbf{Y}; C_1^*, U_2, 0, \gamma_2^*) = \psi_1(\mathbf{Y}) \qquad \text{say}$$
$$\psi_2^*(Y_2) = \psi(\mathbf{Y}; L_1, C_2^*, \gamma_1^*, 1) = \psi_0(\mathbf{Y}) \qquad \text{say,}$$

where U_2 is the largest possible value of Y_2 and L_1 is the smallest possible value of Y_1.

Note that beginning with any test of the form (35.2.4) with $0 \le \gamma_1 < 1$ and $0 < \gamma_2 \le 1$ for which $E_0\psi(\mathbf{Y}) = \alpha$ we can generate a collection of tests having size α by letting γ_1 decrease and γ_2 increase (accordingly) until either γ_1 reaches 0 or γ_2 reaches 1. When either of these occur we use (A.5) to rewrite the test so that again $\gamma_1 > 0$ and $\gamma_2 < 1$. It can be seen that this process will end at $\psi_0(\mathbf{Y})$ and yield a collection of tests satisfying (A.8)–(A.11).

Since $\text{cov}_0(X_{12}, \psi_1^*(\mathbf{Y})) \le 0$ and $\text{cov}_0(X_{12}, \psi_2^*(\mathbf{Y})) \ge 0$ it follows from continuity that $\text{cov}_0(X_{12}, \psi_\beta(\mathbf{Y})) = 0$ for some $\beta \in [0,1]$. Hence (35.4.6) and (35.4.7) are satisfied. ∎

The proof of Theorem 35.4.3 is based on a limiting result of Van Eeden (1965). To state her result precisely let

$x_{11\nu}$	$x_{12\nu}$	$x_{13\nu}$	$n_{1\nu}$
$x_{21\nu}$	$x_{22\nu}$	$x_{23\nu}$	$n_{2\nu}$
$t_{1\nu}$	$t_{2\nu}$	$t_{3\nu}$	n_ν

be a sequence of 2×3 tables with $\lim_{\nu\to\infty} n_\nu = \infty$. Let $\mu_{j\nu} = EX_{1j\nu}$, $\sigma_{j\nu}^2 = \text{Var} X_{1j\nu}$, $\rho_{X_{11\nu}, X_{12\nu}} \equiv$ correlation $(X_{11\nu}, X_{12\nu})$. Assume

$\lim_{\nu \to \infty} \mu_{j\nu} = \mu_j = \infty$, $\lim_{\nu \to \infty} n_{j\nu} = n_j = \infty$, $\lim_{\nu \to \infty} t_{j\nu} = t_j = \infty$, $\lim \rho_{X_{11\nu},X_{12\nu}} = \rho_{X_{11},X_{12}}$ exists. Then $\mathbf{U}_\nu = (U_{1\nu}, U_{2\nu})'$, where $U_{j\nu} = (X_{j\nu} - \mu_{j\nu})/\sigma_{j\mu}$ has a limiting bivariate normal distribution with mean vector $\mathbf{0}$ and covariance matrix $\Sigma_{\mathbf{U}} = \begin{pmatrix} 1 & \rho_{X_{11},X_{12}} \\ \rho_{X_{11},X_{12}} & 1 \end{pmatrix}$.

PROOF OF THEOREM 35.4.3

The normal approximation can be used so that the acceptance region of (35.2.4) becomes $\{\mathbf{Z} : Z_1 \leq K_1, Z_2 \leq K_2\}$ where $Z_i = (Y_i - \mu_{y_i})/\sigma_{y_i}$ and K_1, K_2 are chosen so that (35.4.6) and (35.4.7) are satisfied. To determine the p-value corresponding to such a test procedure, the observed value of (Z_1, Z_2), say (B_1, B_2) must be on the boundary of the acceptance region, in Z-space, of a unique unbiased test region. If it can be established that (35.4.7) is approximated in (35.4.9) with equality, then the size of the unique unbiased test region with (B_1, B_2) on its boundary will be the p-value. This unique unbiased test region will either be (B_1, B_2^*) or (B_1^*, B_2) as outlined in Theorem 35.4.3. Only one of these pairs satisfies (35.4.9) with equality.

Hence the theorem will be proved if we show that (35.4.9) with equality, results by using the normal approximation in (35.4.7). Note (35.4.7) can be expressed as $\operatorname{cov}(X_{12}, \psi(X_{11}, X_{12})) = 0$. In terms of (Y_1, Y_2) this condition becomes

$$E_{\nu=0}\{[(Y_2 - Y_1) - E(Y_2 - Y_1)]\psi(Y_1, Y_2 - Y_1)\} = 0, \qquad (A.12)$$

and in terms of (Z_1, Z_2) it becomes

$$E_{\nu=0}\{(Z_2\sigma_{y_2} - Z_1\sigma_{y_1})\psi^*(\mathbf{Z})\} = 0. \qquad (A.13)$$

Use the fact that

$$\psi^*(\mathbf{Z}) = \begin{cases} 0 & \text{if } Z_1 \leq K_1, Z_2 \leq K_2 \\ 1 & \text{otherwise,} \end{cases}$$

in (A.13) and the fact that $EZ_i = 0$ to rewrite A.13) as

$$\int_{-\infty}^{K_1} \int_{-\infty}^{K_2} \frac{z_2\sigma_{y_2}}{\sqrt{2\pi(1-\rho^2)}} e^{-(z_2-\rho z_1)^2/2(1-\rho^2)} \cdot e^{-z_1^2/2} dz_2 dz_1$$

$$- \int_{-\infty}^{K_1} \int_{-\infty}^{K_2} \frac{z_1\sigma_{y_1}}{\sqrt{2\pi(1-\rho^2)}} e^{-(z_2-\rho z_1)^2/2(1-\rho^2)} \cdot e^{-z_1^2/2} dz_2 dz_1 = 0.$$

$$(A.14)$$

Integrating in (A.14) and substituting (B_1, B_2) for (K_1, K_2) gives (35.4.9) with equality.

Acknowledgements Research of the first author (Arthur Cohen) was supported by NSF Grant DMS-9400476. Research of the second author (H. B. Sackrowitz) was supported by NSA Grant 904-901-1-0506.

References

1. Agresti, A. and Finlay, B. (1997). *Statistical Methods in the Social Sciences*, Second edition, Dellen c/o Macmillan, San Francisco.

2. Berger, V., Permutt, T., and Ivanova, A. (1998). Convex hull tests for ordered categorical data. *Biometrics* **54**, 1541–1550.

3. Berger, V. and Sackrowitz, H. B. (1997). Improving tests of stochastic order. *Journal of the American Statistical Association* **92**, 700–705.

4. Bhattacharya, B. and Dykstra, R. L. (1994). Statistical inference for stochastic ordering. In *Stochastic Orders and Their Applications* (Eds., M. Shaked and J. George Shantikumar), pp. 221–249. Academic Press, New York.

5. Cohen, A. and Sackrowitz, H. B. (1994). Association and unbiased tests in statistics. In *Stochastic Orders and Their Applications* (Eds., M. Shaked and J. George Shantikumar), pp. 251–274. Academic Press, New York.

6. Cohen, A. and Sackrowitz, H. B. (1998). Directional tests for one-sided alternatives in multivariate models. *Annals of Statistics* **26**, 2321–2338.

7. Cohen, A. and Sackrowitz, H. B. (1999a). Testing whether treatment is "better" than control with ordered categorical data: Definitions and complete class theorems. *Statistics and Decisions* (to appear).

8. Cohen, Λ. and Sackrowitz, H. B. (1999b). Testing whether treatment is "better" than control with ordered categorical data: An evaluation of new methodology. Submitted for publication.

9. Eaton, M. L. (1970). A complete class theorem for multidimensional one sided alternatives. *Annals of Mathematical Statistics* **41**, 1884–1888.

10. Emerson, J. D. and Moses, L. E. (1985). A note on the Wilcoxon-Mann-Whitney test for $2 \times k$ ordered tables. *Biometrics* **41**, 303–

309.

11. Grove, D. M. (1980). A test of independence against a class of ordered alternatives in a $2 \times C$ contingency table. *Journal of the American Statistical Association* **75**, 454–459.

12. Hilton, J. F. and Mehta, C. R. (1993). Power and sample size calculations for exact conditional tests with ordered categorical data. *Biometrics* **49**, 609–616.

13. Mehta, C. R., Patel, N. R., and Tsiatis, A. A. (1984). Exact significance testing to establish treatment equivalence with ordered categorical data. *Biometrics* **40**, 819–825.

14. Moses, L. E., Emerson, J. D., and Hosseini, H. (1984). Analyzing data for ordered categories. *New England Journal of Medicine* **311**, 442–448.

15. Moses, L. E., Emerson, J. D., and Hosseini, H. (1992). Analyzing data for ordered categories. In *Medical Uses of Statistics*, Second edition (Eds., J. C. Bailar III and F. Mosteller), NEJM Books, Boston, MA.

16. National Bureau of Standards (1959). *Tables of the Bivariate Normal Distribution and Related Functions*, Applied Mathematics Series **50**, U.S. Government Printing Office, Washington, D.C.

17. Rahlfs, V. W. and Zimmerman, H. (1993). Scores: Ordinal data with few categories – how they should be analyzed. *Drug Information Journal* **27**, 1227–1240.

18. Van Eeden, C. (1965). Conditional limit distributions for the entries in a $2 \times k$ contingency table. In *Classical and Contagious Discrete Distributions* (Ed., G. P. Patil), pp. 123-126. Pergamon, Oxford, England.

19. Zar, J. H. (1996). *Biostatistical Analysis*. Prentice Hall, New Jersey.

36

An Experimental Study of the Occurrence Times of Rare Species

Marcel F. Neuts
The University of Arizona, Tucson, AZ

ABSTRACT We describe a computer experiment to study the waiting times between the first occurrences of rare values in a sequence of independent, identically distributed random variables taking positive integer values. The experiment proceeds in two stages; the first stage requires detailed simulation and lasts only until all the alternatives that have substantial probabilities have occurred. In the second stage, we generate only the times when potentially new, rare values can occur. That stage allows us to look effectively and efficiently at millions of trials. While the proposed methodology is general, we report some findings only for prevalence distributions with a geometric tail.

Keywords and phrases Computer experimentation, coupon collector's problem, waiting times between the occurrences of rare species

36.1 Statement of the Problem

We consider a sequence of independent, identically distributed random variables $\{X_n\}$ taking values in the set of natural numbers. The probabilities $P[X_n = j] = p(j)$, $j \geq 1$, are assumed to be positive and, without loss of generality, non-increasing in j. We are interested in the occurrence times $\{Z(k)\}$ of *new* values in the sequence $\{X_n\}$.

When the density $\{p(j)\}$ concentrates only on the integers $1, \ldots, m$, there is an extensive literature on various related waiting time problems under the heading of as the *Coupon Collector's Problem*. Boneh and Hofri (1997) give a recent survey of that literature. An algorithm for the computation of the probability density of $Z(m)$ and related moments is given in Neuts and Carson (1975) or in Neuts (1995).

Because of the involved dependence on the past of the successive new values in the observed sequence, very little about the coupon collector's problem is amenable to classical analysis. When many of the quantities $p(j)$ are small, when m is large and, *a fortiori* when it is infinite, the capabilities of exact algorithms are limited. The problem is therefore well-suited to exploration by computer experiments.

In this paper, we focus on one experiment only. However, the proposed method is also applicable to other aspects of the coupon collector's problem. For example, in the case of finite m, when many $p(j)$ are small, even in a long string of (simulated) data, the corresponding j-values typically occur only once. It would be interesting to see what can be learned about the relative magnitudes of the corresponding $p(j)$ from the orders of appearance and the times between the occurrences of these isolated rare values. The method proposed here is also useful in that investigation.

Specifically, we experimentally study the relationship between the asymptotic behavior of the sequence $\{p(j)\}$ and that of the inter-occurrence times $Y(k) = Z(k) - Z(k-1)$.

36.2   The Design of the Experiment

For operational purposes, we define two integers N_0 and N, where

$$N_0 = \min \left[\min \left\{ j : \sum_{\nu=1}^{j} p(\nu) \geq 1 - \beta \right\}, \min \{j : p(j) \leq \beta_1\} \right],$$

$$(36.2.1)$$

and

$$N = \min \left\{ j : \sum_{\nu=1}^{j} p(j) \geq 1 - \gamma \right\}, \qquad (36.2.2)$$

where β and β_1 are small positive quantities with typically $\beta = .05, .02$, or .01. The second quantity, β_1, is much smaller than β and typically is set to .001 or .0005. The constant γ is very small, say, $\gamma = 10^{-12}$.

The indices $j \leq N_0$ make up the set of *common indices*. The purpose of β_1 is to avoid including too many alternatives of small probability among the common indices. The index N serves to eliminate alternatives so rare that they are unlikely even to occur in millions of trials. Provided that γ is of a low order of magnitude, our experimental results are insensitive to the specific choice of N. The indices j with $N_0 < j \leq N$ make up the set of *rare indices* .

In the simulation, we use the sequence $\{p(j)\}$ truncated at N. The corresponding computer array is renormalized to sum to one, but the effect of doing so is negligible. An alternative is to place the neglected mass (which is smaller than γ) at an extra alternative $N+1$ corresponding to a *super-rare* event – the sighting of a unicorn, perhaps? Again, that is a matter of choice, but it is of little or no consequence.

The definitions of common and rare species are clearly operational and, therefore, somewhat arbitrary. In some applications, these definitions require revision. For many plausible probability densities $\{p(j)\}$, N_0 is rarely in excess of 75 while N may be a few hundreds (but, exceptionally, can much larger).

In Stage 1 of the experiment, we generate and examine variates with the (truncated) density $\{p(j)\}$ *until all common indices have occurred at least once.* In Stage 2, we generate the first occurrence times of rare indices by an efficient procedure described in Section 3.

Variates from a discrete probability density are generated by the classical *Alias Method*, see e.g., Devroye (1986). The efficacy of that method in generating a large set of variates from a discrete distribution more than offsets the modest overhead computation time needed to set up the alias tables. In Stage 1, we generate variates from $\{p(j)\}$ and we keep a list of the indices that have already appeared. A next variate is generated and checked against those in the list. If it is new, it is added to the list.

After a while, checking recurring common indices wastes too much time. The redundant effort in checking successive variates is precisely the reason for the second stage. Running the code of Stage 1 to check, say, 50,000,000 variates lacks elegance and takes a prohibitively long processing time. In typical runs, Stage 1 requires around 1,000 variates, rarely more. On a good PC, Stage 1 thus takes a negligible processing time. Incidentally, the distribution of the waiting time to see all common events, that is here the number of variates required to complete Stage 1, can be computed by the procedure in Neuts (1995). As discussed there, when the successive trials are performed at the events of a Poisson process of rate one, we obtain a simpler form of the waiting time distribution. That form is better suited to compute, say, an (approximate) high percentile of the waiting time distribution. Given the rapidity of execution of Stage 1, we did not consider a further theoretical study of its duration to be worthwhile.

Rare indices may also occur during Stage 1. We keep these indices in a separate list, say, *RI1*, (*rare indices in Stage 1*). After completion of Stage 1, the probability that a subsequent variate from $\{p(j)\}$ already occurred during Stage 1 is given by

$$\sum_{j=1}^{N_0} p(j) + \sum_{RI1} p(j) = 1 - \beta^*. \tag{36.2.3}$$

Equivalently, β^* is the probability that an index has not shown up during Stage 1.

Before starting Stage 2, we delete all indices up to N_0 or belonging to the set $RI1$. We normalize the resulting array of $p(j)$ by dividing each term by β^*. That is the conditional density of a random index given that it did not appear in Stage 1. We compute the alias table of that density.

36.3   Stage 2 of the Experiment

We denote the successive inter-occurrence times of new labels in Stage 2 by $Y^*(k)$ for $k \geq 1$. In studying asymptotic behavior, we use only these occurrence times. To make that clear, we add an asterisk to the random variables related to Stage 2. The first new label appears at the first success in a sequence on independent Bernoulli trials with success probability β^*. It is therefore found after a geometrically distributed waiting time $Y^*(1)$ with parameter β^*. We place that label in a list $RI2$ of new labels found in Stage 2. The waiting time to the next opportunity for a new label has the same geometric distribution. We generate a variate from the conditional density and, if it is *not* in $RI2$, it is added to that list. The corresponding waiting time is then $Y^*(2)$. If the label belongs to $RI2$, we continue generating geometric waiting times and variates until a new label is found and added to $RI2$. The waiting time $Y^*(2)$ is then a random sum of independent geometric variables with parameter β^*. Subsequent inter-occurrence times $Y^*(k)$ are constructed in the same way. Since the list $RI2$ grows, the successive $Y^*(k)$ are random sums of geometric variables in which the sequence of the numbers of summands is stochastically increasing. The sequence of the interarrival times $\{Y^*(k)\}$ is therefore also stochastically increasing.

Thus, the sequence $Y^*(k)$ is generated by drawing successive geometric variates and, for each, a variate K from the conditional density of a random index that did not appear in Stage 1. To see whether it is new, the variate K needs to be checked against the short list $RI2$ only. Since β^* is typically small, the geometric variates are large. The $Y^*(k)$ are obtained without the onerous checking of repeating common labels. We keep track of their successive partial sums $Z^*(k)$ and we stop when the sum exceeds a specified number such as 20,000,000 or 50,000,000.

36.4 Findings

So far, we have primarily examined the sequence of inter-occurrence times $Y^*(k)$ for distributions with heavy *geometric tails*, that is with a parameter smaller than 0.1. For these cases, we find a sufficiently large number of new indices (typically on the order of two hundred) with the stopping time set equal to 50,000,000. We plotted the logarithms of the $Y^*(k)$ and found that these consistently exhibit a close to linear behavior. The parameters of the linear regression lines fitted to these simulation results show little variation in replications of the experiment. That variation is consistent with the relatively small sample sizes, that is the number of rare indices found even in as many as 50,000,000 trials. The number of rare indices found is remarkably consistent across replications. As an illustration, we display the parameters of the linear fits, the sample sizes, and the stopping times for 25 replications in Table 36.1. In that example, the density $\{p(j)\}$ is *geometric*.

Our results exemplify the challenges in examining asymptotic behavior by simulation. Even if the difficulties associated with simulating rare events can be avoided, the asymptotic parameters must typically be estimated from few data points. The type of sample variability seen in Table 36.1 is representative.

Particularly for short-tailed prevalence distributions, such as Poisson, the combination of rarity and small sample sizes adds to the difficulty of experimentation. On the other hand, the asymptotic behavior may be most interesting for heavy-tailed distributions. We explored cases of the zeta density given by

$$p(j) = [\zeta(\lambda)]^{-1} j^{-\lambda}, \text{for} j \geq 1. \tag{36.4.1}$$

With $\lambda = 3$ and $\gamma = 10^{-11}$, we find that there are $N = 9405$ labels in all of which only $N_0 = 20$ are common. In a particular run, Stage 1 took 6,369 steps and that number does not vary significantly across replications. In that same run, only 7 rare labels showed up in Stage 1 and the smallness of that number is also typical. As is to be expected, heavier tailed distributions are somewhat more generous in producing rare labels. With a stopping time of 50,000,000, that run produced 418 labels during Stage 2 and that number again does not vary significantly across replications.

The logarithms of the $Y^*(k)$ now consistently exhibit a behavior that is well-captured by a regression line of the form

$$y = a + bk^c, \tag{36.4.2}$$

TABLE 36.1
Parameters of the linear regression lines for the logarithms of the
times between occurrences of rare indices in **50,000,000** trials using
a geometric prevalence distribution with success probability **0.05**.
The first column gives the number of the replication; the second,
the intercept, and the third the slope of the linear regression. **J** is
the number of rare indices found in **50,000,000** trials in stage **2**.
The final index gives the time at which the simulations stopped,
the time variable having exceeded **50,000,000**

Nr	a	b	J	Final index
1	-.279E-02	.706E-01	230	50000017
2	-.424E-02	.796E-01	218	50000009
3	-.421E-02	.788E-01	218	50000005
4	-.346E-02	.736E-01	227	50000000
5	-.382E-02	.746E-01	229	50000031
6	-.370E-02	.739E-01	226	50000009
7	-.292E-02	.693E-01	239	50000004
8	-.456E-02	.802E-01	215	50000050
9	-.359E-02	.725E-01	229	50000002
10	-.410E-02	.752E-01	224	50000011
11	-.348E-02	.721E-01	228	50000048
12	-.347E-02	.729E-01	230	50000005
13	-.396E-02	.757E-01	223	50000121
14	-.395E-02	.782E-01	216	50000017
15	-.329E-02	.706E-01	234	50000009
16	-.481E-02	.828E-01	211	50000002
17	-.250E-02	.667E-01	242	50000038
18	-.431E-02	.767E-01	222	50000021
19	-.592E-02	.847E-01	215	50000060
20	-.371E-02	.724E-01	231	50000002
21	-.305E-02	.703E-01	235	50000050
22	-.335E-02	.718E-01	226	50000018
23	-.416E-02	.794E-01	214	50000000
24	-.340E-02	.720E-01	229	50000001
25	-.427E-02	.750E-01	228	50000037

where the variability of the parameter estimates is similar to that in Table 36.1. However, in the interest of brevity, we do not display results for replications of the experiment with the zeta density.

Acknowledgements The author thanks Professor Michael Rosenzweig for stimulating discussions on possible applications. This research was supported in part by NSF Grant Nr. DMI-9306828.

References

1. Boneh, A. and Hofri, M. (1997). The coupon-collector problem revisited – engineering problems and computational methods. *Communications in Statistics — Stochastic Models* **13**, 36–66.

2. Devroye, L. (1986). *Non-Uniform Random Variate Generation.* Springer-Verlag, New York, Berlin.

3. Neuts, M. F. and Carson, C. C. (1975). Some computational problems related to multinomial trials. *Canadian Journal of Statistics* **3**, 235–248.

4. Neuts, M. F. (1995). *Algorithmic Probability: A Collection of Problems.* Chapman and Hall, London.

37

A Distribution Functional Arising in Epidemic Control

Niels G. Becker and Sergey Utev
Australian National University
Institute of Mathematics, Novosibirsk University, Russia

ABSTRACT The properties of a relatively new functional, [Becker and Utev (1998)], are investigated, with emphasis on extreme values with respect to a probability distribution. Scale invariance is established and degenerate distributions are shown to give extreme values. The functional gives, for a certain strategy of vaccination, the fraction of the households that must be immunized to prevent epidemics within a community consisting of a large number of households. Interest lies in seeing how the distribution of household size affects the estimate of the critical immunity coverage. The functional also gives the fraction of individuals that must be immunized to prevent epidemics in a community of individuals who mix uniformly but whose susceptibility is age-specific. Interest then lies in seeing how the distribution of susceptibility affects the estimate of the critical immunity coverage.

Keywords and phrases Epidemic control, functional of distributions, inequalities for functionals

37.1 Introduction

Certain functionals of probability distributions have an important role in probability theory and mathematical statistics. Classical examples are the Fisher information $I(X) = \int [(\ln f)']^2 f dx$ and the Shannon entropy $H(X) = -\int f \ln f dx$. Their study is of interest both for their own sake and for the contribution they make to vital practical problems. See Huber (1981) for applications of Fisher information in statistics, Ellis (1985) for applications of Shannon entropy in statistical mechanics, Barron (1986), Chen (1988), and Cacoullos, Papathanasiou, and Utev (1994) for applications in limit theorems of probability theory.

573

Here we study a distribution functional that is important as an estimate of the immunity coverage required to prevent epidemics in a community of households [see Becker and Utev (1998)]. In part our motivation is to see how the distribution of household size in the community affects this estimate. The same functional is an estimate of the immunity coverage required to prevent epidemics in a community of individuals with varying susceptibility to infection. Our interest then lies in seeing how the distribution of susceptibility over community members affects the estimate.

We first introduce the functional of interest, then establish some of its properties and finally discuss some insights these results provide in the context of epidemic control.

The Functional of Interest

Let X be any nonnegative random variable with finite expectation $EX > 0$, and denote its distribution function by F. Fix a real a, with $0 < a < 1$, and for convenience let $\bar{a} = 1 - a$. The functional of main interest is

$$v(X) := 1 - \frac{1}{\theta(X)\,EX}, \qquad (37.1.1)$$

where the functional $\theta(X)$ is defined by

$$\theta(X): \quad \text{such that} \quad a = E\exp[-\bar{a}\,X\,\theta(X)]. \qquad (37.1.2)$$

It is important to remember that the argument X in $\theta(X)$ and $v(X)$ only reflects their dependence on F. That is, they are not random variables. Sometimes the notation $\theta(F)$ and $v(F)$ is more convenient than $\theta(X)$ and $v(X)$.

Note that the functional $\theta(X)$ exists and is unique when $P(X = 0) < a$. For $P(X = 0) \geq a$, it is convenient to take $\theta(X) = \infty$.

37.2   Properties of the Functional

A substantial part of the theorem below deals with local behaviour with respect to a perturbation of the form $X + \delta_h Y$, for sufficiently small positive $h > 0$, where δ_h is a stochastic process which is independent of the pair (X, Y) and for which we assume that

$$E\delta_h = h \quad \text{for any } h > 0 \qquad \text{and}$$

$$h^{-1}(1 - E\exp[-t\delta_h]) \to Z(t) \text{ as } h \to 0, \text{ for any } t \geq 0. \quad (37.2.1)$$

We consider $Y \geq 0$ and $0 \leq \delta_h \leq 1$. The following parameters will be useful:

$$V(h) = v(X + \delta_h Y), \quad Z_X = \frac{E[X \exp(-\theta \bar{a} X)]}{EX\, E \exp(-\theta \bar{a} X)},$$

$$p_X = \frac{E^2 X}{EX^2} \quad \text{and} \quad \Delta_X = \bar{a} \frac{\text{Var}(X)}{E^2 X}.$$

The main properties of the functional $v(X)$ are as follows:

THEOREM 37.2.1

Using the above notation we have:

1. *For any $c > 0$ we have $v(cX) = v(X)$. In particular, $v(c) = v(1) = 1 + \bar{a}/\ln a$.*

2. *$v(X) \geq v(EX)$, with equality if and only if X has a degenerate distribution.*

3. *$v(X) \leq v(\text{Bin}[1, p_X]) = 1 - (\bar{a} + \Delta_X)/\ln(a - \Delta_X)^+$, where $x^+ = \max(x, 0)$.*

4. *$V'(0)$ has constant sign for all $Y \geq 0$ if and only if X has a degenerate distribution.*

5. *The following statements are equivalent:*
   S1. $V'(0)$ has constant sign for all non-negative Y that are independent of X ;
   S2. X is degenerate or $Z(t)/t \geq Z_X$ for all $t > 0$

Discussion

Assertion 1 indicates scale invariance, while Assertion 2 implies that the distribution functional $v(X)$ assumes its smallest value for degenerate distributions. Assertions 2 and 3 indicate that $v(X)$ can be approximated by $v(EX)$ when the relative variability of X is small; that is, when $\text{Var}(X)/E^2 X$ is small.

To obtain a richer result about extremes of the functional $v(X)$, we consider its local behaviour with respect to a perturbation of the form $X + \delta_h Y$. Assertion 4 states that the only stable points of $V(h)$ with respect to perturbations in the direction Y, where Y is an arbitrary positive random variable, are degenerate distributions. That is, for each strictly positive random variable X with a nondegenerate distribution there exist strictly positive random variables Y_1 and Y_2 such that

$$v(X + \delta_h Y_1) < v(X) < v(X + \delta_h Y_2) \quad \text{for all } 0 < h < h_0. \quad (37.2.2)$$

Two particular types of perturbation promote understanding: the non-random $\delta_h \equiv h$, for which $Z(t) \equiv t$, and Bernoulli chaos with $P(\delta_h = 1) = h = 1 - P(\delta_h = 0)$, for which $Z(t) \equiv 1 - e^{-t}$. Note that the Bernoulli type perturbation is equivalent to a perturbation in the distribution. That is, given distributions F_X and G_U let $Y = U - X$. Then the distribution of $X + \delta_h Y$ is $(1 - h)F_X + hG_U$ and (37.2.2) is read as: for each nondegenerate distributions F there exist distributions H_1 and H_2 such that

$$v[(1 - h)F + hH_1] < v(F) < v[(1 - h)F + hH_2] \quad \text{for all } 0 < h < h_0.$$

Convolutions play a central role in the study of functionals. For example, the inequalities

$$I(X + Y) \geq I(X) \quad \text{and}$$

$$H(X + Y) \leq H(X) \quad \text{for all independent } X \text{ and } Y \quad (37.2.3)$$

are major features of the Fisher information I and Shannon entropy H; see for example Barron (1986). Assertion 5 of the theorem provides a local analogue of (37.2.3) for the functional $v(X)$. In particular, we show that when only perturbations along independent directions Y are considered, the statement (37.2.2) is no longer true in general. For example, for a nonrandom perturbation, we have

$$v(X + \varepsilon Y) < v(X) \quad \text{for all fixed } \varepsilon \text{ such that } 0 < \varepsilon < \varepsilon_0(Y), \quad (37.2.4)$$

for each non-negative Y that is independent of X. In contrast, for a Bernoulli perturbation the local behaviour along independent positive directions Y is approximately the same as local behaviour along arbitrary positive directions Y.

Note that the functional $v(X)$ does not satisfy the property (37.2.4) globally, that is, for all Y independent of X and all $\varepsilon > 0$. Suppose, to the contrary, that

$$v(X + Y) \leq v(X) \quad (37.2.5)$$

were true for all independent random variables X and Y, such that X has a nondegenerate distribution. Choose a Y with a nondegenerate distribution and consider a sequence of random variables $X_1, X_2, \ldots$ such that $X_n \to 1$. From (37.2.5) it would follow that $v(1 + Y) \leq v(1)$ which contradicts Assertion 2.

37.3   Proof of the Theorem

PROOF OF ASSERTION 1.   By direct substitution of $X' = cX$ into (37.1.1).

PROOF OF ASSERTION 2.   From (37.1.1) it follows that we need to show

$$\theta(X) \geq \theta(\mathrm{E}X). \tag{37.3.1}$$

Jensen's inequality gives

$$\exp[-\bar{a}\,\mathrm{E}X\,\theta(\mathrm{E}X)] \;=\; a \;=\; \mathrm{E}\exp[-\bar{a}\,X\,\theta(X)] \;\geq\; \exp[-\bar{a}\,\mathrm{E}X\,\theta(X)]\,,$$

with equality if and only if X has a degenerate distribution. This implies (37.3.1), and hence the result.   ∎

PROOF OF ASSERTION 3.   By Assertion 1 we can assume $\mathrm{E}X = 1$ without loss of generality. Then $p_X = 1/\mathrm{E}X^2 \leq 1$ and the random variable $X_0 = \mathrm{Bin}(1, p_X)/p_X$ is well defined.

Assume that $p_X > \bar{a}$. Then, for X_0 equation (37.1.2) has the unique solution

$$\theta_0 = -\frac{\ln(a - \Delta_X)}{\bar{a} + \Delta_X} = -\frac{\ln[a - \bar{a}(1/p_X - 1)]}{\bar{a} + \bar{a}(1/p_X - 1)}\,.$$

Applying the sharp upper bound $\mathrm{E}\exp(-tX) \leq \mathrm{E}\exp(-tX_0)$, $t \geq 0$, see Stoyan (1983), we obtain

$$\mathrm{E}\exp[-\bar{a}\,X\theta_0] \leq \mathrm{E}\exp[-\bar{a}\,X_0\theta_0] = a = \mathrm{E}\exp[-\bar{a}\,X\theta(X)],$$

which implies $\theta_0 \geq \theta(X)$ and thus the result for the case $p_X > \bar{a}$.

For the case $p_X \leq \bar{a}$ we show that the only possible upper bound is $v(X) = 1$. Define a sequence of random variable $X_\eta = \eta + w\mathrm{Bin}(1, u)$, with u and w chosen so that $wu = 1 - \eta$ and $w^2 u(1 - u) = \mathrm{Var}(X)$, and consider this sequence as $\eta \downarrow 0$. Then equation (37.1.2) implies that $\theta(X_\eta) \to \infty$ and therefore $v(X_\eta) = 1 - 1/\theta(X_\eta) \to 1$ as $\eta \downarrow 0$.

PROOF OF ASSERTION 4.   Define the functions

$$T(h) = \theta(X + \delta_h Y) \quad \text{and} \quad V(h) = v(X + \delta_h Y)\,,$$

where the domains of T and V contain an open interval around the origin. Both T and V are functions of a real variable, and both depend on X and Y.

Applying definition (37.1.2) to $X + \delta_h Y$, replacing $\theta(X + \delta_h Y)$ by $T(h)$ and differentiating at $h = 0$ yields

$$0 = -\mathrm{E}\{[\bar{a}X\,T'(0) + Z(\theta\bar{a}Y)]\exp(-\theta\bar{a}X)\}\,.$$

This gives

$$T'(0) = -\frac{\mathrm{E}[\exp(-\theta\bar{a}X)Z(\theta\bar{a}Y)]}{\mathrm{E}[\bar{a}X\exp(-\theta\bar{a}X)]},$$

and hence

$$V'(0) = \frac{\theta\bar{a}\mathrm{E}Y\,\mathrm{E}[X\exp(-\theta\bar{a}X)] - \mathrm{E}X\,\mathrm{E}[Z(\theta\bar{a}Y)\exp(-\theta\bar{a}X)]}{\bar{a}\theta^2\,\mathrm{E}^2X\,\mathrm{E}[X\exp(-\theta\bar{a}X)]}. \quad (37.3.2)$$

When X is a constant, almost surely, equation (37.3.2) and $Z(y) \leq y$ give

$$V'(0) \geq \frac{\theta\bar{a}\mathrm{E}[Y] - \mathrm{E}Z(\theta\bar{a}Y)]}{\bar{a}\theta^2\mathrm{E}^2X} \geq 0 \quad \text{for all } Y \geq 0.$$

This establishes the necessary condition of Assertion 4.

Now suppose that V' has constant sign. Without loss of generality let $\theta\bar{a} = 1$ and let $H(X,Y)$ denote the numerator of (37.3.2). Take $Y = yI_D$, for a set D, and define $U = \mathrm{E}[X\exp(-X)] - \exp(-X)\,\mathrm{E}X$. Since $Z(y)/y \to 1$ as $y \downarrow 0$, it follows that

$$\lim_{y\downarrow 0} H(X, yI_D)/y = \mathrm{E}(I_D U) \quad \text{has constant sign for every set } D.$$

Substituting the particular choices $D = \{U \geq 0\}$ and $D = \{U \leq 0\}$, in turn, shows that either $\Pr(U \geq 0) = 1$ or $\Pr(-U \geq 0) = 1$. Each of these implies a degenerate distribution for X. For example, suppose $\Pr(U \geq 0) = 1$. That is,

$$\mathrm{E}[X\exp(-X)] \geq \exp(-X)\,\mathrm{E}X, \quad \text{almost surely.}$$

Then

$$\mathrm{E}[X\exp(-X)] \geq \exp(-x_{\mathrm{EI}})\,\mathrm{E}X, \quad (37.3.3)$$

where

$$x_{\mathrm{EI}} = \inf\left\{x : \int_0^x ydF(y) > 0\right\}$$

is the *essential infimum* with respect to the finite positive measure given by $d\mu(x) = xdF(x)$.

Equation (37.3.3) can be written

$$\int_0^\infty [\exp(-x) - \exp(-x_{\mathrm{EI}})]\, x\, dF(x) \geq 0,$$

which implies that F is degenerate.

A similar proof shows that $\Pr(-U \geq 0) = 1$ also implies that F is degenerate. ∎

PROOF OF ASSERTION 5. By Assertion 4, Statement S1 is valid when X is degenerate. Therefore, without loss of generality, we assume that X

is nondegenerate. Since Y is independent of X, it follows from (37.3.2) that Statement S1 is equivalent to the statement

S3. $[EZ(\theta\bar{a}Y)/E(\theta\bar{a}Y)] - Z_X$ has constant sign for all positive Y.

Note that $Z_X \leq 1$, with equality if and only if X is degenerate. We assumed X to be nondegenerate, so that $Z_X < 1$. Since $Z(y)/y \to 1$ as $y \to 0$, it now follows that S3 is equivalent to the statement

S4. $EZ(\theta\bar{a}Y) \geq E(\theta\bar{a}Y)Z_X$ for all positive Y

Since Statement S4 is equivalent to Statement S2 the assertion is proved. ∎

37.4 Application to Epidemic Control

We now point out some insights provided by these results in the context of epidemic control.

Consider an epidemic model describing the spread of infection to susceptible units of a community. By a unit we might mean an individual or a household, depending on the application. Assume that all infectious individuals have the same potential to transmit the disease to susceptible units, but susceptibility to infection may differ among units. In the case of individuals the susceptibility may depend on their age, for example, while in the case of a household the susceptibility to infection depends on its size.

Let $i(t)$ and $r(t)$ denote the proportion of infective and removed individuals at time t, respectively. We classify susceptible units according to type $w \in \mathcal{A}$. The proportion of type-w units that are still susceptible at time t, $s_w(t)$, is related to $i(t)$ and $r(t)$ by

$$\frac{ds_w(t)}{dt} = -\beta f(w)s_w(t)i(t) \quad \text{and} \quad \frac{dr(t)}{dt} = \gamma i(t), \qquad (37.4.1)$$

where $f(w)$ indicates the susceptibility of units of type w. Standard arguments show that there exist limits satisfying

$$s_w(\infty) = s_w(0) \exp[-\theta f(w)r(\infty)],$$

where $\theta = \beta/\gamma$. Let $\mu(w)$ denote a measure for the structure variable w such that $s(t)$, the proportion of all individuals who are susceptible at time t, is given by

$$s(t) = \int s_w(t)d\mu(w).$$

Then $s(\infty)$, the eventual proportion of susceptibles, is given by the solution of

$$s(\infty) = \int s_w(0) \exp\{-\theta f(w)[1 - s(\infty)]\}d\mu(w) \,.$$

This equation can be written in the form of equation (37.1.2) by setting $a = s(\infty)$, provided μ is such that

$$\Pr\{X = f(w)\} = s_w(0)d\mu(w) \qquad (37.4.2)$$

defines a proper probability distribution for the random variable X.

We now introduce two particular settings, to which the model (37.4.1) applies, and associated control strategies for which the critical immunity level is given by (37.1.1).

A Household Structure

Consider a community consisting of a large number of households, in which the vaccination strategy is to select households at random and to immunize every member of each selected household against the infectious disease. Let $f(w) = w$, $w \in \mathcal{A} = \{1, \ldots, n\}$, and let the measure μ be given by $\mu(w) = w/\mu_{\mathrm{H}}$, where μ_{H} is the mean household size. In this setting $s_w(t)$ of (37.4.1) denotes the proportion of households of size w that are still susceptible at time t.

Interest is in knowing how high the immunity coverage needs to be to make the probability of a major epidemic zero. Becker and Dietz (1995) showed that the critical immunity level for a disease that is highly infectious within households is given by (37.1.1), where θ is the mean number of individuals from other households that an infective would infect if everyone were susceptible and the random variable X is the size of the household of an individual selected randomly from the community. The parameter θ needs to be estimated. Suppose that the eventual proportion $s(\infty)$ is observed for an epidemic arising from an introduction of the disease into a totally susceptible community. Then θ can be estimated from estimating equation (37.1.2). Furthermore, the distribution (37.4.2) is proper.

In this setting, Assertion 1 states that the estimate of the critical immunity level v is scale invariant with respect to household size. This means, for example, that we would get the same estimate of v if every household were double its actual size. By Assertion 2 the estimate of v is smallest when every household has the same size. The inequality of Assertion 3 shows that the critical immunity coverage for a community with only two different household sizes may be quite different from that for a community of uniformly-mixing individuals. This conclusion supports some examples considered in Becker and Utev (1998). Assertion 4 considers the effect of perturbations of the household distribution on the functional $v(X)$. Specifically, it indicates that there are no other household distributions with extreme estimates.

An Age Structure

Consider now a community of individuals who mix uniformly, but their susceptibility depends on some characteristic, such as age. Letting w refer to age, $f(w)$ reflects the susceptibility of individuals aged w and $s_w(t)$ of (37.4.1) is the proportion of individuals aged w that are still susceptible at time t. Suppose that we estimate the parameter θ using data from a major epidemic that occurred, over a relatively short period of time, as a result of a few infectives joining a completely susceptible population.

Then an estimating equation for θ is (37.1.2), where $X = f(W)$, the random variable W is the age of an individual selected randomly from the community and $\mu(w)$ gives the age distribution for the community. Suppose the vaccination strategy is to immunize each individual, independently, with probability v. Then the critical immunity level is estimated by (37.1.1).

Assertion 2 states that the estimate of v is smallest when the susceptibility $f(w)$ is constant over age. This implies that it is wise to take the age structure into account when estimating the critical immunity coverage. The upper bound of Assertion 3 suggests that the simple lower bound of Assertion 2 is useful when susceptibility does not change very much with age. Assertions 4 and 5 investigate the local behaviour of the estimate with respect to perturbations in susceptibility of the form $f_\varepsilon(W) = f(W) + \varepsilon Y \equiv X + \varepsilon Y$ with both ε and Y positive. According to Assertion 4 an extreme estimate occurs only when all age groups are equally susceptible. In contrast, Assertion 5 states that each distribution for the susceptibility is stable with respect to perturbations in any "independent" direction.

Acknowledgements Support from the Australian Research Council is gratefully acknowledged.

References

1. Barron, A. (1986). Entropy and central limit theorem. *The Annals of Probability* **14**, 336–342.

2. Becker, N. G. and Dietz, K. (1995). The effect of the household distribution on transmission and control of highly infectious diseases.

Mathematical Biosciences **127**, 207–219.

3. Becker, N. G. and Utev, S. (1998). The effect of community structure on the immunity coverage required to prevent epidemics. *Mathematical Biosciences* **147**, 23–39.

4. Cacoullos, T., Papathanasiou, V., and Utev, S. A. (1994). Variational inequalities with examples and an application to the central limit theorem. *The Annals of Probability* **22**, 1607–1618.

5. Chen, L. H. Y. (1988). The central limit theorem and Poincare type inequalities. *The Annals of Probability* **16**, 300–304.

6. Ellis, R. S. (1985). *Entropy, Large Deviations and Statistical Mechanics*. Springer-Verlag, Berlin.

7. Huber, P. J. (1981). *Robust Statistics*. John Wiley & Sons, New York.

8. Stoyan, D. (1983). *Comparison Methods for Queues and Other Stochastic Models*. John Wiley & Sons, Chichester, England.

38

A Birth and Death Urn for Ternary Outcomes: Stochastic Processes Applied to Urn Models

Anastasia Ivanova and Nancy Flournoy
University of North Carolina, Chapel Hill, NC
The American University, Washington, D.C.

ABSTRACT Consider the situation in which subjects arrive sequentially in a clinical trial. There are K possibly unrelated treatments. Suppose balls in an urn are labeled with treatments. When a subject arrives, a ball is drawn randomly from the urn, the subject receives the treatment indicated on the ball and ball is then returned to the urn. There are three levels of response: success, no response, and failure. If the treatment is successful, one ball of the same type is added to the urn. If a failure is observed, one ball of the same type is taken out from the urn. Otherwise, nothing is done. We describe how the proposed urn process can be embedded in a collection of independent continuous-time birth and death processes. We demonstrate the usefulness of this tool in obtaining limiting results for the continuous-time process that are equivalent to analogous ones for the urn design. Furthermore, important exact statistics are obtained by using a stopping rule to transfer statistics derived from the continuous-time birth and death processes to the urn design. Finally, we give a likelihood ratio test for the comparison of the K treatments.

Keywords and phrases Adaptive designs, clinical trials, experimental design, birth and death process with immigration

38.1 Introduction

An urn model is a useful model for treatment allocation. See Kotz and Balakrishnan (1997) for recent review. Several adaptive designs based on an urn models have been studied [see Ivanova and Rosenberger (2000) for a review]. We consider the situation in which subjects arrive sequentially for a clinical trial in which there are K possibly unrelated treatments with ternary outcomes. Label each ball in the urn with a treatment. A ball labeled i is called a type i ball, $i = 1, \ldots, K$. The collection of all balls in the urn is called the urn composition. In general, when a subject arrives, a ball is drawn randomly from the urn; the subject receives the treatment indicated on the ball, and the ball is then returned to the urn. To facilitate the theory, this discrete time urn process can be embedded into a continuous-time process. The technical justification for this can be found in Athreya and Ney (1972). In this paper, the urn process can be imbedded into a family of continuous-time Markov linear birth and death processes. We emphasize how such embedding permits the use of well-known results for continuous-time Markov processes for characterizing an urn composition and for obtaining other statistics of importance in evaluating the results of a clinical trial.

We present a new adaptive design which is a generalized version of the birth and death urn design for binary outcomes proposed by Ivanova *et al.* (2000). Both exact and asymptotic results are presented. In Section 38.2, we define the birth and death urn design. In Section 38.3, we describe how the urn can be embedded in a continuous-time birth and death process. In Sections 38.4–38.7, we characterize the number of successes and trials in the continuous-time birth and death process. In Section 38.8, we introduce a stopping rule and show how it can be used to transfer statistics derived from the continuous-time birth and death process to the urn design. In Sections 38.9 and 38.10, we use the continuous-time birth and death process to obtain the limiting proportion of trials on each treatment and the limiting proportion of successes on each treatment, respectively. Important asymptotic results are stated specifically for the birth and death urn with ternary responses. However, the proofs of these results are analogous to ones previously reported, and so are not given here. Section 38.11 gives a likelihood ratio test for equal success probabilities. Finally, we make some concluding remarks in Section 38.12.

38.2   A Birth and Death Urn with Immigration for Ternary Outcomes

The term *birth* refers to the addition of a ball to the urn, whereas the term *death* refers to the removal of a ball from the urn. Having an urn process in which balls can be removed from the urn creates the possibility that even ball type(s) with high success probability will become depleted over the course of the experiment. Hence, we define the urn process to include the possibility of randomly replenishing the urn through *immigration*. A ball is called an *immigration ball* if, when it is drawn, it is replaced together with additional balls of other types and no subject is treated.

We define a *birth and death urn design with immigration* for ternary responses as follows:

> An urn contains balls of K types representing K treatments, and aK immigration balls. When a subject arrives, a ball is drawn at random and then replaced. If it is an immigration ball, one ball is added to the urn, with each ball type having probability $1/K$ of being the one added; then the next ball is drawn. The procedure is repeated until the ball representing an actual treatment is drawn (i.e., not an immigration ball). If it is a type i ball, the subject receives treatment i. If a success is observed, a type i ball is added to the urn; if no response is observed the urn composition remains unchanged; if a failure is observed, a type i ball is removed. The process continues sequentially with the arrival of the next subject.

The parameter a is called the *rate of immigration*. Given treatment i, let p_i be the probability of adding a type i ball to the urn; let r_i be the probability that the urn composition remains unchanged entirely, and let q_i be the probability of removing a type i ball from the urn; $q_i + r_i + p_i = 1$. If all $q_i = 0$, $i = 1, \ldots, K$, and $a = 0$, we have the pure birth urn design (called the generalized Pólya urn) considered in Durham, Flournoy and Li (1998). If $r_i = 0$, $i = 1, \ldots, K$, and $a = 0$, we have the birth and death urn design considered in Ivanova (1998). If $r_i = 0$, $i = 1, \ldots, K$, and $a > 0$, we have the birth and death urn design with immigration considered in Ivanova *et al.* (2000). For all these designs, and for the one presented herein as well, a type i ball is added to the urn following a success on treatment i; so $p_i = P\{\text{success}|i\}$. In Ivanova (1998), Ivanova *et al.* (2000), and here as well, a type i ball is removed from the urn following a failure on treatment i; so $q_i = P\{\text{failure}|i\}$. Here we admit the possibility of no response, and given this lack of response the urn composition will remain unchanged entirely, so $r_i = P\{\text{no response}|i\}$. We also permit immigration with rate a.

38.3   Embedding the Urn Scheme in a Continuous-Time Birth and Death Process

To assist in the characterization of urn process defined in Section 38.2, we employ the technique of embedding this urn process into a continuous-time K-type Markov process, which in the present context is a collection of K independent linear birth and death processes with immigration (given by Theorem 38.3.1). Conventionally, continuous-time birth and death processes describe the behavior of particles which split (produce or die) at certain time points. In our context, "particles" correspond to balls in the urn, and the "splitting" of a type i ball corresponds to the addition of a type i ball following a success on treatment i or the immigration of a type i ball, as well as to the removal of a type i ball following a failure on treatment i. We call these *events* instead of splits. When a nonimmigration ball is drawn, a *trial* is initiated with a treatment being selected for a subject. If we start the experiment with the urn having Z_{i0} balls of type i, then $\mathbf{Z}_0 = \{Z_{10}, \ldots, Z_{K0}\}$ is the initial *urn composition*. We assume that the vector $\mathbf{Z}_0$ is fixed. Let $\mathbf{Z}_\omega = \{Z_{1\omega}, \ldots, Z_{K\omega}\}$ denote the composition of the urn after ω consecutive events, where $Z_{i\omega}$ is the number of type i balls. Then $\{\mathbf{Z}_\omega, \omega = 0, 1, 2, \ldots\}$ is a stochastic process defined on K dimensional integer lattice.

Denote the continuous-time analog of the urn process by $\{\mathbf{Z}(t) = (Z_1(t), \ldots, Z_K(t)) \mid t \geq 0\}$, where $Z_i(t)$, $i = 1, \ldots, K$, is a linear birth and death processes with birth rate p_i, death rate q_i and immigration rate a as for the urn process; and $Z_i(t)$ are independent of each other. Consider the event times, τ_ω, $\omega = 0, 1, \ldots$, $\tau_0 = 0$, of $\mathbf{Z}(t)$, and the discrete process $\{\mathbf{Z}(\tau_\omega), \omega = 0, 1, \ldots\}$ defined at the event times. The following theorem is key to characterizing the urn design.

THEOREM 38.3.1
  Given that $\mathbf{Z}(0) = \mathbf{Z}_0$, the stochastic processes $\mathbf{Z}_\omega$ and $\{\mathbf{Z}(\tau_\omega), \omega = 0, 1, \ldots\}$ are equivalent.

PROOF   The proof is similar to one in Athreya and Ney (1972, p. 221). ∎

Characteristics of the number of successes, failures, and trials for the continuous-time process at time t can be obtained through generating functions. In Sections 38.4–38.7, we present results from an artificial experiment conducted according to a continuous-time birth and death process. But time is a construct that has no meaning in terms of the urn process, and hence in terms of the experiment. Therefore, using a stopping rule in Section 38.8, we show how to translate statistics obtained from the continuous-time process back to the urn process at the time the experiment is stopped. Furthermore, we give asymptotic results in Section 38.9 that are the same for both the urn process and its continuous-time analog.

In clinical trials we are interested in estimating the success probability for each treatment. For making inferences concerning the proportion of successes across the various treatments, joint characterization of the numbers of successes and trials is required (Section 38.7). However, we start by obtaining the probability generating function for the number of successes on treatment i.

38.4 The Probability Generating Function for the Number of Successes on Treatment i in the Continuous-Time Birth and Death Process

Because the number of successes at any time depends on the number of type i balls in the urn at that time, to obtain the probability generating function for the number of successes on treatment i, we first derive a differential equation that is satisfied by the joint probability generating function for the number of type i successes and the number of type i balls in the urn. Then the marginal solution yields the generating function for the number of type i successes.

Let $Z_i(t)$ be a linear birth and death process with immigration, that is, $Z_i(t)$ denote the number of balls of type i at time t; $X_i(t)$ denotes the number of successes on treatment i by or at time t; $Y_i(t)$ denotes the number of failures on treatment i, and $N_i(t)$ denotes the number of trials on treatment i. Because the processes corresponding to each ball type in continuous-time are independent, we only consider type i. Also we often suppress the subscript i for notational simplicity when it is not needed for understanding. Let $P^Z_{z_0 \to z}(t)$ $(P^X_{x_0 \to x}(t), P^Y_{y_0 \to y}(t), P^N_{n_0 \to n}(t))$ be the probability of getting z (x, y, n) type i balls (successes, failures, trials) by time t starting with z_0 (x_0, y_0, n_0) type i balls (successes, failures, trials).

$Z_i(t)$ is a Markov process on states $0, 1, \ldots$ with stationary probabilities. The postulates defining a linear birth and death process with immigration rate a are [Karlin and Taylor (1974, p. 189)]:

1. $P^Z_{z \to z+1}(\Delta t) = (p_i z + a)\Delta t + o(\Delta t)$ as $\Delta t \to 0$, $z \geq 0$
2. $P^Z_{z \to z-1}(\Delta t) = q_i z \Delta t + o(\Delta t)$ as $\Delta t \to 0$, $z \geq 1$
3. $P^Z_{z \to z}(\Delta t) = 1 - (p_i z + q_i z + a)\Delta t + o(\Delta t)$ as $\Delta t \to 0$, $z \geq 0$

4. $P^Z_{z \to z}(0) = \delta_{ij}$.

Note that the process $Z_i(t)$ is Markovian, but the processes $X_i(t)$, $Y_i(t)$, $N_i(t)$ are non-Markovian because the transition rates at t depend on $Z_i(t)$. To deal with the dependency of X_i on Z_i, we expand on a hint by Cox and Miller (1965, page 265) and consider the two-dimensional process of the number of successes on treatment i and the number of type i balls, $(X_i(t), Z_i(t))$, having state space the set of nonnegative integer pairs (x, z). Let $P^{XZ}_{x_0,z_0 \to x,z}(t)$ be the probability of having x successes and z balls by time t given there were x_0 successes and z_0 balls at time 0.

Define also the joint probability generating function for the number of balls and the number of successes,

$$G^{XZ}_i(v, s, t \mid Z_{i0}) = \sum_{x=x_0}^{\infty} \sum_{z=0}^{\infty} P^{XZ}_{x_0,z_0 \to x,z}(t) v^x s^z,$$

where $|v| \leq 1$, $|s| \leq 1$. To obtain forward differential equations, assume now that the joint process is at (x, z), $z > 0$, $x > 0$, at time $t + \Delta t$. The following theorem gives the differential equation for the probability generating function. This equation is solved for the case in which the urn is initialized by the immigration process, i.e., $Z_{i0} = 0$. This result is useful for any initial urn composition because with immigration, the effect of the initial urn composition is only substantial in an experiment with a very small sample size. The limiting results do not depend on the initial urn composition, but they do depend on the immigration rate. Note that if the urn initially is not empty so that $Z_{i0} \neq 0$, the process $Z_i(t)$ is the sum of two independent processes: one is initiated by the balls which were in the urn from the beginning; another process is initiated by the balls constantly immigrating into the urn. The first process is a birth and death process without immigration.

THEOREM 38.4.1
Given the initial urn composition $Z_i(0) = Z_{i0}$, the joint probability generating function for the number of successes and balls satisfies the following relation:

$$\frac{\partial G^{XZ}_i(v, s, t \mid Z_{i0})}{\partial t} = (p_i v s^2 - (1 - r_i)s + q_i) \frac{\partial G^{XZ}_i(v, s, t \mid Z_{i0})}{\partial s}$$

$$+ a(s - 1) G^{XZ}_i(v, s, t \mid Z_{i0}). \tag{38.4.1}$$

With $X_i(0) = 0$, the initial condition is

$$G^{XZ}_i(v, s, 0 \mid Z_{i0}) = s^{Z_{i0}}.$$

Furthermore, when the urn is initialized by the immigration process so that $Z_{i0} = 0$, the probability generating function for the number of successes at time t is

$$G_i^X(v, t \mid Z_{i0} = 0)$$

$$= \left[\frac{2\rho}{e^{-\rho t}(2p_i v - 1 + r_i + \rho) - (2p_i v - 1 + r_i - \rho)} \right]^{\frac{a}{p_i v}}$$

$$\times e^{-\frac{ta}{2p_i v}(2p_i v - 1 + r_i + \rho)},$$

where $\rho = \sqrt{(1 - r_i)^2 - 4p_i q_i v}$, $|v| \le 1$.

PROOF Consider the possible transitions which are composite events involving changes in the number of the type i balls and the number of successes on treatment i. There are three that can occur in the time interval Δt and result in x successes and z balls. First is an immigration, i.e., $(x, z - 1) \to (x, z)$, with probability of occurrence $a\Delta t$. The second transition is a trial resulting in a success, i.e., $(x - 1, z - 1) \to (x, z)$. The probability of a success conditional on a type i ball being drawn and on $Z_i(t) = z - 1$ is $(z - 1)p_i \Delta t$. The third transition is a trial resulting in a failure, i.e., $(x, z + 1) \to (x, z)$. The probability of a failure conditional on a type i ball being drawn and on $Z_i(t) = z + 1$ is $(z + 1)q_i \Delta t$. In addition to these composite events, a trial can result in no response, i.e., $(x, z) \to (x, z)$, with probability $zr_i \Delta t$ conditional on a type i ball being drawn and $Z_i(t) = z$. Finally, no composite events occur, i.e., $(x, z) \to (x, z)$, with probability $[1 - (zp_i + zq_i + zr_i + a)\Delta t]$ conditional on $Z_i(t) = z$.

From the postulates defining a birth and death process with immigration, it can be shown that $P_{m,k \to x,z}^{XZ}(\Delta t)/\Delta t = o(1)$ for $k \notin \{z - 1, z, z + 1\}$, where the $o(1)$ term, apart from tending to zero, is uniformly bounded with respect to k and m, and x for fixed z as $\Delta t \to 0$. Then it follows for $z > 0$, $x \ge x_0$, that

$$P_{x_0, z_0 \to x, z}^{XZ}(t + \Delta t) = P_{x_0, z_0 \to x, z}^{XZ}(t)(1 - (zp_i + zq_i + a)\Delta t)$$

$$+ P_{x_0, z_0 \to x-1, z-1}^{XZ}(t)(z - 1)p_i \Delta t$$

$$+ P_{x_0, z_0 \to x, z+1}^{XZ}(t)(z + 1)q_i \Delta t$$

$$+ P_{x_0, z_0 \to x, z-1}^{XZ}(t)a\Delta t + o(\Delta t);$$

whereas for $z = 0$,

$$P_{x_0, z_0 \to x, 0}^{XZ}(t + \Delta t) = P_{x_0, z_0 \to x, 0}^{XZ}(t)(1 - a\Delta t) + P_{x_0, z_0 \to x, 1}^{XZ}(t)q_i \Delta t + o(\Delta t).$$

Taking the derivative and letting $\triangle t \to 0$ for $z > 0$, we obtain

$$\frac{\partial P^{XZ}_{x_0,z_0 \to x,z}(t + \triangle t)}{\partial t} = -(zp_i + zq_i + a)P^{XZ}_{x_0,z_0 \to x,z}(t)$$

$$+[(z-1)p_i + a]P^{XZ}_{x_0,z_0 \to x-1,z-1}(t)$$

$$+(z+1)q_i P^{XZ}_{x_0,z_0 \to x,z+1}(t). \qquad (38.4.2)$$

For $z = 0$, we obtain

$$\frac{\partial P^{XZ}_{x_0,z_0 \to x,0}(t)}{\partial t} = -aP^{XZ}_{x_0,z_0 \to x,0}(t) + q_i P^{XZ}_{x_0,z_0 \to x,1}(t). \qquad (38.4.3)$$

The initial conditions are given by $X_i(0) = x_0$, $Z_i(0) = z_0$ and

$$P^{XZ}_{x_0,z_0 \to x,z}(0) = 1 \quad \text{if } x = x_0, \ z = z_0$$
$$P^{XZ}_{x_0,z_0 \to x,z}(0) = 0 \quad \text{otherwise.}$$

Multiply (38.4.2) and (38.4.3) by v^x and by s^z and sum over all values of x and z. Then rearrange terms and use the fact that

$$s\frac{\partial}{\partial s}G^{XZ}_i(v,s,t \mid Z_{i0}) = \sum_{x=x_0}^{\infty} \sum_{z=0}^{\infty} P^{XZ}_{x_0,z_0 \to x,z}(t)v^x z s^z,$$

to yield the result (38.4.1).

Assuming $Z_{i0} = 0$, apply the algorithm in Anderson (1991) for solving a Lagrange differential equation. The probability generating function for the number of type i successes is thereby obtained from the solution to equation (38.4.1) by putting $s = 1$. It is

$$G^X_i(v,t \mid Z_{i0} = 0)$$

$$= \left[\frac{2\rho}{e^{-\rho t}(2p_i v - 1 + r_i + \rho) - (2p_i v - 1 + r_i - \rho)}\right]^{\frac{a}{p_i v}}$$

$$\times e^{-\frac{ta}{2p_i v}(2p_i v - 1 + r_i + \rho)},$$

where $\rho = \sqrt{(1 - r_i)^2 - 4p_i q_i v}$, $|v| \leq 1$. ∎

38.5 The Probability Generating Function for the Number of Trials on Treatment i in the Continuous-Time Birth and Death Process

Just as we obtained the probability generating function for the number of successes from the joint probability generating function for the number of successes and the number of balls, the probability generating function for the number of trials can be obtained from the joint probability generating function for the number of trials and the number of balls. Let $G_i^{NZ}(w, s, t \mid Z_{i0})$ denote the joint probability generating function for the numbers of successes and trials.

THEOREM 38.5.1

If the urn is initialized by the immigration process, the probability generating function for the numbers of trials is

$$G_i^N(w, t \mid Z_{i0} = 0)$$

$$= \left[\frac{2\rho}{e^{-\rho t}(2p_i w - 1 + r_i w + \rho) - (2p_i w - 1 + r_i w - \rho)} \right]^{\frac{a}{p_i w}}$$

$$\times e^{-\frac{ta}{2p_i w}(2p_i w - 1 + r_i w + \rho)}, \tag{38.5.1}$$

where $\rho = \sqrt{(1 - r_i w)^2 - 4 p_i q_i w^2}$, $|w| \leq 1$.

PROOF Using the same logic as in the proof of Theorem 38.4.1, the probability generating function for the number of trials on treatment i by time t, $N_i(t)$, can be shown to satisfy the differential equation

$$\frac{\partial G_i^{NZ}(w, s, t \mid Z_{i0})}{\partial t}$$

$$= (p_i w s^2 - (1 - r_i w)s + q_i w) \frac{\partial G_i^{NZ}(w, s, t \mid Z_{i0})}{\partial s}$$

$$+ a(s - 1) G_i^{NZ}(w, s, t \mid Z_{i0}).$$

The probability generating function for the number of successes is found by solving this equation and setting $s = 1$. The solution for the case in which $Z_{i0} = 0$ is the result. $\blacksquare$

38.6 The Number of Trials on Treatment i in the Continuous-Time Birth and Death Process

When $Z_{i0} = 0$, the expected number of trials by time t can be obtained from the generating function (38.5.1) as

$$E(N_i(t)) = \left(\frac{\partial G^N(w, t)}{\partial w} \right) \bigg|_{w=1}.$$

Evaluating the derivative yields

$$E(N_i(t)) = \begin{cases} a(e^{(p_i - q_i)t} - (p_i - q_i)t - 1)/(p_i - q_i)^2 & \text{when } p_i \neq q_i \\ at^2/2 & \text{when } p_i = q_i. \end{cases}$$

$$(38.6.1)$$

The variance can be found easily by evaluating

$$Var(N_i(t)) = \frac{\partial}{\partial w}\left(w\frac{\partial G^N(w,t)}{\partial w}\right)\bigg|_{w=1} - E(N_i(t))^2.$$

We show how to translate these moments, from the continuous-time birth and death process to the urn process, in Section 38.8 with the introduction of a stopping rule.

Now we give a limiting result for the (appropriately standardized) number type i balls in the continuous-time birth and death process. In particular, we have

THEOREM 38.6.1
For $a > 0$,

$$N_i(t)/E^*(N_i(t)) \overset{P}{\to} W_i(p_i),$$

where

$$W_i(p_i)$$
$$= \begin{cases} Gamma(a/p_i, p_i/a), & \text{when } p_i > q_i \\ \text{has characteristic function } \left[\cosh\left(2p_i\sqrt{-1}\tau/a\right)\right]^{-\frac{a}{2p_i}} & \text{when } p_i = q_i \\ 1 & \text{when } p_i < q_i \end{cases}$$

and

$$E^*(N_i(t)) = \begin{cases} ae^{(p_i - q_i)t}/(p_i - q_i)^2 & \text{when } p_i > q_i \\ at^2/2 & \text{when } p_i = q_i \\ at/(q_i - p_i) & \text{when } p_i < q_i \end{cases}$$

is the dominant term in $E(N_i(t))$.

PROOF The characteristic function of the random variable

$$\lim_{t\to\infty}\left[N_i(t)/E^*(N_i(t))\right]$$

can be obtained by substituting $w = \exp[i\tau/E^*(N_i(t))]$ in $G_i(w, s, t \mid 0)$ and calculating the limit when t goes to infinity. The rest of the proof is similar to the one for binary responses with $a > 0$ [Ivanova *et al.* (2000)]. ∎

For the pure birth process with $a = 0$, Durham, Flournoy and Li (1998) apply results from Athreya and Ney (1972, pp. 127–130) to show that $N_i(t)/E(N_i(t))$ converges to a gamma distribution with shape parameter z_{i0} and scale parameter $1/z_{i0}$.

38.7 The Joint Probability Generating Function for the Number of Successes and the Number of Trials in the Continuous-Time Birth and Death Process

The observed proportion of successes on treatment i, $X_i(t)/N_i(t)$, is a natural estimator of the success probability of the ith treatment in the continuous-time birth and death process. Therefore, we seek the joint distribution of successes and trials in order to characterize the proportion of successes. But again, in order to obtain this joint distribution function, because the successes and trials depend upon the urn composition, we must first obtain the joint probability generating function for the total numbers of trials, successes and balls, $G_i^{XNZ}(s, w, t \mid Z_{i0})$. We obtain this generating function for the case in which the urn is initialized by the immigration process.

THEOREM 38.7.1

Given the initial urn composition $Z_i(0) = Z_{i0}$,

$$\frac{\partial G_i^{XNZ}(v, w, s, t \mid Z_{i0})}{\partial t}$$

$$= (p_i v w s^2 - (1 - r_i w)s + q_i w)\frac{\partial G_i^{XNZ}(v, w, s, t \mid Z_{i0})}{\partial s}$$

$$+ a(s - 1)G_i^{XNZ}(v, w, s, t \mid Z_{i0}).$$

With $X_i(0) = N_i(0) = 0$, the initial condition is

$$G_i^{XNZ}(v, w, s, 0 \mid Z_{i0}) = s^{Z_{i0}}.$$

When the urn is initialized by the immigration processes, the joint generating function for the numbers of successes and trials is

$$G_i^{XN}(v, w, t \mid Z_{i0} = 0)$$

$$= \left[\frac{2\rho}{e^{-\rho t}(2p_i vw - 1 + r_i w + \rho) - (2p_i vw - 1 + r_i w - \rho)} \right]^{\frac{a}{p_i w}}$$

$$\times \; e^{-\frac{ta}{2p_i vw}(2p_i vw - 1 + r_i w + \rho)},$$

where $\rho = \sqrt{(1 - r_i w)^2 - 4p_i q_i vw^2}$, $|w| \leq 1$, $|v| \leq 1$.

PROOF Proceed in an analogous way to the proof of Theorem 38.4.1 to obtain the differential equation. Then the joint probability generating function for the number of successes and trials, $G_i^{XN}(v, w, t \mid Z_{i0} = 0)$, is the solution of the partial differential equation evaluated at $s = 1$. ∎

The joint distribution of successes and trials is obtained by taking a Taylor series expansion of $G_i^{XN}(v, w, t \mid Z_{i0} = 0)$ around $w = 0$ and $v = 0$. We have been unable to find a closed form solution for the terms in this expansion. A numerical solution can be obtained pointwise and used to evaluate a specific urn design.

38.8 Adopting a Stopping Rule to Convert Continuous-Time Statistics to the Urn Design

Recall that the continuous-time birth and death process is constructed to assist in the analysis of the urn process; time is an artificial construct and is not related to the treatment of subjects. The next subject will receive treatment i with probability proportional to the number of type i balls in the urn. Therefore, we are interested in the number of balls in the urn, Z_{in}, rather than in $Z_i(t)$. One way to relate subjects and time is to introduce a stopping rule. We use the stopping rule introduced by Durham, Flournoy and Li (1998) for the pure birth urn design.

We introduce a fictitious treatment labeled '$K + 1$' having success probability $p_{K+1} = 1$. Balls labeled with this hypothetical treatment are called *control balls*. Thus each time a control ball is drawn and returned to the urn another control ball is added to the urn. Since the control balls are drawn randomly, they mark time, so to speak. The urn starts with one control ball. Use control balls to stop the experiment as follows:

> *stop the sequence of trials when* η, $\eta \geq 1$, *control balls have been drawn from the urn.*

We now show how to derive the expected number of subjects receiving treatment i when the ηth control ball is drawn.

Let $T_\eta = \min\{t : Z(t) = \eta + 1\}$ denote the time the ηth control ball is drawn. The density of T_η was derived by Durham, Flournoy and Li (1998). It is

$$f_\eta(t) = \frac{d}{dt} P\{T_\eta < t\} = \eta(1 - e^{-t})^{\eta-1} e^{\eta-1}. \qquad (38.8.1)$$

Taking an expectation from the continuous-time birth and death process conditional on $T_\eta = t$ and integrating with respect to $df_\eta(t)$ yields the unconditional expectation when the ηth ball is drawn from the urn. For example, using (38.6.1) and (38.8.1), the expected number of trials on ith treatment when the ηth immigration ball is drawn for $p_i \neq q_i$ is

$$E\{N(T_\eta)\} = \int_0^\infty E\{N(T_\eta)|T_\eta = t\} f_\eta(t) dt$$

$$= \int_0^\infty \frac{a}{(p_i - q_i)^2} (e^{(p_i - q_i)t} - (p_i - q_i)t - 1)\eta(1 - e^{-t})^{\eta-1} e^{\eta-1} dt$$

$$= a \frac{p_i + q_i}{p_i - q_i} \left(\frac{\Gamma(1 - (p_i - q_i))\eta!}{\Gamma(\eta - (p_i - q_i) + 1)} - 1 \right).$$

The expected number of successes on treatment i follows from the relationship $E\{X(T_\eta)\} = p_i E\{N(T_\eta)\}$.

In an analogous manner, time can be integrated out of all generating functions and expectations that are obtained for the continuous-time birth and death process.

The observed proportion of successes on treatment i, $X(T_\eta)/N(T_\eta)$, is a natural (but not necessarily unbiased) estimator of the success probability of the ith treatment at the time the experiment is stopped. However, we have to worry about dividing by zero because there is positive probability that $N(T_\eta) = 0$. Therefore, consider the estimator $X(T_\eta)/(N(T_\eta) + 1)$ which will be essentially equivalent for large samples. It is straightforward to write the expectation of this estimator, $X(T_\eta)/(N(T_\eta) + 1)$ from the joint generating function as

$$E\left(\frac{X(T_\eta)}{N(T_\eta) + 1} \right) = \int_0^\infty E\left(\frac{X(T_\eta)}{N(T_\eta) + 1} \bigg| T_\eta = t \right) f_\eta(t) dt$$

$$= \int_0^\infty \left(\int_0^1 \frac{d}{d\eta} \{G_i^{XN}(\eta, w, t|0)\} dw \right) \bigg|_{\eta=1} f_\eta(t) dt.$$

However, we have been unable to find a closed form solution for the conditional expectation $E\left(X(T_\eta)/(N(T_\eta)+1)|T_\eta = t\right)$. It can be obtained pointwise using numerical integration for evaluating a particular design.

38.9 Limiting Results for the Proportion of Trials on Treatment i

For the continuous-time birth and death process, let $\tau_1, \ldots, \tau_n$ denote the trial times (as opposed to the event times considered in Section 38.3). Now for the urn process, define the sample sizes on each treatment after n subjects to be N_{in}, where, in terms of the continuous-time birth and death process, $N_{in} = N_i(\tau_n)$. Note that, by construction, $N_1(\tau_n) + \cdots + N_K(\tau_n) = n$. Consider the proportion of subjects assigned to the ith treatment for each i, $i = 1, \ldots, K$, after n subjects, N_{in}/n. We are interested in the limit of N_{in}/n as $n \to \infty$. It follows from Theorem 38.3.1 that

$$\lim_{n\to\infty} \frac{N_{in}}{n} = \lim_{t\to\infty} \frac{N_i(t)}{\sum_{j=1}^{K} N_j(t)},$$

that is, the limiting proportion of trials on treatment i that is obtained from the continuous-time analog is equivalent to the limiting proportion for the urn design. Therefore, we evaluate this limit in the continuous-time model.

These limits depend on the uniqueness, or lack thereof, of the differences $p_1 - q_1, p_2 - q_2, \ldots, p_K - q_K$. Without loss of generality assume that these differences are ordered: $p_1 - q_1 \geq p_2 - q_2 \geq \cdots \geq p_K - q_K$. Theorem 6 gives the limiting proportion of subjects assigned to treatment i. In Case 1, $p_1 \geq q_1$; in Case 2, $p_1 < q_1$. Let h denote the number of treatments having maximum success probability, and for the treatments with maximum success probability, let Δ denote the difference between the probability of success and the probability of failure, i.e., $\Delta = p_1 - q_1 = p_2 - q_2 = \cdots = p_h - q_h$.

THEOREM 38.9.1
Case 1 *If* $\Delta = p_1 - q_1 = p_2 - q_2 = \cdots = p_h - q_h > p_{h+1} - q_{h+1} \geq \cdots \geq p_K - q_K$; *with* $\Delta \geq 0$, *then*

$$\frac{N_{in}}{n} \xrightarrow[n\to\infty]{P} D_i = \begin{cases} \frac{W_i(\Delta)}{W_1(\Delta)+\cdots+W_h(\Delta)} & \text{when } i \leq h \\ 0 & \text{when } i > h. \end{cases}$$

where $W_i(\Delta), i = 1, \ldots, h$, are random variables defined in Theorem 38.6.1.

Case 2 *If $\Delta = p_1 - q_1 \geq p_2 - q_2 \geq \cdots \geq p_K - q_K$; with $\Delta < 0$, then*

$$\frac{N_{in}}{n} \xrightarrow[n \to \infty]{P} D_i = \frac{\frac{1}{q_i - p_i}}{\frac{1}{q_1 - p_1} + \cdots + \frac{1}{q_K - p_K}}.$$

PROOF The proof is analogous to that of Theorem 2 of Ivanova *et al.* (2000). ∎

Note in Case 1 that the joint distribution of $(D_1, \ldots, D_{h-1})$ is a $(h-1)$-variate Dirichlet distribution with all parameters equal to a/Δ. Furthermore, if $h = 1$, the limiting proportion of subjects receiving the (unique) best treatment is one, and the limiting proportion of subjects on all other treatments is zero.

In both cases, more precise rates of divergence for the treatments subsamples can be derived analogous to those given for binary outcomes by Ivanova *et al.* (2000).

38.10   Limiting Results for the Proportion of Successes on Treatment i in the Urn

Again, limiting results for the proportion of success on treatment i using the urn experiment will be identical to the limiting proportion in the analogous continuous-time process. Hence, we use the continuous-time process to obtain these results. The maximum likelihood estimator of the proportion of successes on treatment i in the continuous-time birth and death model is denoted by $\hat{p}_{in} = X_i(t)/N_i(t)$. Theorem 38.10.1 gives the limiting distribution of $\hat{p}_{in}$, appropriately standardized. Let $p = p_1$, and recall that $\Delta = p_1 - q_1 = p_2 - q_2 = \cdots = p_h - q_h > p_{h+1} - q_{h+1}$.

THEOREM 38.10.1
Case 1. $p_1 \geq q_1$.
Consider the set A, $A = \{1, 2, \ldots, h\}$ of treatments with maximum success probability. The vector with components $\sqrt{N_{in}}(\hat{p}_{in} - p)$, $i \in A$, is asymptotically multivariate normal with mean zero, variances with components $p_i q_i$, and all covariances zero.

Case 2. $p_1 < q_1$.
The vector with components $\sqrt{N_{in}}(\hat{p}_{in} - p_i)$, $i = 1, \ldots, K$, is asymptotically multivariate normal with mean zero, variances with components

$p_i q_i$, and all covariances zero. Furthermore, the vector with components $\sqrt{n}(\hat{p}_{in} - p_i)$, $i = 1, \ldots, K$, is asymptotically a multivariate normal with mean zero, variances with components $p_i q_i / D_i$, and all cross-covariances equal to zero. The constants, D_i, are given in Case 2 of Theorem 38.9.1.

PROOF The proof is analogous to that of Theorem 4.2 in Ivanova *et al.* (2000). ∎

38.11   Asymptotic Inference Pertaining to the Success Probabilities

When the outcome of an experiment is ternary several hypothesis can be tested. One might be interested in testing whether all $p_i - q_i$ are the same. In this paper, we only consider the simplest testing problem that is of interest:

$$H_0 : p_1 = p_2 = \cdots = p_K = p$$

$$H_1 : \text{not all } p_i \text{ are equal.}$$

The likelihood ratio test provides a general method for hypothesis testing. Let $S = \sum_{i=1}^{K} X_{in}$, where $X_{in} = X_i(\tau_n)$. The likelihood ratio test statistic is

$$\lambda_n = n\left((S/n)\log(S/n) + (1 - S/n)\log(1 - S/n)\right)$$

$$- \sum_{i=1}^{K} N_i \left[(X_i/N_i)\log(X_i/N_i) + (1 - X_i/N_i)\log(1 - X_i/N_i)\right].$$

THEOREM 38.11.1
Under H_0,

$$-2\log\lambda_n \xrightarrow{D} \chi^2_{K-1}.$$

PROOF The proof is analogous that of Theorem 5.1 in Ivanova *et al.* (2000). ∎

THEOREM 38.11.2
For $\Delta < 0$,

$$-2\log\lambda_n \xrightarrow{D} \chi^2_{K-1}(\varphi),$$

where the noncentrality parameter φ is

$$\varphi = \sum_{i=1}^{K-1} \frac{D_i \delta_i^2}{p_i q_i)} - \left(\sum_{i=1}^{K} \frac{D_i}{p_i q_i)} \right)^{-1} \left(\sum_{i=1}^{K-1} \frac{D_i \delta_i}{p_i q_i)} \right)^2,$$

where $\delta = \sqrt{n}(p_1 - p_K, \ldots, p_{K-1} - p_K, 0)$.

PROOF The proof is analogous to that of Corollary 5.1 of Ivanova *et al.* (2000). ∎

The limiting distribution for the likelihood ratio statistic in the case $\Delta > 0$ is not known.

38.12 Concluding Remarks

The urn model is useful for treatment allocation in clinical trials. As a randomized procedure, it provides protection from selection bias and from confounding effects of unknown covariates. Strategies for changing the urn composition in response to treatment results can improve the success rate of future subjects. This was the motivation behind the two arm randomized play-the-winner rule [Wei and Durham (1978)] and its generalization to K treatments [Wei (1979)]. The urn composition in these treatment allocation procedures is a K-variate Markov chain, and because the urn composition is Markovian, it can be embedded in a K-variate continuous-time Markov process. However, for these rules, there are dependencies among the K processes in the continuous-time analog of the urn process.

If the urn composition is a K-variate Markov chain, but it is modified after observing a treatment outcome in a way such that only the number of type i balls changes following treatment i, then it can be embedded in a collection of K independent Markov processes. This is the case with the birth and death urn described herein, as well as with those urn models described by Durham, Flournoy, and Li (1998), Ivanova (1998), and Ivanova *et al.* (2000). The advantages are great, in that much existing theory facilitates their analysis. It is easy to obtain limiting results. Also we have shown how a stopping rule can be invoked that permits exact theoretical results to be obtained.

Just as the availability of theoretical results for the continuous-time linear birth and death processes helped to suggest the birth and death urn with immigration, other well-known continuous-time Markov processes can be used to analyze their analogous urn designs, which may be more suitable for a particular application. This paper, used with the references contained herein, provides a template for such an analysis.

References

1. Anderson, W. J. (1991). *Continuous-Time Markov Chains*. Springer Verlag, New York.

2. Athreya, K. B. and Ney, P. E. (1972). *Branching Processes*. Springer Verlag, Berlin.

3. Cox, D. R. and Miller, H. D. (1965). *The Theory of Stochastic Processes*. John Wiley & Sons, New York.

4. Durham, S. D., Flournoy, N., and Li, W. (1998). A sequential design for maximizing the probability of a favorable response. *Canadian Journal of Statistics* **26**, 479–495.

5. Ivanova, A. (1998). A birth and death urn for randomized clinical trials. *Unpublished Doctoral Dissertation*, University of Maryland Graduate School, Baltimore.

6. Ivanova, A. and Rosenberger, W. F. (2000). A comparison of urn designs for randomized clinical trials of $K > 2$ treatments. *Journal of Biopharmaceutical Statistics* **10** 93–107.

7. Ivanova, A., Rosenberger, W. F., Durham, S. D., and Flournoy, N. (2000). A birth and death urn for randomized clinical trials: Asymptotic methods. *Sankhyā, Series B* (to appear).

8. Karlin, S. and Taylor, H. M. (1974). *A First Course in Stochastic Processes*. Academic Press, New York.

9. Kotz, S. and Balakrishnan, N. (1997). Advances in urn models during the past two decades. In *Advances in Combinatorial Methods and Applications to Probability and Statistics* (Ed., N. Balakrishnan), pp. 203–257. Birkhaüser, Boston.

10. Wei, L. J. (1979). The generalized Pólya's urn design for sequential medical trials. *Annals of Statistics* **7**, 291–296.

11. Wei, L. J. and Durham, S. D. (1978). The randomized play-the-winner rule in medical trials. *Journal of the American Statistical Association* **73**, 840–843.

Author Index

Subject Index

Printed and bound by CPI Group (UK) Ltd, Croydon, CR0 4YY

24/10/2024

01778277-0017